Science in the Early Twentieth Century

Other Titles in ABC-CLIO's
History of Science
Series

Science in the Early Twentieth Century

An Encyclopedia

Jacob Darwin Hamblin

A B C · C L I O

Santa Barbara, California   Denver, Colorado   Oxford, England

Library of Congress Cataloging-in-Publication Data

Hamblin, Jacob Darwin.
Science in the early twentieth century : an encyclopedia / Jacob Darwin Hamblin.
    p. cm. — (ABC-CLIO's history of science series)
Includes bibliographical references and index.
ISBN 1-85109-665-5 (acid-free paper)–ISBN 1-85109-670-1 (eBook)
1. Science—History—20th century.   I. Title: Science in the early 20th century.
II. Title. III. Series.
ABC-CLIO's history of science series.
Q121.H345 2005
509—.041—dc22   2004026328

06 05 04 03 10 9 8 7 6 5 4 3 2 1

This book is available on the World Wide Web as an eBook. Visit abc-clio.com for details.

ABC-CLIO, Inc.
130 Cremona Drive, P.O. Box 1911
Santa Barbara, California 93116-1911

This book is printed on acid-free paper ∞.
Manufactured in the United States of America

For my parents, Les and Sharon

Contents

Science in the Early Twentieth Century: An Encyclopedia

Acknowledgments

On July 6, 2003, during a house renovation maneuver marked by stupidity and inexperience, a wall-sized mirror broke in half and slashed open my left leg. Although I hesitate to thank the mirror for this, I must acknowledge that being laid up for the rest of the summer accounts, in part, for my ability to finish this encyclopedia. It took longer than a summer to write, of course, but I established a pace during that time that I tried to enforce once the semester began and I was back to teaching classes. During that time, I relied heavily on the love and support of my wife, Sara Goldberg-Hamblin, who has agreed to keep a blunt object handy should I agree to write any more encyclopedia entries in the course of my career. I also relied on the good humor and encouragement of friends and family. In particular, I thank Houston Strode Roby IV, who routinely refused to denigrate the project. Others include Ben Zulueta and Gladys Ochangco (particularly Ben, who helped muse over the alphabetical interpretation of history), Fred and Viki Redding (prouda ya but miss ya!), Stacey and Branden Linnell (who supplied the mirror but also supplied Rice Krispie treats), Shannon Holroyd (for staying here in Long Beach), Lara and Eli Ralston (for always being around when we need them), and Denny and Janet Kempke (for treating me like family). I also thank my longest-standing friend, my sister Sara, who gave us our niece,

Victoria. Special thanks go to Cathy and Paul Goldberg, whose sunny dispositions could melt an iceberg. My parents, Les and Sharon Hamblin, deserve far more credit and praise than I have ever acknowledged, and I know that my sense of determination came from them. I also owe a debt of gratitude to our dog Truman, who kept me smiling throughout. Last, at least in order of birth, I thank my daughter Sophia for putting everything into perspective.

In a more practical vein, I should mention that the students in my History of Science course at California State University, Long Beach, have been more useful than they ever will know in helping me learn how to communicate ideas. I thank Sharon Sievers for asking me to teach the history of science, Albie Burke for enlisting me to teach it in the honors program, and Marquita Grenot-Scheyer for her efforts to keep us in Long Beach. We are starting to like it. At ABC-CLIO, I thank Simon Mason, the editor in Oxford who guided this project, and William Burns, the series editor, whose comments helped to make the text clearer and more balanced.

Writing an encyclopedia is an enormous task. Although our lives are filled with projects that call to mind our ignorance, this one was very humbling for me. It required me to branch out considerably from my own areas of expertise, to try to do justice to profound and complex ideas, and to capture the sense of the times. I learned a great deal along the way. Having said that, I confess that my notions of the era covered in this book come largely from a few authors, including Lawrence Badash, Daniel J. Kevles, J. L. Heilbron, Peter J. Bowler, Margaret Rossiter, Helge Kragh, John North, Spencer Weart, and the many contributors to *Isis* and other journals. This does not exhaust, by any stretch of the imagination, the list of authors whose work made this encyclopedia possible. It simply acknowledges a deep imprint. This is especially true of my mentor Lawrence Badash, whose expertise in the history of twentieth-century physics I can only hope to approximate.

Jacob Darwin Hamblin
California State University, Long Beach

Introduction

The first half of the twentieth century saw science catapult to the world stage as a crucial aspect of human knowledge and human relations. Revolutions in physics replaced Newton's laws with relativity and quantum mechanics, and biologists saw the birth of genetics and the mainstreaming of evolution. Science was used by racist ideologues to pass discriminatory laws, adapted to political ideology to justify persecution, and used to create the most destructive weapons of modern times. Science was at the forefront of social controversy, in issues related to race, religion, gender, class, imperialism, and popular culture. By mid-century, amidst atomic bombs and wonder drugs, the practice of science had changed dramatically in terms of scale and support, while society tentatively and often begrudging accorded scientists a status in society unprecedented in history.

Reinvigoration of Physics

Although some physicists believed that the vast majority of the great discoveries in their field had already been made and that future work would simply be a matter of establishing greater precision, the last years of the nineteenth century decisively changed that view (see Physics). The fundamental discoveries that reinvigorated physics and shaped its course throughout the twentieth century came from studies of cathode rays, which were produced by the discharge of electricity through a tube of highly rarefied gas. Experiments led to the accidental discovery by German physicist Wilhelm Röntgen of X-rays, later recognized as an intense form of electromagnetic radiation (see X-Rays). The strange "see through" phenomenon of X-rays inspired more studies, resulting in Henri Becquerel's 1896 discovery of "uranium rays," later called radioactivity (see Becquerel, Henri; Radioactivity). The flurry of research on mysterious "rays" sometimes yielded false identifications of new phenomena, but soon researchers realized that the X-rays (resulting from electricity) and radioactivity (emanating from certain substances without needing any "charge" by electricity or by the sun) were rather different. The cathode rays were in 1897 deemed to be streams of particles, or "corpuscles." These little bodies soon came to be called electrons (see Thomson, Joseph John). Around the turn of the century, Marie Curie earned a worldwide reputation for isolating and identifying radioactive elements previously unknown to mankind, such as polonium and radium (see Curie, Marie). These developments indicated avenues for understanding objects even smaller than atoms.

Other fundamental changes in physics were theoretical. In 1900, Max Planck attempted to fix a mathematical problem of energy distribution along the spectrum of light. He inserted a tiny constant into equations measuring energy according to frequency, thus making energy measurable only as a multiple of that tiny number. If theory could be generalized into reality, energy existed in tiny packets, or quanta (*see* Planck, Max; Quantum Theory). Albert Einstein claimed that if this were true, then light itself was not a stream, but rather was made up of tiny "photons" that carried momentum, even though light has no mass. Einstein's famous equation, $E = mc^2$, made energy equivalent to a certain amount of mass (*see* Einstein, Albert; Light). Another of Einstein's well-known ideas, special relativity, was first formulated in 1905. It did away with the nineteenth-century concept of the ether and redefined concepts such as space and time. Ten years later, he published a general theory of relativity that provided explanations of gravitation and argued that light itself sometimes follows the curvature of space (*see* Relativity). Quantum theory and relativity were controversial in the first two decades of the century, and one major vehicle for disseminating both ideas was discussion at the Solvay Conferences (*see* Solvay Conferences). The bending of light, a crucial aspect of general relativity, was observed by Arthur Eddington in 1919 during a solar eclipse (*see* Eddington, Arthur Stanley).

The behavior of electromagnetic radiation, such as X-rays, continued to inspire research during the first few decades of the century. In 1912, German physicist Max von Laue discovered that X-rays were diffracted by crystals (*see* Von Laue, Max). Based on this discovery, Englishmen William Henry Bragg and his son William Lawrence Bragg used X-rays to investigate crystals themselves, because each metal diffracted X-rays differently, yielding unique spectra. This was the beginning of the new field of X-ray crystallography, immensely useful in studying the properties of metals (*see* Bragg, William Henry; Debye, Peter). But still the properties of electromagnetic radiation, such as X-rays and the higher-energy "gamma rays," were poorly understood. Major discoveries about the interplay of electromagnetic radiation and charged particles (such as electrons) were discovered in the 1920s and 1930s. In 1923, Arthur Compton noted that changes in wavelength during X-ray scattering (now known as the "Compton effect") could be interpreted through quantum physics: A photon of radiation strikes an electron and transfers some of its energy to the electron, thus changing the wavelength of both. His work provided experimental evidence for quantum theory and suggested that electrons behave not only as particles but also as waves (*see* Compton, Arthur Holly). In 1928, Indian physicist Chandrasekhara Raman observed that the same is true for ordinary light passing through any transparent medium; it also changes wavelength, because of the absorption of energy by molecules in the medium (*see* Raman, Chandrasekhara Venkata). The importance of the medium was paramount, because the theoretical rules governing light could change, including the notion that light's speed cannot be surpassed. In the Soviet Union, Pavel Cherenkov in 1935 discovered a bluish light emitted when charged particles were passed through a medium. The strange light was soon interpreted by his colleagues to be an effect of the particles "breaking" the speed of light, which is only possible in a transparent solid or liquid medium (*see* Cherenkov, Pavel).

Studies of electrons provoked new questions about the nature of radioactivity and the structure of the atom itself. U.S. experimental physicist Robert Millikan was the first, in 1910, to measure the charge of an electron, and he determined that all electrons were the same. He used the "oil-drop" method, measuring the pull of an electric field against that of gravity (*see* Millikan, Robert A.). New Zealand physicist Ernest Rutherford suggested that radioactivity was the "new alchemy," meaning that some elements were unstable and gradually transformed themselves into other elements. Radioactivity simply was the ejection of one of two kinds of particles, called alpha and beta, in unstable atoms. Beta particles were electrons, and their release would fundamentally change an atom's characteristics. In 1911, Rutherford devised the new planetary model of the atom, based on electrons orbiting a nucleus, which replaced J. J. Thomson's "plum pudding" atom with electrons bathing in positively charged fluid (*see* Atomic Structure; Rutherford, Ernest). Rutherford was a leading figure in experimental physics, and when he became director of the Cavendish Laboratory at Cambridge, England, it became the foremost center for the study of radioactivity and atomic physics. One of the principal tools used at the Cavendish was C. T. R. Wilson's cloud chamber, which allowed scientists to actually see vapor trails of electrons (*see* Cavendish Laboratory; Cloud Chamber).

The Debates on Heredity

While the field of physics was being reborn, scientists were founding the science of genetics and escalating existing debates about heredity (*see* Genetics). Modern theories about the inheritance of traits come from the 1866 work of Gregor Mendel, who was a monk in an area that is now part of the Czech Republic. He hybridized (cross-bred) varieties of garden pea plants to analyze the differences in subsequent generations. He found that, given two opposite plant traits (such as tall or short), one trait was dominant, and all of the first generation had the dominant trait, without any blending of the two. But in the next generation, the less dominant trait came back in about a quarter of the plants. This 3:1 ratio was later interpreted as a mathematical law of inheritance. Although Mendel's work went unappreciated for three and a half decades, a number of plant breeders, such as Dutchman Hugo de Vries, German Carl Correns, and Austrian Erich von Tschermak, independently resuscitated these ideas in 1900 (*see* Rediscovery of Mendel). De Vries proposed a new concept—mutation, the sudden random change in an organism—as the mechanism for introducing new traits, which then would be governed by Mendelian laws (*see* Mutation). The most outspoken

advocate of Mendelian laws was the English biologist William Bateson, who in 1905 coined the term *genetics* to describe the science of inheritance based on Mendel's work. He viewed mutation and Mendelism as a possible way to account for the evolution of species (*see* Bateson, William; Evolution). The Dane Wilhelm Johannsen was the first to speak of *genes* as units that transmit information from one generation to the next, and he differentiated between the genotype (the type of gene, including its dominant or recessive traits) and the phenotype (the appearance, based on which trait is dominant). His theory of "pure lines" established the gene as a stable unit that could be passed down without alteration by the environment (*see* Johannsen, Wilhelm).

The most influential experiments on genetics occurred in the "fly room" at Columbia University, where Thomas Hunt Morgan experimented with fruit flies. Because of its short life span, *Drosophila melanogaster* was the ideal creature upon which to observe the transmission of traits from one generation to the next. Previous theories assumed that any introduction of new traits into a population would be blended so thoroughly that they would not be discernable. But Morgan's fruit flies revealed that random mutations could produce new traits that were indeed inherited through Mendelian laws and, even when phenotypes appeared to show their disappearance, the traits reappeared later. His work also showed that chromosomes, thought to be the physical carriers of genes, were subject to patterns of linkage. This complicated calculations of Mendelian inheritance, because individual traits were not always inherited independently, but instead could only be passed down in conjunction with other genes. Morgan's fruit fly experiments convinced him that individual genes were linked together by residing on the same chromosomes. The chromosome theory of inheritance, devised by Morgan in 1911, implied that genes themselves could be "mapped" by locating them on physical chromosomes. His student, A. H. Sturtevant, was the first to do this, in 1913 identifying six different traits dependent on a single chromosome. The chromosome theory was assailed by other leading geneticists, including William Bateson, for about a decade (*see* Chromosomes; Morgan, Thomas Hunt).

Later, in the 1940s, the chromosome theory received something of a twist. U.S. geneticist Barbara McClintock noted that, although genes were linked together on a single chromosome, the genes were not necessarily permanent residents of any single chromosome. Instead, they could reconfigure themselves. These "jumping genes," as they were called, compounded the difficulties in Mendelian laws even more, because of the impermanence of gene mapping (*see* McClintock, Barbara). Genetics was broadening its horizons in the 1930s and 1940s. For example, embryologists began to look to genetics as a way to understand how tissues organize themselves throughout their development. Also in the 1940s, geneticists George Beadle and Edward Tatum took a biochemical approach to genetics and determined that each gene's function was to produce a single enzyme (*see* Biochemistry; Embryology). Also, scientists revived the notion that chromosomes were made up of deoxyribonucleic acid (DNA), previously disregarded as not complex enough to carry genetic information. Oswald Avery, Colin MacLeod, and Maclyn McCarty approached the problem from a different field altogether: bacteriology. But their work on methods of killing or neutralizing bacteria seemed to suggest that DNA played a major role in making permanent changes in the makeup of bacteria. Continued research on bacteriophages (viruses that destroy bacteria) would result in scientists ultimately agreeing on the importance of DNA in genetics (*see* DNA; Genetics).

We take it for granted today that genetics and evolution are reconcilable, but for many years this was not the case. The leading proponents of Darwinism in the early decades of the century were biometricians such as Karl Pearson, who used large populations to attempt to show how Darwinian natural selection could function, finding a correlation between certain traits and mortality rates (*see* Biometry). Geneticists, on the other hand, initially saw natural selection as a competitor to Mendelian inheritance. Others tried to reconcile the two. R. A. Fisher advocated bringing the two together in a new field, population genetics. J. B. S. Haldane saw natural selection acting on the phenotype, rather than on the genotype, and he noted that natural selection could act far more quickly than most scientists realized. He used a now-famous example of dark-colored moths surviving better in industrial cities, whereas lighter-colored moths were killed off more easily by predators. Sewall Wright introduced the concept of *genetic drift* and emphasized the importance of natural selection in isolated populations. These researchers began the Darwin-Mendel synthesis in the 1920s, and in the next decade, Theodosius Dobzhansky argued that the wide variety of genes in humans accounts for the apparent adaptation to changing environments (*see* Genetics; Haldane, John Burdon Sanderson; Wright, Sewall).

Darwinians and Mendelians were joining forces, but many opponents of evolution held firm. One of the most explicit arguments in favor of evolution came from George Gaylord Simpson, who in the 1940s argued forcefully that the arguments set forth by geneticists agreed very well with the existing fossil evidence (*see* Simpson, George Gaylord). Although this persuaded many scientists, it only fueled antiscience sentiment already burning in society. Darwinian evolution sparked controversy as soon as Darwin published his ideas in 1859, and this continued into the twentieth century. Some objected to evolution in general, disliking the notion of being descended from an ape.

Others directed their attacks specifically at Darwin's contribution to evolution, natural selection, which emphasized random change and brutal competition for resources. Aside from not being a humane mechanism of evolution, its random character left no room for design. Religious opponents claimed that this left no room for God. Evolution had long been a symbol of secularization in the modern world, and even the Pope had condemned it in 1907. Roman Catholics were not the most vocal, however; in the United States, Protestant Fundamentalist Christians cried out against evolution and, more specifically, the teaching of evolution in schools (*see* Evolution; Religion). The 1925 "monkey trial" of John T. Scopes, arrested for teaching evolution in a Tennessee school, showcased the hostility between the two camps. The trial revealed not only hostility toward this particular theory, but toward any scientific ideas that threatened to contradict the Bible. The trial made headlines worldwide and revealed the extremes of antiscience views in America (*see* Scopes Trial).

The New Sciences of Man

The debates about evolution and heredity inspired anthropologists and archaeologists to inspect the fossil record. Was there evidence of linkage between apes and man? Was there a transitional hominid, long extinct, hidden in the dirt somewhere? The search for the "missing link" between apes and man was based on an evolutionary conception that was not particularly Darwinian, but was shared by many scientists and the general public. It assumed a linear progression from inferior to superior beings—making apes an ancestor of man. Darwinian evolution, on the contrary, assumed that both apes and man were species that had evolved in different directions from some common ancestor. Nonetheless, the search for the missing link appeared crucial in proving to skeptics the veracity of evolution (*see* Evolution; Missing Link). The most notorious case was the Piltdown Man, a collection of bones found in 1912 in a gravel bed in England. It served as evidence of the missing link for decades, until it was revealed as a hoax in the 1950s (*see* Piltdown Hoax).

Other discoveries of ancient hominids were more genuine. In the 1920s, workers taking rocks from a quarry near Beijing (typically called Peking at the time) discovered "Peking Man," a collection of the bones of several people, including, despite the name, those of a female. Like Piltdown Man, these bones were used as an argument for the evolution of species, by those using the "missing link" argument, and by those simply looking for an ancient hominid different from man. Unfortunately these bones were lost during World War II; thus many have suspected they were forged in the same fashion as the Piltdown Man (*see* Peking Man). One of the anthropologists involved in both sites was Pierre Teilhard de Chardin, a Catholic priest who saw connections between biological evolution and the evolution of human consciousness. Like many other philosophically minded believers in evolution, he combined Darwinism with Lamarckism, adding a sense of purposefulness. Like species striving for greater complexity, humans as a whole could strive toward higher consciousness (*see* Teilhard de Chardin, Pierre). Another famous bone-hunter was Louis Leakey, who sought to demonstrate that the oldest bones, thought to be in Asia, were actually in Africa. Leakey wanted to show that human beings originated in Africa. Although a number of premature announcements tarnished his reputation in the 1930s, he and his wife, Mary Leakey, continued their search and uncovered some important ancient hominids in the 1940s and 1950s (*see* Leakey, Louis).

Aside from fossil collecting, anthropology made a number of serious changes in the early twentieth century, particularly on the question of race. Nineteenth-century anthropologists were obsessed with classification of races. A few were Social Darwinists, seeing all races in brutal competition. But most were traditionalists who saw all races in a hierarchy, with "superior" European races toward the top and "inferior" African races at the bottom, with most others falling in between. One's place on the scale of civilization could be judged in reference to skin color, skull size, jaw shape, and brain weight (*see* Anthropology). In other words, human society was racially determined. U.S. anthropologist Franz Boas took an entirely different view, arguing instead for cultural determinism. Examining different societies, such as those of the Eskimo, Boas concluded that a complex combination of history, geography, and traditions defined social relations far more than anything biologically determined. Boas's work held an important political message that distressed other anthropologists—that human beings as a species are essentially the same (*see* Boas, Franz; Race). His student, Margaret Mead, took this further and added a caveat. She argued that behaviors that are considered universal and biologically inherent in all humans, such as the attitudes and behaviors of adolescents, are also products of culture. While Boas challenged racial determinism, Mead challenged anthropologists not to assume that what seems "natural" in Western cultures is actually natural for all human beings (*see* Mead, Margaret).

One of the consequences of Darwinian evolution was the application of natural selection to human societies, known generally as Social Darwinism. This outlook encouraged thinking about collective units in competition with one another, which generalized easily to cultural, national, or racial competition. A number of public figures in Europe and the United States voiced concerns about "race suicide," noting that steps should be taken to ensure that the overall constitution of the race was healthy enough to compete. Armed with the data from biometry, scientists argued that laws based on principles of eugenics, or good breeding, ought to be passed to discourage procreation among the unfit and encourage it among the most productive members of society (*see* Biometry;

Eugenics). The birth control movement in Britain and the United States gained support from eugenicists who wanted all women, not simply the most affluent and intelligent ones, to be able to prevent themselves from procreating (*see* Birth Control). Eugenics laws attempting to shape demographics were passed in a number of countries, culminating ultimately in the Nazi racial laws against intermarriage with Jews (*see* Nazi Science). In other countries, most of these laws were directed at the poor or those with mental retardation. In the United States, for example, the Supreme Court protected the involuntary sterilization of people with mental retardation (*see* Mental Retardation). Intelligence testing, developed in France, initially was designed to detect children in need of special education. However, it was used in the United States to classify the intelligence of every individual, and Americans such as Henry Herbert Goddard and Lewis Terman tried to use the intelligence quotient (IQ) to demonstrate the mental inferiority in some immigrant groups (*see* Intelligence Testing).

The first decades of the century saw a renewed interest in studying the individual mind. Sigmund Freud published *The Interpretation of Dreams* in 1900, beginning a new field of psychology known as psychoanalysis. It was based on differences between the conscious and unconscious mind and emphasized the importance of early childhood memories and sexual drive in human thought (*see* Freud, Sigmund; Psychoanalysis). Psychoanalysis had many adherents, though Freud's most famous disciple, Carl Jung, broke away largely because of disagreements about the importance of sexuality. Jung founded his own school, analytical psychology, and began to focus on the idea of the collective unconscious (*see* Jung, Carl). Mainstream psychology underwent a number of theoretical shifts during this period. Holistic approaches to understanding human thought were popular in Europe, particularly Max Wertheimer's *Gestalt* school. In many respects this was a reaction to the prevailing structuralist outlook that saw human psychology as a combination of component physiological activities. *Gestalt* psychology saw the mind as more than the sum of its parts. Another school of thought was functionalism, which denied universality of human minds and emphasized the importance of adaptation to particular environments (*see* Psychology).

Many theorists believed that understanding human psychology was best explored by studying the development of intelligence and consciousness. But by far the most influential movement was behaviorism, which began not in humans but in dogs, through the work of Ivan Pavlov. He became famous in Russia and elsewhere for his concept of signalization, which connected cognitive input to physical output. For example, a dog could be trained to salivate every time it heard a bell, because it associated the bell with being served food. Radical behaviorists believed that all animal actions, including human actions, resulted from learned behaviors rather than any objective, rational thought (*see* Pavlov, Ivan). The most renowned behaviorist was B. F. Skinner, who coined the phrase *operant conditioning* to describe the construction of environments designed to provoke a particular action by an experimental subject. Skinner became notorious for his utopia, *Walden Two,* which advocated a perfectly designed society but struck critics as mind control (*see* Skinner, Burrhus Frederic). Others emphasized the social and cultural aspects of psychology; Lev Vygotsky, for example, was a developmental psychologist whose work on signs and symbols became foundational for studies of special education. Similarly, Jean Piaget saw the development of consciousness in children as a result of assimilation of environmental stimuli and the construction networks of interrelated ideas (*see* Piaget, Jean; Vygotsky, Lev).

Life and Death

Understanding the mind was only one aspect of treating the individual to improve his life or cure disease. Mental conditions such as depression and schizophrenia were objects of study for psychologists; others proposed more radical solutions. People with mental conditions were often confined to sanitoriums, where they were isolated and "treated," which often was merely a euphemism for custodial care. In the 1930s, radical brain surgery was introduced; in Portugal, Egas Moniz invented a device called the leukotome to sever certain nerves in the brain. Psychologists turned increasingly to leukotomies (lobotomies), which entailed surgical severance of nerve fibers connecting the frontal lobes of the brain to the thalamus. Doctors experimented with electric shock treatments, too, but such radical procedures became less important after the development of psychotropic drugs in the 1950s (*see* Mental Health).

Scientists had mixed successes in combating disease in the twentieth century. The famous microbe-hunters at the turn of the century, Paul Ehrlich and Robert Koch, helped to combat major diseases such as tuberculosis (*see* Koch, Robert; Microbiology). Ehrlich hoped to use biochemical methods to develop magic bullets, chemicals that would target only specific microbes and leave healthy human parts alone. This proved difficult, because the side effects of chemical therapy (or chemotherapy) were difficult to predict. One of his drugs, Salvarsan, was useful in combating a major venereal disease, syphilis, but it also caused problems of its own (*see* Biochemistry; Ehrlich, Paul). Alternatives to chemotherapy were antibiotics, which were themselves produced from living organisms and were useful in killing other microorganisms. Among the most successful of these was penicillin, which was developed on a large scale during World War II and helped to bring syphilis and other diseases under control (*see* Antibiotics; Penicillin; Venereal Disease).

Scientists tried to put knowledge to use in promoting health and combating disease. During World War I, many of the young recruits were appallingly unhealthy, leading to new studies in nutrition. One of the most common destructive diseases, rickets, was simply a result of vitamin D deficiency. One of researchers' principal goals was to discover the "essential" vitamins and the "essential" amino acids, from which the body's proteins are made (see Amino Acids; Nutrition). Other scourges, such as cancer, were more difficult to tackle. One of the most rigorous public health programs to combat cancer, which included restrictions on smoking, was put into place by the Nazis as part of their efforts to improve the strength of the Aryan race (see Cancer). Endocrinologists studied the body's chemical messengers, hormones, leading to new medicines to treat diseases stemming from hormone deficiencies. For example, insulin was successfully extracted from the pancreas of a dog and marketed by pharmaceutical companies as a new medicine to combat diabetes (see Endocrinology; Hormones; Industry; Medicine).

Increased knowledge of microbiology also contributed to public health measures. Governments believed that sanitation was crucial for keeping populations healthy, and the importance of education seemed to be crucial. Public health services established regulations, promoted health education programs, and even conducted experiments. One notorious experiment in the United States was the Public Health Service's observation of dozens of African American victims of syphilis over many years. None of them were aware they had syphilis, but were instead told they had "bad blood," and they were not treated for the disease. Experiments on minority groups were not uncommon, and doctors tried to justify them as contributing to the overall health of the public (see Public Health). More well-known cases of human experimentation occurred in the Nazi death camps, in Manchuria under Japanese occupation, and even on unwitting patients by U.S. atomic scientists in the 1940s (see Atomic Energy Commission; Human Experimentation; Nazi Science; Radiation Protection).

Biologists also began to explore the interconnectedness of living organisms. Biologist Ernst Haeckel coined the term *ecology* in the nineteenth century to describe the web of life, emphasizing how species do not exist independently of each other. In the first decades of the twentieth century, researchers broke this giant web into components, noting that there are different systems in any given environment. Arthur Tansley coined the term *ecosystem* to describe these discrete units. Much of the research on ecology was sponsored by the Atomic Energy Commission to study the effects of nuclear reactors on surrounding areas (see Ecology). Research on residual harmful effects would eventually illuminate the dangers of widespread use of pesticides in the 1940s and 1950s (see Pesticides). Aside from research, the ecological outlook also spurred action at the political level, particularly through conservation movements. Although often connected to efforts to protect the environment, conservation was fundamentally an anthropocentric concept—one needs to conserve resources to ensure future exploitation. The U.S. Forest Service was developed, in part, to ensure that lumber supplies in the United States were not depleted. Others emphasized the mutual dependence among humans, animals, and flora. Conservationist Aldo Leopold argued that ecological systems ought to be maintained and that exploitation should take into account not merely man's needs but the survival of the system (see Conservation).

Earth and the Cosmos

Scientists used physics to penetrate the interior of the earth. Knowing that the propagation of waves differed depending on the medium, some reasoned that one could analyze the earth through measuring pressure waves. The science of geophysics is based on using physical methods to extrapolate, without direct visual observation, general principles about the earth's structure (see Geophysics). Doing so by measuring the waves created by earthquakes, seismology, was most common in Europe, through the work of Germans such as Emil Wiechert and Beno Gutenberg (*see* Gutenberg, Beno; Seismology). Andrija Mohorovičić, a Croatian, used such methods to postulate the existence of a boundary between two geologically distinct layers, the crust and the mantle (*see* Mohorovičić, Andrija). The United States became a major center for geophysics for several reasons. Gutenberg moved there in the 1930s, the Jesuits developed a major seismic network, and the California coast is one of the world's most active earthquake zones. U.S. geophysicist Charles F. Richter developed the scale, named after him, to measure the magnitudes of earthquakes (*see* Richter Scale).

One major theory that was rejected firmly by most geologists and geophysicists in the first half of the century was Continental Drift. The German Alfred Wegener proposed that the "jigsaw fit" between South America and Africa was more than a coincidence, and that the two had once been joined. He even proposed that all the earth's land masses once made up a giant supercontinent, which he called Pangaea (*see* Continental Drift; Geology; Wegener, Alfred). Critics such as Harold Jeffreys argued that the theory failed to provide a mechanism for moving such huge blocks of crust that would obey the laws of physics. Jeffreys insisted that the earth's structure was solid, without any major horizontal motions. Theories about the interior of the earth abounded; some supported Wegener by noting that the heat could be transferred through molten convection currents, while others proposed the earth as a honeycomb, with pockets of magma that had not yet cooled (*see* Earth Structure; Jeffreys, Harold).

The idea that the earth was cooling was widely accepted because of the earth's constant loss of heat. In fact, this loss of heat had been measured in the nineteenth century by Lord Kelvin, who used it to calculate the age of the earth. His estimation fell far short of that needed for evolution to occur, lending

antievolutionists a major argument from physics (*see* Age of the Earth). The discovery of a new source of heat, radioactivity, nullified the argument because this vast storehouse of energy implied the possibility of heat loss for much longer periods. Bertram Boltwood and others attempted to calculate the age of the earth based on radioactive decay series, believing that lead was the end-state of all radioactive decay. Thus, the age of rocks could be determined by measuring the proportion of lead they contained (*see* Boltwood, Bertram). Radioactivity also seemed to hold the key to measuring the age of once-living things. Because cosmic rays created radioactive isotopes of carbon, one could calculate the age of carbon-based matter by measuring the amount of radioactive carbon in it (*see* Carbon Dating; Cosmic Rays).

The sciences of the air and sea also saw a major resurgence in the twentieth century. Owing largely to the work of the Norwegian Vilhelm Bjerknes and his Bergen school, meteorology developed some of its basic theoretical premises, such as the existence of weather "fronts." Bjerknes had a widespread influence, particularly among those who wanted to inject a dose of theory and quantification into their subjects (*see* Bjerknes, Vilhelm; Meteorology). One of Bjerknes's students, Harald Sverdrup, became a major figure in oceanography when he took the directorship of the Scripps Institution of Oceanography in 1936. He brought the dynamics-based outlook of his mentor to bear on oceanographic research, giving physical oceanography a major emphasis in the United States (*see* Oceanography; Sverdrup, Harald). The dynamics approach even touched marine biology; the Kiel school of marine biology, based in Germany, saw populations of plankton as large systems interacting. Despite the perceived benefits of quantification, such outlooks also discouraged the study of differences among individual organisms (*see* Marine Biology). Meteorologists and oceanographers often pursued expensive subjects, and thus they needed to attract patrons, and the dynamical approach was attractive to fishery organizations hoping to produce data to enable efficient exploitation of fishing "grounds." Oceanographers also attracted support by making major expeditions to distant waters, such as those around the poles; these occasionally served as national propaganda, as in the case of the American Robert Peary's famous voyage to the North Pole in 1909. Even more famous were the expeditions on foot, such as the race between Norway and Britain to reach the South Pole, ending in a success for Roald Amundsen and the death of Robert Scott's entire team (*see* Oceanic Expeditions; Polar Expeditions).

Turning toward the heavens, the first half of the century saw major efforts toward the marriage of physics and astronomy. One of the most successful astrophysicists was Arthur Eddington, who was among the rare few who understood both the principles of astronomy and Albert Einstein's theories of relativity. Developing theories of the universe that were consistent with relativity was a principal goal of the 1920s and 1930s (*see* Astrophysics). Like most astrophysicists, Eddington owed a great debt to Ejnar Hertzsprung and Henry Norris Russell, who (independently of each other) in 1913 developed a classification system for stars; the resulting Hertzsprung-Russell diagram became the basis for understanding stellar evolution (*see* Hertzsprung, Ejnar). Eddington devoted a great deal of attention to describing the structure of stars, and he analyzed the relationship between mass and luminosity, noting that the pressure of radiation was balanced by gravitational pressures (*see* Eddington, Arthur Stanley). The Indian Subrahmanyan Chandrasekhar famously refuted this in 1930, observing that this equilibrium could only be maintained in stars below a certain mass (*see* Chandrasekhar, Subrahmanyan).

Connecting stars' luminosity to other measurable quantities held the key to many other problems. At Harvard College Observatory, astronomers working under director George Pickering were known by some as Pickering's Harem, because they were predominantly women (*see* Pickering's Harem). One of them, Henrietta Swan Leavitt, observed that in blinking (variable) stars, called Cepheids, the luminosity of stars differed according to the amount of time between blinks (periodicity) (*see* Leavitt, Henrietta Swan). This crucial relationship was then used by others to calculate relative distances of other Cepheid stars whose luminosity-periodicity relationship differed from those observed by Leavitt. Harlow Shapley used such methods to calculate the size of the universe, which he took to be a single unit (*see* Shapley, Harlow). He and Heber D. Curtis engaged in a major debate in 1920, at the National Academy of Sciences, on the question of whether the universe was a single entity or composed of several component galaxies. Edwin Hubble helped to answer this question in 1924 by revealing that the Andromeda nebula (or galaxy) is far too distant to be considered part of our own galaxy. In 1929, Hubble discovered another phenomenon in distant stars, the "red-shift," or their spectral lines; he interpreted this as a kind of Doppler effect, meaning that these stars (in fact, these entire galaxies) were moving away from us (*see* Hubble, Edwin).

Hubble's work on "red-shift" had profound cosmological implications. What was the structure and origin of the universe? No stars appeared to be moving closer to the earth, so the universe seemed to be expanding. But expanding from what? Already Belgian Jesuit Georges Lemaître had proposed his "fireworks" theory of the universe, in which the present world was formed by an explosion at the beginning of time. Now Hubble's finding seemed to demonstrate that Lemaître might be correct. The Big Bang theory, as it was later called, fit well with Lemaître's own religious worldview, because it left room for an act of creation (*see* Big Bang; Religion). Other theories contested it, such as the steady-state theory, which proposed that new matter was being created from nothing virtually all the time. George Gamow and Ralph Alpher lent further support to the Big Bang theory in 1948, when they developed a theoretical rationale for the creation of most of the light elements in the first few moments after the explosion (*see* Cosmology; Gamow, George).

Such theoretical speculation was not limited to the universe as a whole. The creation of the universe begged the question of the creation of life itself on earth. Bacteriologists had debated the possibility of spontaneous creation of life for decades, but the most influential views about the origin of life came from a Russian biochemist, Aleksandr Oparin, who saw Darwinian natural selection acting not just on organisms but on molecules as well. Oparin believed that this would tend to encourage complexity and thus eventually spur the development of living organisms (*see* Origin of Life). The logical assumption, then, was that life did not necessarily exist solely on earth (*see* Extraterrestrial Life). The idea of life on other planets is an old one, but it was a major source of inspiration for one of the most popular twentieth-century genres of literature, science fiction. The fear of alien races invading the earth was the premise of one of H. G. Wells's novels, *The War of the Worlds,* later made into a radio play that scared scores of Americans into believing such an invasion actually was taking place (*see* Science Fiction). Even some astronomers were convinced of the possibility. Percival Lowell, for example, devoted a great part of his career analyzing the "canals" on Mars, claiming that they were built by Martians (*see* Lowell, Percival).

The search for canals on Mars was only one of the many uses for new telescopes in the early twentieth century. Lowell built a major observatory to search for Planet X, which he believed to exist beyond Neptune (Pluto was discovered after his death). The central figure in building new observatories was George Ellery Hale, who had a hand in obtaining the telescopes at the Yerkes,

Mount Wilson, and Palomar Observatories. Hale was an entrepreneur of science and managed to convince rich patrons to fund bigger and more penetrating telescopes (*see* Astronomical Observatories; Hale, George Ellery). Not all astronomical work required access to expensive optical telescopes. Dutch astronomer Jacobus Kapteyn, for example, developed his ideas of "star streams" by analyzing photographic plates with a theodolite, an instrument that only exploits the laws of geometry. One of Kapteyn's students, Jan Oort, became one of the leading astronomers of the twentieth century (*see* Kapteyn, Jacobus; Oort, Jan Hendrik). Indeed, optical light from the stars was only one way to "see" into the heavens. In the 1930s, the basic premise of radio astronomy was discovered, and the capture of radio signals from space became a major tool for astronomy and astrophysics after World War II (*see* Radio Astronomy).

Mathematics and Philosophy

Internal contradictions plagued mathematicians at the end of the nineteenth century, particularly the theory of sets proposed by Georg Cantor, who argued against intuition that a set of all positive numbers has the same number of members as the set of all odd numbers. Most high school geometry students learn that proofs are a fundamental aspect of mathematical reasoning; David Hilbert challenged mathematicians to develop premises that could be proved without inconsistencies. But Kurt Gödel, in his "Incompleteness" theorem, made the disturbing observation that it is impossible to prove any system to be totally consistent. Bertrand Russell tried to develop a method to eliminate inconsistencies by reducing most mathematical problems to logical ones and developed a philosophy of knowledge called logical analysis (*see* Gödel, Kurt; Mathematics; Russell, Bertrand).

Just as Bertrand Russell tried to understand mathematical problems as logical ones, many sought to understand the growth or production of scientific knowledge through philosophical lens. Philosophers of science openly debated the study of knowledge, or epistemology, in order to describe science as it occurred in the past and to prescribe methods for the most effective scientific activity. The most influential of these was Ernst Mach, who was most active toward the end of the nineteenth century. Mach believed that all theories should be demonstrated empirically, or they should not be considered science at all. Mach found many adherents among experimental scientists, and later Mach's philosophy would provide a point of attack for those who wished to assail Einstein's theories of relativity and the quantum theories of Max Planck and Werner Heisenberg. A group called the Vienna Circle, active in the 1920s, became known for "logical positivism," which accepted the positive accumulation of knowledge but insisted that each theory be verifiable, or provable. Karl Popper, influenced by the Vienna Circle, was dissatisfied with this and instead proposed that theories, even if they cannot be "proven" beyond doubt, must be falsifiable, meaning that there must be a conceivable way to demonstrate whether or not they are false (*see* Philosophy of Science; Popper, Karl).

Scientists were active in mixing epistemology with theory. Quantum mechanics challenged fundamental scientific assumptions, particularly the idea that an objective reality is knowable. Werner Heisenberg's first proposals could only be expressed mathematically and seemed to be an abstract representation of reality. His uncertainty principle asserted that at the quantum scale, some variables cannot be known with greater certainty without increasing the uncertainty of others (*see* Heisenberg, Werner; Quantum Mechanics; Uncertainty Principle). He also observed that it is impossible to separate the observer from what is being observed, thus denying the possibility of objective knowledge. In addition, Heisenberg relied on statistics and probabilities to describe the real world, which struck many physicists as fundamentally antiscience. One should, many argued, be able to describe the entire world in terms of cause and effect, the basic notion of determinism (*see* Determinism). Niels Bohr generalized Heisenberg's ideas and asserted that the competing view of mechanics, devised by Erwin Schrödinger, was equally valid. One version saw the world as particles, and the other as waves; they were in contradiction, yet together provided a truer portrait of the world than what either could do alone (*see* Bohr, Niels; Schrödinger, Erwin).

Few ideas were subject to such a strident critique as determinism, which was abandoned not only by physicists, but also by Darwinian biologists who took purpose out of evolution. Determinists preferred a Lamarckian approach, which left some room for the organism's will and action. Austrian biologist Paul Kammerer was discredited in the 1920s for allegedly faking some experiments that "proved" Lamarckian evolution in midwife toads, and he soon after shot himself (*see* Kammerer, Paul). In the Soviet Union, Trofim Lysenko was able to persuade even Joseph Stalin that Lamarckian evolution was friendly to Marxism, because it was a fundamentally progressive view of human evolution that gave power to the organisms, unlike Darwinian evolution, which left progress to chance. This led to a notorious crackdown against Darwinian evolution and Mendelian genetics, leading to the persecution of leading biologists (*see* Lysenko, Trofim). Physicists such as Sergei Vavilov fared better in the Soviet system, largely because by the 1940s Stalin was more interested in acquiring an atomic bomb than in insisting on philosophical correctness (*see* Soviet Science; Vavilov, Sergei).

The Scientific Enterprise

Scientists increasingly sought to make connections between their work and economic and military strength. Patronage strategies shifted dramatically in the first half of the century from reliance on philanthropic organizations to writing proposals for massive grants from governments. Even industries took a serious interest in scientific research. The lucrative possibilities of such investments were seen in the early years of the century with Guglielmo Marconi's development of wireless technology. Communication companies such as the American Telegraph and Telephone Company (AT&T) established corporate laboratories in order to establish patents for new devices. Clinton Davisson's work in physics, for example, was supported by a corporate laboratory. The development of the transistor, and electronics generally after World War II, stemmed from the support of Bell Laboratories, which saw the need to support fundamental research for possible technological exploitation (*see* Davisson, Clinton; Electronics; Industry; Marconi, Guglielmo; Patronage).

But who conducted the research itself? The availability of money from private foundations and the federal government sparked controversy about the equal distribution of funds. Some argued that the best scientists should get all the grants, whereas others pointed out that this would concentrate all scientific activity and simply reinforce the elite status of a few institutions (*see* Elitism). In the United States and elsewhere, racial prejudice barred even the most capable scientists from advancing to prestigious positions, as in the case of the eminent African American marine biologist Ernest Everett Just, who became dissatisfied that he could find a worthwhile position only at a historically black college (*see* Just, Ernest Everett). White women on the whole fared better, but were confined to subordinate roles. One of the most famous examples was Pickering's Harem, a derogatory name given to a group of women astronomers working at the Harvard College Observatory. Although these women did important work, the reason for their employment was that they could be paid less and no one would expect them to advance beyond their technician status (*see* Pickering's Harem; Women).

Some believed that supporting science was the best way to achieve social progress. The technological breakthroughs of the late nineteenth century, particularly in industrializing countries, suggested that scientific knowledge automatically led to the improvement of mankind (*see* Social Progress). This view was also shared by university professors who sought to make their own disciplines seem more like physics and biology by adopting new methods. The term *social sciences* was born from efforts to bring scientific credibility to conventionally humanistic fields (*see* Scientism). Even law enforcement organizations began to adopt scientific methods and techniques, ranging from fingerprinting, to forensic pathology, to the use of infrared light (*see* Crime Detection). Others went so far as to suggest that governments should be controlled by scientists and engineers and that the most progressive societies would be ruled by technocracy (*see* Technocracy). The popularity of the technocracy movement waned considerably during the Great Depression, when there was little money for research and when many people openly questioned the direction of modern, secular, highly technological society. The experience of World War I already contributed to such disillusionment, because leading scientists such as Fritz Haber had put knowledge to use by developing chemical weapons and other more effective ways of killing human beings (*see* Chemical Warfare; Great Depression; Haber, Fritz; Religion; World War I).

Warfare also tarnished the image of science in other ways. Scientists typically believed they belonged to a community that transcended national boundaries, but World War I saw scientists mobilizing for war. Representative national bodies of scientists now became instruments to harness scientific talent for war work. For example, the National Research Council was created in the United States under the auspices of the National Academy of Sciences. Other national bodies created similar units, enrolling scientists in nationalistic enterprises (*see*

Academy of Sciences of the USSR; Kaiser Wilhelm Society; National Academy of Sciences; Nationalism; Royal Society of London). This was not the only way that scientists expressed nationalistic sentiments. Nominations for the symbol of the international scientific community, the Nobel Prize, often fell along national lines (*see* Arrhenius, Svante; Nobel Prize). When World War I ended, international cooperation was stifled, and even international scientific bodies banned membership from the defeated powers (*see* International Cooperation; International Research Council).

Nations used science to exert power in a number of ways. Scientific institutions and practices often served as means to radiate power to colonies. Public health measures, designed to eradicate tropical diseases, also helped to manipulate and control human populations, because Europeans, rather than indigenous peoples, controlled the life-saving treatments (*see* Colonialism). Also, science often was used to justify public policies regarding race. In Germany, biologist Ernst Haeckel endorsed Social Darwinism and lent it his credibility, as did many others (*see* Haeckel, Ernst). Racial competition fueled the appeal of Nazism; when the Nazis came to power in 1933, they instigated a number of racial laws and fired Jewish scientists from their posts. This began the first major *brain drain,* a name given to any large-scale migration of intellectuals. Most of the refugee Jewish scientists moved to Britain or the United States (*see* Brain Drain; Nazi Science).

World War II brought a number of changes to the practice of science. First of all, many technological breakthroughs during the war years indicated the value of science more powerfully than ever before. All of the major combatants attempted to use science. The Germans were the most advanced in the area of rocketry, the British made major strides in the area of radar, and the United States developed the first atomic bombs. In addition, penicillin was developed for widespread use during the war years, saving countless lives from bacterial infections. Scientists also built early computers from designs conceived in the 1930s and developed methods for breaking enemy codes. The research on guided missile technology inspired the first work in cybernetics in the early 1940s. The United States became the first country to attack cities with atomic bombs, ushering in the Atomic Age (*see* Computers; Cybernetics; Radar; Rockets; Turing, Alan; World War II).

The Atomic Age

The period after World War II might be called the Atomic Age, and its origins go back to the early years of the century. Albert Einstein had suggested that very small amounts of mass might be converted into extraordinarily large amounts of energy. Experimentalists who were trying to delve deeply into atoms largely ignored this theoretical background. The development of particle accelerators enabled scientists to force particles to overcome the natural repulsion of the nucleus and to observe how subatomic particles interact with each other. John Cockcroft and Ernest Walton used a linear accelerator to force atoms to disintegrate, thus achieving the first disintegration by artificial means. By the 1930s, Ernest Lawrence's cyclotron was the best tool for atom smashing, and in the 1940s scientists at Berkeley used cyclotrons to create artificial elements such as neptunium and plutonium (*see* Artificial Elements; Cockcroft, John; Cyclotron; Lawrence, Ernest). James Chadwick's 1932 discovery of a particle without charge but with mass, the neutron, gave scientists a kind of bullet with which they could bombard substances without the problem of repulsion by the nucleus and thus no need for particle acceleration. In Italy, Enrico Fermi bombarded all the elements with neutrons to see what reactions occurred. In France, the Joliot-Curies discovered that it was possible to make stable atoms radioactive by bombarding them with alpha particles. Also in the 1930s, Harold

Urey discovered heavy hydrogen, and Hideki Yukawa postulated the existence of mesons inside the nucleus (*see* Chadwick, James; Fermi, Enrico; Joliot, Frédéric, and Irène Joliot-Curie; Urey, Harold; Yukawa, Hideki).

The atomic bomb, however, was based on the phenomenon of nuclear fission. The experimental work on this was done in 1938 by Otto Hahn and Fritz Strassman, whereas the interpretation of the experiments was done by Hahn's longtime collaborator Lise Meitner and her nephew, Otto Frisch (*see* Fission; Hahn, Otto; Meitner, Lise). Bombardment by neutrons had made an element, uranium, split into two atoms of lighter weight, and in the meantime a small amount of the mass of uranium was converted into energy. Here was the fundamental premise of building atomic bombs. If this process could be sustained as a chain reaction, a violent explosion would occur. Shortly after these experiments, World War II began, and secret atomic bomb projects were eventually under way in Germany, the Soviet Union, Britain, the United States, and Japan. Building the bomb required enormous technical, financial, and material resources, including a large supply of uranium. The U.S. project was the only one that succeeded in building a weapon during the war; it was run by the Army and was called the Manhattan Project. In August 1945, two bombs were dropped—one on Hiroshima and one on Nagasaki—with a force equivalent to thousands of tons of dynamite exploding instantaneously (*see* Atomic Bomb; Hiroshima and Nagasaki; Manhattan Project; Uranium).

The advent of an atomic age left many scientists uneasy about the future of the world. During the war a number of scientists urged the U.S. government to reconsider using the atomic bomb. After all, they had envisioned its use against Germany, the country from which many of them had fled. The movement for social responsibility saw its origins toward the end of the Manhattan Project, as scientists developed reasons to limit the military uses of atomic energy. Two of the leading figures in the movement were James Franck and Leo Szilard, who detailed alternatives to dropping atomic bombs on Japan without warning. The movement sparked an organization that emphasized the social responsibility of science called the Federation of Atomic Scientists (*see* Federation of Atomic Scientists; Franck, James; Social Responsibility; Szilard, Leo).

One of the reasons scientists used for not using the atomic bomb was that it would usher in an arms race with the Soviet Union. The Soviet Union had already begun an atomic bomb project of its own, led by Igor Kurchatov. The first years after World War II were confrontational; the world seemed polarized in what became known as the Cold War. Scientists were suspected of helping the Soviets catch up to the United States, and fears of Soviet science and technology seemed justified when the Soviet Union tested its own atomic bomb in 1949. In 1950, one of the Manhattan Project's top scientists was discovered to be a spy for the Soviets. The late 1940s were years of paranoia and fear, with scientists' and other public figures' loyalty opened to inquiry. The nation's top physicist, the director of the National Bureau of Standards, was criticized as being a weak link in national security because of his internationalist views and his opposition to the development of the hydrogen bomb, the next generation of nuclear weapons. The University of California even required that its professors swear an oath claiming to have no affiliations with Communist organizations (*see* Cold War; Espionage; Kurchatov, Igor; Loyalty; National Bureau of Standards). The Cold War mentality led to extreme views. Game theory, for example, was a field connecting mathematics and economics, but increasingly it was used to analyze the most rational decisions that the United States and Soviet Union could make, assuming that a gain for one was a loss for the other. If war was inevitable, some game theorists reasoned, perhaps the United States should launch a preventive war (*see* Game Theory). In 1947, all things atomic fell under the jurisdiction of the newly created Atomic Energy Commission, which tested not only bombs but also the effects of radiation on humans, including some plutonium injection experiments on unwitting hospital patients (*see* Atomic Energy Commission; Radiation Protection).

By mid-century, scientists could say that they were entering the era of "Big Science," with large teams of researchers, instruments such as cyclotrons of ever-increasing size, and vast sums of money from a number of organizations, particularly the Atomic Energy Commission and the Department of Defense. The old questions of elitism remained, but more disturbing were the pitfalls of military patronage. After World War II, leading science administrators such as Vannevar Bush had convinced the U.S. government to support basic research in a number of fields, without precise technological expectations. Yet no civilian agency was

doing it. Young scientists were being enrolled into a military culture, particularly through such funding agencies as the Office of Naval Research. In 1950, those advocating for civilian control of federal patronage got their agency, the National Science Foundation, but it was often highly politicized, and military organizations were not forbidden to sponsor research of their own (*see* National Science Foundation; Office of Naval Research). Observers could agree by 1950 that science had changed considerably over the previous fifty years—in the ideas, in the practices, and in the scale of research. The center of gravity for science shifted decisively after World War II from Europe to the United States, where it enjoyed unprecedented support. But with potentially dangerous context—Cold War competition, military dominance, and the obvious connections between science and national security—few agreed on whether the second half of the twentieth century would see science working for or against the good of society.

Topic Finder

Concepts

Age of the Earth
Atomic Structure
Big Bang
Continental Drift
Determinism
Earth Structure
Eugenics
Evolution
Extraterrestrial Life
Fission
Game Theory
Mental Health
Mental Retardation
Missing Link
Mutation
Nutrition
Origin of Life
Public Health
Quantum Mechanics
Quantum Theory
Rediscovery of Mendel
Relativity
Uncertainty Principle

Fields and Disciplines

Anthropology
Astrophysics
Biochemistry
Biometry
Cosmology
Cybernetics

Ecology
Electronics
Embryology
Endocrinology
Genetics
Geology
Geophysics
Marine Biology
Mathematics
Medicine
Meteorology
Microbiology
Oceanography
Philosophy of Science
Physics
Psychoanalysis
Psychology
Seismology

Institutions/Organizations

Academy of Sciences of the USSR
Astronomical Observatories
Atomic Energy Commission
Cavendish Laboratory
Federation of Atomic Scientists
International Research Council
Kaiser Wilhelm Society
Manhattan Project
National Academy of Sciences
National Bureau of Standards
National Science Foundation
Nobel Prize

A

Academy of Sciences of the USSR

The role of scientists changed dramatically, as did most aspects of Russian life, with the Bolshevik Revolution in 1917. Under the czar, the Academy of Sciences had consisted of eminent scientists, and its role was largely honorific; it was not charged with major tasks by the government. But after the revolution, Vladimir Lenin (1870–1924) demanded that the academy make a study of the distribution of the country's industrial wealth and devise a plan to use it in the most rational way for the benefit of the Soviet state. This was the beginning of a long-standing relationship between the academy and the interests of the state. The Soviet Union's penchant for central planning drew heavily on advice from technical experts, including members of the academy. Lenin's government spent comparatively large sums of money on scientific research, as did Joseph Stalin's (1879–1953). In 1934, the academy sealed its relationship with the state symbolically when it moved its home from Leningrad to Moscow, where it officially was subordinated to the Council of People's Commissars.

Part of the academy's role was to ensure a close fit between its members' scientific theories and the official ideology of the Soviet Union. Its statutes provided for the addition of members only when they were scholars who had proven their ability to "assist the socialist up-building of the USSR." This certainly was a kind of political intrusion into the realm of science; nevertheless, the scientists largely policed themselves, which gave them some control over their own affairs. The academy's decision to embrace this course of action in the 1920s was a result of competition by more ideology-oriented bodies, such as the Communist Academy and the Institute of Red Professoriate, both of which embraced a more explicitly "proletarian" science than had the academy members. To ensure survival, the academy members adapted fully to the Soviet regime.

Nevertheless, the academy came under attack by Stalin toward the end of the 1920s and in the early 1930s. Scientists found themselves under deep suspicion, primarily for spreading theoretical ideas at variance with Marxism. Some members were purged from their positions, and those who were left were compelled to make statements of loyalty to the regime's political and philosophical goals. Scientific qualifications began to take a backseat to political ones, and research programs were integrated into Stalin's state-controlled economic planning. Conformism to Soviet ideology by scientists reached its high point under the influence of agronomist Trofim Lysenko (1898–1976), whose hostility to

1

Mendelian genetics was based on his own interpretation of Marxism-Leninism.

Despite all this, the academy managed to carve a useful niche for itself in the Stalinist state. Although it lost all semblance of autonomy, it achieved high status in a state that increasingly looked toward science and technology as a measure of national strength. The technocracy that emerged under Lenin's and Stalin's leadership, with the former's desire to provide electricity to the cities and countryside and the latter's ambition to develop nuclear weapons and power plants, tended to lend considerable prestige to the academy members. The state's appreciation for science meant new research institutes, not only for practical purposes but also for pure research. One must admit that, despite the inclusion of ideologues in academy membership, the Soviet Union under Stalin valued its scientists highly. It was not interested in pushing out high-caliber scientists, because they held the key to achieving technological goals. This was especially true in the case of physicists in the 1940s. Generally speaking, they were protected from the need to conform to ideology because of the government's desire to develop nuclear weapons.

In 1946, at the dawn of the Cold War, the technological imperative was clear to many Soviets, scientist or not. The academy declared its goal of helping the Soviet Union achieve a new life, recover from the war, and achieve greater technological progress. Pure research still existed, but it did not have the same cachet it had in the West. In 1950, Sergei Vavilov (1891–1951), who had been the academy's president in the 1940s, decried the tendency of some scientists to favor their "personal tastes" over the interests of the nation. But at the same time he acknowledged that the academy needed to maintain some semblance of pure science. Indeed, science in the Soviet Union tried to achieve the ideal of pure science while embracing science in the service of the state. The academy was the symbol of leading scientists' efforts to follow a middle course between the two.

See also Lysenko, Trofim; Soviet Science; Vavilov, Sergei

References

Graham, Loren R. *Science and Philosophy in the Soviet Union.* New York: Alfred A. Knopf, 1972.

Guins, George C. "The Academy of Sciences of the U.S.S.R." *Russian Review* 12:4 (Oct. 1953): 269–278.

Kojevnikov, Alexei. "President of Stalin's Academy: The Mask and Responsibility of Sergei Vavilov." *Isis* 87:1 (1996): 18–50.

Vucinich, Alexander. *Empire of Knowledge: The Academy of Sciences of the USSR, 1917–1970.* Berkeley: University of California Press, 1984.

Age of the Earth

Calculations of the age of the earth have undergone drastic revision during the twentieth century. Even in the nineteenth century, the debate was not merely a religious one. Although those who took the Bible literally insisted on a seven-day creation and a very young earth, most scientists (even religious ones) acknowledged that the earth must be at least a few million years old. But new theories of natural history began to require far greater spans of time than physicists would allow. Charles Darwin's (1809–1882) theory of evolution by natural selection and the proponents of uniformitarian geology suggested long periods of biological history during which time the species of the world evolved and the earth's features formed. Physicists, led by William Thomson (Lord Kelvin) (1824–1907), estimated the age of the earth by measuring its heat loss. Their earth was young and could not accommodate evolution or vast geological time.

The discovery of radioactivity by Henri Becquerel (1852–1908) in 1896 indicated that much of the earth's heat was generated from previously unknown sources. Estimates of the earth's age needed to be revised to account for radioactive processes. This source of heat would provide the earth with a great deal of staying power, far into both the past and the future. But even after Becquerel's

Lord Kelvin reading in his study. Even after Becquerel's discovery of radioactivity, Kelvin estimated the earth's age—based on its rate of cooling—at fewer than 40 million years. (National Library of Medicine)

discovery of radioactivity, Kelvin estimated the earth's age, based on its rate of cooling, at fewer than 40 million years. Although Kelvin has been ridiculed as an elder scientist (he was seventy-three years old) holding on dogmatically to older views, his aim was not to find a way to reject Darwinism. In fact, he was attacking uniformitarian geologists who suggested that the earth's activities were eternal and would go on in perpetuity. To Kelvin, this view violated the second law of thermodynamics. Instead, the release of heat by the earth indicated that it was cooling, and this would not go on forever. But Kelvin's estimates were more than a hundred times smaller than later ones, and his defense of physics has gone down in history as the defense of dogma.

The discovery of new radioactive (and heat-producing) elements, notably by Marie Curie (1867–1934) and Pierre Curie (1859–1906) in the first few years of the twentieth century, convinced most physicists that Kelvin had grossly underestimated the age of the earth. Ernest Rutherford (1871–1937) and Frederick Soddy (1877–1956) proposed that that radioactivity resulted in the "decay" of elements as they transmuted into other elements. Over the next several years, they and other scientists, including Bertram Boltwood (1870–1927), attempted to identify the elements of radioactive decay chains. Where did these decays end? Lead appeared to be the most likely endpoint for radioactive decay, as some proportion of lead, a stable (not radioactive) element, always was present in ores of radioactive elements. Once a decay series was identified and half-lives determined (a half-life is the amount of time for half of the atoms in a sample to decay), one needed only to measure the ratio of lead to uranium in an ore to calculate how long it had been decaying. Boltwood pioneered the measurement of lead in determining the age of rocks and in 1907 published a figure of 2.2 billion years as the age of one of his samples. This figure was later revised because Boltwood had not anticipated that some of the lead came from a thorium decay chain, not a uranium decay chain. By 1915 physicists settled on about 1.6 billion years as the age of the oldest minerals found in the earth's crust.

Geologists were slow to accept these new estimates, because they came from outside their discipline. Geologists' methods for determining the age of the earth were in measuring the deposition of sediments, or in measuring the rate at which sodium is carried by rivers into the oceans. Both of these methods appeared to give an estimate of around 100 million years. The vast difference between geologists' estimates and the radioactivity time scale highlighted a serious problem in earth science.

In 1920, geologist and cosmologist Thomas Crowder Chamberlin (1843–1928) tried to combine radioactive dating with biology and his own planetesimal hypothesis of the earth's original formation, formulated with Forest Moulton (1872–1952) in 1905. The earth, he said, was formed of smaller bodies (planetesimals) coming together, at a temperature fit for the development of life. Biologists estimated the evolution of life after the early Paleozoic era to be about one-tenth the total period of evolution, and radioactive dating put the Paleozoic era at about 400 million years ago. Thus Chamberlin arrived at a figure of about 4 billion years. Other scientists were more conservative; Ernest Rutherford in 1929 suggested a maximum age of 3.4 billion years.

The controversy did not end there. Astronomers disagreed with radioactive estimates because their own estimates of the age of the whole universe was relatively short. From the work of Edwin Hubble (1889–1953) in the 1920s and 1930s, cosmologists believed the universe was expanding at a measurable rate. The birth of the universe, perhaps in a Big Bang, seemed to have occurred about 2 billion years ago. Only in 1950s did astronomers expand their own time scale, helping to resolve a crisis not only in the age of the earth but also in cosmology. Radioactive dating, by measuring the uranium-to-lead ratio, prevailed. This method, which suggested a maximum of 5 billion years, received the blessing of Pope Pius XII (1876–1958) in 1951. In 1953, basing his estimate on meteoritic material, Clair Cameron Patterson (1922–1995) determined the age of the earth at approximately 4.5 billion years. This estimate stands today.

See also Boltwood, Bertram; Carbon Dating; Earth Structure; Radioactivity

References
Badash, Lawrence. "Rutherford, Boltwood, and the Age of the Earth: The Origin of Radioactive Dating Techniques." *Proceedings of the American Philosophical Society* 112 (1968): 157–169.
Brush, Stephen G. "The Age of the Earth in the Twentieth Century." *Earth Sciences History* 8:2 (1989): 170–182.
Burchfield, Joe D. *Lord Kelvin and the Age of the Earth.* New York: Science History Publications, 1975.

Amino Acids

The structures of living organisms—such as skin, tendons, and muscles—are made up of proteins. These proteins are seemingly infinite in number and are the building blocks of the plant and animal worlds. The amino acids are the units that make up protein. The German chemist Emil Fischer (1852–1919) was the first to determine this, in 1902. Fischer was known primarily for his work on sugars in the body, and he was one of the world's leading organic chemists (he was awarded the Nobel Prize in Chemistry in 1902 for his work on sugars). In addition to this work, he noted that each amino acid is composed of at least one amino (base) and one carboxyl (acidic) group. Fischer reasoned that the base of one amino acid could link up with the acidic part of another. He thus proposed that amino acids often joined together, bound together by a peptide bond, creating chains of amino acids called polypeptide chains. As these chains of amino acids grew in length and acquired the requisite kinds of amino acids, they formed proteins. Fischer was able to produce such chains artificially, and his work was later reproduced by other scientists and pursued in other institutions, notably the Rockefeller Institute for Medical Research in New York.

One of the outstanding scientific questions regarding amino acids concerned which of them were indispensable in the animal diet and which of them could be synthesized by the human body without taking them in as food. Scientists believed that there were twenty-one amino acids, but they did not know which of them were "essential" in an organism's diet. Several scientists worked on this problem, including British scientist Frederick Gowland Hopkins (1861–1947), and Lafayette Mendel (1872–1935), Thomas B. Osborne (1859–1929), and William Cumming Rose (1887–1985) in the United States. In the mid-1930s, Rose succeeded in separating the amino acids into ten "essential"— needing to be supplied in food—and eleven "nonessential" amino acids, based on his studies of the diet of laboratory rats. Although these were not precisely the same for all animals, Rose's work provided an invaluable guide to the roles played by the amino acids. He later showed that one of the amino acids, histidine, was an essential amino acid for all animals tested except for humans. These studies, along with Fischer's earlier work, provided the basis for the field of protein chemistry, a major avenue of research within the growing subject of biochemistry.

See also Biochemistry

References
Florkin, Marcel. *A History of Biochemistry.* New York: Elsevier, 1972.
Perutz, M. F. "Proteins: The Machines of Life." *The Scientific Monthly* 59:1 (1944): 47–55.
Van Slyke, Donald D. "Physiology of the Amino Acids." *Science,* New Series 95:2463 (13 March 1942): 259–263.

Anthropology

Anthropology is the study of mankind. Although today anthropology is rarely considered a "hard" science, it has a strong scientific tradition. In the nineteenth century, anthropologists adopted methodologies that enhanced the scientific status of their findings. Skull measuring and brain weighing, for example, for a time lent some credence to claims, made by French anthropologist Paul Broca (1824–1880) and others, about the cognitive superiority and inferiority of different races. Much of the influential work in anthropology dealt with racial classification, and anthropology became an important tool for the apologists for slavery in the United States prior to the Civil War. British anthropologist Edward Tylor (1832–1917) helped to codify an intellectual school of thought, evolutionism, which focused on the development of mankind's civilization through stages. He believed these stages were universal across cultures, but each of the world's peoples existed at a different level. This outlook reinforced preexisting racial attitudes of superiority and inferiority. Because research indicated a wide variety of influences on human societies, however, such as totemism, sexuality, and kinship, early twentieth-century anthropologists such as William H. R. Rivers (1864–1922) tried to augment evolutionism with "diffusionism," noting the separate paths taken by disparate peoples because of differing social structures.

Anthropology became the leading *social science,* a term used to describe fields that explore human society with some methodological tools borrowed from the natural sciences. In the 1920s, a number of anthropologists were dissatisfied with the historical approach of both evolutionism and diffusionism because of its inherently conjectural nature and argued for a view based on more stringent methodology. The result was functionalism, and its leading adherents were British anthropologists Bronislaw Malinowski (1884–1942), Polish-born, and Alfred R. Radcliffe-Brown (1881–1955). Malinowski was particularly keen to make anthropology a strictly empirical science, inspired by the positivist philosophy of Ernst Mach (1838–1916). Functionalism implied that all aspects of society have a purpose that is integral to the whole, and these aspects can be identified and perhaps even quantified to gauge their importance to the whole. Malinowski's other influences included sociologist Émile Durkheim (1858–1917) and psychologist Sigmund Freud (1856–1939), from whom he learned about the importance of social and cultural practices and the individual's internalization of cultural norms. The functionalists emphasized rigorous data collection, requiring extensive fieldwork such as Malinowski's in New Guinea. More than ever before, anthropologists were obligated by their professional standards to leave the comfort of their homes and universities and to travel to the areas they studied and encounter the peoples they analyzed. The impulse to ensure proper "scientific" bases for subjects beyond the natural sciences is often called scientism, and it had a pervasive influence among anthropologists in the twentieth century.

Why did this change from evolutionism to functionalism occur? One reason was certainly tied to the anthropologists' own influences, as in the case of Malinowski. Another was patronage. Funding organizations such as the Rockefeller Foundation provided money for research more readily when the research had sound methodological foundations. In other words, anthropology needed a rigorous, scientific basis to acquire funds for fieldwork. Functionalism asserted precise roles for every aspect of society. Malinowski was skilled at arguing that anthropological research would be of practical benefit to colonial regimes. One important result of this change in anthropologists' attitudes (functionalism replacing evolutionism) was in social control. Evolutionism had emphasized the ranking of different races in a hierarchy of civilization, providing a rationale for imperialism. Functionalism drew out the structures of indigenous cultures, including nascent political relationships, allowing imperial powers (especially the British, where the anthropologists were particularly influential) to understand local institutions and to permit and encourage some degree of autonomous rule.

Probably the most influential of all twentieth-century anthropologists was the German-born American Franz Boas (1858–1942), who was an ardent critic of racial determinism. Boas conducted fieldwork in Canada upon the Kwakiutl, an indigenous tribe. He became convinced that social and even physical differences in peoples resulted not from biological necessity, but rather from a combination of historical, geographic, and social factors, which gave rise to unique cultures. Most of cultural anthropology owes an intellectual debt to Boas, who insisted that culture was the dominant influence on individuals. Cultural anthropology struck a blow against universalism; it asserted that societal differences are not racially determined and have nothing to do with any "stage" of civilization; instead, they are a product of the specific cultural environment into which individuals are born and raised. Some of the most prominent anthropologists were disciples of Boas, including Ruth Benedict (1887–1948) and Margaret Mead (1901–1978)—both of whom tried to shatter universalist notions. Mead, for example, asserted that what most Westerners thought were normal, natural experiences of adolescents were absent in some populations in the Pacific region. Though some of Mead's findings later were questioned seriously, the outlook of cultural anthropology became a particularly "American" tradition by mid-century.

See also Boas, Franz; Leakey, Louis; Mead, Margaret; Missing Link; Peking Man; Piltdown Hoax; Scientism

References

Goody, Jack. *The Expansive Moment: The Rise of Social Anthropology in Britain and Africa, 1918–1970.* New York: Cambridge University Press, 1995.

Harris, Marvin. *The Rise of Anthropological Theory: A History of Theories of Culture.* New York: Cromwell, 1968.

Kuper, Adam. *Culture: The Anthropologists' Account.* Cambridge, MA: Harvard University Press, 1999.

Stocking, George W., Jr. *After Tylor: British Social Anthropology, 1888–1951.* Madison: University of Wisconsin Press, 1995.

Antibiotics

The term *antibiotic* was first used during World War II to describe any biologically produced substance that was harmful to other living organisms. Since then, its meaning has narrowed to refer to drugs that combat bacterial infections. The control of infectious diseases was addressed, at the beginning of the twentieth century, by public health measures and new vaccines for traditional human enemies such as tuberculosis. Around the turn of the century, assessing disease as a product of specific microbes helped German scientists such as Robert Koch (1843–1910) and Paul Ehrlich (1854–1915) to develop chemical therapies for specific diseases. Ehrlich was famous for having proposed the possibility of a "magic bullet" for each disease—a specific substance that would destroy another specific substance.

Aside from chemical agents, microbiologists recognized that some living organisms had natural, deleterious effects on others. At Rutgers University in 1915, Selman A. Waksman (1888–1973) turned the Department of Soil Microbiology into a center for research on the interrelationships among microbes, particularly their harmful effects on one another. In 1939, influenced by the work at Rutgers, Rockefeller Institute scientist René Dubos discovered tyrothricin, an antibiotic useful for fighting pneumonia. It became the first antibiotic to be commercially exploited.

The most well-known antibiotic, penicillin, was recognized in 1928 by British scientist Alexander Fleming (1881–1955). He had been researching influenza, and he accidentally allowed one of his culture plates to become contaminated by the mold penicillium. He later discovered that penicillium dissolved the bacteria that had existed prior to contamination by the mold. He named the antibacterial substance penicillin. Some years later, the need for therapeutic penicillin during World War II jump-started research. Ernst Chain (1906–1979) and Howard Florey (1898–1968) began in 1939 to explore the therapeutic effects of antibacterial substances, including penicillin. Financed by the Rockefeller Foundation, these two headed a team of British scientists who eventually produced small quantities of penicillin that killed certain germs in the human body.

Manufacturing penicillin on a large scale was one of the great industrial challenges during the war. Although the Manhattan Project (the effort to build the atomic bomb) claimed the lion's share of notoriety for scientific achievement, penicillin was no less of an accomplishment. Coordinated by the Office of Scientific Research and Development (OSRD) in the United States, scientists collaborated with several drug manufacturers and put the drug into widespread use during the war to help prevent infections from wounds and to cure rampant diseases such as syphilis. For their role in discovering penicillin and developing it as a therapy, Fleming, Chain, and Florey shared the Nobel Prize in Physiology or Medicine in 1945.

Dr. Selman A. Waksman shown at work in his Rutgers University laboratory in 1952 in New Brunswick, New Jersey. In 1943, Waksman announced his development of streptomycin, which he hoped would help in the fight against resistant strains of tuberculosis. (Bettmann/Corbis)

Despite the seemingly miraculous effects of penicillin for preventing infections and curing venereal diseases, it was not the cure-all that some hoped. Scientists continued research on other antibiotics, and Waksman's group at Rutgers assumed a leadership role in developing drugs to combat bacteria that resisted penicillin and tyrothricin. In 1943, Waksman announced his development of streptomycin, which he hoped would help in the fight against resistant strains of tuberculosis. After the war, antibiotic medicine dominated health care as a panacea for innumerable dangers, from preventing infection in major wounds to treating infectious diseases. Because no single antibiotic served as a true panacea, research on new antibiotics continued in order to address new or resistant strains of diseases.

See also Penicillin; World War II

References

Dowling, Harry F. *Fighting Infection: Conquests of the Twentieth Century.* Cambridge, MA: Harvard University Press, 1977.

Epstein, Samuel, and Beryl Williams. *Miracles from Microbes: The Road to Streptomycin.* New Brunswick, NJ: Rutgers University Press, 1946.

Parascandola, John, ed. *The History of Antibiotics: A Symposium.* Madison, WI: American Institute of the History of Pharmacy, 1980.

Sheehan, John C. *The Enchanted Ring: The Untold Story of Penicillin.* Cambridge, MA: MIT Press, 1982.

Arrhenius, Svante

(b. Vik, Sweden, 1859; d. Stockholm, Sweden, 1927)

Svante Arrhenius, a Swedish chemist, was an exemplar of interdisciplinary science. Although he published his most significant work in the late nineteenth century, his methods, combining physics and chemistry, were emblematic of the blurred boundaries between these disciplines at the turn of the century. He wanted to study physics for his doctorate at the University of Uppsala, but became dissatisfied with his mentor and chose chemistry instead, working under Erik Edlund (1819–1888). His interest in both fields would serve him well, because he became a principal founder of physical chemistry.

His initial aim was to find a way to determine molecular weight of dissolved compounds by measuring how well they conducted electricity. A compound that conducts electricity when dissolved is an electrolyte. He experimented with electrolytes of varying levels of dilution in water, and his work turned toward measuring how such dilution affected conductivity. Increasingly he found that the most interesting aspect was the state of the electrolyte itself, part of which seemed to break down into its constituent parts. The extent to which this occurred depended on how much it was diluted.

In 1884 he published his results as his Ph.D. dissertation and in a memoir to the Swedish Academy of Sciences. He revealed that electrolytes assume an active form and an inactive form, and in fact only the active part conducts electricity. He later claimed that this was a precaution designed to ensure that he receive his degree. He did not yet feel confident enough (at least in a doctoral dissertation) to say that he believed that dilution was in fact breaking down the compounds, or dissociating them. Most chemists were not willing, for example, to say that by diluting sodium chloride in water, some of it would dissociate into sodium and chlorine.

His dissertation met a cautious reception and was awarded only a fourth-class distinction, or *non sine laude* (not without praise). Few knew what to make of his theories. Chemists did not recognize them as chemistry, and physicists did not recognize them as physics. Over the next few years, Arrhenius traveled to other European laboratories while developing his ideas into a mature theory of electrolytic dissociation. This he published in 1887, when he specified that the electrolytes, under greater dilution, in fact became dissociated, meaning that the compound was breaking down into its constituent parts. This work won him the Nobel Prize in Chemistry in 1903. In 1905, he became director of the Nobel Institute for Physical Chemistry. Thus, as an "elder statesman" of science, Arrhenius was positioned to play a crucial role in the awarding of Nobel Prizes in the early twentieth century. Historian Elisabeth Crawford credits Arrhenius with the reorientation of the prize committees toward the "avant-garde of international physics and chemistry," praising the new work in quantum physics and atomic structure.

Looking back on these events in light of the rapid changes in physics at the dawn of the twentieth century, Arrhenius said that his work showed that atoms charged with electricity played an important role in chemistry. He pointed to a general tendency for scientists to attach more and more importance to electricity, "the most powerful factor in nature." Arrhenius was very much aware that studies of electricity had led to J. J. Thomson's (1856–1940) electron theory, and he predicted rapid progress along similar lines. The boundary between physics and chemistry was becoming less clear, and the nature of the atom itself would become crucial for both fields in the first decades of the twentieth century.

See also Nobel Prize; Thomson, Joseph John

References

Arrhenius, Svante. "Development of the Theory of Electrolytic Dissociation" (Nobel Lecture, December 11, 1903). In *Nobel Lectures, Chemistry, 1901–1921*. Amsterdam: Elsevier, 1964–1970, 45–58.

Crawford, Elisabeth. "Arrhenius, The Atomic Hypothesis, and the 1908 Nobel Prizes in Physics and Chemistry." *Isis* 75:3 (1984): 503–522.

———. *Arrhenius: From Ionic Theory to the Greenhouse Effect.* Canton, MA: Science History Publications, 1996.

Snelders, H. A. M. "Arrhenius, Svante August." In Charles Coulston Gillispie, ed., *Dictionary of Scientific Biography,* vol. I. New York: Charles Scribner's Sons, 1970, 296–302.

Artificial Elements

Experiments in nuclear physics escalated dramatically with the discovery of the neutron and the invention of the cyclotron in the 1930s. British physicist James Chadwick (1891–1974) discovered the neutron in 1932; because it lacked an electric charge but had considerable mass, scientists immediately saw that it would not be repelled by other atomic nuclei; thus it could be used to bombard other elements and observe the results. Scientists working under Ernest Lawrence (1901–1958) at the University of California–Berkeley already had been accelerating particles with Lawrence's invention, the cyclotron, to conduct such bombardments. Accelerating elements such as deuterium (a heavy isotope of hydrogen) provided physicists with fast-moving atoms that, upon collision, could disturb the makeup of other atoms and create interesting results.

Claims of having discovered artificial elements, which were predicted from empty slots on the periodic table of elements, dated from the 1920s. But definitive evidence of these missing elements did not appear until the late 1930s. From a sample created at the cyclotron in Berkeley, physicists in Italy, Carlo Perrier (1886–1948) and Emilio Segrè (1905–1989), first discovered an artificially created element, with atomic number 43, in 1937. Because it was the product of human technology, they called the new element technetium. Francium, atomic number 87, was discovered in France (hence the name) in 1939. Promethium, atomic number 61, was thought to have been discovered in the 1920s, but was identified in 1947 by a U.S. team at the Oak Ridge National Laboratory in Tennessee. Astatine, atomic number 85, was identified in 1940, at Berkeley. These elements were considered artificial because they were not discovered in nature but rather were produced in the laboratory. Scientists suspected, however, that they did occur in nature but simply had not yet been found. Both francium and astatine were found to be part of the natural radioactive decay series of actinium, and thus their "artificial" status comes purely from their historical context.

The more celebrated artificial elements were the transuranic elements, the name given to heavy elements with atomic numbers higher than uranium on the periodic table. None of them exist in nature, but instead must be produced artificially. One of them, plutonium, proved invaluable in the construction of atomic bombs. The first transuranic element, with atomic number 93 (neptunium), was discovered in 1940 at Berkeley, by Edwin McMillan (1907–1991) and Philip Abelson (1913–). Their announcement of the discovery was one of the last papers about uranium to appear in *The Physical Review* before the research went under a veil of secrecy during wartime. Plutonium, element 94, was discovered in 1941 by Glenn T. Seaborg (1912–1999), but it was not announced publicly until 1946. This element became crucial for the atomic bomb project because, like the rare isotope uranium 235, it was fissile: Its nucleus could split easily with the addition of a neutron, and thus could be assembled into a critical mass and sustain a fission chain reaction. It provided an alternate route to the bomb. The bomb dropped on Hiroshima was made from uranium 235, whereas the bomb dropped on Nagasaki was constructed from plutonium.

During the war, Seaborg worked at the University of Chicago as part of the atomic bomb project. During that period, he and his team discovered elements 95 and 96, naming them americium and curium (the former was named for its place of discovery and its chemical similarity to another element, europium, also named for geographical reasons; the latter was named after Marie Curie [1867–1934] and Pierre Curie [1859–1906], and indeed could also apply to Irène Joliot-Curie [1897–1956] and Frédéric Joliot [1900–1958], all of whom contributed to studies of radioactivity). Americium was a decay product of an isotope of plutonium, whereas curium was produced through alpha-particle bombardment. When the war ended, the element hunting continued unabated. Bombardments yielded elements 97, 98, and 101—berkelium, californium, and mendelevium, respectively. The first two of these were named after the locations of discovery, and the last was named after the inventor of the periodic table, Dmitry Mendeleyev (1834–1907). Elements 99 and 100 were not discovered through cyclotron bombardment, but as the residue of a more powerful force: the hydrogen bomb. Named einsteinium and fermium—after Albert Einstein (1879–1955) and Enrico Fermi (1901–1954)—these elements were found after the 1952 thermonuclear test on the Pacific atoll Eniwetok.

This flurry of discovery centering on the U.S. nuclear weapons research resulted in the identification of elements 93 through 101, and Seaborg dominated these activities. He shared the Nobel Prize in Physics with Edwin McMillan for these efforts in 1951. But soon others got involved, including Swedish scientists who would name element 102 (nobelium) after Alfred Nobel (1833–1896), discovered in 1957. The next element, discovered in 1961, was discovered in Berkeley and was named for the cyclotron's inventor (lawrencium).

See also Cyclotron; Physics; World War II
References
Badash, Lawrence. *Scientists and the Development of Nuclear Weapons: From Fission to the Limited Test Ban Treaty, 1939–1963.* Atlantic Highlands, NJ: Humanities Press, 1995.
Heilbron, John L., and Robert W. Seidel. *Lawrence and His Laboratory: A History of the Lawrence Berkeley Laboratory,* vol. 1. Berkeley: University of California Press, 1989.
Kragh, Helge. *Quantum Generations: A History of Physics in the Twentieth Century.* Princeton, NJ: Princeton University Press, 1999.

Astronomical Observatories

Many new astronomical observatories were built in the first half of the twentieth century. The most important of these were built specifically to house large telescopes to enhance astronomers' ability to see deeper into space. These telescopes were difficult to build, requiring great technical precision from both science and industry. Although the United States took the lead in building them, astronomical observatories existed all over the world. Some of the leading ones were quite old: The observatory at the University of Leiden was established in 1633, the Paris Observatory was established in 1667, and the Royal Observatory in Greenwich, England, was established in 1675. Over the centuries these housed sextants, quadrants, and other instruments, giving way in the nineteenth century to large telescopes. By the early twentieth century, telescopes of increasing size dominated the work of major observatories. Observatories used telescopes in combination with photographic technology to resolve and record the visible light in space.

Most observatories around the turn of the century were using photometric techniques. Under the directorship of Edward C. Pickering (1846–1919), for example, the Harvard College Observatory used photographs to catalogue the sky, collecting photographic plates from its own telescope and others throughout the world. These efforts led to Henrietta Swan Leavitt's (1868–1921) work on Cepheid variables, which helped astronomers judge distances. The Harvard College Observatory declined in importance in the twentieth century; however, one of Pickering's students, George Ellery Hale (1868–1938), became the principal advocate of astronomical observatories in the United States.

Although the history of twentieth century astronomical observatories was dominated by U.S. efforts, this was because of economic and political reasons rather than lack of talent. Many leading astronomers were trained in Europe, only to be attracted later to the new large U.S. observatories. One leading European center was the Hamburg Observatory, in Germany, but in 1931 one of its leading figures, Walter Baade (1893–1960), decided to move to the United States. The Estonian Bernhard Schmidt (1879–1935) designed the astronomical camera, giving his name to the Schmidt corrector plate and the Schmidt telescope, the designs of which were copied in many countries. He worked at the Hamburg Observatory in the 1920s and 1930s, and his designs allowed very large regions of the sky to be photographed. Owing to internal political problems and the advent of war in 1939, the Hamburg Observatory was unable to capitalize on Schmidt's design. The Mount Wilson Observatory in the United States, under the guidance of former Hamburg astronomer Walter Baade, began to build a 48-inch Schmidt telescope in the late 1930s.

The United States became the focal point of efforts to build the largest telescopes in the world. That was James Lick's (1796–1876) goal toward the end of the nineteenth century. A quirky millionaire, Lick paid for an observatory with the biggest-yet refractor telescope to be built near San Jose, California, dedicated in 1888 and named after him. Another center of astronomical research sprung up at the University of Chicago. In 1892, Hale and the university's president convinced businessman Charles T. Yerkes (1837–1905) to finance an observatory with a 40-inch refractor telescope. The University of Chicago developed a cooperative relationship with the University of Texas, which had built the McDonald Observatory in Mount Locke, Texas. In the 1930s, both Yerkes and McDonald came under the directorship of

Hooker telescope with 100-inch reflector at the Mount Wilson Observatory. (Hale Observatories, courtesy AIP Emilio Segrè Visual Archives)

Russian immigrant Otto Struve (1897–1963), who recruited many of Europe's top astronomers, including Gerard Kuiper (1905–1973) of the Netherlands and Subrahmanyan Chandrasekhar (1910–1995), the astrophysicist from India (then at Cambridge) who had transformed physicists' understanding of the evolution of stars.

Hale was a key figure behind some of the most advanced telescopes in the United States. Aside from his role in founding the Yerkes Observatory, he acquired grants from the Carnegie Institution of Washington to build at Mount Wilson (near Los Angeles, California) a 60-inch telescope. Completed in 1908, it was the largest telescope lens in use at the time. While it was being built, Hale already was planning another, even larger telescope. This one, with a 100-inch lens, was less reliable because of it sensitivity to temperature, but with far greater resolution of distant objects. Leading astronomers flocked to Mount Wilson; it was there that scientists such as Harlow Shapley (1885–

1972) and Edwin Hubble (1889– 1953) calculated the size of the galaxy and the radial velocities of nebulae. In 1928, Hale convinced the Rockefeller Foundation to pay for an even larger lens, a $6 million, 200-inch reflecting telescope, to be installed at Palomar, a hundred miles from Pasadena, California. The Palomar Observatory was built in the 1930s and was operated by the young campus of the California Institute of Technology, another of Hale's many projects. The 200-inch lens was a long-term project, with the difficult task of grinding, polishing, and aligning the massive chunk of Pyrex. It was installed and dedicated, finally, in 1948, and the first photographic exposure from the lens was taken in early 1949. Hale had died in 1938, and the new telescope was named after him.

See also Hale, George Ellery; Hubble, Edwin; Kapteyn, Jacobus; Leavitt, Henrietta Swan; Lowell, Percival; Pickering's Harem

References

Donnelly, Marion Card. *A Short History of Observatories.* Eugene: University of Oregon Books, 1973.

Meadows, A. J. *Greenwich Observatory, vol. 2, Recent History (1836–1975).* New York: Scribner, 1975.

Osterbrock, Donald E. *Yerkes Observatory, 1892–1950: The Birth, Near Death, and Resurrection of a Scientific Research Institution.* Chicago: University of Chicago Press, 1997.

Wright, Helen. *James Lick's Monument: The Saga of Captain Richard Floyd and the Building of the Lick Observatory.* New York: Cambridge University Press, 1987.

Astrophysics

Astrophysics is a branch of astronomy that deals primarily with dynamic relationships and physical structures. It came to prominence before the twentieth century and was substantially helped by the development of new technology. The first of these was the incorporation of photographic methods. By the late nineteenth century, photography was almost entirely integrated into astronomical observation, and the great observatories used photographic plates to compile comprehensive star catalogues and to measure (with the use of a theodolite) the coordinates of those stars. This led to the involvement of women technicians to conduct the more rudimentary calculations for lower pay than men, as was the case at the Harvard College Observatory under Edward Pickering's (1846–1919) leadership at the turn of the century. The innovator of this method, Dutch astronomer Jacobus Kapteyn (1851–1922), employed men from a local prison near Groningen University. The study of photographic plates led to major compilations such as the *Henry Draper Catalogue,* published between 1918 and 1924, based largely on the work of Annie Jump Cannon (1863–1941). It also led to the study of the Cepheid variables, a useful tool in judging distances between stars, pioneered by Henrietta Swan Leavitt (1868–1921). Aside from photography, the construction of observatories with increasingly large lenses proved to be a boon to astrophysics in the twentieth century. U.S. astronomer George Ellery Hale (1868–1938) was one of the principal figures in these developments; he was a great fund-raiser and had a hand in the construction of the Yerkes, Mount Wilson, and Palomar Observatories in the United States.

One of the controversies in twentieth-century astrophysics was the nature of the galaxy. Scientists had conjectured about "island universes" for years, because of the presence of cloudlike masses (nebulae) in space. But the notion that our own universe is only one of many was hotly contested. A famous debate at the (U.S.) National Academy of Sciences in 1920, between Harlow Shapley (1885–1972) and Heber Curtis (1872–1942), centered around the size of the Milky Way and the possibility of there being other galaxies separate from ours. Curtis argued for a small Milky Way that was one of many; Shapley argued for a large Milky Way that encompassed the nebulae. Only the work of Edwin Hubble (1889–1953) appeared to settle the argument. In the 1920s,

Hubble used Cepheid variables to judge distance and noted (from stellar spectra) that some of the most distant nebulae were receding. Not only were they far, far away—and thus separate galaxies—but they were moving away from us, indicating that the universe was expanding. This finding proved astounding, and it lent support to what Belgian physicist Georges Lemaître (1894–1966) had proposed in 1927, namely, the idea that became known as the Big Bang theory. This theory of the origin of the universe, with its stars and galaxies the result of an enormous explosion, sat well with many religious-minded people (including Lemaître, a Jesuit), who sought a theory of the universe that left room for an act of creation.

A number of astrophysicists devoted their work to developing hypotheses of stellar evolution. Their primary task was to use recent revolutions in physics—namely, Max Planck's (1858–1947) quantum theory and Albert Einstein's (1879–1955) theories of relativity—to construct a meaningful model of stars. The German Karl Schwarzchild (1873–1916) was one of the first to propose such a model, based on radiation transfer according to the laws of quantum physics, but his theory (1906) was replaced by that of Arthur Eddington (1882–1944) in the 1920s. A lasting contribution of Schwarzchild's, however, was the concept of a black hole, which was simply a star that had collapsed to a radius below a certain threshold determined by its mass; such a star could not emit any radiation at all—in other words, even light could not escape, so these stars are invisible. Another limit on stellar structure was proposed by Subrahmanyan Chandrasekhar (1910–1995) in 1928. He noted that there is a limit to the mass a star can have and still behave according to the prevailing model. Some stars, he observed, will never reach the state "white dwarf," the supposed end stage of stars of any mass that have lost most of their energy. According to Chandrasekhar, at masses larger than 1.44 solar masses, the gravitational load on the outer layers of the star will cause it to collapse, even when theoretically there should be a balance of repulsion and gravitational attraction. This idea was immediately criticized by most, including Eddington, but eventually Chandrasekhar won the Nobel Prize for it (1983). It gave rise to research on alternative end stages, such as neutron stars and black holes.

For the evolution of stars, astrophysicists relied on the simultaneous yet independent work of the Dane Ejnar Hertzsprung (1873–1967) and the American Henry Norris Russell (1877–1957). Correlating spectrum type with absolute magnitude, they proposed the existence of two series of stars, one "normal," and the other composed of very bright, giant stars. The graphical representation of the evolutionary sequence, first published in 1911, was called the Hertzsprung-Russell diagram. From these ideas, British astrophysicist Arthur Eddington worked out a mass-to-luminosity relationship in stars and constructed a model of stars based on the balance of internal and external pressures. He published it as *The Internal Constitution of Stars* (1926), calling the Hertzsprung-Russell diagram not simply a classification scheme but an indicator of stellar evolution.

The 1930s were critical years in the history of astrophysics. In 1932, Dutch astrophysicist Jan Hendrik Oort (1900–1992) postulated the existence of "dark matter." He calculated that the prevailing beliefs about the universe's dynamic action required there to be two or three times as much mass in the universe as could be seen. Because the galaxy did not appear to be coming apart, but rather rotating, there had to be enough mass to account the gravitational attraction needed to keep the galaxy in place. In 1937, Caltech astrophysicist Fritz Zwicky (1898–1974) determined that most of the universe's mass is this mysterious dark matter. Zwicky also played a crucial role, along with Walter Baade (1893–1960), in research on neutron stars and supernovae. Also in the 1930s, radio astronomy was born, which allowed scientists to "see" in space without relying on the visible light necessary for optical telescopes. Like the development of photographic methods at the end of the nineteenth century, radio astronomy became the principal technology to transform the fields of astronomy and astrophysics after World War II.

See also Chandrasekhar, Subrahmanyan; Eddington, Arthur Stanley; Hertzsprung, Ejnar; Hubble, Edwin; Oort, Jan Hendrik; Radio Astronomy

References

Gingerich, Owen, ed. *Astrophysics and Twentieth-Century Astronomy to 1950.* Cambridge: Cambridge University Press, 1984.

Lang, Kenneth R., and Owen Gingerich, eds. *A Source Book in Astronomy and Astrophysics, 1900–1975.* Cambridge, MA: Harvard University Press, 1979.

North, John. *The Norton History of Astronomy and Cosmology.* New York: W. W. Norton, 1995.

Shapley, Harlow. *Through Rugged Ways to the Stars.* New York: Scribner, 1969.

Wright, Helen. *Explorer of the Universe: A Biography of George Ellery Hale.* New York: Dutton, 1966.

Atomic Bomb

The atomic bomb was developed during World War II as a weapon that could produce, in a single explosion, devastation equivalent to thousands of tons of conventional explosives. It was based on the possibility that the nuclei of atoms could be split, releasing particles and energy at the same time; if enough such reactions occurred, a fireball of immense proportions would be the result. Several countries attempted to develop the atomic bomb in the first half of the twentieth century, and two (the United States and the Soviet Union) were successful. Atomic bombs have been used twice in war, both times by the United States against Japan in 1945.

Long before it became a reality, the atomic bomb was a source of speculative fiction. Celebrated novelist H. G. Wells (1866–1946) wrote *The World Set Free* in 1914, predicting the use of atomic bombs in the 1950s. The bombs described by Wells were continuously explosive, requiring lumps of "pure Carolinum" to generate an unstoppable radiation of energy, like a miniature active volcano. To Hungarian physicist Leo Szilard (1898–1964), who had read Wells's book in the early 1930s, the possibility of such weapons seemed genuine after the discovery of nuclear fission in 1939. This was the name given to the division of an atomic nucleus into two parts; for example, a single atom of a heavy element could split into two atoms of a lighter element. The process was catalyzed by the bombardment of uranium with free neutrons, and it released a small amount of energy. Szilard became convinced that the process of fission could be sustained in many atoms at once, because of the ejection of a neutron during fission. That neutron might catalyze more fission reactions, which would free more neutrons; these in turn could instigate still more fission reactions. The possibility of a chain reaction meant that an extraordinary amount of energy release was possible—an atomic bomb could be built.

The discovery of fission by Germans Otto Hahn (1879–1968) and Fritz Strassman (1902–1980) occurred on the eve of World War II (the experiments were conducted in late 1938 and early 1939). After the war began, several of the main combatant countries attempted to build atomic bombs, with varying degrees of commitment. Germany put some of the best theoretical physicists to work on the project, including Hahn and the world-famous pioneer of quantum mechanics, Werner Heisenberg (1901–1976). Led by Heisenberg, the German project never determined an adequate means of building a nuclear reactor, the crucial test of a fission chain reaction. As the war drew to a close, Heisenberg's team was captured by Allied forces and transported to England, where they stayed under surveillance in a country manor called Farm Hall. Their conversations were recorded in secret, and they appeared to admit their shortcomings and lack of knowledge about how the bomb had been built by the Americans. However, Heisenberg later made the controversial claim that his team deliberately avoided making an all-out effort, not wanting German leader Adolf Hitler (1889–1945) to possess a weapon of such power. Germany's ally, Japan, had a strong nuclear physics community as well and had bought several cyclotrons (particle accelerators) from the United States prior to the war. Historians have taken an increasing interest in the Japanese atomic bomb in recent years, but it appears that the Japanese physicists did not pursue lines of inquiry that provided feasible solutions to the problem of building a bomb.

The atomic bombs that were dropped on the Japanese cities of Hiroshima and Nagasaki were developed from a cooperative effort of Britain and the United States. The United States took an interest in the possibility in 1939, especially after celebrated refugee physicist Albert Einstein (1879–1955), prodded by Szilard, described the destructive potential of the bomb to President Franklin D. Roosevelt (1882–1945). But more important, German refugee physicists Rudolf Peierls (1907–1995) and Otto Frisch (1904–1979) conducted theoretical work in Britain in 1940 and 1941 that suggested if a sufficient amount of uranium—a "critical mass"—could be brought together at a high speed, it would yield an explosion equivalent to thousands of tons of dynamite. A British committee, code-named MAUD, soon issued a report with similar findings (1941).

With the encouragement and recommendations of the British scientists, the U.S. and Britain combined their efforts in 1941 to try to develop the weapon. The first great success of the project occurred on 2 December 1942, when Italian immigrant Enrico Fermi (1901–1954) led a team of scientists (including Szilard) to achieve the first controlled fission chain reaction. They called it a "pile," but it was in fact the first working nuclear reactor. In the next several months, the United States devoted vast sums of money to develop ura-

nium and plutonium production plants in Tennessee and Washington and set up a bomb-building site in Los Alamos, New Mexico. The project officially began in 1942 and was code-named Manhattan Engineer District, though typically it is called the Manhattan Project. The scientific leader would be U.S. physicist J. Robert Oppenheimer (1904–1967), but the overall director was a general in the Army Corps of Engineers, Leslie Groves (1896–1970). The first successful test of a weapon occurred in Alamogordo, New Mexico, in July 1945. The next month, the first atomic bombs used in war were dropped on Japan. The cooperative relationship with Britain ended, particularly after it was discovered that some of the important security leaks had been from British participants. The French were also excluded after the war ended, although some participated as part of a Canadian branch of the project.

The Soviet Union also developed atomic bombs. There were a number of scientists on the British and U.S. teams who either sympathized with communism or wanted to ensure that the United States did not have a monopoly on the bomb. Spies such as German immigrant (part of the British team) Klaus Fuchs (1911–1988) transferred valuable information from Los Alamos to the Soviet Union during the war. However, the Soviet government did not make a commitment to building the bomb until its most pressing problem—the advance of German tanks and infantry—was taken care of. In early 1943, after the German army's decisive defeat at the battle of Stalingrad, Germany's advance was halted and the Soviet Union began its counteroffensive. Around the same time, Soviet leader Joseph Stalin (1879–1953) decided to hand over everything the government knew about atomic weapon development to leading physicists. Physicist Igor Kurchatov (1903–1960) led the Soviet bomb project. Despite the expectations of other countries, the Soviets produced an atomic bomb in 1949, thus breaking the U.S. monopoly.

See also Atomic Energy Commission; Cold War;
Fermi, Enrico; Fission; Hiroshima and
Nagasaki; Kurchatov, Igor; Manhattan Project;
Szilard, Leo

References

Clark, Ronald. *The Greatest Power on Earth: The
International Race for Nuclear Supremacy from
Earliest Theory to Three Mile Island.* New York:
Harper & Row, 1980.

Gowing, Margaret. *Britain and Atomic Energy,
1939–1945.* London: Macmillan, 1964.

Hewlett, Richard G., and Oscar E. Anderson Jr.
*The New World: A History of the United States
Atomic Energy Commission, vol. I, 1939–1946.*
Berkeley: University of California Press, 1990.

Holloway, David. *Stalin and the Bomb: The Soviet
Union and Atomic Energy, 1939–1956.* New
Haven, CT: Yale University Press, 1994.

Wells, H. G. *The World Set Free.* New York:
Dutton, 1914.

Atomic Energy Commission

The creation of the Atomic Energy Commission (AEC) was one of the first major political battles involving scientists, military leaders, and politicians after World War II. During the war, science had benefited handsomely from generous government patronage, particularly from the military services. The Manhattan Project in particular had brought an international team of scientists together under military control, putting enormous resources at scientists' fingertips, for the purpose of building a weapon of unprecedented magnitude. When atomic bombs were dropped on the Japanese cities of Hiroshima and Nagasaki, some of these scientists believed that the new weapon should be taken out of military hands and put into civilian ones.

At the core of the argument was that the atomic bomb could not be considered a purely military weapon. Its use against Japan had required an order by the U.S. president; unlike conventional weapons, its use would have important political ramifications as well as military ones. Scientists were surprised when, less than two months after the atomic attacks against Japan, they learned that plans

for the postwar control of atomic energy had resulted in a bill being sent to Congress without their input. The bill originated in committees oriented toward military affairs. Named the May-Johnson bill (after its initial sponsors in Congress), it was widely perceived as a blatant effort to use the momentum of victory in war to keep atomic weapons in the hands of the military. Active-duty officers of the military would lead the commission for atomic energy, according to this bill, and civilian scientists would play little role.

Scientists who recently had formed the Federation of Atomic Scientists virulently attacked the bill, which led to a new one sponsored by Senator Brien McMahon (1903–1952). The McMahon Bill gave the government a monopoly on bomb material, established a full-time commission, and prohibited active-duty military officers from being employed. Although there were numerous policy questions to be resolved between the bills, the most significant issue appeared to be the question of control: Should atomic energy rest in civilian or military hands? The Atomic Energy Act was finally passed in 1946, taking shape from the McMahon Bill and acknowledging the need to establish civilian control. It also provided a new policy role for scientists by creating within the AEC a general advisory committee made up of top scientists who had taken part in the U.S. bomb project.

The AEC officially took control of the country's atomic energy establishment on 1 January 1947, with David E. Lilienthal (1899–1981) as the first AEC chairman. Although there was some expectation that the AEC would attempt to pursue nuclear power plants for peaceful applications, the vast majority of its attention was focused on weapons development and production. Growing antagonism with the Soviet Union compelled AEC leadership to focus on making the atomic arsenal stronger. The number of bombs was a closely guarded secret. Only in the 1980s did the government reveal that the total number of bombs possessed by the

United States had been rather small: A total of nine had been produced in 1946. Also the Hiroshima and Nagasaki bombs had used a mere fraction of the material available for atomic fission, so the AEC's technical agenda seemed clear: Build more efficient bombs.

Aside from these tasks of production and development, the AEC also conducted numerous tests on how to integrate atomic weapons into battle tactics and to gauge the effects of atomic weapons on humans. Under a veil of secrecy, the AEC acted in concert with the U.S. military services in the Sandstone tests of 1948, conducted in the Marshall Islands of the Pacific Ocean. Techniques included handing out badges to individual soldiers to measure the doses they were receiving while carrying out conventional military operations. Radiation safety proved problematic for the AEC, because individuals often misunderstood or disobeyed orders, or tests unfolded in entirely unexpected (and unpredictable) ways, leading to many cases of accidental overexposure to the harmful effects of atomic radiation. The AEC also conducted experiments on unwitting patients in civilian hospitals to test their bodies' tolerance to injections of plutonium.

See also Atomic Bomb; Federation of Atomic Scientists; Manhattan Project; Radiation Protection

References

Badash, Lawrence. *Scientists and the Development of Nuclear Weapons: From Fission to the Limited Test Ban Treaty, 1939–1963.* Atlantic Highlands, NJ: Humanities Press, 1995.

Hacker, Barton C. *Elements of Controversy: The Atomic Energy Commission and Radiation Safety in Nuclear Weapons Testing, 1947–1974.* Berkeley: University of California Press, 1994.

Hewlett, Richard G., and Oscar E. Anderson, Jr. *The New World: A History of the United States Atomic Energy Commission, vol. I, 1939–1946.* University Park: Pennsylvania State University Press, 1962.

Hewlett, Richard G., and Francis Duncan. *Atomic Shield: A History of the United States Atomic Energy Commission, vol. II, 1947–1952.* University Park: Pennsylvania State University Press, 1969.

Atomic Structure

For many centuries, *atom* was the word used to describe the smallest particles. The ancient Greeks Leucippus and Democritus had proposed, in the fifth century B.C., that atoms were the building blocks of the world. In subsequent centuries, atom was an abstract concept that meant little more than something small and indivisible. In the twentieth century, however, the atom lost its basic character of indivisibility, though physicists agreed that there were indeed atoms, that all material was composed of them, and that they share similar properties. Several models of atomic structure were devised.

The first modern model to consider the atom as a unit with subatomic components was devised by British physicist J. J. Thomson (1856–1940) in 1897. His model was called the "plum pudding" atom because of the similarity between his model and the soupy culinary concoction. That year, Thomson already had made a serious breakthrough by discovering the electron, a negatively charged particle. He believed that the atom was composed of these electrons (Thomson called them corpuscles), which swam about randomly in a positively charged medium. The charges canceled each other out, and the atom itself had no charge at all. Thomson was aware that his model posed some problems, particularly his positively charged medium, which many (including himself) had a difficult time believing could have no mass and cause no friction. By 1910, experiments with radioactive decay showed the scattering of both beta and alpha particles (the first negative and the other positive), which discredited the basic assumption of the model that only the negative charges were corpuscular in nature.

The above alpha particle experiments were conducted at Britain's Manchester University under New Zealander Ernest Rutherford (1871–1937), who in 1911 devised another model to replace Thomson's. Rutherford agreed with Thomson that the atom was composed of negatively charged particles (electrons). But he also believed

Two elderly physicists and Nobel Laureates, Sir Joseph John Thomson (1856-1940), of England, and Lord Ernest Rutherford (1871-1937), of New Zealand, speaking together. (Bettmann/Corbis)

that each atom had a very tiny center with a positive charge. The whole atom was about ten thousand times the size of the nucleus, and yet almost all of the atom's mass was contained there. Initially this model was proposed simply to explain his scattering experiments. In order to explain the deflection of alpha particles by angles of more than ninety degrees, Rutherford insisted that there must be a point-particle of positive charge in the target medium. It soon became the main model of the atom. The negative electrons orbited the positive nucleus like the earth and other planets orbit the sun.

Like the Thomson model, Rutherford's atom had many problems; but in Rutherford's case, they did not lead to the model's demise. Instead, they required a theoretical explanation along the lines of the new scien-tific paradigm, namely, quantum physics. One serious problem with Rutherford's atom was the energy released by electrons: If the electrons are in orbit, why do they not emit energy, as required by the laws of thermodynamics? And if they do in fact release energy, should the electrons simply collapse into the nucleus? Danish physicist Niels Bohr (1885–1962) "saved" Rutherford's atom from these tough questions in 1913 by developing the concept of quantum orbits. The atom only gains or releases energy when electrons jump from one quantum orbit to another; otherwise, the electrons can orbit without concern about energy release. In addition, the electrons can never jump closer to the nucleus than the innermost quantum orbit. Bohr justly became known as the leading theoretical nuclear physicist from this work, and it won him the Nobel Prize in 1922. The structure of the atom as constructed by Rutherford and Bohr was a crucial foundation of modern physics.

Physicists continued to debate about the structure of the atom, usually in regard to the nucleus itself. Nuclear physics deals with the structure of atomic nuclei, which was studied intensely in the 1930s by observing nuclear reactions and through theoretical prediction. Particle accelerators such as the cyclotron, developed at Berkeley by Ernest Lawrence (1901–1958) in the 1930s, probed into the nucleus by speeding up particles and smashing them into nuclei. Outside the laboratory, Hideki Yukawa's (1907–1981) 1935 "discovery" of mesons, a crucial aspect of the nucleus, was entirely theoretical. His was a mathematical prediction of the existence of a particle in the nucleus that physicists soon identified with observations made by U.S. cosmic ray researchers Carl Anderson (1905–1991) and Seth Neddermeyer (1907–1988) in 1937. Although this view was revised a decade later (Yukawa had predicted what came to be called pions, whereas the Americans had found the muon), the research on mesons demonstrated that the atom itself was no "atom" at all in the sense envisioned by the ancient Greeks; instead, there were many more subatomic particles to be identified.

See also Bohr, Niels; Chadwick, James; Fission; Gamow, George; Quantum Theory; Rutherford, Ernest; Thomson, Joseph John; Yukawa, Hideki

References

Conn, G. K. T., and H. D. Turner. *The Evolution of the Nuclear Atom.* London: Iliffe Books, 1965.

French, Anthony P., and P. J. Kennedy, eds. *Niels Bohr: A Centenary Volume.* Cambridge, MA: MIT Press, 1985.

Kragh, Helge. *Quantum Generations: A History of Physics in the Twentieth Century.* Princeton, NJ: Princeton University Press, 1999.

B

Bateson, William

(b. Whitby, England, 1861; d. London, England, 1926)

According to his wife, William Bateson was reading a long-neglected paper by Gregor Mendel (1822–1884) while on a train to London, on the way to give a lecture at a meeting of the Royal Horticultural Society. In the course of the journey, he decided to change the lecture to include a reference to the laws set forth in that long-forgotten paper. The truth is less sensational (the paper was probably by Hugo De Vries [1848–1935] and was about Mendel, not by him), but it conveys a sense of the impact of Gregor Mendel's work on hybridization, and its rediscovery at the turn of the century, on the life work of William Bateson.

By the time of the rediscovery of Mendel in 1900, Bateson already was an accomplished biologist, though out of step with most of his contemporaries. Early on he was convinced that the prevailing view of evolution did not provide an adequate description of positive change over time. He attacked Charles Darwin's (1809–1882) theory, evolution by means of natural selection, for its focus on small but continuous changes and its reliance on adaptation as the guiding force behind evolution. He refused to believe that minor variations could ever provide a useful change for an organism. Rather than constant, gradual accumulation of an organism's modifications, Bateson proposed major modifications occurring at once, in sudden steps.

These views brought Bateson into direct conflict with the biometrical school of biology, led by Karl Pearson (1857–1936) and Walter F. R. Weldon (1860–1906). These biometricians correlated death rates to measurable characteristics, such as the width of a crab's shell. They were using statistical analysis to demonstrate natural selection in species. Bateson's hostility to continuous change provoked a heated debate, but certainly not one that was resolved in his favor. His efforts to counter the arguments of the biometricians led him to conduct breeding experiments of his own, to identify the process by which features were inherited. His views were reinforced by the 1900 rediscovery of Mendel. Bateson came to view Mendel's work as the mathematical cornerstone of a new science. He soon became Mendel's most vocal champion and even published the first translation of his work in English. Bateson saw in Mendel the possibility of establishing universal laws of heredity through discontinuous variation for both plants and animals.

Bateson published *Mendel's Principles of Heredity* in 1902 and began to trumpet Mendelism as a new science. In 1905 he called it "genetics." More than half of his research staff was composed of women, who he employed to conduct breeding experiments on various plants and animals to prove the laws of Mendelian inheritance. Lacking a secure position (he finally became a professor in 1908), he was particularly ruthless in his critiques of others, trying to establish a name for himself. He waged against the biometricians an embittered and increasingly personal battle that, while doing credit to neither side, helped to publicize the new science. He cofounded *The Journal of Genetics* in 1910. In 1914, he became president of the British Association for the Advancement of Science, clear evidence that he had brought his new science into the mainstream.

Genetics soon outstripped Bateson's intentions, somewhat to his dismay. Instead of providing a mechanism for positive evolution—for new changes in organisms—heredity more and more appeared to be confined to establishing laws by which already existing traits were passed from one generation to the next. The process of introducing new characteristics seemed less clear. The mutation theory of Hugo De Vries provided the kind of saltations, or significant and discontinuous changes, that Bateson long wished to identify. But by 1914, Bateson openly expressed doubt that evolution ever produced genuinely new characteristics. Mutations, in his view, were degenerative and destructive, not positive and constructive. Perhaps, he suggested, what we see as "new" characteristics are simply the result of a mutation destroying a gene that previously had masked another.

Bateson also opposed another development in genetics: the chromosome theory of inheritance. This theory posited a material element within the cell's nucleus as the agent of heredity. But Bateson could find no direct, visible relationship between any given chromosome and a physical trait in an organism. He proffered in place of the chromosome theory a "vibratory" theory of inheritance based on motion. This mechanistic theory seemed, to Bateson, preferable to the materialistic one provided by chromosome theory. Bateson's career after about 1915 took a conservative turn, as his life's creation took a turn toward the chromosome theory he distrusted. Perhaps for him it evoked the same materialistic devils that had troubled him about natural selection. His "vibratory" theory of inheritance found few adherents. In the final decade of his life, Bateson found himself on the periphery of the science he had helped to found.

See also Biometry; Chromosomes; Genetics; Mutation; Rediscovery of Mendel

References
Bowler, Peter J. *The Mendelian Revolution: The Emergence of Hereditarian Concepts in Modern Science and Society.* Baltimore, MD: Johns Hopkins University Press, 1989.
Coleman, William. "Bateson, William." In Charles Coulston Gillispie, ed., *Dictionary of Scientific Biography,* vol. 1. New York: Charles Scribner's Sons, 1970, 505–506.
Richmond, Marsha L. "Women in the Early History of Genetics: William Bateson and the Newnham College Mendelians, 1900–1910." *Isis* 92:1 (2001): 55–90.
Sturtevant, A. H. *A History of Genetics.* New York: Harper & Row, 1965.

Becquerel, Henri

(b. Paris, France, 1852; d. Le Croisie, Brittany, France, 1908)

Henri Becquerel was born into a dynasty of French physicists. His father and grandfather had both been members of the French Academy of Sciences and both were professors of physics at Paris's Museum of Natural History. Henri would do the same, all before achieving more lasting distinction as the discoverer of radioactivity. In the final years of the nineteenth century, Becquerel's active research seemed to be behind him. He was in his mid-forties, had established his reputation through studies in optics, and recently had succeeded his father's two chairs of physics at the museum and at the Conservatoire National des Arts et Métiers. Yet his singular contribution to twentieth-century science still lay ahead.

This next phase was catalyzed by Wilhelm Röntgen's (1845–1923) discovery of X-rays in 1895. These strange rays, which passed easily through many kinds of solid material and could produce photographs of bones underneath the skin of living people, astonished the world and the scientific community alike. The rays appeared to originate in the region of a vacuum tube struck by cathode rays (experiments with cathode rays had been widespread among physicists for decades). That region appeared to glow, or "phosphoresce," leading Becquerel and others to believe that perhaps all luminescent bodies emit something similar to X-rays.

Becquerel's subsequent experiments are now famous for leading to an accidental discovery that could only be made by a prepared mind. He exposed samples of potassium uranium sulfate to sunlight and then set them near photographic plates wrapped with metallic objects, such as coins, in thick black paper. He found that, despite the presence of the paper, the photographic plates had been exposed, and the resulting photograph showed a silhouette of the metallic object. To Becquerel, this seemed to demonstrate that phosphorescence—the lingering glow emitted after absorbing light from the sun—indeed led to the emission of some kind of ray, like X-rays. But he soon noticed something puzzling. He put a sample of the mineral into a dark drawer next to photographic plates, again wrapped in thick black paper with metal objects. Then he decided to develop the plates. Why did Becquerel develop these plates if the point of his experiments was recording phosphorescence? He later claimed that he expected to find weak images, perhaps as a result of long-lived phosphorescence in the mineral from previous exposure to the sun. Had he found such results, he would have found evidence that the emitted rays diminished in intensity over time after being removed from sunlight.

Henri Becquerel, discoverer of radioactivity. (Bettmann / Corbis)

However, when he developed these plates, Becquerel saw that the silhouettes of the metallic objects were just as pronounced as before, and thus the "ray" was just as intense as it had been immediately after the mineral was exposed to the sun. He had to conclude that the exposure was not a result of phosphorescence as it was then understood. Instead, something entirely different was occurring that might have nothing to do with a lingering glow from previous exposure to sunlight. The "ray" came from the mineral itself, regardless of external stimulation.

Becquerel had discovered radioactivity, although this name came into use only after Marie Curie (1867–1934) coined it. But at the time, it was far from clear what these new rays truly were. They were not X-rays, but still Becquerel thought perhaps they could be explained by some particular manner in which metals phosphoresce. Soon he, followed by

Pierre Curie (1859–1906) and Marie Curie, began to study this new phenomenon in earnest, finding that pure uranium emitted "rays" several times more intense than the mineral first used. Soon the Curies would identify other radioactive metals and even discover new radioactive elements. The three of them shared the 1903 Nobel Prize in Physics. The scientific problems presented by the discovery of radioactivity in the twilight of the nineteenth century engaged the best minds in chemistry and physics during the first half of the twentieth century. Ultimately radioactivity would transform human understanding of nature, especially the structure of the atom.

> *See also* Curie, Marie; Radioactivity; Uranium; X-rays
>
> *References*
> Badash, Lawrence. "Becquerel's 'Unexposed' Photographic Plates." *Isis* 57:2 (1966): 267–269.
> Romer, Alfred. "Becquerel, [Antoine-] Henri." In Charles Coulston Gillispie, ed., *Dictionary of Scientific Biography,* vol. I. New York: Charles Scribner's Sons, 1970, 558–561.
> Romer, Alfred, ed., *The Discovery of Radioactivity and Transmutation.* New York: Dover, 1964.

Big Bang

The Big Bang was the generic phrase given to theories that the universe originated in an explosion. More specifically, the introduction of the Big Bang theory into cosmology in the 1930s provided yet another replacement of Newtonian physics with a new idea brought about by Albert Einstein's (1879–1955) theory of general relativity. In place of the stable, mechanistic clockwork came an unstable, fast-moving, expanding universe. The term originally was used derisively by Fred Hoyle (1915–2001) in 1950 to describe the cosmological ideas set forth by Georges Lemaître (1894–1966).

Georges Lemaître was a Belgian physicist whose interest in creation (he was also a Jesuit) ultimately led to the Big Bang theory. His cosmological preconceptions convinced

him that a beginning state could be found, and he proposed the existence of a "primal atom" long ago that exploded to form the present universe. Lemaître had studied under Arthur Eddington (1882–1944) at Cambridge University in the early 1920s and spent nine months at the Harvard College Observatory. His ideas were built on Einstein's general theory of relativity and his own critique of existing cosmologies, such as that of Willem De Sitter (1872–1934). He published about his theory in 1927, but it was not widely read until it was translated and published by Eddington in 1931. By that time, the interpretation of red shifts in nebular spectra by Edwin Hubble (1889–1953) in 1929 had convinced most astronomers of the possibility that the universe is expanding. The question arose: Expanding from what?

Many physicists were hostile to the idea of the Big Bang. Einstein at first rejected it when he heard of Lemaître's idea, but he was also one of the first to change his mind. Eddington objected to the notion of a "beginning" to the present order of things in the universe. But in 1933 Lemaître argued persuasively that the quantum interpretation of thermodynamic principles seemed to lead inexorably toward his own view, that of a "fireworks" theory that sees the present universe as a very different one from the way things once were. As he put it: "The last two thousand million years are slow evolution: they are ashes and smoke of bright but very rapid fireworks" (Danielson 2000, 410). In the late 1940s, Russian-born physicist George Gamow (1904–1968) and his colleague Ralph Alpher (1921–) extended Lemaître's theory by envisaging a primordial state of neutrons, protons, and electrons surrounded by radiation. A major explosion, they argued, would have given birth to the light elements in the first few moments after the "bang," as subatomic particles began to adhere together in (relatively) cooler temperatures.

British cosmologist Fred Hoyle noted in 1950 that "this big bang idea" was unsatisfactory, and the name stuck. Hoyle's criticisms in the 1950s gave the Big Bang a competing

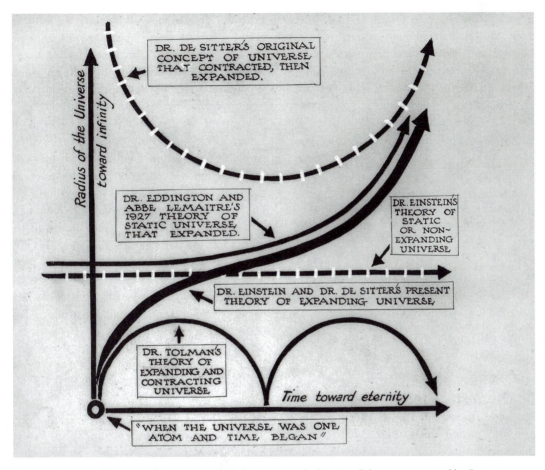

A new explanation of the origin of the universe, dubbed by critics as the "Big Bang" theory, was proposed by Georges Lemaître. (Bettmann/Corbis)

cosmology. Hoyle did not argue for the eternality of the world, but rather the continuous creation of matter. Hermann Bondi (1919–), Thomas Gold (1920–), and Hoyle proposed their Steady State theory in 1948. While Lemaître and Gamow wrote of an event at the beginning of time, they proposed that new matter spontaneously was created constantly in a steady state (hence the name).

The Big Bang theory proved controversial. Some religious critics objected to the contradictions with the revelations of the Bible, while some atheists felt that the theory left far too much room for God, because it located the universe's origins in a specific event. Certainly this aspect of the theory was appealing to theologically minded thinkers such as Lemaître. But astronomers were not convinced by mid-century. The theory had some serious problems, such as its inability to explain galaxy formation. In addition, although the explanation by Gamow and Alpher of the creation of light elements was impressive, they had not done anything close to a convincing job of explaining the creation of heavy elements. No one could find a mathematically sound solution that could explain elements heavier than helium. Although the theory was not abandoned, it clearly was incomplete. The theory's architects appeared to abandon it: Gamow moved on to other scientific fields, including molecular biology, while Alpher took a job in private industry. The Steady State theory enjoyed much more notoriety and support in the 1950s. However, the Big Bang theory enjoyed a renaissance in the 1960s when discoveries of cosmic background radiation, relics of the universe's original fireworks, seemed to back up predictions made by the Big Bang theorists. Lemaître, who died in 1966, lived long enough to see his "fireworks" theory revived.

See also Astrophysics; Cosmology; Gamow, George; Religion

References
Danielson, Dennis Richard, ed. *The Book of the Cosmos: Imagining the Universe from Heraclitis to Hawkins.* Cambridge: Perseus, 2000.
Isham, Chris J. *Lemaître, Big Bang, and the Quantum Universe.* Tucson, AZ: Pachart, 1996.
Kragh, Helge. *Quantum Generations: A History of Physics in the Twentieth Century.* Princeton, NJ: Princeton University Press, 1999.
North, John. *The Norton History of Astronomy and Cosmology.* New York: W. W. Norton, 1995.

Biochemistry

Biochemistry is concerned with chemical processes in living matter. In the first half of the twentieth century, some of the major work in this field was related to metabolic processes in the living cell and the role of enzymes. By mid-century, biochemistry permeated many fields, including fruitful interdisciplinary studies in areas such as morphology and genetics.

Germany was the major focal point of chemistry around the turn of the century. German scientists often worked to aid that country's robust dye industry. From this community came the work of Adolf Von Baeyer (1835–1917), who had determined the structure of a series of organic dyes, including indigo. Another was Otto Wallach (1847–1931), whose work on oils proved very useful to German industry. One of Von Baeyer's students, Emil Fischer (1852–1919), also worked on dyes. But Fischer studied other organic compounds, such as those in tea and coffee, and he found that many organic substances contained chemicals that appeared to be related, in a category he called

purines. His work on purines and his syntheses of sugars in the 1890s provided a foundation for the chemistry of organic substances. Fischer, Von Baeyer, and Wallach all were recognized in the fist decade of the century by being awarded Nobel Prizes in Chemistry.

Another winner of the Nobel Prize was Eduard Buchner (1860–1917), whose work on enzymes became fundamental for the study of biochemistry. Buchner focused on the fermentation of sugar as it turned into alcohol, noting that the presence of yeast cells was not always necessary. This appeared to contradict the older notion that the basic life force comes from living cells. Scientists such as Louis Pasteur (1822–1895) in the nineteenth century had argued that fermentation required the presence of living yeast cells. By contrast, Buchner demonstrated that fermentation resulted from the action of the enzymes (a protein substance) produced by the cells. Buchner made this discovery in 1897, and this date is often hailed as the birth of biochemistry.

Englishman Arthur Harden (1865–1940) had begun work on glucose but soon was fascinated by the chemistry of the yeast cell. Extending Buchner's work, he and his colleague William John Young (1878–1942) showed that fermentation also required another substance, the co-zymase. Harden's systematic chemical study of the fermentation process of sugar in yeast proved useful for numerous other scientists interested in biochemical processes. He demonstrated the important role played by phosphates in alcoholic fermentaion. This resulted in his book, *Alcoholic Fermentation* (1911). He became an influential figure in the field, editing the *Biochemical Journal* from 1913 to 1938. Nearly two decades after his pioneering work on fermentation, he was awarded the Nobel Prize in Chemistry (1929), an honor shared with the German Hans von Euler-Chelpin, who also had contributed to understanding alcohol fermentation.

Studies of enzymes culminated in the 1920s in the work of Cornell University's

James B. Sumner (1887–1955), who claimed that enzymes could be crystallized. This work was not accepted immediately, however, and work on organic crystals did not accelerate until after World War II. Most of the important work in biochemistry before the war was related to the chemistry of metabolic processes. Scientists sought to discover how living cells broke down fats, proteins, and carbohydrates into smaller molecules. The relationships between these processes proved elusive to scientists. The Germans Hans Krebs (1900–1981) and his colleague Kurt Henseleit (1907–1973) at the University of Freiburg were the first to identify the crucial metabolic "pathway," or the underlying process taking place by the action of the living cell. The "Krebs cycle," as it became known, accomplished the breakdown of complex molecules, releasing energy that was then used by the cell itself for its own sustenance. In 1933, the Nazis fired Krebs, and he moved to England, where he continued his work and more fully developed his conception of the cycle, also called the citric acid cycle, by 1937. German-born Fritz Lippman (1899–1986) provided evidence that affirmed Krebs's model of metabolic processes. Both men were awarded the Nobel Prize in Physiology or Medicine in 1953.

Biochemistry, already an interdisciplinary study, expanded further in the 1930s and 1940s. British biologist Joseph Needham (1900–1995), known as both scientist and historian of science, helped to unify biochemistry and morphology during these years. In 1931, he published his book, *Chemical Embryology,* which treated the development of the embryo in terms of chemical processes. In 1942, he made more explicit connections between these two subfields of biology, resulting in the book *Biochemistry and Morphogenesis.* Also in the early 1940s, Americans George Beadle (1903–1989) and Edward Tatum (1909–1975) brought genetics to the study of biochemistry. Although genetics typically had been useful in studying heredity—or developmental genetics—Bea-

dle and Tatum used it to identify biochemical pathways in propagation of molds and fungi, particularly *Neurospora,* a bread mold. After World War II, studies of biochemistry increasingly focused on genetic relationships.

See also Amino Acids; DNA; Genetics
References
Fruton, Joseph. *Proteins, Enzymes, Genes: The Interplay of Chemistry and Biology.* New Haven, CT: Yale University Press, 1999.
Kohler, Robert E. *From the Medical Chemistry to Biochemistry: The Making of a Biomedical Discipline.* New York: Cambridge University Press, 1982.
Teich, Mikuláš, with Dorothy M. Needham. *A Documentary History of Biochemistry, 1770–1940.* Rutherford, NJ: Fairleigh Dickinson University Press, 1992.

Biometry

Biometry, or statistical biology, came to prominence in the early twentieth century as a means of representing patterns of heredity. Francis Galton (1822–1911) was one of the first to make significant use of biometry in the nineteenth century, as a way to demonstrate that saltative change—or sudden, discontinuous change—took place and that such changes were inherited. He believed that a study of change in a large population, analyzed through statistics, could show the spread of variation. His disagreement with the views of his cousin Charles Darwin (1809–1882) hinged upon Darwin's insistence on continuous, nonsaltative variation in organisms. Ironically, the twentieth-century biometricians took the opposite view, that Darwin was correct and that statistics could prove it. The argument about continuous change and discontinuous change created serious discord between Darwinian evolution and Mendelian genetics in the first decades of the twentieth century.

Statistics had the power to lend scientific credence to vague and perhaps intangible ideas by providing them with a semblance of precision and quantitative accuracy. Biometry in particular gave authority to the eugenics movement, and leading eugenicists often constructed their work in such a way that the results could be represented easily with numerical values. Galton himself was the founder of eugenics. He believed that Darwin's view of natural selection indicated that "artificial" selection could be implemented in order to improve the race as a whole, at least in the short term (but in the long term, he believed, only sudden mutations could have a permanent effect). One of Galton's admirers, the mathematician Karl Pearson (1857–1936), carried on the biometric methodology to demonstrate Darwin's view of continuous variation. Karl Pearson also believed that his statistical data provided a strong scientific justification to urge governments to guide or shape human breeding in society. Because he believed, unlike Galton, in continuous change, Pearson also believed that long-term racial improvement was possible through social policy.

Two principal biometricians were Pearson and the biologist Walter F. R. Weldon (1860–1906). Pearson founded *Biometrika: A Journal for the Statistical Study of Biological Problems* in 1901. Acting as editor, Pearson used this journal as the mouthpiece for his own ideas and attracted others outside his laboratory to use biometric methods and principles and publish them. Like Pearson, Weldon was a Darwinian biometrician and a convinced eugenicist. Weldon conducted studies of crabs and snails, hoping to show a correlation between death rates and a particular trait; in both cases, the trait was the size of the shell. The criterion seemed arbitrary to some opponents, particularly because Weldon made little effort to identify precisely what the utility of a larger shell would be if the Darwinian mechanism of natural selection was supposed to "select" crabs with larger shells over crabs with smaller ones. Weldon's work succeeded in showing that in some cases, larger crabs survived more often than smaller crabs, but these results convinced few of the skeptics.

Pearson and Weldon became bitter opponents of Mendelian geneticists such as William Bateson (1861–1926). The geneticists, taking for granted the mathematical "laws" of inheritance posited by Gregor Mendel (1822–1884) in the nineteenth century, believed that inheritance should be discontinuous. Bateson did not believe that natural selection had any significant power, certainly not when compared with genetic inheritance. The emphasis of the biometricians on continuous change and the emphasis of the geneticists on discontinuous change created disharmony between the Darwinian and Mendelian views. The dispute, which turned into personal animosity between Bateson and Weldon, separated the Mendelians and Darwinians into hostile camps, making scientific reconciliation difficult.

Biometry itself evolved as scientists of a younger generation accepted a synthesis of Mendelian and Darwinian ideas, and as they adopted the tools of chromosome analysis. The field more properly was known as population genetics, as researchers such as Ronald A. Fisher (1890–1962) adapted biometrical methods to understanding Mendelian inheritance. Dealing with large numbers of organisms, population genetics embraced the statistical approach pioneered by the biometricians but without the hostility toward Mendelian inheritance. In the 1920s, researchers such as Fisher, J. B. S. Haldane (1892–1964), and Sewall Wright (1889–1988) provided the theoretical synthesis to ameliorate the problems that had plagued the field when it was dominated by the antagonistic voices of Weldon and Bateson. The new findings emphasized the primacy of Mendel's laws of inheritance but included other factors as well, such as mutation as an agent of change and Darwinian natural selection as a mechanism for the evolution of the whole population.

See also Bateson, William; Eugenics; Evolution; Genetics

References

Bowler, Peter J. *Evolution: The History of an Idea.* Berkeley: University of California Press, 1989.

Edward, A. W. F. "R. A. Fisher on Karl Pearson." *Notes and Records of the Royal Society of London* 48:1 (1994): 97–106.

MacKenzie, Donald A. *Statistics in Britain, 1865–1930: The Social Construction of Scientific Knowledge.* Edinburgh: Edinburgh University Press, 1981.

Provine, William B. *The Origins of Theoretical Population Genetics.* Chicago: University of Chicago Press, 1971.

Margaret Sanger, president, National Committee on Federal Legislation for Birth Control. (Underwood & Underwood / Corbis)

Birth Control

Human reproduction was a difficult political issue in the first half of the twentieth century. In the United States, the birth control movement was coupled with efforts to reclaim women's control over their own bodies during a time when both science and law seemed poised to deprive them of it. Many states passed laws in the late nineteenth century forbidding not only abortions but also any form of contraception, while male professional physicians were marginalizing traditional women's roles in childbirth. Meanwhile, efforts to promote birth control were tied very closely to the eugenics movement in both the United States and Britain.

At the turn of the century, about half of the births in the United States involved the use of midwives. The percentage was even greater among immigrants, the working class, and in rural areas. Over the next generation or so, however, midwives were replaced as more Americans put their trust in doctors or as doctors gradually pushed midwives out of the process. The change sparked lively debate in the 1920s about the necessity of "modern" medical knowledge in ensuring healthy childbirth. Midwives argued that their own expertise was sufficient for most births. But for the most part, the doctors won the argument, consolidating their control over childbirth and associating it increasingly with hospitals, anesthetics, and even surgery. It was also an example of gender conflict, as childbirth once was the province of female expertise, but by the first half of the twentieth century it was firmly under the control of a profession dominated by men.

The birth control movement became radical between 1910 and 1920, at the same time that the women's suffrage movement saw its great successes in the United States. Birth control became a major issue largely through the efforts of the American Margaret Sanger (1883–1966), whose mother had died of tuberculosis at an early age; Sanger attributed her mother's weakness to the exhaustion of giving birth to eleven children and enduring several miscarriages. Margaret Sanger advocated birth control as a means of achieving feminine autonomy, believing that unwanted or unplanned pregnancies were a serious burden for women. In 1914 she began using the term *birth control,* and she promoted contraception despite the 1873 Comstock Act, which had made contraception illegal. She opened a birth control clinic in 1916 and was arrested; two years later a federal court ruled that birth control sometimes could be permitted for health reasons. In 1936, a more important decision was reached in a circuit court of appeals that contraceptive devices could be sent through the mail. This had at least two effects: It provided a way to circumvent the many state laws against birth control, and it proved to the medical establishment and industry that there might be a profitable market for improved methods of contraception. Capitalizing on this, Sanger spearheaded the effort to develop an oral contraceptive. Her efforts sparked movements elsewhere also. Before the rise of the Nazis in Germany, for example, Sanger's work informed a major movement of sexual liberation, including an effort to decriminalize abortions. The emphasis on women's sexual freedom was effectively halted by the Nazis, who used birth control in quite a different way—to shape the reproductive patterns of the entire population to ensure racial purity.

Although the objective to empower women typifies historical understanding of the birth control movement, there were many motivations for it. Part of the rationale for birth control was the role it could play in eugenics, or selective breeding. U.S. and European physicians and biologists saw birth control as a way to contain the reproduction of the "unfit" in society. Part of the problem, they believed, was that the limitations in birth control technology forced educated or conscientious women to self-select themselves and their potential offspring out of society. Birth control technology remained relatively stagnant during the first half of the century. Douches, diaphragms, and condoms were available before the twentieth century. They already were fairly effective; by 1900

U.S. women on average bore about half as many children as they had a century before (7.04 in 1800 and 3.56 in 1900). The search for a pill, pursued from the 1920s onward, came largely from the belief that these traditional forms of birth control often were too complex for all classes of society and that poorly educated women needed a simple alternative that would still give them reproductive control. The term *race suicide* embodied fears that the racial composition of society would be dominated by those least likely to use birth control—the poor, the uneducated, or the supposedly inferior ethnic minorities. Eugenicists were ardent supporters of early birth control clinics in the 1920s and 1930s, and they also financed scientific efforts to find an alternative to traditional birth control technology. In return, birth control advocates like Sanger and others used the potential eugenic benefit of birth control to encourage biologists, geneticists, and doctors to take a more active role in studying contraceptive technology. Because of this potential well of support, the argument for eugenics was more commonplace among birth control advocates during the 1920s and 1930s than was the argument for feminism.

See also Eugenics; Women

References
Gordon, Linda. *Woman's Body, Woman's Right: A Social History of Birth Control in America.* New York: Grossman, 1976.
Grossman, Atina. *Reforming Sex: The German Movement for Birth Control and Abortion Reform, 1920–1950.* New York: Oxford University Press, 1995.
Kennedy, David M. *Birth Control in America: The Career of Margaret Sanger.* New Haven, CT: Yale University Press, 1970.
Litoff, Judy Barrett. *American Midwives, 1860 to the Present.* Westport, CT: Greenwood Press, 1978.
Reed, James. *From Private Vice to Public Virtue: The Birth Control Movement and American Society since 1830.* New York: Basic Books, 1978.
Soloway, Richard A. "The 'Perfect Contraceptive': Eugenics and Birth Control Research in Britain and America in the Interwar Years." *Journal of Contemporary History* 30 (1995): 637–664.

Bjerknes, Vilhelm

(b. Christiania [later Oslo], Norway, 1862;
d. Oslo, Norway, 1951)

Vilhelm Bjerknes was born into science through his father, Carl Bjerknes (1825–1903), who already had conducted important work in hydrodynamics, the branch of physics concerned with fluids in motion. Although he made a decision in the 1880s to cease collaboration with his father, Vilhelm Bjerknes never was able to divorce himself from his father's work. Ultimately the younger Bjerknes, too, would make major contributions to hydrodynamics, and in fact he would become one of the principal founders of modern meteorology.

A native of Norway, Bjerknes traveled widely during his studies, including a two-year stint in Germany as Heinrich Hertz's (1857–1894) assistant at the University of Bonn. There he conducted work on Hertz's specialty, electrodynamics, a branch of physics that attempts to understand how electric currents behave, using magnets and other electric currents. When he returned to Norway, he wrote his Ph.D. dissertation on that topic, and his work provided him with the scientific credentials to obtain an appointment as professor at the University of Stockholm in 1895.

But Bjerknes did not stray from his father's field of specialization forever. While at Stockholm, he brought his own education in physics to bear on the problems of hydrodynamics. He generalized the ideas of two leading physicists, William Thompson (Lord Kelvin) (1824–1907) and Hermann Helmholtz (1821–1894), who had discussed the velocities of circulation and the conservation of circular vortices in fluids. Bjerknes conceived of the fluid not just as a hydrodynamic system, but also as a thermodynamic system, where the presence of heat and its conversion to other forms of energy had to be taken into account. Such an approach, Bjerknes reasoned, would enable a better understanding of the atmosphere and the ocean. This broadened the scientist's understanding of what was involved in dynamic processes but made

the systems themselves much more complex for the theoretician.

If theories of circulation, combining hydrodynamic and thermodynamic principles, could be applied to the atmosphere and ocean, scientists could use knowledge of present conditions to predict future ones. Bjerknes soon found himself being hailed as a leading geophysicist, and eventually he was invited to the University of Bergen to start a geophysical institute. There, Bjerknes promoted the practical value of his research for fisheries, weather forecasting, and military uses. His work at the University of Bergen resulted not only in some efforts to develop reliable meteorological services, but also in a research program in theoretical meteorology. While at Bergen, he published one of his classic works, *On the Dynamics of the Circular Vortex with Applications to the Atmosphere and to Atmospheric Vortex and Wave Motion* (1921).

The influence of Bjerknes's "Bergen school" went far beyond his native Norway. His approach to the study of the sea and atmosphere inspired interdisciplinary research in geophysics in Europe and the United States and also provided a model for modern weather services.

See also Meteorology; Sverdrup, Harald
References
Friedman, Robert Marc. *Appropriating the Weather: Vilhelm Bjerknes and the Construction of a Modern Meteorology.* Ithaca, NY: Cornell University Press, 1989.
Pihl, Mogens. "Bjerknes, Vilhelm Frimann Koren." In Charles Coulston Gillispie, ed., *Dictionary of Scientific Biography*, vol. II. New York: Charles Scribner's Sons, 1970, 167–169.

Boas, Franz

(b. Minden, Germany, 1858; d. New York, New York, 1942)

Franz Boas was the most influential anthropologist in the United States during the first decades of the twentieth century. His efforts to make anthropology more "scientific" had mixed results: On the one hand, he emphasized rigorous methodology, yet on the other hand, his hostility to speculation often blocked evolutionary approaches to anthropological problems.

Boas began his career by familiarizing himself with the anthropology community in Berlin, and he was influenced by Adolf Bastian (1826–1905) and Rudolf Virchow (1821–1902). In 1883, he participated in an expedition to the arctic regions of North America, where he studied local geography and Eskimo culture. This began a career-long fascination with the linguistics and ethnology of the native peoples of the Northwest Coast. Boas returned to Germany and lectured on geography at the University of Berlin. But his early career was clouded by anti-Semitism, which made him the object of discrimination and insults. A visit to the United States, for a meeting of the American Association for the Advancement of Science, led to his taking a position as assistant editor of the journal *Science*. He became a lecturer in anthropology at Clark University and, in 1892, conferred North America's first Ph.D. in anthropology on A. F. Chamberlain. He then worked in museums for several years before taking both a position in physical anthropology at Columbia University (1899) and the post of curator of anthropology at the American Museum of Natural History (1901).

One of the key ideas behind Boas's work was that reality has some underlying, knowable structure that is masked by the variations apparent in nature. The way to discover this structure in humans was through empirical investigation to determine the range of differences; these differences had many external origins, based on environment and historical experience. In his introduction to the *Handbook of American Indian Languages* (1911), he noted that the aim of a scientific anthropology was to outline the psychological laws that govern human behavior. To do so, one needed to understand and identify the variations of them.

The differentiation between a stable inner core and a relatively plastic external form echoed biological thinking on heredity at the dawn of the twentieth century. This is no accident; Boas's goal was to make his discipline as rigorous and exact—as "scientific"—as possible. For him, this meant a rejection of speculative theories, including evolution. He believed that investigators should proceed inductively, confining themselves to their data and avoiding large-scale inferences. He advocated the "historical" method, which would analyze processes of cultural change—such as in language, myths, folklore, and art—to measure individual forms and variations of them. In this way, "elements" of culture could be plotted through time and space. Because of the multitude of possible variations, Boas cautioned strongly against making broad inferences and preferred as strict an empirical method as possible.

The result of this approach was that Boas opposed evolutionary approaches and any interpretations that proposed cultures being diffused en bloc, without alteration. He strongly opposed the notion that societies are defined purely by their racial composition (in other words, he was against racial determinism), but instead were a complex mix of various environmental and historical factors. He explicitly critiqued the notion of African American inferiority and praised African cultural achievements. His statements in this regard proved influential on African American intellectuals such as W. E. B. Du Bois (1868–1963). Cultural traditions were external representations molded by complicated historical processes, without possibility of uniformity. Yet for Boas, human types seemed permanent. The job of the anthropologist was to demonstrate the range and historical sources of variations, to allow these permanent features to be recognized. Only then could anthropologists identify genuinely "genetic" human types. In his *Materials for the Study of Inheritance in Man* (1928), Boas showed how the distinctive head shapes of two different immigrant groups (Jews from Eastern Europe and southern Italians) seemed less pronounced after the groups arrived in New York and shared the same urban environment. For Boas, this illustrated his point that most variation in nature was environmental; the true differences would be much more difficult to identify.

Boas and his ideas were very influential in the United States, not only among anthropologists, but among scientists as a whole. He served as president of the New York Academy of Science (1910) and the American Association for the Advancement of Science (1931). His hostility to focusing on evolutionary problems and his great influence, not only on colleagues but also on a generation of students, seemed to tarnish his legacy somewhat. Some argue that he inhibited studies of anthropology, particularly in the United States, where he exercised extraordinary influence in publication and organizations. But the reverse is also true: His rejection of evolutionary approaches, which he deemed overly speculative, facilitated an embrace of methodological rigor and an acknowledgment of the limitations and ambiguity of evidence.

See also Anthropology; Determinism; Mead, Margaret; Race; Scientism

References
Herskovits, Melville J. *Franz Boas.* New York: Scribner's, 1953.
Stocking, George W., Jr. *The Ethnographer's Magic and Other Essays in the History of Anthropology.* Madison: University of Wisconsin Press, 1992.
Voget, Fred W. "Boas, Franz." In Charles Coulston Gillispie, ed., *Dictionary of Scientific Biography,* vol. II. New York: Charles Scribner's Sons, 1970, 207–213.
Williams, Vernon J., Jr. *Rethinking Race: Franz Boas and His Contemporaries.* Lexington: University Press of Kentucky, 1996.

Bohr, Niels

(b. Copenhagen, Denmark, 1885; d. Copenhagen, Denmark, 1962)

Niels Bohr was a giant of twentieth-century physics, rivaled only by Albert Einstein (1879–1955) in reputation. He brought quantum theory into the atom and established a fundamental, even philosophical, principle in quantum mechanics. He also helped to build the atomic bomb and tried, unsuccessfully, to prevent a nuclear arms race. Bohr stands against Einstein as the symbol of the new physics, with its rejection of determinism and its abandonment of the dream of understanding the fundamental reality of nature.

After finishing his doctorate at the University of Copenhagen in 1911, Bohr traveled to England to work with J. J. Thomson (1856–1940) at Cambridge. He hoped to extend his dissertation work on electrons in metals. But Thomson had given up his interest in electrons, so Bohr moved to Manchester to work with Ernest Rutherford (1871–1937). Here he took up the problem of atomic structure. Rutherford's model of the atom, with electrons in orbit around a nucleus, had recently supplanted Thomson's "plum pudding" atom. But Rutherford's model was incomplete because mechanical laws predicted that the electrons would collapse into the nucleus. Bohr set out to account for the atom's apparent stability in spite of the prevailing laws of physics.

Three major papers by Bohr appeared in the *Philosophical Magazine* in 1913 addressing the problem of atomic structure. He formulated two postulates. One was that atomic systems possessed stationary states, in which the laws of mechanics apply. The other was that, in addition to these, there were transition states, in which these classical laws of physics did not apply. Bohr adopted Max Planck's (1858–1947) quantum theory to account for the energy difference between these stationary states. The transition state is accompanied by the emission or absorption of a discontinuous quantum of radiation.

Bohr's theory was tremendously influential because it cast the atom into the language of quantum physics. His description of quantum orbits provided the needed stability in Rutherford's atom. He moved back to

Danish physicist Niels Bohr, pictured in his laboratory, received the 1922 Nobel Prize in Physics for his work on the structure of atoms. (Bettmann / Corbis)

Denmark and became director of the new Institute for Theoretical Physics in Copenhagen. Scientists abroad, such as Henry Moseley, James Franck (1882–1964), and Gustav Hertz (1887–1975), demonstrated his atomic model through experiments in X-ray spectroscopy and electron collision. Subsequent years saw theoretical physicists obsessed with generalizing Bohr's work into a theory of quantum mechanics. In doing so, physicists such as Werner Heisenberg (1901–1976) found themselves abandoning dearly held notions beyond mechanics. In 1927, Heisenberg announced his uncertainty principle (or indeterminacy principle), which stated that it was meaningless to describe the exact position of an electron. Its position was governed by quantum relationships and sta-

tistics and could never be objectively observed. The philosophical implication was that the classical concept of causality was untenable, because no physical system could ever be defined precisely; instead, one had to be satisfied with probabilities.

Bohr reacted to the crisis in quantum mechanics by accepting Heisenberg's uncertainty principle and developing his own principle of complementarity in 1927. The standard example of complementarity is the wave-particle duality in quantum mechanics. The wave description and the particle description, each a classical concept on its own, are complementary ways to understand reality. They are each valid; they cannot, however, be used together, for they are contradictory explanations. Because the physicist himself chooses how to frame his questions and decides what to measure (for example, he frames his observation from a particle outlook), he eliminates the possibility of resolving questions that the other outlook might raise. What, then, is really being observed? Surely they cannot both be valid. Should investigations of physical properties be framed in terms of wavelength and frequency, or rather energy and momentum? This important question is cast aside by the principle of complementarity, which accepts different avenues toward understanding phenomena even if one avenue contradicts the other. Bohr claimed that the aim of physics was not necessarily to discover reality, but rather to find adequate means to explain and predict phenomena.

Complementarity became the basis of what was later dubbed the Copenhagen interpretation of quantum mechanics. But Bohr's embrace of complementarity, and its philosophical implications, put him at intellectual odds with Albert Einstein. The two engaged in personal debate about whether or not quantum mechanics was a complete theory if it set such clear limitations on the ability to observe physical phenomena. Einstein disliked the implication, based on Heisenberg's uncertainty and Bohr's complementarity, that reality was not discoverable and that, at a certain point, the physical world could only be described by statistics. Einstein said that God "does not throw dice." But at two Solvay Congresses—in 1927 and 1930—Bohr's arguments appeared to counter all of Einstein's objections; although Einstein never abandoned his faith in the ability to determine causality, the debates brought many leading physicists into Bohr's camp.

By 1936 Bohr's model of the atom took on a form that helped leading scientists conceptualize the process of fission. Bohr's "droplet model" suggested a plasticity that led Lise Meitner (1878–1968) and Otto Frisch (1904–1979) to believe that an atom might elongate and divide itself. In 1939, they announced this interpretation—nuclear fission—of an experiment by Otto Hahn (1879–1968) and Fritz Strassman (1902–1980). Bohr was the one to carry news of the discovery to the United States, where he was attending a conference. Soon scientists in Europe and the United States began experiments on uranium fission, and Italian émigré Enrico Fermi (1901–1954) presided over the first nuclear chain reaction in 1942. Bohr and other eminent scientists from Europe became leaders in the Manhattan Project, the U.S. effort to build the atomic bomb.

Although Bohr participated in the project, he was alarmed at the geopolitical consequences of the bomb. In 1944, he urged President Franklin Roosevelt (1882–1945) to put atomic weapons under international control. The existence of the bomb, because of its devastating effects, had the potential to forge a lasting peace if handled with diplomatic skill. Even before the war ended, Bohr wanted the United States to announce its possession of atomic bombs to the rest of the world, as a measure of good faith. Were it to be kept secret, and the United States was to keep a monopoly on the atomic bomb, surely an arms race would ensue. All of Bohr's efforts to forestall such an arms race failed. He continued his efforts after World War II, to little effect.

See also Atomic Structure; Determinism; Philosophy of Science; Physics; Quantum Mechanics; Rutherford, Ernest; Solvay Conferences

References
Faye, Jan. *Niels Bohr: His Heritage and Legacy.* Dordrecht: Kluwer, 1991.
Kragh, Helge. *Quantum Generations: A History of Physics in the Twentieth Century.* Princeton, NJ: Princeton University Press, 1999.
Petruccioli, Sandro. *Atoms, Metaphors, and Paradoxes: Niels Bohr and the Construction of a New Physics.* New York: Cambridge University Press, 1993.
Rosenfeld, Leon. "Bohr, Niels Henrik David." In Charles Coulston Gillispie, ed., *Dictionary of Scientific Biography,* vol. II. New York: Charles Scribner's Sons, 1970, 239–254.

Boltwood, Bertram

(b. Amherst, Massachusetts, 1870; d.
Hancock Point, Maine, 1927)

After the discovery of radioactivity in 1896, scientists became enthralled with the properties of radiation and the materials that produced it. Bertram Boltwood, as a U.S. scientist, worked outside the mainstream of these scientific activities. Yet from afar, he made a number of important contributions to understanding radioactive decay series. Eventually he used such knowledge to estimate the age of the earth, adding vastly to the relative youthfulness attributed to it by nineteenth-century physicists.

Boltwood began his own research on radioactivity in 1904, after reading about a new interpretation by Ernest Rutherford (1871–1937) and Frederick Soddy (1877–1956). These scientists claimed that radioactive atoms "decay" as they emit radiation, thus transforming (or "transmuting" as they wrote) into different elements. Boltwood set out to prove a family relationship among radioactive elements often found together in ore. He tried to "grow" radium, recently discovered by Marie Curie (1867–1934), from uranium.

Boltwood's efforts proved more difficult than he expected, mainly because he was not yet aware of the complexity of the decay series. There were in fact several steps between uranium and radium. In addition, he was not yet aware of the vast amount of time that such transmutation would require. After more than a year of observing, he did not observe any change from uranium's known daughter product, "uranium X," to radium. But rather than reject the whole premise, he concluded that there must be a very long-lived product still between them. Over several years he and others, including Soddy, searched for the missing steps in the decay series. In 1907 Boltwood identified a substance he called "ionium" as the immediate parent of radium, and more than a decade later Soddy showed the family relationship between uranium and ionium. Their work identified the whole decay series from uranium to radium.

Where did the decay series lead? Boltwood hypothesized that because lead was always present with uranium, it might be the stable end product after a long process of decay. He noted that geologically older minerals contained higher proportions of lead, which is easily explained if one assumes that lead is accumulating over time from the decay of radioactive elements. On a practical level, this pointed to a new method for dating the age of rocks. To determine the age of an ore, one needed only to determine the rate of formation of lead and measure the total amount found. This method, developed in 1907 by Boltwood, resulted in a startling calculation of the age of the earth. In the nineteenth century, William Thompson (Lord Kelvin) (1824–1907) had calculated an age in the tens of millions of years. Now, with Boltwood's method, scientists had to revise these estimates upward to the billion-year range. Although Boltwood's initial calculation later was revised, his method forced scientists in all fields to accept a much longer time span for the origin of the planet and the evolution of its life forms.

Boltwood's life and work had a decisive impact on the study of radioactivity, despite his distance from the main centers of activity in Europe. But his personal relationship with eminent radiochemist Ernest Rutherford, who treated him as a scientific equal as well as a friend, ensured that his ideas found a receptive audience. Boltwood's own life took a tragic turn in 1927, when he lost his battle against depression and committed suicide.

See also Age of the Earth; Radioactivity;
    Rutherford, Ernest; Uranium

References

Badash, Lawrence. *Radioactivity in America: Growth and Decay of a Science.* Baltimore, MD: Johns Hopkins, 1979.

———, ed. *Rutherford and Boltwood: Letters on Radioactivity.* New Haven, CT: Yale University Press, 1969.

Bragg, William Henry

(b. Westward, England, 1862; d. London, England, 1942)

William Henry Bragg published very little for the most part of his career, which he spent as professor of mathematics and physics at the University of Adelaide, Australia. Educated at Cambridge University in England, he rose to prominence in Australia largely through his administrative talents, teaching abilities, and involvement in the Australian Association for the Advancement of Science. In the dawn of the new century, he was on the periphery of scientific activity in physics to say the least. His position in Australia drew inevitable comparisons with the upstart New Zealander Ernest Rutherford (1871–1937), who was making a name for himself in England with his model of the atom and was fast gaining disciples as a leading experimental physicist.

After formulating a scathing critique of others' work on the scattering and absorption of ionizing radiation, Bragg began his own series of experiments in 1904. He published several papers over the next few years and began to question the nature of gamma rays and X-rays. Were they "ether pulses" destined to be scattered widely? Were they waves? Were they of a material nature? Were they made up of tiny bodies? Bragg did not believe that X-rays acted as waves, and he formulated a high-energy corpuscular theory that would keep most of an X-ray in a straight beam. His views seemed to be confirmed by the development of the cloud chamber by C. T. R. Wilson (1869–1959) in 1911. The chamber showed that a beam of X-rays, when exposed in the chamber to a gas, did not diffuse into a fog. Instead, the cloud chamber showed many short lines. Still, Bragg's premise, that X-rays did not behave as waves, would soon be refuted by his own son.

Bragg returned to England in 1909 and, three years later, began work with his son William Lawrence Bragg (1890–1971) on a new phenomenon discovered in 1912 by Max Von Laue (1879–1960). The German scientist had found that X-rays were diffracted by crystals. The younger Bragg, William Lawrence, noted that some of Von Laue's observations could be interpreted as resulting from the reflection of X-rays. The Braggs soon came to the conclusion that their previous view, that X-rays were not waves, had to be rejected. The elder Bragg then developed an instrument to examine X-rays more thoroughly. The X-ray spectrometer allowed a crystal face to reflect X-rays at any angle and measure its spectrum. Using the spectrometer, William Henry Bragg made another major discovery: Each metal used as a source of X-rays gave its own unique X-ray spectrum of definite wavelengths, just as elements give unique optical spectra. Bragg had discovered a new means to analyze not only X-rays but also crystal structure. By examining the various faces of the crystal, one could note the angles and intensity with which they reflected X-rays, thus determining how the atoms were arranged in the crystal.

The two men joined forces in earnest in 1913 to experiment with the X-ray spectrometer. The elder Bragg concentrated on studying the X-rays produced by different metals, whereas the younger was most interested in the crystal structure. The younger Bragg summarized the importance of this work as demonstrating the fundamental principles of X-ray analysis of atomic patterns. In recognition of their accomplishment, father and son shared the 1915 Nobel Prize in Physics. The younger Bragg was twenty-five years old, the youngest Nobel laureate to date. Unfortunately, World War I interrupted their work together. William Henry Bragg worked on problems of submarine detection and had relatively little time for the X-ray spectrometer. After the war ended, he took on additional responsibilities as head of the Royal Institution. From 1935 to 1940 he was president of the Royal Society as his country entered another world war. He did not live to see it end.

See also Physics; Von Laue, Max; X-rays
References
Ewald, P. P., ed. *Fifty Years of X-Ray Diffraction.*
    Utrecht: International Union of
    Crystallography, 1962.
Forman, Paul. "Bragg, William Henry." In
    Charles Coulston Gillispie, ed., *Dictionary of
    Scientific Biography,* vol. II. New York: Charles
    Scribner's Sons, 1970, 397–400.
Karoe, G. M. *William Henry Bragg, 1862–1942:
    Man and Scientist.* New York: Cambridge
    University Press, 1978.

Brain Drain

Brain drain is a term used to describe the loss of scientific manpower from a particular profession, scientific field, or geographic region. It usually implies a "brain gain" somewhere else, as manpower gets siphoned from one area and moved into another. The term has been used particularly to describe the shift of focus in scientific activity, particularly physics, from Europe to the United States during the 1930s, culminating in the large-scale scientific projects of World War II and leading to an unprecedented level of support for scientific research by the United States in the postwar period. The loser in this brain drain was Germany and its allies, while the United States and (to a lesser extent) Britain became the world's scientific powerhouses.

Physics in the United States grew rapidly in the 1920s, owing to patronage by philanthropic organizations such as the Rockefeller Foundation. Americans were keen to compete with their European colleagues and establish their community as a strong one; many young scholars traveled to European institutions in the 1920s and 1930s for graduate or postdoctoral work before taking up positions in U.S. universities. For physicists, the most attractive countries were Germany and England, both with strong research communities (especially at Göttingen, Munich, Berlin, and Cambridge), though some chose to go to Denmark to be part of Niels Bohr's (1885–1962) institute in Copenhagen. These young scholars returned to the United States and made it fertile ground for top-level physics research, culminating in Nobel Prizes for Americans such as Ernest Lawrence (1901–1958) and Arthur Compton (1892–1962). Such developments made the United States, with its many universities and its desire to compete for the best minds, very attractive to European scholars, not only the established ones who could demand a great deal of money, but also the ones who felt marginalized in their own countries.

The rise of fascism in some countries of Europe in the 1930s facilitated the intellectual migration (or flight) from Europe to the United States. Although most of them were German Jews who had lost their positions because of the racial policies of the Third Reich, others left for different but related reasons. The Italian Enrico Fermi (1901–1954), for example, was not a Jew, but his wife was. All told, about a hundred physicists came to the United States from Europe between 1933 and 1941. Many leading physicists remained; Werner Heisenberg (1901–1976), Max Von Laue (1879–1960), and Otto Hahn (1879–1968), for example, became the physicists charged with building the Nazi atomic bomb. A great many of their friends and colleagues, however, had left. In fact, the U.S. atomic bomb project owed its existence and success to émigré scientists who had fled fascist countries. The irony is that the racial policies of the Third Reich undercut its scientific strength, draining its brains and literally giving them to a future enemy.

Not all of those leaving the European continent fled to the United States. Others went to Britain, such as Germans Max Born (1882–1970) and Paul Peter Ewald (1888–1985), and Austrian Erwin Schrödinger (1887–1961). Some who initially went to Britain ultimately settled in the United States, such as Hungarians Leo Szilard (1898–1964) and Edward Teller (1908–2003) and the German Klaus Fuchs (1911–1988). The last would spy for the So-

Albert Einstein receiving from Judge Phillip Forman his certificate of American citizenship (1940). (Library of Congress)

viet Union while working on the U.S. atomic bomb project. The most famous immigrant was Albert Einstein (1879–1955), who decided early in the 1930s to make the United States his home rather than live in Nazi Germany. Other U.S.-bound leading scientists were Enrico Fermi, the German James Franck (1882–1964), and a host of others. Many of the émigré scientists initially stayed closer to home, finding positions in Switzerland, the Netherlands, Denmark, or France, but ultimately they moved farther away, mostly to the United States.

If the names of eminent scientists were not enough to demonstrate that continental Europe was losing its scientific strength, one could consider the quantitative loss of those who were recognized internationally by receiving the Nobel Prize in Physics. Among those migrating to Britain and the United States from fascist countries were five Nobel laureates—Erwin Schrödinger, Peter Debye

(1884–1966), Albert Einstein, Enrico Fermi, and Gustav Hertz (1887–1975). In retrospect, we can consider the future benefits of this "brain gain" as well. Two émigrés to Britain, Max Born and Dennis Gabor (1900–1979), would later win Nobel Prizes; seven to the United States would—Felix Bloch (1905–1983), Max Delbrück (1906–1981), Gerhard Herzberg (1904–1999), Emilio Segrè (1905–1989), Eugene Wigner (1902–1995), Hans Bethe (1906–), and Otto Stern (1888–1969). In addition to these were many others whose contributions were never recognized by this prestigious prize.

The United States and Britain benefited handsomely from the wave of migrations. In 1933, Leo Szilard helped to set up in Britain the Academic Assistance Council, whose purpose was to find temporary posts for refugee scientists, to facilitate their transition to their new homes. It envisioned the United States as the ultimate endpoint for them. This and other programs for displaced scientists eased some of the difficulties brought on by the "brain gain." With the limited number of jobs, for example, competition was still fierce, and the migrations were resented by some native-born scientists in both Britain and the United States. In addition, anti-Semitism in these countries (especially the United States) caused difficulties in placing scientists. Displacement programs helped to address this problem by creating special posts for recent immigrants. By the late 1930s, the United States led the world in physics. That leadership remained during World War II and long after.

See also Nationalism; Nazi Science; Szilard, Leo; World War II

References

Kevles, Daniel J. *The Physicists: The History of a Scientific Community in Modern America.* Cambridge, MA: Harvard University Press, 1995.

Kragh, Helge. *Quantum Generations: A History of Physics in the Twentieth Century.* Princeton, NJ: Princeton University Press, 1999.

Rider, Robin E. "Alarm and Opportunity: Emigration of Mathematicians and Physicists to Britain and the United States, 1933–1945." *Historical Studies in the Physical Sciences* 15 (1984): 107–176.

C

Cancer

For centuries, cancer was part of a pantheon of deadly ailments that threatened human lives. Although the development of vaccines and public health measures helped to conquer traditional enemies such as cholera and tuberculosis during the first half of the twentieth century, cancer remained an obstinate foe. *Cancer* was a widely inclusive term used to describe abnormal, often malignant growth in various parts of the body: skin, brain, colon, breast, and myriad other areas. The word *cancer* also became a widely used metaphor for a diseased blight on anything pure. Often it was used to describe some deleterious factor in modern, industrial society.

Lacking a cure, cancer was addressed most successfully by measures of prevention. In the United States, the American Society for the Control of Cancer was founded in 1913; its battle against cancer was composed mainly of public education to enable early detection of symptoms that might lead to malignant forms of cancer. Such efforts met with only limited success, because health care coverage was not universal, and thus even early detection did not necessarily translate into prevention. By the 1930s, cancer mortality rates were on the rise. The U.S. government attempted to address the problem by funding, in 1937, the National Cancer Institute, to be advised by the National Advisory Cancer Council. The federal government funneled considerable funding into biomedical research in the hopes that scientists would eradicate cancer as they had other diseases. This early optimism faltered when confronted by many years without a decisive cure. Soon a debate emerged over which should be most emphasized for the good of public health: education about prevention or scientific research for a cure.

Although the exact cause of cancer eluded scientists, there were some obvious contributors. Many of these were connected to the alleged benefits of science and technology. Certainly the pollution of industrial cities contributed, as did the widespread use (and overuse) of X-rays for a host of medical problems. Radiation in general was little understood in the first half of the twentieth century, and it appeared to yield some negative effects. For example, the famed Marie Curie (1867–1934) became feeble in her later years and developed strange lesions on her hands after a career of working with X-rays and radioactive substances. A group of watch-dial painters, casually licking their radium-paintbrushes, developed cancers in their mouths. In such cases, better knowledge of the cause of cancer might have sufficed in place of a cure. The most serious, widespread cause of

Technician standing at laboratory table, preparing organic compounds used in cancer research, 1930s. (National Library of Medicine)

cancer was tobacco smoke; because smoking was addictive and had a powerful corporate lobby behind it, cancers developed despite the existence of an ostensibly simple means to prevent them (namely, not smoking).

Once cancer was diagnosed, treatments for it were radical. Among them were chemotherapy and radiation therapy, which aimed to arrest the spread of cancer by killing cancerous cells. These often were accompanied by dangerous side effects. To treat breast cancer, the surgeon William Halsted (1852–1922) first performed a radical mastectomy (removal of the breast, chest wall muscles, and lymph nodes) in 1882. Despite the psychological trauma of forever losing

one's breast, Halsted's surgery saved many women's lives. This procedure for treating breast cancer was the most popular in the United States throughout the first half of the twentieth century.

One of the bizarre twists of history is the connection between cancer research and racial hygiene, epitomized by the Nazi regime in Germany during the 1930s and 1940s. Adolf Hitler (1889–1945) and many of his Nazi subordinates were concerned not only with purifying the Aryan race but also with eliminating other biological "threats" to the health of the race, including cancer. Historian Robert Proctor has argued that the Nazi war against cancer had much to recommend it:

The Nazis collected statistics intensively and centralized the effort to combat the disease, passing laws to prevent harmful carcinogens in public areas. The Nazis focused on prevention, partly because the removal of high-caliber Jewish scientists made a cure for cancer unlikely in the near future. Government regulations against asbestos, smoking, radiation exposure, and harmful dyes saved German lives. During the war, many of these efforts were abandoned, particularly those seeking to eliminate cigarettes and alcohol from soldiers' routines. But in general, the number of people developing lung cancer in Germany in postwar years fell below that of other countries, including the United States. These achievements were understandably overshadowed in the postwar years by the atrocities of the Nazi regime; Hitler had wanted to purify the race and destroy anything he considered a pollutant, even humans. After the war, efforts to curb cancer by stigmatizing smokers became easy prey for accusations of prejudiced, Nazi-like policies.

The development of the atomic bomb ushered in a new era of cancer-related fears. Because of the capacity of radiation to damage body cells and pose long-term genetic problems, governments possessing atomic weapons came under scrutiny. Before 1949, only the United States possessed the bomb; however, the mining of uranium for research, bombs, or future nuclear power plants was widespread. The exact nature of the links between radiation exposure and cancer were not well known; the only existing data was taken of the survivors of the bombings of Hiroshima and Nagasaki during the war. In addition, the Atomic Energy Commission conducted human exposure experiments in the 1940s and 1950s, sometimes without the knowledge of the subjects, to help determine acceptable limits of radiation exposure. By the 1950s, the threat of cancer from long-term radiation exposure erupted into a full-scale controversy about the risks of radioactive fallout from nuclear tests.

See also Medicine; Nazi Science; Radiation Protection; X-rays
References
Lerner, Barron H. *The Breast Cancer Wars: Hope, Fear, and the Pursuit of a Cure in Twentieth-Century America.* New York: Oxford University Press, 2001.
Lindee, M. Susan. *Suffering Made Real: American Science and the Survivors at Hiroshima.* Chicago: University of Chicago Press, 1994.
Patterson, James T. *The Dread Disease: Cancer and Modern American Culture.* Cambridge, MA: Harvard University Press, 1987.
Proctor, Robert N. *The Nazi War on Cancer.* Princeton, NJ: Princeton University Press, 1999.
Welsome, Eileen. *The Plutonium Files: America's Secret Medical Experiments in the Cold War.* New York: Delta, 1999.

Carbon Dating

Carbon dating, developed in the late 1940s by chemist Willard F. Libby (1908–1980) and his colleagues, provided a means to date ancient organic material. Using carbon 14, or radiocarbon, Libby's technique has empowered archaeologists to date historical sites and uncover the mysteries of human settlements at various locations throughout the world, going back thousands of years. It has also helped to resolve controversies about the age of artifacts and human and animal remains. In 1960, Libby won the Nobel Prize in Chemistry for this contribution.

University of Chicago's Institute of Nuclear Studies was one of many postwar research centers studying radioisotopes, the radioactive forms of common elements. These typically were produced artificially, using instruments such as the cyclotron. But some radioisotopes were created through natural processes. Carbon is not normally radioactive; radiocarbon (as carbon 14 was called) is created, as New York University's Serge Korff (1906–1989) noted in 1939, by powerful cosmic rays. High-energy particles entering the atmosphere from space liberate neutrons from the upper atmosphere, which in turn react with the omnipresent nitrogen in the atmosphere and create unstable (radioactive) isotopes of carbon, namely, carbon 14. These isotopes decay over time and become nitrogen, but this is a long process; carbon 14 has a half-life of some 5,600 years. One of the Institute of Nuclear Studies' scientists, chemist Willard Libby, found that carbon 14 could be useful in determining the age of organic materials, and he developed carbon dating by 1948.

Libby determined that humans were part of a giant pool containing radiocarbon produced by cosmic rays. Radiocarbon exists in all organic bodies, including humans. He assumed that, while alive, organic material assimilates carbon 14 (by consuming other organic material filled with it) at the same rate as it decays to form the stable isotope, nitrogen 14. That is, during life, the decay of radiocarbon is compensated by the intake of new radiocarbon. Quantitatively, living organisms do not acquire or lose radiocarbon during their lives. But when they die, radioactive decay no longer is compensated by the intake of radiocarbon-filled organic material. The process of decay is quantifiable. Thus, after 5,600 years, dead organic material should be about half as radioactive as living material. Libby realized that measuring the amount of radiocarbon in any dead organic substance should give some indication of how long ago it died.

Measuring the carbon 14 in organic substances to determine their age (carbon dating) became an important tool for tracing human history. Libby compared his results with material of historically known origin and found acceptable agreement. Consequently, with carbon dating, archaeologists could compare less sure historical estimates. The technique initially was deemed valid for organic material up to 50,000 years old. One of the oldest sites in France, for example, was the Lascaux Cave, in which there are paintings on the walls depicting ancient animals. Measuring the carbon 14 in the charcoal in the caves has helped archaeologists to deter-

mine their age at some 15,000 years. Most geological events are beyond the reach of carbon dating, but (relatively) recent phenomena, such as the arrivals and departures of ice ages, can be tracked with the technique. Other uses were in oceanography and meteorology, where carbon dating helped to determine the rate of mixing. But the most notable uses have been archaeological. As Libby noted in his Nobel speech, carbon dating "may indeed help roll back the pages of history and reveal to mankind something more about his ancestors, and in this way perhaps about his future."

See also Age of the Earth; Anthropology; Cosmic Rays; Radioactivity

References

Libby, W. F. *Radiocarbon Dating.* Chicago: University of Chicago Press, 1955.
Libby, Willard F. "Radiocarbon Dating." Nobel Lecture, 12 December 1960. In *Nobel Lectures, Chemistry, 1942–1962.* Amsterdam: Elsevier, 1964–1970, 593–610.

Cavendish Laboratory

In mid-nineteenth-century England, both Oxford University and Cambridge University wanted to create special areas for practical knowledge such as the sciences. Cambridge's efforts to do so proved difficult because of financial hardships. The Cavendish Laboratory came into being through the efforts of William Cavendish (1808–1891), the seventh Duke of Devonshire, who saved the plans from failure and financed the laboratory. In 1874, the first director of the laboratory received the title "Cavendish Professor," a position that was supposed to terminate after the death of the first person to hold it. But Cambridge University saw the value of the laboratory and its director, and the position continued; under the guidance of its world-renowned directors, the Cavendish Laboratory became a principal center of physics in the twentieth century.

The directors of the Cavendish Laboratory were several of the leading physicists of their day. The first, James Clerk Maxwell

(1831–1879, director 1871–1879), was the chief figure in electrodynamics during the nineteenth century. Following him were John William Strutt, better known as Lord Rayleigh (1842–1919, director 1879–1884), who studied the noble gases and electricity; J. J. Thomson (1856–1940, director 1884–1919), who discovered the electron; Ernest Rutherford (1871–1937, director 1919–1937), who proposed the nuclear atom and made fundamental contributions to radioactivity; and William Lawrence Bragg (1890–1971, director 1938–1953), who pioneered the study of X-ray crystallography.

The Cavendish was primarily an experimental laboratory, not a theoretical one. It attracted many students from abroad who wished to learn experimental techniques, particularly after J. J. Thomson's work had led to the discovery of the electron. The laboratory became a major center not only on cathode rays, which had led to the discovery, but on radioactivity and atomic structure. After Ernest Rutherford took over the directorship in 1919, young physicists from Europe and North America flocked to the laboratory to study the behavior of subatomic particles, by observing radioactive "disintegrations" (decay) and by bombarding elements with alpha particles (now recognized as helium atoms). In the 1930s, scientists at the Cavendish built the first particle accelerator—John Cockcroft (1897–1967) and Ernest Walton (1903–1995)—and discovered the neutron—James Chadwick (1891–1974)—a particle that was of particular importance in understanding atomic interactions and later atomic fission.

After Rutherford's death, the Cavendish passed into the hands of Sir William Lawrence Bragg. He and his father, William Henry Bragg (1862–1942), had won the Nobel Prize for their work on the interactions between X-rays and crystals. Bragg turned the focus of the laboratory toward his own interests, as his predecessors had done. The laboratory's work was severely curtailed during World War II, falling into relative dormancy as its scientists moved to war-related work, in radar, nuclear physics, and other fields.

Despite its pivotal role in experimental nuclear physics, the Cavendish Laboratory was weak in an area of profound importance to twentieth-century physical theory, namely, quantum physics. This was likely owing to the fact that the emphasis of the laboratory tended to follow the interests of the director. Although Rutherford had not shied from theory—he conceived of the nuclear atom—the quantum descriptions of this work were done largely by others (for example, Niels Bohr [1885–1962] was the first to explain the nuclear atom in terms of quantum orbits). The laboratory was also slow to make use of sophisticated instrumentation, although this was sometimes a badge of honor; nostalgic scientists reminded their colleagues that Rutherford accomplished more with sealing wax and string than the scientists in many of the best-equipped laboratories. The Cavendish Laboratory was slow to move into the age of "Big Science," with large teams of researchers and expensive equipment.

See also Chadwick, James; Cloud Chamber; Physics; Radioactivity; Rutherford, Ernest; Thomson, Joseph John

References

Cambridge University Physics Society. *A Hundred Years and More of Cambridge Physics.* Cambridge: Cambridge University Physics Society, 1974.

Crowther, J. G. *The Cavendish Laboratory, 1874–1974.* New York: Science History Publications, 1974.

Chadwick, James

(b. Cheshire, England, 1891; d. Cambridge, England, 1974)

James Chadwick, English physicist, discovered the neutron in 1932, transforming the understanding of subatomic physics and nuclear interactions. Chadwick spent much of his career under the guidance of Ernest Rutherford (1871–1937), the leading experimental physicist in the field of radioactivity and subatomic physics of his day. Working with Rutherford, he received his master of science degree in 1913 at the University of Manchester. After graduating, he traveled to Berlin and worked with Hans Geiger (1882–1945) at the Physikalisch-Technische Reichanstalt. He was there when war broke out, and he spent World War I interned at a racetrack in Ruhleben, Germany. When the war ended, he joined Rutherford at the Cavendish Laboratory, at Cambridge University, where he conducted experiments bombarding various elements with alpha particles. He became assistant director of research at the Cavendish in 1927.

Chadwick's discovery of the neutron came from his analysis of the mysterious "beryllium radiation." German physicist Walther Bothe (1891–1957) and his student Herbert Becker, at the Physikalisch-Technische Reichanstalt, had discovered this odd radiation when they bombarded the element beryllium with alpha particles. They believed the radiation was akin to gamma rays. In Paris, Frédéric Joliot (1900–1958) and Irène Joliot-Curie (1897–1956) found that the radiation could facilitate the ejection of protons in other material. Chadwick, aware of these studies, provided a different interpretation, observing that the radiation was in fact a particle with mass and a neutral charge, the neutron. After repeating the experiments at the Cavendish, he determined that the interaction between alpha particles and beryllium caused a nuclear transformation resulting in an atom of carbon and a lone neutron.

The director of the Cavendish, Ernest Rutherford, already had predicted the existence of a particle with mass, but neither negative nor positive charge. He envisioned it not as a separate particle but rather as a composite of an electron and proton, each negating the other's charge. Although Chadwick discovered evidence for the existence of the neutron, he continued to interpret it much as Rutherford had, as a composite of two oppositely charged particles. According to that interpretation, the electron was still present in the atom's nucleus. Ambiguity about whether the neutron was indeed an elementary particle, rather than a composite, lingered for some years. For his discovery of the neutron, Chadwick was awarded the Nobel Prize in Physics in 1935.

Chadwick made this discovery in 1932, often called the annus mirabilis (year of miracles) of physics. Although the neutron was the most important discovery of this year, it was accompanied by a couple of others around the same time: In late 1931, U.S. chemist Harold Urey (1893–1981) discovered heavy hydrogen, or deuterium. In 1932, Cavendish scientists John Cockcroft (1897–1967) and Ernest Walton (1903–1995) developed a device to accelerate protons and used these high-energy particles to bring about the first nuclear disintegration (ejection of a particle) achieved by artificial means, proving that particle acceleration could facilitate experimentation on atoms. Meanwhile, Ernest Lawrence (1901–1958) in Berkeley, California, was perfecting another accelerator, the cyclotron, which would revolutionize experimental nuclear physics. The discovery of the neutron came at an opportune moment to explain the interactions among atoms, analyses previously confined to alpha and beta particles.

In 1935, Chadwick accepted the Lyon Jones Chair of Physics at the University of Liverpool. In an effort to turn it into a first-rate research center, he built a cyclotron with money from the Royal Society. Rutherford, still at the Cavendish, had built a reputation for achieving a great deal without the need of expensive equipment; Chadwick's efforts at Liverpool marked a break with his mentor's own distaste for accelerators.

During the war years, Chadwick became deeply involved in efforts to build the atomic bomb. The cyclotron at Liverpool proved useful in this regard, helping to develop feasibility studies on nuclear reactions. Chadwick and other British scientists, members of the so-called MAUD Committee, reported in

1941 that that the amount of U–235 (an isotope of uranium) needed to achieve a critical mass and sustain a fission reaction was less than previously believed. Their findings helped to stimulate hopes that a bomb could be made during the war, thus making it a potentially decisive weapon. Later the British team was consolidated with the U.S. one, and Chadwick became a part of the Manhattan Project. He lived in the United States from 1943 to 1946 and was chief of the British team of researchers on the project.

After the war, Chadwick returned to Liverpool. He took part not only in physics research but also served in a policy-advising role to the British government in matters of atomic energy. In 1948, Chadwick retired to become master of Gonville and Caius College, Cambridge, a post he held for ten years; in 1969 he and his wife moved back to Cambridge. He received numerous honors in his lifetime, including a knighthood in 1945.

See also Atomic Bomb; Atomic Structure;
   Cavendish Laboratory; Physics; Radioactivity
References
Brown, Andrew P. *The Neutron and the Bomb: A Biography of Sir James Chadwick.* Oxford: Oxford University Press, 1997.
Gowing, Margaret. "James Chadwick and the Atomic Bomb." *Notes and Records of the Royal Society of London* 47:1 (1993): 79–92.
Peierls, Rudolf. "Recollections of James Chadwick." *Notes and Records of the Royal Society of London* 48:1 (1994): 135–141.

Chandrasekhar, Subrahmanyan

(b. Lahore, India [later Pakistan], 1910; d. Chicago, Illinois, 1995)

Subrahmanyan Chandrasekhar was born and raised in India, receiving his education at the University of Madras. He came from a successful Brahman family, and he was inspired by the achievements of Indian intellectuals who then were achieving world renown in numerous fields. His uncle, Chandresekhara Raman (1888–1970), won the Nobel Prize in Physics in 1930, bringing that honor to India for the first time. That same year, Chandrasekhar left India at the age of twenty to begin his doctoral studies under physicist Ralph H. Fowler (1889–1944) at Cambridge University in England. Ultimately his interests in stellar structure brought him to work instead under the astrophysicist Edward Milne (1896–1950). While a graduate student, he also visited Niels Bohr's (1885–1962) institute in Copenhagen for six months working on quantum statistics. During the five years at Cambridge, he developed the concept that came to be called the "Chandrasekhar limit" in stellar mass, providing the key to understanding differences in the end stages of stars.

At Cambridge, Chandrasekhar was challenged both personally and professionally. His devotion to vegetarianism, part of his identity as an Indian, was particularly difficult in a community with no familiarity with his dietary needs. More difficult was the cold reception to his scientific ideas. He had attempted to improve on Fowler's theory of white dwarf stars by including relativistic physics in his conception of stellar structure. The kernel of his idea occurred to him during his long voyage from India in 1930. He proposed that the equilibrium between electron repulsion and gravitational attraction in stars simply would not hold for stars with mass greater than about 1.44 solar masses. Such massive stars, instead of becoming white dwarfs, would collapse into a previously unknown kind of star of enormous density. But the well-known astrophysicist Arthur Stanley Eddington (1882–1944), who had initially encouraged Chandrasekhar, turned against him in 1935. He roundly criticized these ideas at a meeting of the Royal Astronomical Society, and unfortunately his prestige convinced other scientists not to interfere in their disagreements. Chandrasekhar, a young, foreign, and unproven scientist, could not hope to challenge Eddington. But in the ensuing years Chandrasekhar was vindicated, and others identified black holes as the end stage of those massive stars.

Astrophysicist Subrahmanyan Chandrasekhar talks to reporters in 1983, shortly after being awarded the Nobel Prize in Physics for work done half a century earlier. (Bettmann/Corbis)

Chandrasekhar did not limit his work to stellar structure. His contributions to astrophysics included subjects such as the dynamics of stars, atmospheres of planets and stars, hydrodynamics, and black holes. He became an authority on relativistic astrophysics. Chandrasekhar's many publications seemed, according to some, to follow a pattern. He would write highly technical papers within a particular subject and then wrap them up after some years with a detailed treatise that would serve as a benchmark for the whole subject.

After receiving his doctorate, Chandrasekhar went against the wishes of his family in India in at least two ways. First, he married a woman of his choosing (and her choosing) rather than having a marriage arranged. Second, he decided not to return to India and instead embarked on a career as a research scientist. He took a position at the University of Chicago in 1937, where he helped transform the astronomy department toward relativistic astrophysics and supervised some fifty doctoral students over the years. Beginning in 1952, he became editor of the *Astrophysical Journal,* a responsibility he kept for nearly twenty years. Chandrasekhar and his wife, Lalitha, became U.S. citizens in the 1950s. Many of his colleagues were astonished that he was never awarded the Nobel Prize in Physics, a situation that was rectified in 1983, nearly fifty years after he conducted the work on the structure of stars.

See also Astrophysics; Eddington, Arthur Stanley

References

Chandrasekhar, S. *Truth and Beauty: Aesthetics and Motivations in Science.* Chicago: University of Chicago Press, 1991.

North, John. *The Norton History of Astronomy and Cosmology.* New York: W. W. Norton, 1995.

Wali, Kameshwar C. *Chandra: A Biography of S. Chandrasekhar.* Chicago: University of Chicago Press, 1991.

Chemical Warfare

One of the notorious hallmarks of the first half of the twentieth century was the development and use of weapons of mass destruction. Chemical weapons, specifically poison gas, were the first to be recognized internationally as special kinds of weapons needing special rules. Poison gas was developed and used by both belligerent sides during World War I, and the image of soldiers clad in gas masks has become virtually synonymous with that conflict. When the war ended, the role of poison gas in future wars became a subject of fierce debate.

Scientists had suggested the possibility of chemical attacks as early as the Napoleonic wars, but their novelty, along with skepticism about their usefulness, precluded development on a significant scale. During the U.S. Civil War, the United States might have developed "asphyxiating shells" intended to spread fire and noxious fumes in areas occupied by the enemy. Despite these possibilities, chemical warfare did not develop in earnest until the early twentieth century. During World War I, German chemists Fritz Haber (1868–1934) and Walther Nernst (1864–1941) developed a means to flush enemy soldiers out of their trenches by saturating them with toxic gas. This was only one of Haber's plans to put chemists to work in service of German Kaiser Wilhelm II (1859–1941); although he became notorious for his role in developing poison gas, he also helped develop fertilizers and explosives.

Chemical weapons made their wartime debut in an experimental chlorine gas attack by the Germans at Ieper, Belgium, in April 1915. Released from canisters, the gas was carried by the wind over no-man's-land and into the trenches on the other side. Gagging and coughing, caught by surprise, the French forces retreated. But the weapons were not fully integrated into German fighting forces, and despite this victory, they were not able to turn poison gas into the desired decisive weapon. Soon both sides developed the capa-bility to make such attacks and issued gas masks to their troops, and chemical weapons lost any decisive role they might have had. In fact, gas attacks killed (relatively) few soldiers; its primary role was to unsettle and harass the enemy, producing casualties that later would recover.

Chemical warfare sparked one of the first major working relationships between military groups and scientists. Typically, the application of scientific knowledge to military technology was done by engineers or technical specialists. But Haber's group had worked directly with military officers, and the situation was soon replicated in the Allied countries. One important consequence of chemical weapons was the identification of scientists with a weapon of destruction; the postwar disillusionment with progressive Western society also provoked doubts about the role of science in social progress. Allied scientists, who opposed such linkages between them and what they considered the atrocities of war, were appalled when Haber was selected in 1919 as the recipient of the Nobel Prize in Chemistry.

In society as a whole, chemical warfare provoked an abiding sense of dread. Would poison gas be the future of war? Activist groups in Europe and the United States pointed to poison gas as the precursor of an apocalyptic future. Poison gas came to symbolize the inhumanity of the war as a whole. It was indiscriminate in its destruction, and the use of poison gas on civilian populations, using the recently invented airplane as a delivery vehicle, seemed just around the corner.

The question of whether chemical weapons posed a special menace and therefore should be banned provoked sharp disagreement. In the United States, scientific lobbies such as the American Chemical Society convinced Congress to support the Chemical Warfare Service to keep the country abreast of the latest weapons technology, in the interests of preparedness. Leading scientific figures in Britain spoke up in support

A posed photograph from World War I showing the risks of failing to wear gas masks. (Library of Congress)

of chemical weapons. One of these was J. B. S. Haldane (1892–1964), whose *Callinicus: A Defense of Chemical Warfare* appeared in 1925. He noted that asphyxiation, however terrible, was preferable to a shell wound from a conventional weapon. Haber himself asked whether choking in a cloud of gas was really different from being killed by a bullet. Given that roughly 90 percent of gas victims during the war had survived, some newspapers in the United States even called the weapon a humane alternative to killing.

After the war, the United States made poison gas an important component of the arms control negotiation. The recently created League of Nations followed suit, pursuing dialogue among nations to address chemical weapons. In 1925, diplomats met to sign the Geneva Protocol for the Prohibition of the Use in War of Asphyxiating, Poisonous or Other Gases, banning the use of chemical weapons in war. The U.S. delegation signed it, but the Senate did not ratify it until some fifty years later. Although peace activists had succeeded in creating the Geneva Protocol, they misjudged the opposition to ratifying such a document. But whether countries rat-

ified it or not, this legal infrastructure did little to stop the use of chemical weapons. Instead, chemical warfare served as a model for deterrence in a prenuclear world. Edward Spiers has argued that fear of retaliation, not international law, prevented gas warfare during World War II. Meanwhile, in conflicts where one side had the capability and the other did not, such weapons were used freely. For example, in the Abyssinian War of 1935–1936, Italian troops broke with the Geneva Protocol (ratified by Italy in 1928) and gassed their Ethiopian foes. Likewise, the Japanese used poison gas during their conquest of China in the 1930s. In both cases, the victim lacked a deterrent.

Between the industrialized belligerents of World War II, outright chemical warfare in the form of gas attacks was avoided. However, bodies such as the Chemical Warfare Service (U.S.) provided useful armaments such as chemical mortars (designed to deliver gas shells but used for high explosives and smoke shells) and flamethrowers. Mortar battalions organized by the Chemical Warfare Service saw extensive combat in the invasion of Sicily and in Italy, where the terrain was better suited to mortar units than the larger artillery units. In the Pacific region, in addition to mortar shells, portable flame-throwers (and in some cases, flame-throwing tanks) provided U.S. troops with a means to penetrate bunkers and caves that had proven impervious to conventional artillery attacks. The chemists also helped develop effective smoke shells to provide cover for troops during assault. The most devastating weapons were incendiary bombs, or "fire bombs," which were the most effective destroyers of cities before the development of the atomic bomb.

See also Haber, Fritz; Haldane, John Burdon Sanderson; Hormones; Social Progress; World War I; World War II

References

Haber, L. F. *The Poisonous Cloud: Chemical Warfare in the First World War.* New York: Clarendon Press, 1986.

Kleber, Brooks E., and Dale Birdsell. *The Chemical Warfare Service: Chemicals in Combat.* Washington, DC: Government Printing Office, 1966.

Miles, Wyndham D. "The Idea of Chemical Warfare in Modern Times." *Journal of the History of Ideas* 31:2 (1970): 297–304.

Slotten, Hugh R. "Human Chemistry or Scientific Barbarism? American Responses to World War I Poison Gas, 1915–1930." *Journal of American History* 77:2 (1990): 476–498.

Spiers, Edward M. *Chemical Warfare.* Urbana: University of Illinois Press, 1986.

Cherenkov, Pavel

(b. Voronezh region, Russia, 1904; d. Moscow, USSR, 1990)

Pavel Cherenkov was one of several outstanding young physicists working in the Soviet Union in the 1930s. He was a graduate student under Sergei Vavilov (1891–1951), who was busy creating the Physics Institute of the Academy of Sciences (FIAN). Cherenkov received his doctorate there, in physico-mathematical sciences, in 1940. Among his colleagues were Vavilov, Il'ia Frank (1908–1990) and Igor Tamm (1895–1971). At Vavilov's urging, Cherenkov began to ob-

serve how gamma rays interact with uranium salt solutions. He soon noticed, in 1935, a bluish light being emitted by a beam of charged particles as it passed through a medium. This soon became known as "Cherenkov radiation," or simply as the "Cherenkov effect." (It is often spelled Cerenkov.) Soviet scientists referred to it as "Vavilov-Cherenkov" radiation to emphasize Vavilov's role in discovering it.

Tamm and Frank distinguished themselves in 1937 by providing a theoretical explanation for Cherenkov radiation. The blue glow was radiation emitted when electrons or other charged particles pass through a transparent (solid or liquid) medium at velocities faster than that of light passing through the same medium. After arriving at this conclusion through mathematical calculations, Tamm and Frank were eager to find an alternative, because, according to Albert Einstein's (1879–1955) theory of relativity, light sets the speed limit of the universe. "Superlight" velocities seemed out of the question. But they soon determined that relativity has a limited validity: It applies only to the velocity of light in a vacuum. Any medium can change the velocities of light or other particles in different ways. Thus the Cherenkov radiation was, for light, something like a sonic boom, the effect of an airplane exceeding the speed of sound.

The work of these three researchers, and the efforts of Vavilov to create a strong physics community, provided a major boost to the prestige of science in the Soviet Union. Recognition by the international community came slowly, however. Only after the death of Joseph Stalin (1879–1953) in 1953 did Soviet scientists have the (relative) freedom to forge international contacts and promote their stature abroad. The recognition of the efforts of 1930s physicists was a major stride in this direction. Cherenkov, Frank, and Tamm shared the Nobel Prize in Physics in 1958 (Vavilov died in 1951 and thus was not eligible). They were the first Soviet scientists to do so.

See also Academy of Sciences of the USSR; Light; Physics; Radioactivity; Relativity; Soviet Science; Vavilov, Sergei

References
Kojevnikov, Alexei. "President of Stalin's Academy: The Mask and Responsibility of Sergei Vavilov." *Isis* 87 (1996): 18–50.
Tamm, I. E. "General Characteristics of Vavilov-Cherenkov Radiation." *Science*, New Series, 131:3395 (22 January 1960): 206–210.

Chromosomes

In the nineteenth century, biologists looked to the study of cells (cytology) to provide information about the transmission of characteristics from one generation to the next. When cells divided, small bodies appeared in the cell nucleus. These rodlike bits were called chromosomes because they readily absorbed coloring stains needed to view them clearly under a microscope. The work of German anatomist Walther Flemming (1843–1905) described the splitting of chromosomes during cell division, which produced identical chromosomes in each of the new cells. German theoretical biologist August Weismann (1834–1914) also took an interest in cytology. Rather than focus on populations and statistics, Weismann urged the study of living cells in order to understand the material agent of heredity. He predicted that some process would be discovered that reduced chromosomes by half, allowing for fertilization to restore cells to the full number. His prediction was confirmed in 1888.

By the twentieth century, scientists were familiar with some of the basic features of the chromosome, such as its splitting during mitosis (normal cell division) and meiosis (cell division in which the number of chromosomes are reduced by half). Also, chromosomes appeared to carry very specific information; for example, U.S. cytologist Clarence McClung (1870–1946) suggested in 1902 that certain pairs of chromosomes were responsible for determining sex. Scientists were beginning to view chromosomes as the key to understanding heredity. In 1903,

one of McClung's students, Walter S. Sutton (1877–1916), argued that work on chromosomes would prove that their interactions and separation during cell division constituted the physical basis of Mendelian laws. At that time, Gregor Mendel's (1822–1884) work on genetics had resurfaced and many biologists accepted it as a new mathematical law of heredity. Chromosomes, Sutton believed, were crucial in transmitting Mendelian characters. Aside from uniting cytology and genetics, this belief put genetic theory on material footing, opening a path for laboratory experimentation.

Despite the clear relationship between chromosomes and heredity, it was equally clear that there were far more characteristics to pass on than there were chromosomes. Sutton noted that there must be many different genetic characters associated with a single chromosome. Characters were not always independent; in fact, it seemed that some of them always were inherited together, perhaps lumped onto a single chromosome. This seemed to suggest that "genes," as the unit of heredity was called, were arranged into linkage groups. U.S. biologist Thomas Hunt Morgan's (1866–1945) breeding experiments with *Drosophila melanogaster*—the fruit fly—showed clear deviations from the Mendelian laws if one assumed that characters all were inherited independently. Although he initially rejected the notion that genetics characters were carried by chromosomes, these experiments convinced him that "linkage" indeed was taking place. For example, he noted that the trait "white eyes" was always passed on with another, "rudimentary wings."

In 1911, Morgan's chromosome theory of inheritance suggested that multiple genes were carried on chromosomes and that these genes could be identified, or "mapped." Morgan's student A. H. Sturtevant (1891–1970) published the first such chromosome map in 1913 in the *Journal of Experimental Zoology*. It showed the arrangement of six characters on the chromosome. Over the next several years, the chromosome theory of inheritance increasingly gained supporters, until finally its greatest opponent, British geneticist William Bateson (1861–1926), abandoned his criticisms and embraced it in 1922. Morgan became the leading geneticist and his students, among them Sturtevant and H. J. Muller (1890–1967), helped more fully to establish the chromosome as the material carrier of heredity.

See also Bateson, William; Genetics; Morgan, Thomas Hunt

References

Bowler, Peter J. *The Mendelian Revolution: The Emergence of Hereditarian Concepts in Modern Science and Society.* Baltimore, MD: Johns Hopkins University Press, 1989.

Magner, Lois N. *A History of the Life Sciences.* New York: Marcel Dekker, 1979.

Cloud Chamber

The cloud chamber was developed by Charles T. R. Wilson (1869–1959) in 1911. This instrument, improved over subsequent years, was designed to trace the paths of ionizing particles by observing them in a chamber packed with vapor. Wilson, who was then working at Cambridge University's Cavendish Laboratory, arrived at the idea from his interest in meteorological studies. The principle of the chamber was simple: As charged particles (such as protons and electrons) passed through vapor, the ionized droplets became visible because of their sudden expansion. Wilson discovered this principle in the 1890s and began work on a chamber to turn the phenomenon into an experimental tool.

In 1911 Wilson created the first working chamber and took the first photograph of the ionization traces. By photographing the traces that appeared in the cloud chamber, physicists had a means to observe the ionization process and track the paths of charged particles in any experimental environment. The cloud chamber, like the ionization counter developed afterward by German physicist Hans Geiger (1882–1945), provided the detection methods for the intensive

experimental research on particle physics in the 1920s and 1930s. The cloud chamber was improved in the 1920s by Japanese physicist T. Shimizu and Patrick M. S. Blackett (1897–1974), both working at the Cavendish. The small cloud chambers typically were used to detect the particles released during the decay of radioactive minerals. Large cloud chambers proved useful for detecting the presence of cosmic rays.

See also Cavendish Laboratory; Physics; Radioactivity

References

Kevles, Daniel J. *The Physicists: The History of a Scientific Community in Modern America.* Cambridge, MA: Harvard University Press, 1995.

Kragh, Helge. *Quantum Generations: A History of Physics in the Twentieth Century.* Princeton, NJ: Princeton University Press, 1999.

Wilson, J. G. *The Principles of Cloud Chamber Technique.* New York: Cambridge University Press, 1951.

Cockcroft, John

(b. Todmorden, Yorkshire, England, 1897; d. Cambridge, England, 1967)

John Cockcroft, who shared the 1951 Nobel Prize in Physics with his collaborator, Ernest T. S. Walton (1903–1995), led a scientific career at the center of high-energy physics. His journey in this field began when his studies at the University of Manchester happened to coincide with Ernest Rutherford's (1871–1937) tenure there, and he attended some of the latter's lectures on the atom. But another decade would pass before he would begin a more lasting relationship with Rutherford. Like so many of his countrymen, Cockcroft went into active service during World War I. He was a signaler for the Royal Field Artillery. He began to take an interest in electrical engineering, and when the war ended, he spent several years in that field and in studying mathematics. After these many years of training, in 1924 Cockcroft joined Rutherford's group at Cambridge University, at the Cavendish Laboratory.

Cockcroft fit in well with the people at the Cavendish Laboratory, whose fascination with atomic structure he came to share. One of these colleagues was the Russian George Gamow (1904–1968), who had provided the theoretical basis for the escape of alpha particles from atomic nuclei. Gamow's statistical calculations endowed alpha particles with wave properties, allowing some of them to penetrate the energy barrier that seemed to trap them. Cockcroft reversed this and calculated that protons of high energy would be likely to penetrate the same barriers to enter the nucleus. This could, he reasoned, be achieved in a laboratory if a contraption could be devised to accelerate protons.

After receiving Rutherford's blessing to work on this problem at the Cavendish Laboratory, Cockcroft and Ernest T. S. Walton constructed a machine to accelerate protons by 1930. It consisted of a vacuum tube provided with a proton source attached to a voltage multiplier. With the multiplier they managed to achieve more than 700 kilovolts. When turning the proton beam on a lithium target, they detected alpha particles being emitted after raising the multiplier to 125 kilovolts. As they increased the voltage, the amount of alpha particles increased. Cockcroft and Walton soon realized that they were witnessing the disintegration of elements. Bombarding a lithium element with a proton resulted in a transformation into two alpha particles (ions of helium).

The results of these experiments, announced in 1932, were astounding. Already in 1919 Rutherford had demonstrated the possibility of transmuting atoms with protons, but Rutherford had confined himself to protons at natural speeds. Now, through acceleration, Cockcroft and Walton had achieved the first nuclear transformation from artificial means. They continued their work on accelerators in the subsequent years, although by this time U.S. physicist Ernest Lawrence (1901–1958) was constructing an even more powerful device for particle acceleration, the cyclotron. Cockcroft then took his cue from Lawrence, building a cyclotron for the Cavendish Laboratory in the late 1930s.

Cockcroft entered into war work again during World War II. He contributed to Britain's radar projects, to help detect submarines and aircraft. This wartime work on radar would have a decisive impact on the course of the war, particularly in the struggle for air supremacy in the Battle of Britain. His government sent him to the United States in 1940 as part of Sir Henry Tizard's (1885–1959) effort to share military secrets with the United States. For the next several years, Cockcroft worked primarily on radar before becoming the director of atomic energy research for the British and Canadians in 1944.

When the war ended, Cockcroft became the dominant scientific voice in British nuclear affairs, as director of the Atomic Energy Research Establishment at Harwell from 1946 to 1959. He continued to support research on particle acceleration, helping to establish not only a series of accelerators at Harwell but also the Rutherford Laboratory at Chilton, and by supporting the European Center for Nuclear Research (CERN).

See also Cavendish Laboratory; Cyclotron; Gamow, George; Physics; Radioactivity; World War II

References

Cockcroft, John D. "Experiments on the Interaction of High Speed Nucleons with Atomic Nuclei." (Nobel Lecture, 11 December 1951). In *Nobel Lectures, Physics, 1942–1962.* Amsterdam: Elsevier, 1964–1970, 167–184.

Hartcup, Guy, and T. E. Allibone. *Cockcroft and the Atom.* Bristol: Adam Hilger, 1984.

Spence, Robert. "Cockcroft, John Douglas." In Charles Coulston Gillispie, ed., *Dictionary of Scientific Biography,* vol. III. New York: Charles Scribner's Sons, 1971, 328–333.

Walton, Ernest T. S. "The Artificial Production of Fast Particles." (Nobel Lecture, 11 December 1951). In *Nobel Lectures, Physics, 1942–1962.* Amsterdam: Elsevier, 1964–1970, 187–194.

Cold War

As World War II drew to a close in 1945, scientists began to play an important role in the mounting "Cold War" between the United States and the Soviet Union. Because of the primacy of the atomic bomb in both U.S. and Soviet global strategy in the years following the war, supremacy in science and technology became the centerpiece of the Cold War confrontation. Although the countries did not fight each other directly, they antagonized each other politically, while pouring money into scientific institutions and military establishments in a race to prevent the other from maintaining either a diplomatic advantage or a strategically superior position in the event of an actual war.

The United States and the Soviet Union were allies during the war, but the most important scientific projects were not shared between them. The United States collaborated with Britain on both radar and the atomic bomb project, but the Soviet Union was excluded from these arrangements. Some atomic scientists, such as Leo Szilard (1898–1964) and Niels Bohr (1885–1962), favored sharing information with the Soviet Union as a peaceful gesture and argued forcefully for the international control of atomic energy as the only means of averting a confrontation and a costly arms race. But political leaders—especially British prime minister Winston Churchill (1874–1965)—were adamantly opposed to it, believing that a postwar Anglo-American alliance would need the bomb as diplomatic leverage to prevent the aggression of the Soviet Union. Scientific know-how, atomic or otherwise, increasingly became a commodity to be protected; in the final months of the war, U.S. and Soviet special intelligence teams sought out and "captured" German scientists for their own use in various scientific programs, such as nuclear physics and rocketry. After the war, the occupation forces deported a number of German scientists and technicians. The U.S. effort to do this,

dubbed Project Paperclip, captured leading scientists such as Wernher Von Braun (1912–1977), who became instrumental in U.S. rocket programs.

Although the Americans and British did not want to share the secrets of the atomic bomb, the Soviets already knew a great deal about it through a sophisticated intelligence network. The Federal Bureau of Investigation learned that Manhattan Project scientists Alan Nunn May (1911–2003) and Klaus Fuchs (1911–1988) had sold secrets to the Soviets throughout the war, and Fuchs had continued to do so when he became a leader of Britain's postwar nuclear establishment. Because of such revelations, scientists came under suspicion for Communist sympathies or for their efforts to share information with the Soviet Union. The "internationalist" proclivities of leading scientists such as National Bureau of Standards director Edward Condon (1902–1974), for example, led to hearings about his loyalty. Their access to national secrets, combined with some left-leaning political sentiments, made scientists common targets for congressional investigations led by the House Un-American Activities Committee (HUAC) or leading anticommunist politicians such as Joseph McCarthy (1908–1957) in the Senate.

The Soviet Union began its own atomic bomb project after the battle of Stalingrad, in early 1943. Already it had received intelligence material, chiefly from the activities of Klaus Fuchs. Igor Kurchatov (1903–1960) headed the project. Most officials in the United States could not conceive of the Soviet Union being able to make the enormous effort necessary to build the bomb, at least for a decade or so. But Soviet leader Joseph Stalin (1879–1953) made it a priority, believing that his country would have no diplomatic power against the dominance of capitalist countries such as the United States or Britain if it did not also possess the bomb. This provided the physics community in the Soviet Union some protection from ideologues who wished to enforce conformity of thought, as occurred in the biological sciences. In his desire to build a bomb, Stalin kept physicists above that dangerous political fray. This, combined with a highly successful espionage effort, allowed the Soviet Union to test its first device in 1949, years ahead of others' expectations.

Both the Soviets and Americans made science and technology central aspects of their national security strategies, first in the area of nuclear energy, but also in numerous other fields related to military technology. The Soviet success made the decision by the United States to pursue the hydrogen (fusion) bomb much easier; some scientists had argued that this weapon, which could be made a thousand times more powerful than the fission bomb dropped on Hiroshima, was not a military weapon, but rather an instrument of genocide. But President Harry Truman (1884–1972) reasoned, as did many others, that the Soviets would likely pursue it (he was right), and that the United States could not afford to be out-leveraged diplomatically. He made the decision to go ahead with the hydrogen bomb in 1950, beginning a new phase in the nuclear arms race.

The complexity of the Manhattan Project, with its reliance on scientific expertise, industrial production, political support, and efficient management, convinced many of its participants that science in the postwar world would be equally complex. In light of the potential confrontation with the Soviet Union, new projects would be secret and funded by the military, necessitating new partnerships. The Navy, for example, would need to forge strong bonds with scientific institutions, to see new weapon systems move from the theoretical stage to the implementation stage. The Air Force, which became an independent branch of the armed services in 1948, became the backbone of U.S. defense because its strategic bombers would be the delivery vehicles of atomic weapons in the event of war. The Cold War thus created situations that some called Big Science, with teams of well-paid researchers coordinating

research and techniques between universities, corporations, and the armed services, to achieve large-scale, complex projects. It fostered close relationships that, a decade later, President Eisenhower would warn against and call the military-industrial complex.

The United States also wanted to create scientific links with its allies, to ensure the existence of an anticommunist bloc well grounded in science and technology. Aware of the importance of science in U.S. power, the U.S. Department of State took an active role in carving a place for science within its own structure. It established science liaison offices overseas, to have U.S. scientists attend conferences and meetings, to keep watch over the latest scientific and technological developments in other countries, in an effort to keep U.S. scientists abreast of new ideas and achievements. The first liaison office, in London, was designed to report on British activities while also fostering cooperative relationships. Led by Lloyd Berkner (1905–1967), a State Department committee issued a report in 1950 entitled "Science and Foreign Relations," outlining the need to keep apace of science and technology for the security of all free peoples of the world.

In 1949, President Truman issued a directive that assigned to the Department of State the responsibility of collecting and disseminating ideas about the basic sciences from other countries. The Department of Defense also developed an intelligence network to ensure that military technology in the United States was equivalent or superior to that of any other nation. At the same time, Truman declared in his 1949 inaugural address that the United States wanted to pool the world's scientific talent to help countries of the developing world. Because it was the fourth point in his address, the U.S. effort became known as Point Four. Like the Marshall Plan, which provided grants and loans to Europe, it was designed to help the recovery of poor nations by providing them with technical expertise. Behind Point Four, however, was Truman's strategy of containing communism; to strengthen poor, nonaligned countries was a way to win them over to the "free world" without pouring vast sums of cash into them, as had occurred with the Marshall Plan in Europe.

See also Atomic Bomb; Atomic Energy Commission; Espionage; Game Theory; Loyalty; Office of Naval Research; Patronage; Soviet Science

References

Holloway, David. *Stalin and the Bomb: The Soviet Union and Atomic Energy, 1939–1956.* New Haven, CT: Yale University Press, 1994.

Kevles, Daniel J. *The Physicists: The History of a Scientific Community in Modern America.* Cambridge, MA: Harvard University Press, 1995.

Needell, Allan A. *Science, Cold War and the American State: Lloyd V. Berkner and the Balance of Professional Ideals.* Amsterdam: Harwood Academic Publishers, 2000.

Sherwin, Martin J. *A World Destroyed: The Atomic Bomb and the Grand Alliance.* New York: Alfred A. Knopf, 1975.

Wang, Jessica. *American Science in an Age of Anxiety: Scientists, Anticommunism, and the Cold War.* Chapel Hill: University of North Carolina Press, 1999.

Colonialism

Modern science and European colonialism were closely connected after the fifteenth century, with newly discovered species of flora and fauna requiring reevaluation of old systems of biological classification and unprecedented access to the world's physical and human resources, transforming scientists' ability to study and observe. By the end of the nineteenth century, "colonial science" was an activity with special characteristics, decidedly different from that taking place in metropolitan, industrialized Europe. Its most salient characteristic was its reliance on Europe or North America for authority and guidance. But Europeans depended on the colonies, too, as a source of previously unknown forms of "exotic" knowledge, and they used their own knowledge to exercise imperial control over distant lands.

One view of the differences between metropolitan and colonial science, proffered by historian George Basalla, emphasized the process by which colonial science became more autonomous and capable of independent scientific activity. He saw this as a progression of the spread of Western science to other parts of the world, and he devised a three-stage approach to the development of scientific communities. First, outlying territories became resources for scientists in advanced countries (in Europe or the United States). Second, they developed scientific communities of their own, but they were dependent on Europeans. And third, they became wholly independent communities. This model did not necessitate an official colonial status but referred instead to the relationship between scientific communities. The model is instructive but limited in explanatory power, as it does not help to understand how colonies shaped science in metropolitan areas, and it does not provide ways to understand how colonial science facilitated the consolidation of imperial power.

Scientific knowledge often helped to reinforce imperialism. This was particularly true in the field of medicine, as Europeans tried to eradicate tropical diseases in their colonies. In combating diseases such as malaria, sleeping sickness, bilharzia, and smallpox, Europeans defined the critical problems themselves and imposed solutions upon indigenous populations; local peoples played little or no role in using knowledge to develop public health measures. Although this was a means of promoting sanitation and fighting fatal or debilitating illnesses, it provided imperial authorities—armed with the tools of science—with extraordinary power over indigenous populations, because they were the sole bearers of life-saving knowledge. In addition, such efforts typically began as a means to prevent widespread deaths among Europeans, who died in larger numbers from such diseases because of lack of previous contact. Keeping white Europeans alive in distant lands, through knowledge of medicine, helped to reinforce control over colonies.

The feedback of information and impressions by colonial Europeans to their homelands transformed many scientific fields and created others. One of the strongest of these influences was in anthropology. This field developed in the nineteenth century, often motivated by the desire to rank the races of the world in a hierarchy. Britain's vast empire was used as a collecting ground for specimens of flora and fauna, but also of different races of men. After the first two decades of the twentieth century, however, by the end of World War I, many British intellectuals abandoned their progressive view of racial hierarchy, which had placed British blood and civilization at the top. The war itself radically altered that belief, casting doubt on whether Western civilization was truly progressing at all. "Functionalist" anthropologists began to study the operations of individual societies for their own sake, rather than to place aboriginal peoples at some rung on a ladder in the climb toward greater—that is, more "Western"—civilization.

Also after World War I, colonial science often was reoriented to encompass more strategic goals. For years, French scientists looked to the colonies in North Africa and elsewhere as sources of employment and as a vast source of data, making climate, seismic, magnetic, astronomical, and other kinds of observations and classifying exotic species. As in other national contexts, colonial authorities looked toward experts at home for guidance on scientific and other matters. Thus, argues historian Lewis Pyenson, French scientists could spread not only their ideas but also their cultural values to the colonies. Yet after the war, French scientists focused less on the collection of data and specimens for the benefit of scientists in France and more on the development of resources in the colonies. Scientists argued that their research was of strategic value, making up an important part of French strength not only at home but abroad in the colonies. This sparked debate over who should set agendas for scientists in the colonies, whether the colonial scientists or

strategic planners in France. Germany also developed colonial science after World War I for strategic reasons. In Western Samoa, for example, scientists established a geophysical station, and scientists in other disciplines were active in Argentina and China. In the case of Germany, much of its influence had no political status (the countries were independent). Yet through scientific knowledge, Germany exerted cultural influence; its power to do this helped scientists convince the German government to fund their activities for prestige value.

Longstanding imperial relationships led to deep divisions among the peoples in colonial territories about the value of Western science. Was it a tool of control, or was it simply a universal kind of knowledge to which no culture or country could claim ownership? Widespread decolonization after World War II provoked such difficult questions about the role of Western science in an era when the atomic bomb appeared to make science an even more important tool of great power diplomacy. In India, activist Mohandas Gandhi (1869–1948) had spent many years, long before the war, challenging his countrymen to avoid becoming yet another European or U.S. state. Like many other colonies, India's economy was transformed to attempt industrialization after World War I, because the British government wanted to ensure that its crown jewel (the colony itself) would provide war goods if fighting broke out again. Gandhi was highly critical of technology and industrialization, which he believed were used by a few individuals to control the masses. Scientific knowledge could do this as well, and Gandhi insisted that India pursue science only for productive, humanitarian ends. When India underwent its independence movement in the late 1940s, science and technology were highly valued, but specifically for their role in catalyzing economic and social development. Other former colonial territories, now forced to define explicitly their attitudes toward "Western" science—a legacy of empire—typically took a similar road, believing that they should harness it when useful for development, but also that they should be wary of the power of such knowledge in the hands of those wanting to manipulate and control others.

See also Geophysics; Medicine; Patronage; Scientism; Social Progress; Technocracy

References

Basalla, George. "The Spread of Western Science." *Science,* New Series 156:3775 (5 May 1967): 611–622.

Bonneuil, Christophe. *Des Savants pour l'Empire: La Structuration des Recherches Scientifiques Coloniales au Temps de la 'Mise en Valeur des Colonies Francaises,' 1917–1945.* Paris: ORSTOM, 1991.

Farley, John. *Bilharzia: A History of Imperial Tropical Medicine.* Cambridge: Cambridge University Press, 1991.

Kuklick, Henrika. *The Savage Within: The Social History of British Anthropology, 1885–1945.* New York: Cambridge University Press, 1992.

Kumar, Deepak. "Reconstructing India: Disunity in the Science and Technology for Development Discourse, 1900–1947." *Osiris* 15 (2001): 241–257.

Pyenson, Lewis. *Civilizing Mission: Exact Sciences and French Overseas Expansion, 1830–1940.* Baltimore, MD: Johns Hopkins University Press, 1993.

———. *Cultural Imperialism and the Exact Sciences: German Expansion Overseas, 1900–1930.* New York: Peter Lang, 1985.

Compton, Arthur Holly

(b. Wooster, Ohio, 1892; d. Berkeley, California, 1962)

Arthur Compton is known, as the Nobel Prize committee put it in 1927, "for his discovery of the effect named after him." Compton lent his name to science as few scientists have, and in the 1920s he became a major figure in the world of physics. As an American, Compton was an anomaly during a period when Europeans dominated both experimental and theoretical physics. He later helped to transform his country into a center of scientific activity while directing the best scientific minds under the Manhattan Project.

Compton received a Ph.D. from Princeton in 1916, writing a thesis on X-ray diffraction and scattering. After the United States entered World War I, he worked as a research engineer for the Westinghouse Electric and Manufacturing Company and helped to develop aircraft instruments for the Army Signal Corps. While at Westinghouse, Compton continued his studies of X-rays, formulated a conception of an electron with finite size, and conducted studies of magnetization on the reflections of X-rays by crystals.

After the war, Compton received a fellowship to study at Cambridge University's Cavendish Laboratory, where physicists flocked during the interwar years to work with the leading investigators of the atom, J. J. Thomson (1856–1940) and Ernest Rutherford (1871–1937). However, at Cambridge, the equipment Compton needed to continue high-voltage X-ray experiments was not available, so he shifted temporarily to the study of the scattering of gamma rays. During his year abroad he held fast to his notion of an electron with finite size, which he hoped would account for the intensity of scattering and the change of wavelength with scattering angle. But his work increasingly only enhanced a growing conceptual disparity between the laws of physics and the results he observed.

When he returned to the United States, Compton took a job in the physics department at Washington University, in St. Louis, Missouri. He turned again to X-rays, using the crystal spectrometer developed by William Henry Bragg (1862–1942). With the aid of this instrument, he measured the change in wavelength of monochromatic X-rays scattered from a target at various angles. The shift in wavelength itself, initiated by striking electrons at rest in the target, has since become known as the Compton effect.

Compton's achievement in 1923 was not merely in describing the effect, but also in explaining it in the context of quantum theory. Although Compton was well acquainted with quantum theory, it was only after he read a paper by Albert Einstein (1879–1955) on the linear momentum of photons that he saw a way to demonstrate it using X-rays. A photon, according to quantum theory, was the basic unit (or quantum) of electromagnetic radiation. If an X-ray photon carried linear momentum as well as energy, Compton could treat the interaction in terms of momentum and its conservation, as an X-ray photon collides with an electron in the target substance. Assuming the conservation of energy (a fundamental principle of physics), Compton had to account for all of the energy after impact. He showed that the collision resulted in a new photon of less energy (and thus greater wavelength) being scattered after contact, while the target electron took on some of the energy as well. The total energy was conserved. The shift in wavelength depended on the mass of the electron and the angle of scattering. This work was crucial in establishing experimental evidence for conceiving of electromagnetic radiation (such as light and X-rays) as composed of quanta, with both energy and directed momentum.

The Compton effect was important not only for its description of photon scattering, but also for its ramifications for understanding electrons. In the interaction just described, the electron was at rest. After a collision, however, the electron recoiled. Compton calculated the wavelength of the electron in motion after striking a photon, and the result became known as the Compton wavelength. Compton's results, which support the notion that radiation behaves as both wave and particle, precipitated a flurry of fundamental work in quantum physics in the 1920s. The quantum mechanics that emerged at the end of the decade can be viewed in part as the theoretical explanation of the experimental evidence found in Compton's laboratory. Even Werner Heisenberg's (1901–1976) uncertainty principle, asserting the impossibility of locating the electron with accuracy, can trace its origins to the problems of electron recoil described by Compton in 1923.

Compton's 1927 Nobel Prize, shared with C. T. R. Wilson (1869–1959), demonstrated the international recognition of his work and cemented his leading position in the U.S. community of physicists. He turned his research from X-rays to cosmic rays, for which he led expeditions throughout the world to measure their intensity. This work ended abruptly during World War II, when Compton entered the project to build the atomic bomb.

Compton achieved a position of leadership in the project from an early date. Even before the attack on Pearl Harbor, which brought the United States into the war, he chaired the National Academy of Sciences Committee on Uranium, for which he worked with Ernest Lawrence (1901–1958) in determining the military possibilities of uranium and the recently discovered element plutonium. The two men took a leading role in initiating the Manhattan Project to build the bomb. Compton directed the Metallurgical Laboratory at the University of Chicago, where he recruited scientists, many of them émigrés from Europe, to create a nuclear chain reaction. Italian physicist Enrico Fermi (1901–1954) led this particular project, along with Hungarian-born Leo Szilard (1898–1964) and other scientists working under a veil of secrecy. When Fermi succeeded, it was Compton who phoned his colleague James Conant with the cryptic news, "The Italian navigator has landed in the New World." Compton went on to play a leading role in creating the crucial nuclear facilities in Oak Ridge, Tennessee, and Hanford, Washington. When the war ended, he left the world of research and returned to Washington University in St. Louis to become its chancellor.

See also Atomic Bomb; Manhattan Project; Physics; Quantum Theory; X-Rays

References

Compton, Arthur Holly. *Atomic Quest: A Personal Narrative.* New York: Oxford University Press, 1956.

Shankland, Robert S. "Compton, Arthur Holly." In Charles Coulston Gillispie, ed., *Dictionary of Scientific Biography,* vol. III. New York: Charles Scribner's Sons, 1971, 366–372.

Stuewer, Roger. *The Compton Effect: Turning Point in Physics.* New York: Science History Publications, 1975.

Computers

The word *computer* once referred to a person charged with the task of making routine calculations. This computer's job was eased by the development of various calculating machines, which had cumbersome names like Charles Babbage's (1791–1871) nineteenth-century "difference engine." The modern computer, typically viewed as a revolutionary new device, might also be seen as evolutionary. Computing technology grew from pre-existing needs and equipment. The modern computer's ancestors were the typewriter, the cash register, punch-card equipment, and, most important, the human worker.

The word *computer* itself appears to exclude a widely perceived element of such technology, namely, artificial intelligence. Do machines simply "compute," or is intelligence in machines possible? In 1936, Alan Turing (1912–1954) developed a theoretical "machine" that could serve the same function as a human brain in calculating a range of mathematical problems. Turing's machine, based on sequential calculations, provided a limitless supply of space for basic calculations (a "tape" as long as needed). He went on to write some of the earliest programs in artificial intelligence, including not only mathematical problem solving but also chess.

The transition from analog to digital computers was a fundamental change in modern computing. Analog computers, like Vannevar Bush's (1890–1974) differential analyzer, created in various incarnations beginning in 1931, could address complex problems but required a great deal of effort in setting up the precise parameters of the problem to be solved. Bush's machine essentially modeled the variables in the problem, simulating as precisely as possible expected conditions. The results were not exceptionally accurate, but they often were useful. His machine was first envisioned in 1927, and it took some $25,000 to build. Turing's theoretical machine sparked interest in digital computers that could solve complex problems not through simulation but through

long, sequential, basic, but highly accurate calculations. Although analog computers had several advantages over digital ones, the increasing speed and accuracy of digital calculations appeared to outweigh them.

The first machines to demonstrate the superiority of electronic digital computers over analog (or even mechanical digital computers) were developed in the United States at the University of Pennsylvania's Moore School of Electrical Engineering and at Princeton University's Institute for Advanced Study. The Moore School's Electronic Numerical Integrator and Computer (ENIAC) was born during World War II under a veil of secrecy and became known to the public in 1946. ENIAC could make 5,000 addition and 300 multiplication calculations per second, at least a hundred times faster than any existing computer. The Electronic Discrete Variable Computer (EDVAC) soon followed, largely based on the ideas of John Von Neumann (1903–1957). The EDVAC stored the calculating program in its memory, an innovation that would soon set the norm in computers.

During the war, computers served a decisive role. For example, Turing and others developed machines for the British government that were used to break German codes. British leadership in the field was soon taken over by the Americans. Project Whirlwind, begun in 1944, aimed to produce an aircraft simulator that used a computer to respond quickly to pilots' actions. The computer itself soon became the focus of the project, and the simulator was neglected in favor of developing a real-time computer, Whirlwind I. This computer, useful for design projects and simulation, became a critical part of U.S. national defense systems in the 1950s.

Although the computer can be seen as an end product of research and development, it should also be viewed as a beginning. As Von Neumann noted, not only was the computer an important tool, it was an object of scientific study. The dynamics of numerical processes had fascinated Von Neumann since the 1920s. Information exchanges and

Computer operators program ENIAC, the first electronic digital computer, by plugging and unplugging cables and adjusting switches. (Corbis)

processes seemed akin to neural processes; like other cybernetics enthusiasts, Von Neumann drew a great deal of inspiration from the apparent parallels among computing, biology, and neurology. His Silliman Lectures, "The Computer and the Brain," delivered in the 1950s, explored some of these connections. He proposed a theory of automata, to develop the logical parameters by which machines independently could regulate themselves. But many of his ideas remained undeveloped, because he died of cancer in 1957.

In the first decade after World War II, computers did not seem commercially viable. Typically they were large, one-of-a-kind computers developed for specific purposes, usually funded through government grants. They were unreliable, and they were huge. The initial phase of implementation of computers on a large scale would occur in the 1950s, when banks and accounting firms saw

them as potentially time- and money-saving devices. In addition, military sponsors looked increasingly to computers to solve the defense problems of the Cold War, such as the SAGE early warning system, a network of radar stations positioned to alert the United States of the presence of enemy bombers.

See also Cold War; Electronics; Gödel, Kurt; World War II

References
Aspray, William. *John Von Neumann and the Origins of Modern Computing.* Cambridge, MA: MIT Press, 1990.
Campbell-Kelly, Martin, and William Aspray. *Computer: A History of the Information Machine.* New York: Basic Books, 1996.
Cortada, James W. *Before the Computer: IBM, NCR, Burroughs, and Remington Rand and the Industry They Created, 1865–1956.* Princeton, NJ: Princeton University Press, 1993.
Flamm, Kenneth. *Creating the Computer: Government, Industry, and High Technology.* Washington, DC: Brookings Institution, 1988.

Goldstine, Herman H. *The Computer from Pascal to Von Neumann.* Princeton, NJ: Princeton University Press, 1972.

Hodges, Andrew. *Alan Turing: The Enigma.* New York: Simon & Schuster, 1983.

Conservation

Conservation has roots in economics, owing to concerns about the possible depletion of resources brought about by overhunting, overharvesting, and overfishing. In the early twentieth century, the world's wildernesses were shrinking and governments began to recognize the need to manage their resources more efficiently. Scientists played a leading role in helping to establish guidelines for sensible resource exploitation with long-term conservation in mind. By mid-century, these efforts were complemented by a relatively new approach to conservation that incorporated ecological ideas.

Early conservation policies in the United States owed a great deal to the efforts of Gifford Pinchot (1865–1946), forester and politician. Between 1898 and 1910, Pinchot developed the U.S. Forest Service and expanded the number of national forests considerably, while publicly noting that the United States needed to conserve its natural resources. He was particularly concerned about the lumber supply, but he also argued for the conservation of other nonliving resources such as water, soil, and minerals. His brand of conservation was utilitarian; Americans, he believed, should conserve in the interest of making their resources last. Although Pinchot was not particularly interested in wildlife preservation, he played a major role in influencing major conservation efforts because of his political connections. Pinchot's conservationist attitudes held sway over his friend Theodore Roosevelt (1858–1919), the president of the United States (1901–1909), who was an avid hunter. In addition to helping to establish the U.S. Forest Service, Roosevelt set aside land for the creation of some 150 national forests,

more than 50 wildlife refuges, and 5 major national parks. The national parks also benefited from the handsome donations of philanthropists, especially the Rockefeller family, which contributed to the Grand Teton, Great Smoky Mountains, and Acadia parks.

Another leading U.S. conservationist was Aldo Leopold (1887–1948), who spent part of his career working for the U.S. Forest Service. Unlike Pinchot, Leopold was adamant about protecting wildlife. Leopold was among the first to draw attention to the fact that killing wolves and other animals, a routine matter for pioneers in the nineteenth century, was disturbing the ecological balance of nature. Leopold himself became a pioneer in developing a less human-centered vision of nature, based on interdependence. Like many early conservationists, Leopold was a hunter concerned with diminishing populations (fishermen had similar concerns). But he urged land-use policies that conformed to the relationships of ecological systems rather than the resource needs of humans. In the 1930s, Leopold promoted the view that the land itself was an organism, and human action should work to keep it alive, rather than simply to strip it of flora, fauna, and minerals. He developed the concept of a "land ethic," tying land-use utilitarianism to the moral imperative of keeping an ecological community alive. Leopold's writings, especially the posthumously published *A Sand County Almanac* (1949), became classics of ecology-minded conservationists.

Conservation as a scientific and political movement took hold most prominently in the United States, where it consistently was challenged by the apparent demands of industrialization and territorial development, but efforts to conserve resources took place elsewhere as well. One of these was at sea; just as hunters in the United States hoped to prevent the demise of bison and other beasts, fishermen hoped to manage the exploitation of the sea's herring, cod, and plaice. Countries with mutual interests in the fish of the North Atlantic and surrounding waters, particularly Britain and Norway, formed the International Council (later called the International Council for the Exploration of the Sea) in 1902, whose goal was to adopt sensible management policies based on scientific research on marine biology and physical oceanography. Conservation also was promoted by scientists in Russia and, after the Bolshevik Revolution, in the Soviet Union. As in the case of the United States, however, the Soviet Union's conservation goals often conflicted with its development agenda. The Soviet Union's plans for accelerated development, embodied in Joseph Stalin's (1879–1953) "five-year plans" during the 1930s, overpowered these efforts and took a serious toll on the natural environment. Although Soviet scientists had been among the leaders in ecology and nature conservation, Stalin's programs to force economic growth upon the country cast many of these concerns aside. Overexploited resources and unchecked pollution made the Soviet case stand as one of the great ecological catastrophes of the twentieth century.

See also Ecology; International Cooperation; Oceanography; Soviet Science

References

Meine, Curt. *Aldo Leopold: His Life and Work.* Madison: University of Wisconsin Press, 1988.

Pinkett, Harold T. *Gifford Pinchot: Private and Public Forester.* Urbana: University of Illinois Press, 1970.

Rozwadowski, Helen. *The Sea Knows No Boundaries: A Century of Marine Science under ICES.* Seattle: University of Washington Press, 2002.

Weiner, Douglas R. *Models of Nature: Ecology, Conservation, and Cultural Revolution in Soviet Russia.* Bloomington: Indiana University Press, 1988.

Winks, Robin W. *Laurence S. Rockefeller: Catalyst for Conservation.* Washington, DC: Island Press, 1997.

Continental Drift

Continental Drift was the name given to the idea that continents were horizontally mobile, and that the apparent congruence of major continental coastlines was evidence that the continents once were assembled differently than they are at present. The idea that the continents once had been joined was not entirely new and already had been a subject of speculation for years, as soon as the jigsaw fit of South America and Africa was noted on modern maps. The first scientific effort to demonstrate it, however, came in the twentieth century. Although one effort to do so was made by U.S. geologist Frank Taylor (1860–1939) in 1907, the first that was widely discussed was that done by German meteorologist Alfred Wegener (1880–1930). In 1910, Wegener noted that the continental margins, visible only when viewing a bathymetric map (which reveals water depth, providing more detailed information than a coastline map), were strikingly congruent between South America and Africa, while other areas seemed to fit as well, given a little creative thinking.

Wegener's idea challenged the basic assumptions about the earth at the time. These were based largely on the work of Austrian scientist Eduard Suess (1831–1914), whose monumental late-nineteenth-century work *Das Antlitz der Erde* (*The Face of the Earth*) had proposed that the earth was contracting, and that any major motion in the earth was due to vertical sinking or uplift. According to Suess, the earth had been in a constant state of contraction since the time of its formation. Because the inner portions of the earth contracted faster, most of the tensions in the earth were found in the crust. Such tensions were resolved from time to time by structural shifts, resulting in the disparate features (hills, mountains, etc.) on the face of the earth. It should be emphasized that Suess did not argue for a stable earth; like Wegener, he believed that the earth's features were impermanent. In particular, Suess believed that in the distant past, there were two vast continents—he named them Atlantis and Gondwanaland, separated by a sea he called Tethys.

Wegener first announced his own interpretation of the origins of landmasses in

1912, claiming that there had once been a giant continent, which he called Pangaea, but over time it had broken up and the pieces had drifted apart. He broke with Suess, who claimed that the ocean basins and continents were simply uplifted or sunken versions of each other. Wegener, following the principle of isostasy (the tendency toward gravitational equilibrium), noted that the difference in average elevation between them must have meant they were distinct in composition. The continents were lighter than the ocean basins. He also observed that his theory could do more than Suess's in incorporating recent discoveries of radioactive material in the earth's crust, and it avoided the serious flaw in Suess's theory, namely, the lack of ocean sediment on the continents, which would be required if the modern continents had once been ocean basins. Most of the supporting bits of evidence for his theory were geological and paleontological, citing similar formations and fossil evidence on both sides of the Atlantic. These and other ideas in support of continental drift were published in his 1915 work, *Origins of Continents and Oceans.*

Wegener's theory of continental drift had a few supporters. The primary reason for rejecting continental drift was the inability of Wegener to convince the physicists that such motion was possible. Despite the evidence in its favor, continental drift simply did not have an adequate mechanism to be considered a possibility. What physical forces were strong enough to move such huge blocks of rock across the face of the earth? The earth, scientists such as British physicist Harold Jeffreys (1891–1989) argued, was fixed. Jeffreys was particularly influential upon English-speaking scientists through his textbook, *The Earth: Its Origin, History and Physical Constitution,* first published in 1924. Many U.S. geologists were very critical of Wegener. Nonetheless, Wegener's ideas found some fertile ground. The most outspoken supporter of continental drift, especially after Wegener's death in 1930, was South African geologist Alexander Du Toit (1878–1948), who expanded the store of paleontological and geological evidence in his comparisons between South America and his homeland in Africa.

One of the more helpful supporters of continental drift, particularly in light of the later development (in the 1960s) of the theory of sea-floor spreading, was British physicist Arthur Holmes (1890–1965). If finding an adequate mechanism was the main difficulty facing continental drift, Holmes provided a possible answer in the 1930s. He proposed that the loss of heat from the earth, intensified by the presence of radioactive substances, could be considered a process of convection rather than conduction. This would require not merely the movement of heat, but also of physical material from deep within the earth toward the crust. Continuous convection would create a kind of current in the earth, with hot material moving toward the crust and cooler material being forced back down. This cycle, Holmes argued, might provide a force strong enough to push continents around on the surface of the earth. Even die-hard advocates of a fixed crust such as Jeffreys agreed that the theory could indeed provide a possible mechanism, but few were willing to accept it as anything more than a theoretical possibility. It was upon these convection currents that the theory of continental drift rested, having persuaded few, until it was revived in the early 1960s.

See also Earth Structure; Geology; Geophysics; Wegener, Alfred

References

Frankel, Henry. "Continental Drift and Plate Tectonics." In Gregory A. Good, ed., *Sciences of the Earth: An Encyclopedia of Events, People, and Phenomena.* New York: Garland, 1998, 118–136.

Hallam, A. *A Revolution in the Earth Sciences.* Oxford: Oxford University Press, 1973.

LeGrand, H. E. *Drifting Continents and Shifting Theories.* Cambridge: Cambridge University Press, 1988.

Oreskes, Naomi. *The Rejection of Continental Drift: Theory and Method in American Earth Science.* Oxford: Oxford University Press, 1999.

Cosmic Rays

Cosmic rays were discovered during a spate of scientific work on "rays" of various kinds, some of them real and others not, around the turn of the century. After the discovery of radioactivity in 1896, most measurements of ions in the atmosphere were assumed to originate in the earth's radioactive minerals or gases; few speculated that some of them came from outer space. In 1912, Austrian physicist Victor Hess (1883–1964) conducted some observations aboard a balloon at a height of more than 5,000 meters. To his surprise, at altitudes higher than about 1,500 meters, radiation intensity spiked sharply upward. He concluded that powerful rays must enter the earth's atmosphere from above. This first record of cosmic rays suggested their existence, but their nature was almost entirely unknown.

By the 1920s, most research on radiation was centered around radioactive minerals or man-made electromagnetic radiation such as X-rays. Cosmic radiation was far less studied or understood. California Institute of Technology (Caltech) physicist Robert Millikan (1868–1953) replicated Hess's and others' work and coined the term *cosmic rays*. He gave little credit to the early work. Many believed he had discovered them, which is why they were sometimes called "Millikan rays." This oversight was later corrected when, in 1936, Hess was awarded the Nobel Prize in Physics for his discovery. Because cosmic ray particles typically possessed large energies, they were a source of fascination by physicists. The problem was that, unlike alpha particle sources from radioactive decay, it was very difficult to study cosmic rays in a controlled, laboratory environment.

Because cosmic rays defied laboratory control, scientists who studied them developed a reputation as adventurers, as they traveled to locations all over the world to gather data. Instruments to detect cosmic rays were placed in balloons and atop mountains. Apart from the cost of travel, research on cosmic rays was initially relatively cheap, because nature provided radiation of higher energies than anything that man-made particle accelerators could produce at the time.

U.S. scientists led the world in studying cosmic rays. Millikan believed initially that the rays consisted of high-energy photons. Under this conception, the rays were simply an intense variety of electromagnetic radiation, probably some kind of super gamma ray. Millikan argued that because no one had detected deflection of the cosmic rays by the earth's magnetic field, they could not consist of charged particles. Research on these cosmic "photons" was productive. Caltech's Carl Anderson (1905–1991) credited them with so much power that he believed they were responsible, upon collision with atomic nuclei, for ejecting a previously unknown particle: the positive electron, or positron. Anderson's discovery of the positron in 1932, which would lead to a Nobel Prize in Physics, shared with Victor Hess in 1936, only heightened interest in the cosmic "photons." But soon Arthur Compton (1892–1962), of Washington University in St. Louis, Missouri, observed the deflection of cosmic rays by the earth's magnetic field and concluded that the rays must include charged particles. Rather than gamma rays, the cosmic rays seemed to be made up of penetrating particles, a fact that was generally accepted by 1935.

Physicists disagreed on exactly what kind of particles—electrons or protons—the cosmic rays really were. Most thought they were electrons, but they behaved differently than the typical electrons studied in the laboratory. For example, they were more penetrating and were not as readily absorbable. This feature led Anderson to call them "green" electrons, as opposed to the more easily stoppable "red" electrons. In 1937 he and his colleague Seth Neddermeyer (1907–1988) announced, from their cosmic ray data, the existence of these "green" electrons, with a mass larger than normal electrons and much smaller than protons.

Across the Pacific Ocean, Japanese physicist Hideki Yukawa (1907–1981) took notice of Anderson and Neddermeyer's discovery, and correlated it with a prediction that he had made (although few in the West noticed it) two years earlier. On theoretical grounds, he had proposed the existence of an intermediate particle that was necessary to bind the protons and neutrons to the nucleus. Yukawa's work soon found a wide audience, and some even toyed with the idea of naming the Caltech group's particle after him (the "yukon"). Instead, the particle was dubbed "meson," indicating its middle status between electron and proton. However, as it turned out, Yukawa's particle and the Caltech group's particle were not one and the same. Yukawa's theoretical particle was later called the charged pion, while the particle that the Californians found in cosmic rays was named the muon. By 1950, its precise mass was determined, though much about it remained unknown. Still, the discovery of the muon established the identity of some of the particles from cosmic rays.

The existence of cosmic rays has had an enormous impact on considerations of risk in radiation protection, a serious political and public health issue in the years following the development of nuclear weapons and power. The natural "background" radiation was often cited after World War II as justification for playing down the dangers of radiation from nuclear power plants, from radioactive waste, and from nuclear fallout. Scientists and laypersons alike pointed out the risks of such man-made sources of radiation and the need to eliminate them; however, their opponents (also scientists and laypersons alike) pointed out that the omnipresence of cosmic rays showed that radiation was not all bad, and it has been part of people's lives since the beginning of human existence.

See also Millikan, Robert A.; Physics; Radioactivity; Yukawa, Hideki

References

Hanson, Norwood Russell. "Discovering the Positron." *The British Journal for the Philosophy of Science* 12:47 (November 1961): 194–214.

Kargon, Robert H. "The Conservative Mode: Robert A. Millikan and the Twentieth-Century Revolution in Physics." *Isis* 68:4 (1977): 508–526.

Kopp, Carolyn. "The Origins of the American Scientific Debate over Fallout Hazards." *Social Studies of Science* 9:4 (1979): 403–422.

Kragh, Helge. *Quantum Generations: A History of Physics in the Twentieth Century.* Princeton, NJ: Princeton University Press, 1999.

Sekido, Yataro, and Harry Elliot, eds. *Early History of Cosmic Ray Studies: Personal Reminiscences with Old Photographs.* Dordrecht: D. Reidel, 1985.

Cosmology

Understanding the structure of the universe—its composition, its origins, and its dynamics—is one of the oldest problems in the history of science. An essential feature of the "scientific revolution" in the sixteenth and seventeenth centuries was the replacement of an earth-centered universe with a sun-centered one. Questions still lingered over whether it was a finite universe, as presented by the ancient Greeks, or an infinite one, as proposed by Giordano Bruno in the sixteenth century. Modern cosmology owes much of its origins to certain findings of the early twentieth century and can be divided into observational cosmology and theoretical cosmology.

Cosmology based on astronomical observations dominated thinking in the early twentieth century. Star counting and cataloguing brought father and son William Herschel (1738–1822) and John Herschel (1792–1871) widespread recognition in the nineteenth century. Between 1890 and 1920, Dutch astronomer Jacobus Kapteyn (1851–1922) and German astronomer Hugo Von Seeliger (1849–1924) led the efforts to assess the distribution of stars and to understand the universe as a system, subject to the mutual effects of Newtonian gravitation laws. Albert Einstein's (1879–1955) publication of the general theory of relativity in 1915 did a great deal to disrupt cosmology. Most astronomers believed they lived in "Kapteyn's universe," based on his understanding (published in 1921) of the Milky Way as a flattened disc about 40,000 light-years in size, with the sun close to the center.

New techniques in determining distances changed Kapteyn's universe considerably. Henrietta Swan Leavitt (1868–1921), while working at Harvard College Observatory, found a direct relationship between period and luminosity in certain variable stars called Cepheids. She suggested in 1912 that, because the stars she studied were all in the Magellanic Clouds, they were all about the same distance from the earth. Astronomers realized that they could be compared with other variable stars in the universe to help judge distances (same apparent luminosity but differing periods would reveal a difference in distance). One of the first to use Cepheid variables successfully in this way was Harlow Shapley (1885–1972), then working at the Mount Wilson Observatory. Shapley increased the estimates of the size of the Milky Way to take into account the distances revealed by Cepheid variables, resulting in a galaxy some ten times larger than previously believed. He also determined the location of the Milky Way's center and revised Kapteyn's belief that the sun was near the center (Shapley put the sun far on the periphery).

A hotly contested topic was the status of nebulae, the cloudlike bodies in space. Shapley and others argued that the Milky Way was alone in the universe, and that the nebulae were like eddies in the distance, integrated (however distantly) to the Milky Way. In the nineteenth century, some speculated that the nebulae were not part of the Milky Way at all, but rather were *island universes* with star systems of their own, very far away. They took their cue from eighteenth-century philosopher Immanuel Kant, who coined the phrase. Shapley and Lick Observatory astronomer Heber D. Curtis (1872–1942) debated this question openly at the National Academy of Sciences in 1920. Shapley insisted that the nebulae belonged to a single system shared by all, whereas Curtis favored island universes.

Their opinions were put into print, and eventually Curtis's view prevailed because of measurements of distance made by Edwin Hubble (1889–1953). By 1924, Hubble found thirty-six variable stars in the Andromeda nebula, and he calculated a distance of approximately 900,000 light-years. But the maximum estimate of the Milky Way's diameter was accepted at about 100,000 light-years. These results indicated that the nebula was very distant and well beyond the reaches of our galaxy. His results were announced at a meeting of the American Astronomical Society in December 1924. When Shapley received a letter from Hubble with the results, he reportedly said that the letter destroyed his universe. Curtis's views were vindicated, and the galaxy became a new basic structural unit for understanding the universe.

In 1929, Hubble made another remarkable discovery that changed assumptions about how to conceptualize the universe. For more than fifteen years, scientists had been measuring radial velocities of stars (radial velocity is the rate at which the distance between object and observer is changing) by observing the displacement of spectral lines from starlight. The displacement was attributed to the Doppler effect. This effect usually is understood in relation to sound, like a police siren changing tones as it comes closer or moves farther away from the listener, as sound waves are compressed or elongated. It can also be applied to light, and the compressions and elongations are evident in light's spectrum. Hubble noted that radial velocities increased with distance, and he calculated a ratio establishing the proportionality of radial velocity to distance. Hubble and his colleague Milton L. Humason (1891–1972) examined spectra from the most distant observable stars, using a fast lens to take photographs of faint spectra. Through this work, Hubble showed (in what became known as Hubble's Law) that the previously stated proportionality of radial velocity to distance could be extended to bodies at a distance of more than 100 light-years.

Spectral lines indicating radial velocity

were displaced, or "shifted," always toward the red end of the spectrum (as opposed to the violet end). Distant stars had a greater "red shift" than the closer stars, indicating greater radial velocity. The conclusion was staggering for cosmologists, who knew that red shifts indicated objects moving away from each other, whereas violet shifts (if they existed) would indicate objects moving closer. Because only red shifts were observed, all galaxies must be moving away from each other. The most distant galaxies are moving the fastest. The unavoidable conclusion, that the universe must be expanding, stimulated renewed interest in cosmological theories and cast doubt on ones that perceived the universe as stable and unchanging.

Theoretical cosmology had already been a contested field for years. After Einstein published his general theory of relativity in 1915, cosmology had a new paradigm with which to develop theory. Relativistic cosmology equated gravity with acceleration, making the universe an unstable place, and light itself was subject to being bent. Still, for the next fifteen years or so, cosmologists assumed that the universe was static. Two rival cosmologies vied for influence: Einstein proposed a closed (finite) universe, whereas Dutch physicist Willem De Sitter (1872–1934) proposed an infinite one. They argued over the importance of gravitational attraction and how to calculate the average density in infinite universes. De Sitter's cosmology made the counterintuitive claim that the universe was devoid of matter because its average density had to be zero; Einstein's finite universe solved this problem but could not account satisfactorily for expected gravitational relationships. To resolve the issue, he introduced a "cosmological constant," an arbitrary mathematical fix that he later regarded as a monumental mistake. In 1922, a Russian mathematician, Alexander Friedmann (1888–1925), suggested that the outstanding discrepancies between De Sitter's and Einstein's views could be addressed if one accepted an infinite universe whose properties changed in relation to time. In other words, if

the universe's theoretical difficulties might be solved by a non-static model.

The first persuasive non-static model was proposed by a Belgian Jesuit, Georges Lemaître (1894–1966), who connected his cosmology very explicitly to time. He suggested that the universe might have had a beginning—a violent explosion that led to an unstable, expanding universe. The present universe was simply the remnant of past fireworks. Later derided by physicist Fred Hoyle (1915–2001) as the "big bang idea," Lemaître's concept became the dominant view of cosmology in the twentieth century. The work of Hubble and others, showing observational evidence of an expanding universe, seemed to affirm the ideas of both Friedmann and Lemaître. The theory received a boost in the late 1940s when George Gamow (1904–1968) and Ralph Alpher (1921–) calculated that, from an initial soup of subatomic particles and radiation, a massive explosion could create the light elements in the first moments of the universe. The theory gained further orthodoxy beyond the scientific community when Pope Pius XII (1876–1958) noted in 1951 that the Big Bang was essentially harmonious with Christianity.

The Big Bang theory retained its orthodoxy, but it had many outspoken critics. In the late 1940s, Hoyle and his colleagues Hermann Bondi (1919–) and Thomas Gold (1920–) issued a challenge to the Big Bang with their "Steady State" theory. It was also a non-static model, but proposed not one creation but many, in a continuous (steady) state of spontaneous creation of matter throughout the universe. Hoyle attacked the Big Bang theory as nonscientific, with its distant beginnings unobservable and thus not testable. The Big Bang as explained by Gamow and Alpher had other problems as well: inability to explain the formation both of galaxies and of the heavier elements. Later, in the 1960s, the discovery of cosmic background radiation, predicted by Alpher and others as the after-effects of the Big Bang, established renewed consensus about the theory.

See also Astronomical Observatories; Astrophysics; Big Bang; Gamow, George; Hubble, Edwin; Leavitt, Henrietta Swan; Origin of Life; Relativity; Religion; Shapley, Harlow

References

Bertotti, B., R. Balbinot, S. Bergia, and A. Messina, eds. *Modern Cosmology in Retrospect.* New York: Cambridge University Press, 1990.

Harwit, Martin. *Cosmic Discovery: The Search, Scope, and Heritage of Astronomy.* New York: Basic Books, 1981.

Kragh, Helge. *Cosmology and Controversy: The Historical Development of Two Theories of the Universe.* Princeton, NJ: Princeton University Press, 1996.

————. *Quantum Generations: A History of Physics in the Twentieth Century.* Princeton, NJ: Princeton University Press, 1999.

Lightman, Alan, and Roberta Brawer. *Origins: The Lives and Worlds of Modern Cosmologists.* Cambridge, MA: Harvard University Press, 1990.

North, John. *The Norton History of Astronomy and Cosmology.* New York: W. W. Norton, 1995.

Paul, Erich Robert. *The Milky Way Galaxy and Statistical Cosmology, 1890–1924.* New York: Cambridge University Press, 1993.

Crime Detection

Just as science entails a great deal of problem solving, the tools of science proved useful to law enforcement agencies in the first half of the century to detect numerous crimes, including espionage. The most celebrated application of scientific methods to crime detection was the use of fingerprinting. Because human beings tend to have distinctive fingerprints, matching prints from a crime scene to the suspects of a crime was a successful means of identifying and prosecuting alleged criminals. The famed eugenicist Francis Galton (1822–1911) proposed in the 1890s that fingerprints could be used for identification; although Galton was interested in using them to study racial composition, he observed that fingerprints do not change over one's lifetime and that each person's were unique. Thus, they could be useful for identification.

In 1901, fingerprinting techniques were adopted in England and Wales, using a classification system devised by Edward Richard Henry (1850–1931). The next year, fingerprinting was used systematically for the first time in New York. Its prison system was among the first to subject all of its convicts to fingerprinting. Within a few years, the U.S. military services used fingerprints as a means for personal identification, and over the next several decades, fingerprinting became a standard procedure in law enforcement.

Science was also useful in understanding the nature of crimes. Particularly in urban areas where the identity of victim and culprit could be equally anonymous, forensic pathology (the scientific study of the cause of death) was useful in solving crimes by letting corpses speak for themselves. For example, the police could call in a forensic pathologist to provide evidence based on expertise: cause of death and the physical specifics such as angle of entry of the murder weapon. Perhaps surprisingly, scientific evidence of this kind entered the law enforcement community very slowly. The political power of coroners, who were seldom required to have any special training, began to erode in favor of forensic evidence during the first couple of decades of the twentieth century. To a large degree, this was because of the efforts of progressive-minded reformers who saw coroners as symbols of the corruption of municipal bureaucracies. New York City, for example, abolished the coroner's office in 1915, in the wake of a corruption scandal involving the mayor; it was replaced by a medical examination system and was praised widely as the beginning of a new progressive era in the city. When a series of murders were mishandled in the 1920s by the coroner's office in Essex County, New Jersey, the county copied New York's system. In other cities and counties, coroners' lack of competence to judge whether a death was a result of bludgeoning, suicide, or disease gradually gave way to increasing emphasis on forensic pathology.

A forensic scientist analyzes the clothing of Nancy Titterton who was strangled with a garment she was wearing (1936). (Hulton- Deutsch Collection / Corbis)

In addition to the new emphasis on scientific methods, the decline of the one-man expert began to occur in the 1930s. At Glasgow University, in Scotland, forensic medicine had been used to solve crimes since the mid-nineteenth century, and individuals were trained to analyze blood, fingerprints, and body fibers, and to understand the nature of disease, psychology, and even ballistics. During the widely publicized case of Buck Ruxton, who killed his wife and maid, chopped up the body parts, and deposited them in different parts of Scotland, a group led by Glasgow professor John Glaister (1892–1971) used forensic methods to incriminate Ruxton (1899–1936). The variety of methods, from blood samples to X-ray techniques, helped to promote the use of multidisciplinary forensic teams rather than single individuals.

The physical and chemical sciences helped to determine the properties of materials used in crimes. In 1939, the Federal Bureau of Investigation (FBI) in the United States received a directive from the president to coordinate activities against espionage and subversive activities, thus widening the scope of investigation to include a host of cases in which crimes were secret or had not yet been committed. This resulted in an extraordinary volume of information that consistently poured into the FBI from various sources.

Among the ways to incorporate twentieth-century scientific discoveries in physics was the use of X-rays to examine materials, particularly in cases of industrial fraud in which evidence of faulty craftsmanship was necessary. Even more useful than X-ray recorders were spectrographs. The spectrograph recorded the distinctive dark bands evident in the spectra of light from incandescent objects; the bands differed according to chemical composition of the source of light. This instrument, which had proved useful to astronomers by helping them to understand the composition of stars, also became a favored tool for analyzing the evidence in crime investigations. Any piece of evidence could be burned, in whole or in part, to determine its chemical composition.

In 1942, the FBI captured eight German agents when they landed on the shores of Long Island. The tools they found on the culprits helped the FBI to shape its own methods of crime detection. It discovered, for example, that the spies carried boxes of matches tipped with quinine; such matches could be used for writing in invisible ink. The widespread use of such practices in espionage cases led the FBI to develop the means of detecting "invisible" ink through the use of ultraviolet light. The FBI also used infrared light to detect forged documents; the spectrophotometer used such light to discern chemical and physical properties of various materials.

Metallurgical techniques to determine composition of metals and the possible presence of cracks and other breakage were useful in uncovering evidence of sabotage. The FBI made extensive use of petrographers (petrography deals with the classification of rocks), whose expertise in abrasives, minerals, and soils facilitated comparisons of materials. Grease and oil in a crime scene or a piece of evidence could be analyzed in a laboratory and the origins more readily determined. Finding abrasives in oil often was a sure sign of industrial sabotage. By the end of World War II, the FBI laboratory conducted thousands of such investigations looking into the possibility of sabotage; according to FBI director John Edgar Hoover (1895–1972), the use of these new methods led to the discovery of nearly 2,000 genuine cases of sabotage during the war.

See also Espionage; X-rays
References

Crowther, M. Anne, and Brenda White. *On Soul and Conscience: The Medical Expert and Crime: 150 Years of Forensic Medicine in Glasgow.* Aberdeen, Scotland: Aberdeen University Press, 1988.

Forrest, D. W. *Francis Galton: The Life and Work of a Victorian Genius.* New York: Taplinger, 1975.

Hoover, John Edgar. "FBI Laboratory in Wartime." *The Scientific Monthly* 60:1 (1945): 18–24.

Johnson-McGrath, Julie. "Speaking for the Dead: Forensic Pathologists and Criminal Justice in the United States." *Science, Technology, and Human Values* 20:4 (1995): 438–459.

Curie, Marie

(b. Warsaw, Poland, 1867; d. Sancellemoz, France, 1934)

As a national hero, Marie Curie is rivaled in Poland only by Copernicus, who put the sun at the center of the universe in the sixteenth century. Born Marie Sklodowska in Warsaw, Poland, Marie Curie has become a symbol of both French and Polish science (although her scientific life was confined to France). She also became a symbol and inspiration for women in science, as proof that women were capable of making significant contributions to knowledge. She moved to France in 1886, worked as a governess, then returned to Poland. When she returned in the early 1890s, she studied physics and mathematics and met Pierre Curie (1859–1906). They married in 1895. Their relationship was both romantic and scientific, and together they revolutionized science.

Curie's scientific life coincided with the discovery of X-rays by Wilhelm Röntgen (1845–1923) and the discovery of radioactivity by Henri Becquerel (1852–1908). In 1896, when the Curies took up the subject, the two seemed interrelated. Becquerel had determined that uranium salts emitted some kind of ray without needing to be exposed to sunlight beforehand. Although Becquerel suspected that the rays were due to some long-lived residual effect of previous exposure to the sun (i.e., the rays were due to phosphorescence, not something within the material itself), he had in fact discovered a new property in some minerals. Marie Curie sought to discover if there were other elements, in addition to uranium, capable of emitting these mysterious rays.

Curie herself coined the term *radioactive* to describe the existence of rays coming from the elements she was trying to isolate and identify. She noted that pitchblende, the ore containing uranium, was more radioactive than pure uranium. This led her to believe that another highly radioactive element was present. The process of chemically treating pitchblende was laborious and time consuming, but necessary to identify any new element that pitchblende might contain. Pierre Curie set aside his own research to join her quest, and in 1898 the two of them announced that they had identified a previously unknown element. They called it polonium, after Marie's homeland. A few months later, they announced yet another new element, which they called radium. They soon were assisted in their work by André Debierne (1874–1949), who discovered actinium. Marie Curie's obsession became radium, and in 1902 she isolated a decigram of pure radium and determined its atomic weight as 225 (now accepted as 226).

The work of Marie and Pierre Curie was recognized as fundamental throughout the scientific community. Marie won national prizes such as the Berthelot Medal and the Gegner Prize. The French scientific community was unaccustomed to honoring women, and it continued to treat Marie Curie according to the customs of the time. For example, Marie Curie found out she had won the Gegner Prize when the Academy of Sciences wrote not to her but to Pierre, "with respectful compliments to your wife." International acclaim came when the Curies were awarded the Humphrey Davy Medal by England's Royal Society and, more importantly, when they shared with Henri Becquerel the 1903 Nobel Prize in Physics. Her research led her to accept and help demonstrate the theory,

Marie Curie (1867-1934), the Polish-born French physicist and winner of the Nobel Prize in Physics (1903) and Chemistry (1910), at work in her laboratory. (Hulton-Deutsch Collection / Corbis)

proposed by Frederick Soddy (1877–1956) and Ernest Rutherford (1871–1937), that radioactive elements decayed and transmuted into other elements. She became not only the leading radiochemist in France but also an international scientific celebrity. When Pierre died in 1906, Marie succeeded his chair in physics, thus becoming the first woman to teach at the Sorbonne.

Curie's fortunes in the scientific community were not always so pleasant. After suffering the tragedy of Pierre's death, she began an affair with another scientist, Paul Langevin (1872–1946). Because Langevin was married, Marie Curie became the object of a newspaper campaign to impugn her name and reputation. Revolted by such intrusions into her private life, many eminent scientists came to her defense, including Albert Einstein (1879–1955), who urged her to "simply stop reading that drivel." The French public and some French scientists, however, treated her with scorn, especially when Langevin's wife published letters that Curie and Langevin had written to each other. The scandal dredged up hostility toward Curie that might have had other origins: She was a "foreigner" destroying a French family, and she had the audacity to challenge time-honored traditions by seeking to be the first woman admitted to the French Academy of Sciences (the request was denied).

Despite the scandal, Marie Curie became not only the first woman, but the first person, to receive two Nobel Prizes in science. She already had won the 1903 prize in physics, which she shared with her husband, Pierre, and Henri Becquerel. But her work also had provided new foundations for chemistry. Thus the 1911 prize in chemistry, awarded for the discovery of radium and polonium, as well as her subsequent isolation and studies of radium, was Marie Curie's alone. Yet the scandal that rocked France also prompted Swedish Nobel laureate Svante Arrhenius (1859–1927) to notify her that, had the Swedish Academy of Sciences known the truth about her affair with Langevin, it would never have awarded her the prize. He even urged her not to come to the award ceremony in Stockholm. But Curie insisted that the prize was based on scientific accomplishments, not on her private life, and she refused to take Arrhenius's advice.

In subsequent years Curie's prestige regained itself and she continued to work with radioactive materials and X-rays. Curie helped to equip army medical teams with X-ray apparatuses and created courses on medical radiology during World War I. An organization of admiring U.S. women presented her with an expensive gift: a gram of radium to be used for research purposes. She traveled to the United States in 1921 to receive the radium directly from President Warren G. Harding (1865–1923). Marie Curie had achieved a stature unprecedented in either France or the United States. But

what kind of impact did Marie Curie have for women in science? Certainly, most cheered her accomplishments. Some heralded her as evidence that women could and should enter into science, a profession dominated by men. Others have argued that Curie's genius had the opposite effect: A woman had to be a "Madame Curie" to consider a career in science.

The many years of research on radium carried a heavy price. The 1920s saw an increased awareness of the dangers of radioactivity, symbolized by the deaths of several women in New Jersey who had been exposed while painting the dials of watches with luminescent radium paint. For Marie Curie, extended physical contact with the dangerous substance produced lesions on her fingers and other health problems. Even an enthusiastic handshake could put her arm in a sling. She lived long enough to see her daughter and son-in-law, Irène Joliot-Curie (1897–1956) and Frédéric Joliot (1900–1958), create an artificially radioactive isotope in a laboratory. This work earned them the Nobel Prize in Chemistry in 1935, awarded the year after Marie Curie died.

> *See also* Becquerel, Henri; Cancer; Joliot, Frédéric, and Irène Joliot-Curie; Nobel Prize; Physics; Radioactivity; Rutherford, Ernest; Uranium; Women; X-rays
>
> *References*
> Quinn, Susan. *Marie Curie: A Life*. Reading, MA: Addison-Wesley, 1996.
> Reid, Robert. *Marie Curie*. New York: Saturday Review Press, 1974.
> Wyart, Jean. "Curie, Marie (Maria Sklodowska)." In Charles Coulston Gillispie, ed., *Dictionary of Scientific Biography*, vol. III. New York: Charles Scribner's Sons, 1971, 497–508.

Cybernetics

Cybernetics is the comparative study of information processes in machines controlled by mechanical or electronics means and animal nervous systems. The term *cybernetics* comes from the Greek word *kybernētēs*, meaning "pilot" or "steersman." It is the same root as the word *governor,* which helps to explain the purpose of cybernetics: understanding the control of complex systems. Although the term had already been used in the nineteenth century by André Ampère to describe political science, and James Clerk Maxwell (1831–1879) had discussed governors in relation to thermal systems, cybernetics as a new field of study was born during World War II.

U.S. mathematician Norbert Wiener (1894–1964) and his colleagues Julian Bigelow (1913–2003) and Arturo Rosenblueth (1900–1970) published the first paper on cybernetics in 1943. The inspiration was Wiener and Bigelow's war work on guided missile technology, which necessitated studying the real-time assessment of wind direction, altitude, and target trajectory to help determine firing times. The complexity of this war work did not lend itself readily to practical application, but it did lead them to consider the actions of communication and feedback of data in increasingly complex machines. In their paper, published in the journal *Philosophy of Science,* they made the bold claim that men and machines behave in parallel, despite drastic differences in their composition.

The new field of cybernetics, based on the notion of parallelism between humans and machines as complex information-processing systems, was interdisciplinary. Beginning in 1946, leading figures in science, mathematics, psychology, and anthropology began to meet in a series of conferences sponsored by the Macy Foundation. The first meeting described its subject under the title, "Feedback Mechanisms and Circular Causal Systems in Biology and Social Sciences." But soon it adopted Wiener's term, *cybernetics.* The field's early major figures were Wiener, Claude Shannon (1916–), and Warren Weaver (1894–1978).

Norbert Wiener published the technical book *Cybernetics* in 1948, and he followed it in 1950 with his more popularly written *The Human Use of Human Beings.* In these works, Wiener classed communication and control together. If a person communicates with another, he is sending a message. The response is a unit of information somewhat related to the initial message. In control, the only difference is that the initial message is a command; otherwise, control and information transmission are similar in that they can be studied through feedback. Wiener wrote that society could only be studied through such message processes; further, the involvement of machines in that process would only increase with time, as humans developed more sophisticated ones capable of more complex messaging systems. Thus cybernetics should play a major role in facilitating that process.

Because of the connection to social sciences, cyberneticists claimed that they were creating a new universal science, not just another field of inquiry. But Wiener also made strong connections to physics, particularly the law that entropy tends toward a maximum. Similarly, information becomes disorganized or distorted as it passes from source to recipient, never fully retaining the wholeness of its initial meaning. The purpose of cybernetics, then, was to develop language to enable greater control and, more broadly, to study information processes. "In control and communication," Wiener argued, "we are always fighting nature's tendency to degrade the organized and to destroy the meaningful." Communication and feedback of information were the fundamental elements of cybernetics. Wiener drew out some of the social implications of this by noting: "To live effectively is to live with adequate information. Thus, communication and control belong to the essence of man's inner life, even as they belong to his life in society" (Wiener 1988, 17).

The reception of cybernetics was mixed. Scientists in the Soviet Union were hostile to it, dubbing it a "reactionary pseudoscience" that reflected the values of the capitalist world, namely, "its inhumanity, striving to transform workers into an extension of the machine." (Bowker 1993, 111). Others found it quite appealing, as in the case of French physicist Pierre Auger (1899–1993), who noted that humankind was moving from a world of material and energy to a world of forms. Conflating humans with machines stirred up fantasies of a world run by machines, an idea that has inspired countless science fiction stories. The *New York Times* even editorialized in 1948 that man's only role in the future would be to try to prevent his own destruction at the hands of machines. In the 1950s and 1960s, cybernetics enthusiasts would find themselves at the center of controversy, as Wiener himself confronted the religious implications of equating humans with machines in his book, *God and Golem, Inc.*

See also Computers; Mathematics; Religion; Science Fiction; World War II

References

Bowker, Geof. "How to Be Universal: Some Cybernetic Strategies, 1943–1970." *Social Studies of Science* 23:1 (1993): 107–127.

Heims, Steve Joshua. *The Cybernetics Group.* Cambridge, MA: MIT Press, 1991.

Wiener, Norbert. 1950. *The Human Use of Human Beings: Cybernetics and Society.* New York: DaCapo Press, 1988.

Cyclotron

The cyclotron, developed in the early 1930s by Ernest Orlando Lawrence (1901–1958) and his colleagues, provided a powerful tool for physics and revolutionized laboratory life in the United States. Its purpose was particle acceleration, as scientists sought particles of increasingly higher energies to facilitate nuclear experiments. The importance of the cyclotron was recognized immediately by the widespread construction of them in the United States and elsewhere. Lawrence won the Nobel Prize in Physics in 1939, one of the few times that prize has been awarded primarily because of an invention rather than a scientific discovery.

Early particle accelerators were developed in England, at Cambridge University's Cavendish Laboratory. In 1930, John Cockcroft (1897–1967) and Ernest Walton (1903–1995) created a device to impart energy to protons by passing them through a vacuum tube attached to a voltage multiplier.

In 1932, they announced the first nuclear transformation by means of an accelerated particle, revealing a major avenue of research in which physicists could exploit particle acceleration. In 1931, chemist Harold Urey (1893–1981) and his colleagues at Columbia University isolated and identified the heavy isotope of hydrogen and dubbed it deuterium; they soon realized that it would make an excellent projectile to facilitate nuclear reactions if accelerated. The deuteron (so the nucleus of deuterium was called) was poised to provide the ammunition for even more effective particle accelerators.

The new particle accelerators were developed in the United States. In 1931, for example, Robert Van de Graaf (1901–1967) constructed one with a maximum voltage of 1.5 million volts, and the same year Ernest Lawrence created a linear accelerator of slightly less voltage. Lawrence, a physicist at the University of California–Berkeley since 1928, soon devised a method to build a much more powerful accelerator. His experimental accelerator used magnetic fields to force particles to spiral with an increasing radius as they achieved higher speeds. Although the first version, built in 1932, accelerated protons to a mere 1.2 million volts of energy, Lawrence and his colleague M. Stanley Livingston (1905–1986) predicted new models pushing protons to 10 million volts and beyond. Their hopes were soon confirmed, and the laboratory slang term *cyclotron* came into official usage by 1936. Livingston built his own cyclotron at Cornell University, and the cyclotrons began to proliferate throughout the United States. By the late 1930s, scientists in Japan and Europe had begun to build them, modeled on Lawrence's successes.

Cyclotrons accelerated charged particles to high energies, making them useful for research experiments. Studies in nuclear physics often involved bombarding atomic nuclei with particles in the hope of producing an observable effect, such as the ejection of a particle from the nucleus. Acceleration seemed necessary because of the repulsion of the nu-

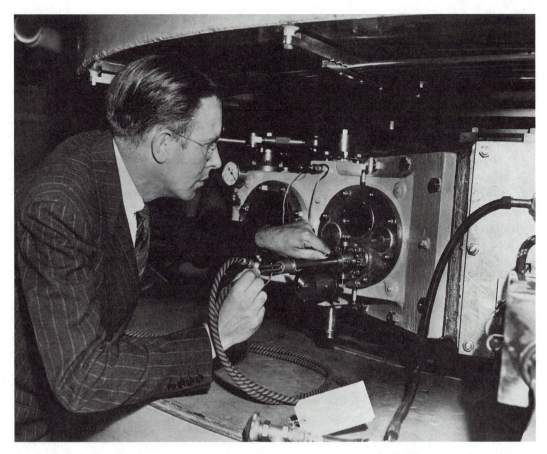

Ernest Orlando Lawrence (1901–1958), the American physicist with the cyclotron he designed, for which he was awarded the 1939 Nobel Prize. (Hulton-Deutsch Collection/Corbis)

cleus; protons, if accelerated sufficiently, could overcome such repulsion and collide with the nucleus. The 1932 discovery of the neutron, an uncharged particle, provided a means to conduct such experiments without worrying about repulsion. But cyclotrons nonetheless promised a future of research in high-energy physics. In addition, the cyclotron served the needs of researchers in the United States and abroad by providing them with a source of radioisotopes, artificially radioactive isotopes of elements not typically associated with radiation. Because radioactivity was readily detectable, radioisotopes could be used as "tracers," helpful in understanding a variety of natural processes from human physiology to ecosystem dynamics.

Ironically, chemists (not physicists) reaped the most noteworthy rewards of the cyclotron initially. Biochemist Martin Kamen (1913–2002) oversaw the production of radioisotopes in Lawrence's laboratory in the late 1930s. His experiments led to the identification and separation, with colleague Sam Ruben, of the long-lived radioisotope carbon 14. Also, Glenn T. Seaborg (1912–1999) used the cyclotron profitably through chemical research and discovered man-made transuranic elements (elements of higher atomic number than uranium, such as plutonium). Part of the reason for the lack of scientific breakthroughs in physics, in a time when such breakthroughs were commonplace in other laboratories, was that

Lawrence devoted most of his energies to expanding his laboratory and designing cyclotrons of ever-increasing power. Especially embarrassing was the 1934 discovery of artificial radioactivity (by the Joliot-Curies) using instruments far less sophisticated than the cyclotrons at Berkeley.

Cyclotron research has become nearly synonymous among historians with the term *Big Science*. In his Radiation Laboratory (or Rad Lab) at the University of California, Lawrence pursued bigger accelerators throughout the 1930s and managed, even in the lean years of the Great Depression, to find a great deal of money for research. For example, by advertising the benefits of the cyclotron for medical research, he acquired funds from the Rockefeller Foundation. The U.S. government provided funding as well, sending scientists, electricians, and other laborers to Lawrence's Rad Lab under the auspices of New Deal agencies such as the Works Progress Administration. So many scientists and technical experts worked at the Rad Lab that the cyclotron appeared to spell the end of science's heroic age of individualism and to herald in the days of team research. The scope of research was interdisciplinary, including not only physics and chemistry, but also engineering, biology, and medicine. During World War II, the cyclotron was harnessed to help build the first atomic bomb. By 1944, Lawrence's staff rose to some 1,400 people, and his instruments were converted for use in developing methods of electromagnetic separation of isotopes, a crucial step in providing the fissionable materials for the bomb.

The cyclotron's role in creating radioisotopes decreased after World War II, because nuclear reactors provided more of these than anyone could use. But experimental research with the cyclotrons themselves continued. Already the cyclotrons had made an enormous contribution not only to science but also to the geographic base of research. The United States became the center of world physics after the war for numerous reasons, and the cyclotron was one of them. Its early development there, along with the wave of immigrant scientists fleeing Europe during the 1930s, helped to create a strong nuclear physics community in the United States before World War II.

See also Artificial Elements; Atomic Bomb; Lawrence, Ernest; Manhattan Project; Physics

References
Baird, Davis, and Thomas Faust, "Scientific Instruments, Scientific Progress and the Cyclotron." *British Journal for the Philosophy of Science* 41:2 (1990): 147–175.
Childs, Herbert. *An American Genius: The Life of Ernest Orlando Lawrence, Father of the Cyclotron.* New York: Dutton, 1968.
Heilbron, John L., and Robert W. Seidel. *Lawrence and His Laboratory: A History of the Lawrence Berkeley Laboratory,* vol. 1. Berkeley: University of California Press, 1989.
Kamen, Martin D. *Radiant Science, Dark Politics: A Memoir of the Nuclear Age.* Berkeley: University of California Press, 1985.
Kevles, Daniel J. *The Physicists: The History of a Scientific Community in Modern America.* Cambridge, MA: Harvard University Press, 1995.
Kragh, Helge. *Quantum Generations: A History of Physics in the Twentieth Century.* Princeton, NJ: Princeton University Press, 1999.

D

Davisson, Clinton
(b. Bloomington, Illinois, 1881; d. Charlottesville, Virginia, 1958)

Clinton Davisson is best known for providing experimental evidence, through electron diffraction, for wave mechanics. After taking a Ph.D. in physics from Princeton University in 1911, he joined the Carnegie Institute of Technology. But in 1917, during World War I, he left to join the Western Electric Company Laboratories (later Bell Telephone Laboratories) on a military communications project. After the war, he decided to stay on rather than return to teaching duties at Carnegie. Although working in a commercial laboratory, Davisson enjoyed the freedom to pursue his own research interests. One of these was the emission of electrons from metals from electron bombardment. He and C. H. Kunsman found in 1919 that some of the secondary electrons had the same energy as the primary electrons; they measured energy and angles of bombardment but found no theoretical explanation.

Davisson and Lester H. Germer (1896–1971) continued these investigations and, after a particularly fortuitous accident in 1925 that required their equipment to be cleaned by heating, found that the results were entirely different after the extreme heat had recrystallized the metal target. Now the target consisted of a few large crystals rather than many tiny ones; observing the electrons striking the large crystals showed a strong dependence on the crystal's direction. In 1927, Davisson showed that the nickel diffracted the electrons in exactly the kind of way that theoretical physicists, notably Louis De Broglie (1892–1987) and Erwin Schrödinger (1887–1961), had postulated. These physicists had argued that, despite the work on quantum mechanics, which seemed to emphasize light's particulate nature, light still behaved like a wave. Wave mechanics appeared vindicated by Davisson's work. Independently, George Paget Thomson (1892–1975) arrived at the same results in England. The two later shared the 1937 Nobel Prize in physics.

Davisson's experience reveals a great deal about patronage for science in the United States. The 1920s saw increased interest by large corporations, such as General Electric and the American Telephone and Telegraph Company (AT&T), in paying for research in science and technology. Davisson's research was related to the development and manufacture of vacuum tubes by AT&T. In 1925, AT&T transformed its industrial research

Clinton Davisson, co-recipient of the 1937 Nobel Prize in Physics, in his laboratory. (Bettmann/Corbis)

department into the Bell Telephone Laboratories. It was at Bell that Davisson conducted his research on the emission of electrons from metals. When Davisson's work provided observational evidence for wave mechanics, the results were twofold. On one hand, it was a prestige plum for Bell Labs; but on the other hand, it suggested that these industrial laboratories were allowing their scientists too much freedom to pursue arcane academic problems. What did wave mechanics have to do with making telephones? Bell Labs was quick to point out that few scientists enjoyed that kind of liberty. Some also complained that researchers like Davisson were not the kind of scientist that corporations ought to employ. Davisson's discovery drew attention to the fact that, although industry was becoming a patron for science, it would be an unlikely place to find long-term acquiescence of academic values, such as the freedom to pursue ideas without thinking of economic consequences.

See also Industry; Patronage; Physics; Quantum Mechanics

References

Kevles, Daniel J. *The Physicists: The History of a Scientific Community in Modern America.* Cambridge, MA: Harvard University Press, 1995.

Koizumi, Kenkichiro. "Davisson, Clinton Joseph." In Charles Coulston Gillispie, ed., *Dictionary of Scientific Biography,* vol. III. New York: Charles Scribner's Sons, 1971, 597–598.

Debye, Peter

(b. Maastricht, Netherlands, 1884; d. Ithaca, New York, 1966)

The Dutch chemical physicist Peter Debye had diverse interests, contributing to a number of different scientific questions. He had an equally diverse career, moving from one university to the next and working in the Netherlands, Germany, Switzerland, and finally the United States. Debye studied with mathematical physicist Arnold Sommerfeld (1868–1951) in Aachen before following his mentor in 1906 to the University of Munich. He received his Ph.D. in 1908, and in 1911 he briefly assumed a post, recently vacated by Albert Einstein (1879–1955), as professor of theoretical physics at the University of Zurich. Years of wanderlust followed, with jobs in three different countries. Yet these years, up to and including World War I, saw some of Debye's most productive work.

Debye's abiding interest was in molecules. One lasting contribution was his method for investigating molecular structure using electric fields. Scientists such as Debye expected that the properties of electric charges held the key to the structure of molecules. This motivated him, in 1912, to study the behavior of molecules when subjected to an electric field. He began to see problems with the conventional way of viewing the dielectric constant of a substance (a measure of its ability to resist the formation of an electric field). This constant was supposed to decrease very slowly with rising temperature. But Debye noted that some substances showed a much more rapid decrease, and polarization occurred faster than expected. The existing formula had to be revised. He conjectured that molecules of some substances had permanent dipole moments that contributed to the overall polarization in the presence of an external field. His concept accomplished two things. Debye created a revised equation for dielectric constant, and he demonstrated the existence of permanent electric dipole in some molecules, along with a means to determine its moment (its tendency to rotate around an axis). Debye had devised an effective method to analyze the structure of molecules. The unit of measurement for dipole moment is called the debye.

In 1912, Max Von Laue (1879–1960) and the Braggs—father and son, William Henry Bragg (1862–1942) and William Lawrence Bragg (1890–1971)—had demonstrated that X-rays were diffracted by crystals, and that diffraction patterns revealed a great deal about the arrangement of atoms in the crystals. This work opened up a new field of inquiry using X-ray spectra as a means to study the structures of molecules. In the years following the initial discovery of X-ray diffraction, Debye published a series of papers in this new field. The most influential of these was published with Paul Scherrer (1890–1969) on X-ray interference patterns of randomly oriented particles. They took photographs of X-ray spectra diffracted through a powder sample of lithium fluoride. Since this happened to be a substance with very good diffracting properties, they really had a stroke of luck. They found that even fine powders could be analyzed through X-ray diffraction, with no need for large crystals.

Debye's work in the 1920s continued to be of fundamental value. One of these was the 1923 Debye-Hückel theory of electrolytes, which improved upon Svante Arrhenius's (1859–1927) theory of electrolytic dissociation and replaced it with one that took into account thermodynamic relationships. He worked in Germany in the 1930s, first at the University of Leipzig and then at the University of Berlin. His fame attracted some of the best young physical chemists to his laboratories. In 1936, his work on molecular structure was recognized by the award of the Nobel Prize in Chemistry. A few years later, his celebrity status seemed to have been cemented: His native city constructed a bust of him in its town hall.

Dutch chemical physicist Peter Debye lecturing in a classroom. One of Debye's lasting contributions was his method for investigating molecular structure using electric fields. (Division of Rare and Manuscript Collection, Carl A. Kroch Library, Cornell University)

Like many scientists of his time, Debye's career was transformed by Nazism in Germany. During this Nazi period, he retained his Dutch citizenship. But when the war began, German officials demanded he become a German citizen. Debye refused, more out of distaste for the Nazis than loyalty to the Netherlands. He moved to the United States and took a position at Cornell University in 1940. He became a citizen in 1946. Debye was among the most respected scientists of his time. In addition to his Nobel Prize, he was elected to more than twenty academies of science throughout the world and held eighteen honorary degrees. He con-tinued to produce work of the highest order at Cornell, such as his research on light scat-tering, but his wanderlust years had ended.

See also Physics; X-rays

References
Debye, Peter J. W. *The Collected Papers of Peter J. W. Debye.* Woodbridge, CT: Ox Bow Press, 1988.
Ewald, P. P., ed. *Fifty Years of X-Ray Diffraction.* Utrecht: International Union of Crystallography, 1962.
Smyth, Charles P. "Debye, Peter Joseph William." In Charles Coulston Gillispie, ed., *Dictionary of Scientific Biography,* vol. III. New York: Charles Scribner's Sons, 1971, 617–621.

Determinism

Determinism is a term that refers to the belief that physical relationships have precise, knowable, and predictable relationships. The mechanical world picture established by Isaac Newton (1642–1727) and others in the seventeenth and eighteenth centuries, which describes the world as if it were a machine obeying mathematical laws, is the foundation of deterministic thinking. Determinism was a powerful tool for scientists who believed that the project of science was to unveil the mysteries of nature, to provide an increasingly accurate and predictable understanding of it. Thus determinism was an important aspect of scientists' philosophy about the world. In the twentieth century, many fields of scientific inquiry saw determinism challenged, raising important scientific and epistemological questions (regarding the nature of knowledge itself).

One of the attacks on determinism originated in the nineteenth century, when Charles Darwin (1809–1882) published his theory of evolution, by means of natural selection. Unlike other evolutionary views, such as Lamarckian evolution, natural selection was based on competition among species whose traits had developed randomly. Lamarckian evolution, by contrast, proposed that species develop through the use or disuse of their parts, and that changes made in the course of a species' life could be inherited. This gave power to the organism itself, and it had attractive philosophical implications. Perhaps man could improve himself based on his self-knowledge and his will to change. Darwinian evolution denied this and left any change in species to chance. This view was opposed in many quarters in the first decades of the twentieth century, but nowhere more vehemently than in the Soviet Union, where many biologists spoke out against Darwinian natural selection. Although plant breeder Trofim Lysenko (1898–1976) was notorious for his hostility toward genetics, one of his most virulent criticisms was directed at natu-ral selection. Science, he announced to a gathering of agronomists in 1948, is the enemy of chance. Scientists should not be confined to waiting for nature to bestow its gifts randomly; instead, scientists should be active, able to make improvements (such as agricultural ones) based on their knowledge of causality. Soviet biology was vociferously deterministic, disliking the implications of indeterministic natural selection.

A famous controversy about determinism arose in the field of physics, from some of the epistemological ramifications of quantum mechanics. German theoretical physicist Werner Heisenberg (1901–1976), who was a leading figure in developing quantum mechanics in the 1920s, in 1927 announced the uncertainty principle, which drew out some of the disturbing implications of trying to unite the notion of the quantum (an indivisible packet of energy) with Newtonian mechanics. Relying on the quantum would require physicists to abandon their notions of causality. Heisenberg noted that, in order to predict future conditions, one must first describe accurately the initial conditions. His uncertainty principle stated that initial conditions would always be veiled in a certain amount of uncertainty. At the quantum scale, increased certainty about some variables would result in decreased certainty about others, making it impossible to understand conditions with precision. Although many physicists were accustomed to describing the future in terms of probabilities, they did not believe that one must necessarily conceive of the future in this way. The uncertainty principle was inherently indeterministic because it denied the possibility of calculating cause and effect, because the present itself could never be understood precisely. Although most physicists accepted this, and indeed much of modern technology is built on the principles of quantum mechanics, physicist Albert Einstein (1879–1955) made a famous objection to indeterminacy when he argued that God does not play dice.

See also Anthropology; Bohr, Niels; Einstein, Albert; Evolution; Heisenberg, Werner; Philosophy of Science; Physics; Quantum Mechanics; Uncertainty Principle

References

Bowler, Peter J. *Evolution: The History of an Idea.* Berkeley: University of California Press, 1989.

Cassidy, David C. *Uncertainty: The Life and Science of Werner Heisenberg.* New York: W. H. Freeman, 1993.

Cassirer, Ernst. *Determinism and Indeterminism in Modern Physics: Historical and Systematic Studies of the Problem of Causality.* London: Oxford University Press, 1957.

DNA

Material in the living cell's nucleus, called nuclein or nucleic acid, was discovered by the 1870s. From this early period, nuclein appeared to hold the key to understanding the material agent of heredity. In Germany, Oscar Hertwig (1849–1922) proposed in 1885 that these acids played the critical role in heredity. Albrecht Kossel (1853–1927) demonstrated that there were two kinds of nucleic acid, each one composed of four bases. His work on proteins and on the chemical nature of these bases won him the Nobel Prize in Physiology or Medicine in 1910. Although this strong work seemed to locate the chemical basis of heredity in nucleic acid, it would be nearly half a century before the scientific community accepted it.

Although chromosomes were widely regarded as the carriers of genetic characters in the early twentieth century, scientists were less sure about its constituents. What was the fundamental genetic material? By 1930, nucleic acids faded from possibility, largely owing to the work of Russian physician-turned-chemist Phoebus A. Levene (1869–1940), who moved to the Rockefeller Institute for Medical Research in New York in 1905. Levene published more than 700 papers on nucleic acids and was a major authority on the subject. In the 1920s, he discovered the presence of ribose and deoxyribose

in the nucleic acids, which forms the basis of the names ribonucleic acid (RNA) and the deoxyribonucleic acid (DNA). Levene formulated the "tetranucleotide interpretation" of the acids, which stated that their four bases were always present in equimolar amounts. This led him and others to dismiss the notion that, although nucleic acid molecules were large, they were complex enough to carry genetic information.

In the 1940s, the primacy of DNA found advocates in a different field, bacteriology. Oswald Avery's (1877–1955) research on bacteria indicated that the transfer of genetic material could make bacteria passive, rather than active. Avery and his colleagues Colin MacLeod (1909–1972) and Maclyn McCarty (1911–) published a paper in 1944 proposing that DNA was the agent of genetic change. At the time, bacteriology seemed far afield from research on heredity, yet Avery's work raised the question of whether DNA might be the genetic material after all. Research on bacteriophages, or viruses that destroy bacteria, continued under the "phage group" centered on Max Delbrück (1906–1981), Alfred Hershey (1908–1997), and Salvador Luria (1912–1991). Tracking DNA with radioactive substances, Hershey and Martha Chase (1928–2003) showed in 1952 that the active part of the virus was indeed DNA, whereas proteins merely protected it.

Despite the findings by Avery and his colleagues in the 1940s, most scientists still believed that proteins were the true genetic material. Levene's tetranucleotide interpretation held considerable sway. Proteins seemed complex, thus well suited for storing information. By contrast, the nucleic acids did not seem complex at all, nor did they seem to differ much from one cell to the next. But the Austrian chemist Erwin Chargoff (1905–2001) changed this view during his work at Columbia University in the 1940s. Chargoff analyzed nucleic acids from beef thymus, spleen, liver, yeasts, human sperm, and even the tubercle bacillus. He

In the 1920s, Phoebus A. Levene (1869-1940) discovered the presence of ribose and deoxyribose in the nucleic acids, which form the basis of the names ribonucleic acid (RNA) and the deoxyribonucleic acid (DNA). (National Library of Medicine)

different organs within the same body did share the same DNA.

The renewed agency given to DNA by Chargoff and the "phage group" sparked new studies of the structure and role of DNA. Among them was the American James Watson (1928–), a member of the "phage group" who traveled to Cambridge, England, to team up with Englishman Francis Crick (1916–). The two of them used X-ray diffraction techniques to investigate the structure of DNA molecules. Together, they constructed a model for the structure of DNA, based on X-ray diffraction patterns collected by Maurice Wilkins (1916–) and Rosalind Franklin (1920–1958). Although Wilkins favored a spiral model, Watson and Crick soon conceived of the double helix as the basic model of DNA. This kind of structure, they argued, not only accounted for the X-ray diffraction evidence but also provided an explanation of how the DNA molecule can generate an exact replica of itself. Their 1953 paper became a classic of genetics, explaining how genetic information gets encoded through a sequence of base-pairs, which is transmitted from one cell to the next by the duplication of chromosomes.

used new techniques developed during and after World War II, such as paper chromatography, ultraviolet spectrophotometry, and ion-exchange chromatography. These enabled him to demonstrate that the four bases of nucleic acids were not equal as Levene had insisted. Further, DNA was not identical from one organism to the next, although

See also Amino Acids; Biochemistry; Chromosomes; Genetics; X-rays

References

Bowler, Peter J. *The Mendelian Revolution: The Emergence of Hereditarian Concepts in Modern Science and Society.* Baltimore, MD: Johns Hopkins University Press, 1989.

Magner, Lois N. *A History of the Life Sciences.* New York: Marcel Dekker, 1979.

E

Earth Structure

Knowledge of the earth's structure came primarily from studies of gravity, heat flow, and seismology. Although the later twentieth century would see the rise of plate tectonics, the notion that continents "drifted" was explicitly rejected in the first half of the century. Before 1900, conceptions of the earth varied, some viewing it as a giant sponge with myriad caverns. Depending on one's viewpoint, such caverns could be filled with violent winds, thus causing earthquakes, or filled with water or fire, explaining the discovery of such things as lava and rivers flowing beneath the earth's surface. Before the 1950s, most scientists believed in a solid earth whose movements, such as earthquakes, resulted from the settling of large blocks of crust due to gravitational forces.

At the beginning of the century, knowledge about the structure of the earth came largely from the late-nineteenth-century synthesis of geological ideas set forth in the work of the Austrian Eduard Suess (1831–1914), especially his book *The Face of the Earth,* first published in 1883 but also updated in the first decade of the twentieth century. Suess introduced the concept of a supercontinent, which he named Gondwana, which over time had changed considerably owing to the action of the oceans and the vertical motion of the

crust's masses (sinking and rising). Alfred Wegener (1880–1930) proposed an alternative hypothesis about the earth in 1915. He also proposed a primitive supercontinent, which he called Pangaea, but with a crucial difference: He believed that the great motions of the earth were horizontal (rather than vertical), and that the extraordinary coincidence of the apparent puzzle-fit of South America and Africa was no coincidence at all. These continents, he believed, had been joined in the distant past, only to drift apart over time. Continental drift, as the theory became known, was rejected by most geologists. This was in part because of the objections of physicists such as Harold Jeffreys (1891–1989), who argued that the force required to move continents was too great to attribute to any known physical process in the earth.

Seismology provided the tools to penetrate the earth's interior. Because certain waves—shear waves—could be transmitted through a rigid medium, seismologists could use them to judge what parts of the earth's interior were solid and liquid. By the late 1920s, such waves were believed to travel through the earth's crust at a relatively constant speed, only to change speed dramatically at a certain depth, indicating a different kind of material. Because the waves still traveled through this material—known as the mantle—seismologists concluded that it, too, was solid. But the earth's core was decidedly different, and the waves seemed to disappear at that depth. Thus scientists believed that the earth's core was liquid, but it was surrounded by a solid mantle and crust.

The prevailing theory of the earth's structure came from beliefs about its origins. As William Thompson (Lord Kelvin) (1824–1907) had done in the nineteenth century, physicists such as Jeffreys argued in the 1920s that the earth probably solidified from a liquid origin by gradual cooling. Kelvin's conception was now modified to include other sources of heat such as radioactivity. As the earth cooled, a thin crust formed on the surface, which would then contract and thus break up and sink, only to melt again; the process would repeat until a permanent, honeycombed solid crust was formed around a denser, liquid core. The honeycombs themselves were filled with liquid, but through gradual conduction of heat to the surface of the earth, these would harden and provide an even more solid earth.

One of the principal problems of earth structure was posed by mountains. In the late nineteenth century, George Airy (1801–1892) and John Henry Pratt (1809–1871) developed hypotheses claiming that the variations in mass over the surface of the earth—such as mountains—must be neutralized by opposite variations (for Airy, this meant deep roots for mountains; for Pratt, this meant less density of material in mountains) in order to keep the density of the crust uniform. Called "isostasy" in 1889 by Clarence E. Dutton (1841–1912), the concept presumed that the apparent differences in density presented by hills and mountains were compensated somehow, ensuring uniform distribution of mass over the earth's surface. Investigations of such "compensation" constituted much of the geophysical work in the first half of the twentieth century. For example, The United States Coast and Geodetic Survey (C&GS) conducted a wide-ranging study of isostasy along coastlines, in nonmountainous interior regions, and in mountainous areas, to test various hypotheses about the extent to which the principal of isostasy held. William Bowie (1872–1940), the C&GS's chief of geodesy (the science of measuring the earth), published some of the results of this survey in 1917, providing a data set of the comparative gravitational forces in different parts of the earth.

Several scientists noted from Bowie's and others' data that there were widespread inequalities over the earth's surface that must somehow be compensated. Harold Jeffreys took note of this in his 1924 textbook, *The Earth,* which was one of the most influential textbooks on earth structure in subsequent decades. He believed that the continents, for example, had to be the same mass as the earth beneath the oceans; but because of their obvious difference in size (continents being taller), the ocean floor must be made of denser material than that of the continents. The conjecture made by Jeffreys in his 1924 book *The Earth* was tested in subsequent years by geologists aboard submarines. Felix Andries Vening Meinesz (1887–1966) of the Netherlands and Harry Hess (1906–1969) of the United States both made gravitational studies at sea during the 1920s and 1930s, "weighing" the earth—in other words, taking gravitational measurements to confirm or refute the principal of isostasy in relation to the ocean floors. These studies and others, especially after World War II, helped to confirm that the sea floor was made of decidedly different material than the continents. The basic assumptions about earth structure, based on these findings about the ocean floor, would be questioned intensively in the 1960s in light of new interpretations about the mobility of the earth's crust.

See also Age of the Earth; Continental Drift; Geophysics; Jeffreys, Harold; Seismology; Urey, Harold; Wegener, Alfred

References
Adams, Frank Dawson. *The Birth and Development of the Geological Sciences.* New York: Dover, 1954.

Jeffreys, Harold. *The Earth: Its Origins, History, and Physical Constitution.* Cambridge: Cambridge University Press, 1924.

Oreskes, Naomi. *The Rejection of Continental Drift: Theory and Method in American Earth Science.* Oxford: Oxford University Press, 1999.

Ecology

Ecology is the branch of biology that examines organisms in relation to one another and to their environment. It is the study of the "web" of life, taking the interconnectedness of living things as a foundation of scientific inquiry. The term was coined by the German Ernst Haeckel (1834–1919) in 1866, although the first major figure in the history of the subject was the Danish botanist Eugenius Warming (1841–1924), whose 1895 treatise *Plantesamfund* laid out the parameters of ecology as the complex relationships of plants and animal communities. Warming emphasized the state of flux in these communities. He and other ecologists at the turn of the century were inspired by the work of Charles Darwin (1809–1882), who described the world of competitive nature as an entangled bank, hiding laws such as natural selection. These early ecologists debated whether this chaos hid a fundamental stability, or if it was a manifestation of the complexity of continuous change.

One of the key figures of early-twentieth-century ecology was University of Chicago botanist Henry C. Cowles (1869–1939), who developed the concept of plant succession. His investigations of vegetation in the Indiana Dunes in 1897 led him to trace a route of simple forms, such as brush, to more complex forms, such as beech and maple trees. These trees, for Cowles, were the "climax" plants allowable under those environmental conditions. Cowles in 1914 helped to found the Ecological Society of America and later served as its president. Another American, Frederic Clements (1874–1945), studied the Nebraska countryside and proposed that theories of plant communities ought to consider vegetation as large-scale, complex organisms. He developed this idea with his colleague Victor Shelford (1877–1968) in their 1939 book, *Bio-Ecology,* which called that organism the *biome,* a single unit that consists of all plants and animals within a habitat.

The term *ecosystem* was introduced in 1935 by British ecologist Arthur George Tansley (1871–1955), who had also been a founder of the British Ecological Society in 1913. Tansley accepted the concept of separate units, or ecological systems, inside of which all things are interdependent. His view was mechanistic: The ecosystem could be understood as the exchange of energy and matter. Studies of ecosystems were not limited to vegetation: Scientists such as George Evelyn Hutchinson (1903–1991) and Charles Elton (1900–1991) pioneered ecological approaches to soil and lake chemistry and animal ecology.

Among the most significant developments for theoretical ecology was Raymond Lindeman's (1915–1942) 1942 work on trophic (nutritional) relations in lake ecosystems. Over five years, he and his wife, Eleanor, took samples of all identifiable organisms in a shallow body of water at the boundary of two succession groups. Their labors helped to integrate the dynamics of food cycles into prevailing notions of ecological succession. Lindeman argued that one had to understand short-term nutritional functions in an ecosystem in order to trace its long-term changes, namely, succession. Lindeman cast nutritional considerations in terms of energy flow, which, when expressed mathematically, could help calculate new theoretical concepts such as the biological efficiency of an ecosystem. By reducing trophic relationships to energy, Lindeman provided ecologists with a theoretical unit with which to explore fundamental problems.

Scholar Donald Worster has argued that ecology has never been far removed from human values, and thus its history should be understood in the context of the political and social aims of the participants. This is probably true of most scientific activity, but especially so for ecology, which has an overtly utopian foundation: interdependent communities. It is difficult to judge whether this is a value that has shaped ecology, or whether ecological studies have reinforced it as a social value. By mid-century, the emphasis of ecology on the balance of nature perhaps lent it a reputation for favoring such a balance at the expense of disruption, such as the exploitation of natural resources, the introduction of chemical wastes, or any other radical change to the environment.

After World War II, many ecologists attempted to distance themselves from the growing environmental movement, lest their science be associated simply with political or social preference. At Britain's Nature Conservancy and the U.S. Oak Ridge National Laboratory, scientists developed communities of "ecosystem ecologists" who buttressed their scientific legitimacy through close affiliation with physicists. They emphasized pure research rather than environmental advocacy, though privately or publicly many of them did the latter. Ironically, much of their important work in the 1950s and 1960s was done under the auspices of agencies such as the Atomic Energy Commission, which came under heavy scrutiny for its role in environmental degradation.

See also Atomic Energy Commission;
    Conservation; Haeckel, Ernst; Patronage

References
Bocking, Stephen. *Ecologists and Environmental Politics: A History of Contemporary Ecology*. New Haven, CT: Yale University Press, 1997.
Cook, Robert Edward. "Raymond Lindeman and the Trophic-Dynamic Concept in Ecology." *Science,* New Series 198:4312 (7 October 1977), 22–26.
Hagen, Joel B. *An Entangled Bank: The Origins of Ecosystem Ecology*. New Brunswick, NJ: Rutgers University Press, 1992.
Worster, Donald. *Nature's Economy: The Roots of Ecology*. San Francisco, CA: Sierra Club Books, 1977.

Eddington, Arthur Stanley
(b. Kendall, England, 1882; d. Cambridge, England, 1944)

Arthur Eddington was a celebrated astrophysicist who provided an early model of stellar structure and helped to spread the ideas of general relativity. Eddington was educated at Cambridge University, where he studied physics initially before turning to mathematics. In 1905, he began work at the Royal Observatory in Greenwich and focused more intensively on astronomy. Later, in 1914, he became director of the Cambridge Observatory. One of his first achievements was to shatter the age-old notion that the stars are fixed in the heavens, permanently destined to retain the patterns that the ancients and moderns embellished with zodiacal signs. He showed that they do in fact move, albeit almost imperceptibly, relative to each other.

A more significant of his important contributions involved the structure of stars. His work partly was based on variable stars, or Cepheids. These stars oscillate in brightness, turning dimmer or brighter in periodic cycles. In the early 1920s, Eddington established the relationship between mass and luminosity in stars and used the increasing knowledge of Cepheid variables to construct a model of stellar structure. In *The Internal Constitution of Stars* (1926), he described an equilibrium state in stars resulting from the opposing forces of radiation pressure and gas pressure.

Eddington's conception of the structure of stars met with an early critique by a younger and less experienced colleague. In 1930, the Indian Subrahmanyan Chandrasekhar (1910–1995) arrived in Cambridge to begin his studies with Eddington, but had determined (en route, aboard ship) that if a star's mass exceeded a certain limit, it could not reach the expected state of equilibrium. Chandrasekhar believed that if the mass of a star were greater than 1.44 solar masses, it would collapse under its own weight. Although it was subsequently accepted and known as the "Chandrasekhar Limit"—leading to a Nobel Prize in 1983—the idea was severely and publicly criticized by Eddington, providing Chandrasekhar with an inauspicious career start.

One of Eddington's goals was to promote Albert Einstein's (1879–1955) theory of general relativity, and indeed he became one of Einstein's principal evangelists. One of the main features of the theory was that space is curved by gravitational mass and that the shortest distance between two points is occasionally a curve. Eddington and others determined that the way to prove this would be to observe a massive object actually bending the light from a star. In 1919, he traveled to the island of Principe, off the coast of Africa, to find an ideal location to witness a solar eclipse, when the sun's light would be obscured enough to trace the path of starlight as it passed the massive body of the sun. To his delight, Eddington found that the light of the star was indeed affected by the sun's gravity, thus lending support to Einstein's theory.

Eddington was a Quaker and had been a pacifist during World War I, as had Einstein. As a conscientious objector, he avoided the fight and continued his research at Cambridge. He hoped that his eclipse expedition might be an expedition to heal the wounds caused by the war—the two scientists came from countries on opposite sides of the battlefield. Here was a British scientist making an expedition to test and support a German theory. He worked to get British scientists to accept general relativity, although some British nationalists objected to a German theory replacing a previously British theory of gravity—Isaac Newton (1642–1727), the founder of the classical laws of physics, had been English. Eddington's efforts to persuade the Royal Society to recognize the accomplishments of Einstein with its prestigious Gold Medal failed initially, although after several years the medal was indeed awarded to Einstein.

English astrophysicist Arthur Stanley Eddington helped to provide empirical evidence for Einstein's general theory of relativity. (Hulton-Deutsch Collection/Corbis)

Eddington also shared Einstein's dream of discovering a theory to unify relativity and quantum mechanics. Also like Einstein, his efforts set him apart from the mainstream physics community in the 1930s. His 1936 *Relativity Theory of Protons and Electrons* found few sympathetic readers. Most physicists were not interested in his highly mathematical physics, which seemed to have little regard for experiment and focused on algebraic equations. He attempted to demonstrate the harmony between the fine structure constant in quantum theory and the ratio of masses between protons and electrons. Despite his reputation, Eddington's later work was scorned by most and labeled mystical, as if he were a superstitious numerologist. Some believed that this work was pseudoscientific. He died before achieving his goal of a fundamental theory, though his ideas were published posthumously in his *Fundamental Theory* (1946).

See also Astrophysics; Chandrasekhar, Subrahmanyan; Einstein, Albert; International Cooperation; Relativity
References
Chandrasekhar, S. *Eddington: The Most Distinguished Astrophysicist of His Time*. New York: Cambridge University Press, 1983.
Kilmister, C. W. *Eddington's Search for a Fundamental Theory: A Key to the Universe*. Cambridge: Cambridge University Press, 1994.
North, John. *The Norton History of Astronomy and Cosmology*. New York: W. W. Norton, 1995.
Stanley, Matthew. "'An Expedition to Heal the Wounds of War': The 1919 Eclipse and Eddington as Quaker Adventurer." *Isis* 94 (2003): 57–89.

Ehrlich, Paul

(b. Strehlen, Germany [later Strzelin, Poland], 1854; d. Bad Homburg, Germany, 1915)

Much of Paul Ehrlich's career belongs to the nineteenth century. But at the turn of the twentieth century, he began a program of research that became the basis for all future work in immunology and chemotherapy. His chemical treatment of syphilis, certainly his most famous accomplishment, opened up new possibilities for using chemicals to cure disease. At the same time, chemotherapy created a world of potentially destructive unknowns in medical practice.

In the 1880s, Ehrlich was the head physician at the medical clinic in Berlin's Charité Hospital. While there, he conducted experiments in histology (the study of tissue structure) and biochemistry. In particular, he was interested in blood cells. Hematology, a field in which Ehrlich made major contributions, is the study of blood. During this period, he studied and reported on the forms and functions of blood cells, and the causes and nature of diseases in them. In other words, he was studying the morphology, physiology, and pathology of blood cells. His achievement was in establishing ways to detect and differentiate leukemias and anemias. In observing the behavior of white blood cells, or leucocytes, he began to believe that chemical affinities tend to govern all biological processes. Many of his later experiments were directed at determining the affinities between various tissues and dyes, to improve staining methods necessary for histological research.

Ehrlich pioneered the use of dyes in research and therapy. He showed that bodily organs can be classified according to their oxygen avidity (avidity is chemical affinity, the force that attracts certain elements to others, keeping them together). Ehrlich used dyes to detect the presence of chemicals, such as a color test to find bilirubin in urine, a sign of typhoid fever. He moved with his wife to Egypt after discovering he had tuberculosis. He returned a year later and began the tuberculin treatment recently discovered by Robert Koch (1843–1910). He spent most of the 1890s conducting research on antitoxins.

In 1899 Ehrlich became director of the Royal Prussian Institute for Experimental Therapy in Frankfurt. This "Serum Institute" became the center for government control of chemical agents used in immunotherapy, such as tuberculin and diphtheria antitoxin. It also became a major center of research. A few years later, the Research Institute for Chemotherapy was erected next to it. Ehrlich's community became a mecca for international researchers, including visitors from other European countries, the United States, and Japan.

His years at Frankfurt were extraordinarily fruitful. He elaborated his "side-chain" theory of immunity, which identified two distinct attributes in diphtheria toxin. One was toxic, and the other had the power to bind with nontoxins. The nontoxic parts attached to a cell to which they had a chemical affinity, and they locked in the toxin to the healthy cell's side chains (later termed *receptors*), exposing it to damage through them. If the cell survived, its side chains would be its efficacy, the demand for this kind of treatment by chemicals, or chemotherapy, increased dramatically. Ehrlich tried to restrict its distribution, but the level of need spurred large-scale manufacture under the name of Salvarsan. For his work in chemotherapy, he was nominated again for the Nobel Prize, but the controversial status of the chemical made a second award impossible. Ehrlich had to combat many unforeseen complications from injecting the chemical into human bodies. Experiments with the chemical continued amidst high demand. Yet he found himself the target of accusations of charlatanism and brutal human experimentation, including an allegation that prostitutes were forced into Salvarsan treatment against their will. Although these sensational activities abated when World War I erupted, Ehrlich's health deteriorated during these years. After a slight stroke, he entered a sanatorium; a second stroke soon thereafter killed him.

See also Biochemistry; Koch, Robert; Medicine; Microbiology; Patronage; Venereal Disease

References
Baümler, Ernest. *Paul Ehrlich: Scientist for Life.* New York: Holmes & Meier, 1984.
Dolman, Claude E. "Ehrlich, Paul." In Charles Coulston Gillispie, ed., *Dictionary of Scientific Biography,* vol. V. New York: Charles Scribner's Sons, 1972, 295–302.
Marquardt, Martha. *Paul Ehrlich.* New York: Schumann, 1951.

At the turn of the twentieth century, Paul Ehrlich began a program of research that became the basis for future work in immunology and chemotherapy. (Hulton-Deutsch Collection / Corbis)

rendered inert, and the injured tissue would regenerate without them. Ehrlich wrote that this created blood cells that were immune to damage from that particular toxin. Also in these years, Ehrlich conducted thousands of experiments grafting malignant tumors into mice, hoping (in vain) to find a means to immunize against cancer. His efforts in immunity resulted in sharing the Nobel Prize in Physiology or Medicine in 1908.

In 1906, Ehrlich predicted the creation of substances that could be directed exclusively against parasites and not the surrounding tissue, acting as "magic bullets" to combat diseases. A few years later, in 1909, he announced his discovery of a synthetic compound of arsenic that seemed to be an effective magic bullet to cure syphilis. The chemical was injected into animals and human patients. After Ehrlich demonstrated

Einstein, Albert
(b. Ulm, Germany, 1879; d. Princeton, New Jersey, 1955)

Albert Einstein was the best-known scientist of the twentieth century. His theories of relativity challenged the most fundamental notions of physics, revising considerably the concepts of Newtonian mechanics. His arrival to worldwide scientific celebrity was unorthodox, because he was not a star pupil. Einstein was born in southern Germany and spent his early life in Munich, where his father and uncle operated a company that manufactured plumbing and electrical appliances.

During his teens, the business collapsed and the family moved to Milan, Italy, leaving the young Albert behind to finish his schooling. But instead of completing his studies, he quit school and abandoned Germany, renouncing his citizenship, possibly to avoid compulsory military service. He enrolled in school in Switzerland, closer to his parents, and later would acquire Swiss citizenship. He entered the Swiss Federal Polytechnic, intending to gain a credential to become a math or physics teacher in secondary schools; he gained his diploma in 1901. An early effort to gain a doctorate from the University of Zurich failed (he was awarded it in 1906). In 1902, he took a full-time government job as a patent examiner, providing the financial stability needed for a wife and newborn son.

During his time at the patent office, and shortly after his first son's birth, Einstein formulated the works that would do more to revolutionize physics than anyone since Isaac Newton's (1642–1727) work in the seventeenth century. Inspired by Max Planck's (1858–1947) quantum theory, recently proposed, Einstein wrote a paper in 1905 on the quantization of light itself, arguing that light behaved like a stream of particles, despite seeming also to behave like a wave. The properties of light that he described explained what was known as the photoelectric effect, and light quanta were termed *photons*. In the same year, 1905, he explained the behavior of tiny bodies suspended in liquids, known as Brownian motion, in terms that seemed to proffer a persuasive case for the existence of atoms.

Despite these important works—the photoelectric effect won Einstein the Nobel Prize in Physics in 1921—Einstein's lasting fame rested largely on his theories of relativity. He devised the special theory of relativity in the same year, 1905. His goal was to find a way to reconcile Newtonian mechanics with the recent findings in the nineteenth century in electromagnetic fields by scientists such as James Clerk Maxwell (1831–1879). Einstein was dissatisfied with the fact that it

was difficult to fit field theory into the mechanical outlook of the previous era. His motivation for devising theories of relativity was to provide a foundation for physics with greater explanatory power, encompassing both Newtonian mechanics and Maxwellian electrodynamics.

Before Einstein, scientists believed in the presence of the ether, a medium that we cannot see or feel but that connects all things. Because physicists believed that, in a purely mechanical universe, there can be no action at a distance through hidden forces, there must be an intervening medium—an ether—through which some subtle mechanical action is taking place, allowing for gravitational attraction, magnetism, or the propagation of light waves (like the ripples in a pond). Einstein abandoned the concept of the ether, which had never actually been observed, and developed instead his theory of special relativity, which rejected the entire concept of any fixed medium connecting all objects in space. By discarding the ether, he rejected the notion of absolute rest, forcing physicists to assess any physical object in relation to some other physical object. Concepts of space and time had to be framed in relative terms, rather than in terms of some notion of fixed points in the ether. Fixed points, Einstein argued, do not exist in reality. The only constant is the speed of light. According to the special theory, the laws of physics appear the same to all observers, no matter how fast they are traveling, because one's own speed appears to be a state of rest from one's own position. However, as speed approaches that of light, one perceives (in the outside world) time moving more slowly, space being contracted, and mass increasing.

The implications of special relativity seemed mind-boggling. Special relativity set the velocity of light as the speed limit of the universe and opened up questions about the nature of time itself. Einstein also tried to embed electrodynamics more firmly in mechanical terms. Electrodynamics could be incorporated into a relativistic mechanics by

endowing radiation itself with inertia; photons, for example, have mechanical properties. Assuming some kind of equivalent relationship between radiation energy and mass,
Einstein developed a famous equation:
$E = mc^2$, or energy equals mass multiplied by
a huge constant, namely the square of the
speed of light.

Special relativity applied only to objects
moving at uniform velocities (constant
speeds). What would happen if an object
were accelerating or slowing down? This required a generalization of the theory, or the
theory of general relativity, devised by Einstein in 1915. Here he made an original observation: Gravitational mass (the mass that
attracts other masses) and inertial mass (the
mass that resists acceleration) are equivalent.
Just as moving objects crossing the path of an
accelerating object seem to follow a curved
path, objects near a very massive object
should be expected to follow a curved path.
Isaac Newton had never been able to explain
the cause of gravitational attraction, despite
setting forth explicit laws governing its behavior. According to Einstein, gravitational
attraction was no mysterious force; instead,
it was simply the curvature of space brought
about by mass.

Beginning in the 1920s, Einstein turned
his intellectual energies toward unified field
theories. Initially his goal remained the same
as before, namely, unifying mechanics with
electrodynamics. But increasingly another
contender arose: quantum mechanics. He
was skeptical of the implications of quantum
mechanics as it unfolded during the 1920s.
Werner Heisenberg's (1901–1976) uncertainty principle, in particular, struck him as
too reliant on statistical understandings of nature. Einstein was a determinist who was unwilling to believe that certain facets of nature
were unknowable and thus unpredictable.
God, he famously asserted, does not play
dice. But Einstein was increasingly alone in
his intransigence, as the majority of physicists
embraced quantum mechanics and its philosophical implications of indeterminacy. Part

*Albert Einstein's theories of relativity challenged the most
fundamental notions of physics, revising considerably the
concepts of Newtonian mechanics. (Library of Congress)*

of the reason for this was that relativity provided few tools for new physical ideas,
whereas quantum mechanics did. Einstein's
famous energy equation was an exception; it
hinted at the possibility of unleashing great
amounts of energy by converting it from
mass, a process that would become fundamental to nuclear weapons and power.

After working in academic posts in Bern,
Zurich, and Prague, Einstein became a professor at the University of Berlin in 1914, thus
returning to his homeland and becoming Germany's chief scientific celebrity. He tried to
dissociate himself from militarism and war in
general. His Swiss citizenship helped him in
this regard, obviating the need to take a stand
for or against Germany during the World
War I. Outspokenly pacifistic, he was one of
the few German scientists who managed to
avoid being vilified by U.S., French, and
British scientists. After the war, anti-Semitism tarnished the enjoyment of his status as a

German celebrity. Beginning in the 1920s, he and his ideas were occasionally derided by anti-Semitic colleagues. Einstein's theories, many of them incomprehensible to experimental physicists, appeared to some as quintessential examples of how theoretical physicists lived in a dreamworld, completely divorced from reality. In Germany, theoretical physics came under attack as "Jewish science," eroding the strength of German experimental physics. When the Nazis came to power in 1933, Einstein was abroad in the United States. Just as he had been vocal in his pacifism during World War I, he was openly critical of the Nazis' aims. He did not wish to return, and he accepted a generous job offer from Princeton University.

Einstein's celebrity status gave him the power to be very effective when lending his support to certain causes. For example, a letter from Einstein, crafted by Hungarian physicist Leo Szilard (1898–1964), warned U.S. president Franklin Roosevelt (1882–1945) about the potential dangers of a weapon harnessing atomic energy. He urged Roosevelt to take into account the fact that the Nazis were pursuing such a weapon and to take care to requisition sufficient supplies of uranium—needed for such a weapon—for the United States. Einstein's efforts catalyzed the U.S. project to build an atomic bomb. Later, after the war, Einstein turned his influence toward more peaceful pursuits. He criticized U.S. scientists for allowing the military to dominate scientific research, and in the 1950s he lent his name to many disarmament and antinuclear organizations. He was offered the presidency of Israel in 1952, but declined.

See also Atomic Bomb; Determinism; Light; Manhattan Project; Nazi Science; Physics; Quantum Mechanics; Race; Relativity; Social Responsibility; Solvay Conferences

References
Bernstein, Jeremy. *Einstein.* New York: Viking, 1973.
Cassidy, David. *Einstein and Our World.* Atlantic Highlands, NJ: Humanities Press, 1995.
Clark, Ronald W. *Einstein: The Life and Times.* New York: World Publishing Co., 1971.
Frank, Philipp. *Einstein: His Life and Times.* New York: Alfred Knopf, 1947.
Pais, Abraham. *"Subtle Is the Lord . . .": The Science and Life of Albert Einstein.* New York: Oxford University Press, 1982.
Pyenson, Lewis. *The Young Einstein: The Advent of Relativity.* Boston: Adam Hilger, 1985.

Electronics

The field of electronics is concerned with electric currents transmitted as pulses through devices of various kinds that manipulate the current's behavior to make it behave like a signal. Early work in this field was conducted in the 1820s and 1830s by the British physicist Michael Faraday (1791–1867), who constructed an electric motor and a device for inducing electricity. Manipulation of electric currents in the nineteenth century improved early forms of communication, such as telegraphy; Samuel Morse's "code," developed in the 1840s, for long-distance communication still retains his name. In the 1880s, Thomas Edison (1847–1931) discovered that electricity flowed from a hot filament to a metal wire *through a vacuum.* This became known as the Edison effect.

The study of electric currents in vacuum tubes became the starting point of twentieth century electronics. Through vacuum tube experiments, British physicist J. J. Thomson (1856–1940) in 1897 discovered the electron, a charged subatomic particle. He identified the cathode rays inside vacuum tubes as being electric currents composed of these tiny electrons. British inventor John Ambrose Fleming (1849–1945) developed the first electronic valve in 1905, an "oscillation valve" that allowed electric current to be converted to a signal. Shortly thereafter the American Lee De Forest (1873–1961) added a grid to Fleming's valve in order to amplify the signals. He called his invention the Audion and patented it in 1907, to Fleming's dismay. The applications of the two inven-

Lee De Forest demonstrating his oscillating tube. (Bettmann / Corbis)

tions seemed clear: Devices for signals and amplification of signals could be very useful to the communications industries. In fact, soon AT&T acquired De Forest's patent and made improvements.

Vacuum tubes continued to be the focus of study in electronics research until the 1930s, although the tubes were not very efficient. Around that time the field of solid-state physics was teeming with interest in the relationship between the new quantum mechanics and the behavior of electrons in crystalline materials. Scientists such as Princeton physicists Eugene Wigner (1902–1995) and Frederick Seitz (1911–) turned their attention to semiconductors, crystals whose properties of conducting electricity were inconsistent but promising. If semiconductors could be used

to amplify electronic signals, they would be a vast improvement over the vacuum tubes.

William Shockley (1910–1989), a scientist at Bell Telephone Laboratories, tried a number of times in the 1940s to devise a solid-state amplifier using semiconductors, but failed. His frustration was exacerbated in 1947 when two scientists from his team of researchers, John Bardeen (1908–1991) and Walter Brattain (1902–1987), succeeded. Their invention, soon dubbed the "transistor," ensured the future of solid-state amplification. Although disappointed that his role was not greater, Shockley shared the Nobel Prize in Physics with Bardeen and Brattain in 1956 for this work. The first transistor used a slab of germanium crystal, without a single vacuum tube. Soon more efficient transistors would be composed of silicon as the semiconductor crystal. Electronics appeared to demonstrate that pure science and technology were not terribly far apart. In his Nobel speech, Shockley noted that most of his research originated in the goal of producing a useful device. He applauded his colleagues at Bell Telephone Laboratories for their long-standing appreciation of the role of fundamental research in developing industrial technology.

Applications of the transistor did not appear immediately for general public consumption. Televisions and radios continued to be produced with bulky, heavy tube amplifiers. But in the mid-1950s, "transistor radios" and other similarly named products flooded the markets, produced by various companies such as the Japanese-based Sony. Shockley himself moved to California and started a semiconductor company in the region that eventually would be called Silicon Valley.

See also Industry; Physics; Thomson, Joseph John

References

Braun, Ernest, and Stuart MacDonald. *Revolution in Miniature: The History and Impact of Semiconductor Electronics.* Cambridge: Cambridge University Press, 1978.

Finn, Bernard, Robert Bud, and Helmuth Trischler, eds. *Exposing Electronics.* Amsterdam: Harwood, 2000.

Riordan, Michael, and Lillian Hoddeson. *Crystal Fire: The Invention of the Transistor and the Birth of the Information Age.* New York: W. W. Norton, 1997.

Seitz, Frederick, and Norman G. Einspruch. *Electronic Genie: The Tangled History of Silicon.* Urbana: University of Illinois Press, 1998.

Shockley, William. "Transistor Technology Evokes New Physics" (Nobel lecture 11 December 1956). In *Nobel Lectures, Physics, 1942–1962.* Amsterdam: Elsevier, 1964–1970, 344–374.

Tyne, Gerald F. J. *Saga of the Vacuum Tube.* (Indianapolis, IN.: Sams, 1977.

Elitism

Science has almost always been an elite activity, a pursuit possible only for those with the time and resources to devote to it. In the twentieth century, however, science was pursued by a variety of people and institutions, owing to the development of salaried positions in universities and private laboratories. Although governments took roles as patrons in their attempts to promote science, new institutions often lacked the prestige of older ones. In Britain, for example, the Imperial College of Science and Technology was founded in 1907, but initially it lacked the prestige of older institutions not focusing primarily on science. Elites already in academic institutions resented the notion that government patronage might make academic careers in science more lucrative than in other fields, and they claimed such patronage infringed on academic freedom. In Britain, however, the government's desire to see its populace better trained in scientific and technical matters ensured that government patronage would make scientists into new elites.

As governments took an interest in science, some began to resent the practice of providing money only to those institutions that had the best talent. Critics argued that such funds would only reinforce these institutions' dominance, making it impossible for smaller, less-known institutions to grow and compete with them. This was an especially pressing problem in the United States, where the disparity among institutions could be understood geographically—there were regions in the United States that lacked well-known institutions and thus would receive very little funding in any "best-science" scheme. To ease this geographical conflict, the National Research Council decided in 1918 to recommend funding the best projects within each state. Of course, this did not eliminate the institutional conflicts within the states, but it did appear to resolve a major political question among states. But in the 1930s, National Bureau of Standards director Lyman Briggs (1874–1963) attempted unsuccessfully to convince the government to earmark funding for science; some of it would be spent in the Bureau of Standard's own laboratories, and the National Academy of Sciences would decide where to spend the rest. This proposal would have given the country's elite scientists the job of deciding which projects to fund. It met with considerable criticism from members of Congress, such as Texas's Fritz Lanham (1880–1965), from states located outside of the Northeast and the Midwest, and California. This frustrated scientists who wanted the government to help them solve problems rather than simply throw money at science. The issue was also political: Best-science elitists typically favored laissez-faire approaches more generally—not only in science but in industry—and had favored the conservative policies of Republican presidents in the 1920s. They disliked efforts of the government to equalize opportunity and wanted the government to base its decisions upon merit. Others, however, were New Dealers who hoped that the Democratic president Franklin Roosevelt (1882–1945) would pursue more equitable policies. Government patronage in the 1930s led to the creation of the National Cancer Institute in 1937 and earmarked a considerable sum each year to the National Institutes of Health. Both were part of Roosevelt's New Deal, yet they also supported best-science elitism, leaving the question of elitism without a definitive answer. When the United States entered World War II in 1941, the issue of elitism evaporated temporarily, because funding on the basis of merit went unchallenged owing to short-term demands for results.

Elitism again became an issue in the United States during the efforts to create a national science foundation. As World War II came to a close, leading science administrators such as Vannevar Bush (1890–1974) believed that a permanent body was needed to support science, in the best interests of the state. Bush knew about this firsthand, as he had managed U.S. wartime scientific projects as director of the Office of Scientific Research and Development and had authored a report to the president of the United States entitled *Science—the Endless Frontier* (1945), arguing for continued federal patronage. The

motivation for a national science foundation had been, in part, to provide a civilian basis for science rather than to let the military continue to bear the responsibility for supporting science. But a civilian agency would be vulnerable to the travails of politics. Immediately the question was raised: In an era of increased federal funding for science, who would get the money? Especially if the rationale for supporting science was its ability to strengthen the nation as a whole, it might make sense to ensure that only the best scientists received money. If that occurred, then only the elite universities would benefit. The issue did not get resolved in the formation of the National Science Foundation in 1950, although its formulators urged that the government should avoid concentrating funding in a few elite centers; at the same time, they resisted the efforts of some state universities to apply a rigid formula to ensure equitable distribution of money.

See also Great Depression; National Academy of Sciences; National Bureau of Standards; National Science Foundation; Patronage; Technocracy

References
Argles, Michael. *South Kensington to Robbins: An Account of English Scientific and Technical Education since 1851.* London: Longmans, 1964.
Dupree, A. Hunter. *Science in the Federal Government: A History of Policies and Activities.* Cambridge: Belknap Press, 1957.
England, J. Merton. *A Patron for Pure Science: The National Science Foundation's Formative Years, 1945–57.* Washington, DC: National Science Foundation, 1982.
Kevles, Daniel J. *The Physicists: The History of a Scientific Community in Modern America.* Cambridge, MA: Harvard University Press, 1995.

Embryology

Embryology is the study of the growth of animals from the moment of fertilization to the full development of organs enabling independent life. In addition to its inherent interest, embryology has been one of the key fields in debating some of the larger questions in biology, such as the eighteenth- and nine-teenth-century controversy between vitalists and mechanists. As the embryo develops, does it simply pass through configurations of some preformed mass (preformation, a mechanist idea), or is each stage a separate, spontaneous growth based on some ill-defined, life-giving—thus vital—force (epigenesis, a vitalist idea)? The crucial question of embryology remained relatively unchanged in the twentieth century: how to explain the patterns of development in its earliest stages.

Many embryologists by mid-century believed that eventually the science could be reduced to a branch of biochemistry. The seeds of this idea were present from the dawn of the century. German zoologist Curt Herbst proposed in 1901 that experimentalists eventually would discover that the embryo developed through induction, a process by which some formative stimulus "induces" the growth in neighboring cells. Herbst never fully demonstrated this, but the idea was strongly indicative of a reductionist mentality; in other words, he believed that development could be reduced to mechanical action.

The most renowned researches in embryology were conducted by German biologist Hans Spemann (1869–1941) at the University of Freiburg. The concepts of "organizers" in the process of induction came from Spemann's work. Spemann had begun experiments prior to World War I at the Kaiser Wilhelm Institute for Biology in Berlin-Dahlem, but his momentous discoveries were made after moving to Freiburg in 1919. Spemann's experiments used the embryos of amphibians, most often the common striped newt (*Triton taeniatus*). According to Spemann, in a theory developed by the early 1920s, there were certain parts of the embryo that acted as the center of organization. Its tissue cells were the organizers, exerting influence on nearby tissue, to stimulate development. Even if parts of this center were transplanted to a different part of another embryo, they would "induce" the beginnings of a secondary embryo. One of Spemann's

students, Hilde Proescholdt (1898–1924) (later Hilde Mangold), conducted a series of experiments in 1921 in which a secondary embryo was "induced" in this way. From this work, Spemann in 1924 developed his "organizer" theory, and later he determined that the different parts of the organization center were responsible for different parts of the embryo's development. This work's importance was recognized in 1935 with the Nobel Prize in Physiology or Medicine.

Even with Spemann's work, however, embryology seemed to carry vitalistic overtones, with the "organizer" as the mysterious agent of change. One of Spemann's former students, Johannes Holtfreter (1901–1992), attempted to reduce embryology even further to mechanistic relationships. This "reductionist" vision of embryology was essentially a biochemical one. During his experimental program, he developed a saline solution in which the embryos could live, and developed methods to deter infection. Holtfreter's work, much of it accomplished in the 1930s, cast the embryo less as a holistic organism and more as a population of interacting cells. Did the cell tissue surrounding the organizers simply follow instructions, or did they self-instruct to some extent? In 1938, he found the latter to be true and thus diminished the importance of the organizer as the vital agent of change; Holtfreter claimed that in most cases it simply released the capacity to develop already inherent in the tissue. Another of his principal findings, also demonstrated by 1938, was that organizer material that was no longer living could still induce embryonic growth when transplanted. This further diminished the vital property of the organizer, because it did not itself need to be alive in order to induce! In addition, the organizer material in the embryo was not the only material that seemed capable of induction. The ramifications were astounding: Embryonic growth was not necessarily induced by living tissue, and organizers could come in different varieties. Because of Holtfreter's discovery, much of the work after 1938 was directed toward finding other kinds of inducers, beyond the organizer material identified by Spemann. British embryologist Joseph Needham (1900–1995) was one of the key searchers for such inducers (in fact, the two were close colleagues, and Needham provided refuge to Holtfreter in 1939, when the latter fled the Nazi regime).

Because work in genetics also was concerned with the transmission of organizing information, much of the subsequent history of embryology was marked by efforts to close the perceived gaps between it and genetics. Few scientists until the 1930s attempted this, mainly because the leading embryologists (including Spemann) distrusted geneticists as interlopers and did not believe in the usefulness of genetics in understanding development. Two scientists who tried to bridge the gap were British biologist Conrad H. Waddington (1905–1975) and German biologist Salome Glücksohn-Schönheimer (1907–), a former student of Spemann's. Instead of experimenting on embryos by altering them, Glücksohn-Schönheimer believed that the genes already were altering them—"experimenting" on them, and that it would be useful to study the alterations in development caused by mutated genes. Glücksohn-Schönheimer began to call herself and others with similar goals *developmental geneticists,* a term that endured in the postwar period.

See also Genetics

References

Gilbert, Scott, ed. *A Conceptual History of Modern Embryology.* New York: Plenum Press, 1991.

Hamburger, Viktor. *The Heritage of Experimental Embryology: Hans Spemann and the Organizer.* New York: Oxford University Press, 1988.

Willier, Benjamin H., and Jane M. Oppenheimer, eds. *Foundations of Experimental Embryology.* Englewood Cliffs, NJ: Prentice-Hall, 1964.

Endocrinology

Endocrinology is a branch of physiology concerned with internal secretions by glands. Endocrinologists studied the effects and functions of such secretions. Twentieth-century endocrinology largely was built upon the nineteenth-century work of experimenters such as Claude Bernard (1813–1878) of France and Ivan Pavlov (1849–1936) of Russia. Bernard had introduced the concept of glands and had begun to identify the role of the secretions in the liver and pancreas. Pavlov studied digestion in dogs by analyzing internal fluids, which he gathered by surgically implanting fistulas (collecting tubes) into living dogs.

In the early twentieth century, scientists came to understand internal secretions as regulators of the body. In 1902, British physiologist Ernest Starling (1866–1927) introduced the word *hormone* to describe the regulating "chemical messengers" inside the body. Hormone research became the cornerstone of endocrinology, and scientists hoped to isolate and define the roles of the myriad kinds of such secretions. Researchers in 1921, led by Frederick Banting (1891–1941) and J. J. R. Macleod, successfully extracted the hormone in a dog's pancreas that regulated the body's sugars. That hormone, soon marketed as "insulin," became a veritable wonder drug in treating diabetes, a disease caused by a lack of the hormone in humans. The two men were rewarded with the Nobel Prize in Physiology or Medicine in 1923. Over the next decades, new drugs were developed for "hormone therapy," designed to augment existing secretions or compensate for their absence. Much of the impetus for research in endocrinology came from drug companies during the 1920s and 1930s; that situation changed after World War II, when governments (particularly that of the United States) began to fund research on a much larger scale.

Endocrinology's scope widened in the twentieth century, because hormones were recognized as the body's regulators, with the power to cause (or cure) disease but also to define one's sexual identity. For example, those advocating political freedoms for homosexuals turned to endocrinology for evidence that homosexuality should not be considered a disease or a criminal act, but rather a consequence of hormone differences. German physician Magnus Hirschfeld (1868–1935) developed this idea, describing homosexuality as a physiological phenomenon that could be traced back to the sex glands. Hirschfeld was the founder of the Institute for Sexual Science, and he advocated this interpretation of homosexuality in the years prior to World War I, culminating in the 1914 work, *Homosexuality in Man and Woman*. His efforts to use science to end discrimination against homosexuals in Germany succeeded in influencing other scientists and activists, but ultimately failed in its political goal, particularly in the 1930s during the rise of the Nazis.

Research on the body's "juices" continued through mid-century. In 1936, Carl Cori (1896–1984) and Gerty Cori (1896–1957) collaborated to study the role of insulin and epinephrine. Their studies of the starch glycogen and its conversion into glucose, an essential process in the body, led to a Nobel Prize in Physiology or Medicine in 1947. They shared it with Bernardo Houssay (1887–1971), whose work on pituitary glands helped to assess the causes of diabetes and trace the path of carbohydrate metabolism. After the war, endocrinology became an essential component of neurophysiology, as more scientists began to study the role of fluids in regulating neural processes in the brain.

See also Hormones; Patronage

References

Bliss, Michael. *The Discovery of Insulin.* Chicago: University of Chicago Press, 1982.

Hughes, Arthur F. "A History of Endocrinology." *Journal of the History of Medicine and Allied Sciences* 32:3 (1977): 292–313.

Sengoopta, Chandak. "Glandular Politics: Experimental Biology, Clinical Medicine, and Homosexual Emancipation in Fin-de-Siècle Central Europe." *Isis* 89:3 (1998): 445–473.

Espionage

As scientists took increasingly important roles in national security in World War II, science itself became the centerpiece of intrigue. Between belligerents, the most important kind of intelligence was in code-breaking. Radio intercepts between military commanders provided U.S. intelligence officers with far more information than they otherwise would have had. But it was interalliance spying, specifically on scientific projects, that became the greatest concern by the late 1940s. As the Soviet Union and the United States grew colder toward each other and began to base conceptions of national security increasingly on scientific and technological knowledge, fear of spies in the scientific community gripped society.

Scientific intelligence during World War II was gathered through a variety of means, from publication browsing to code breaking. Both sides of the conflict devoted intensive efforts to gleaning as much information as possible from published and unpublished documents, aided by people and new technology. Paul Rosbaud (1896–1963), an Austrian science editor, smuggled information about the Nazi rocketry and atomic bomb projects, in addition to a detailed report about conventional incendiary bombs. British knowledge of Wernher Von Braun's (1912–1977) rocketry work at Peenemünde, for example, came largely from the efforts of Rosbaud, and British bombers attacked the site in 1943. As for new technology, Britain's intelligence center at Bletchley Park managed to decrypt Germany's military codes and laid the foundations for the field of modern cryptanalysis, which proved immensely useful in knowing German positions in advance. The Soviet Union stepped up its own intelligence efforts in 1942 to assess troop concentrations as the Germans advanced deeper into Russia. The British technology proved useful to the Soviets; one of the officials at Bletchley Park, John Cairncross (1913–1995), was a Soviet spy.

Espionage also helped the Soviet Union to learn about, and make the decision to pursue, the atomic bomb. Although Soviet physicists knew that a bomb was a theoretical possibility, the resources simply could not be spared on such an "experiment" while the Germans were advancing. But after the battle of Stalin-grad, in early 1943, the Red Army began a major counteroffensive against the Germans called Uran. Perhaps the reason Joseph Stalin (1879–1953) chose this name was that he also made another important decision at the time to encourage his scientists to pursue a uranium bomb. The Soviets knew that the British and Americans already had begun to build one, and Stalin did not want a powerful weapon to be in the hands of his only competition in a postwar world. His foreign minister, Viacheslav Molotov, handed over intelligence materials to physicist Igor Kurchatov (1903–1960) in February 1943. After that time, Kurchatov and others knew that a bomb was possible. The materials were based on the espionage activities spearheaded by a physicist named Klaus Fuchs (1911–1988).

Fuchs, a German, had joined the Communist Party in the 1930s, in opposition to the Nazis. Persecuted, he fled the country and settled in England, then moved to Scotland, where he joined Max Born's (1882–1970) physics laboratory. When the war began, he became part of Britain's wartime scientific establishment, which, among other things, worked on nuclear fission. By early 1942, he was spying for the Soviet Union. His connections with the Communist Party raised few eyebrows, because it was a fairly popular alternative to fascism in the 1930s. When this secret project was moved to the United States, Fuchs joined the British contingent at Los Alamos, the top-secret site where scientists were designing the first atomic bomb. He rose to a high position and had knowledge of myriad aspects of the bomb project, despite the compartmentalization instigated by project leaders to keep everyone on a need-to-know basis. He continued his espionage activities through the war and returned to Britain in 1946, where he headed the theoretical physics division for developing the atomic bomb in Britain. In 1949, British intelligence officers, armed with new information from the Federal Bureau of Investigation about wartime messages from the Soviet con-

sulate in New York, arrested Fuchs and made him confess. He was imprisoned for nearly a decade (his sentence initially was for fourteen years), and in 1959, he moved to Communist-controlled East Germany.

The Fuchs debacle had many detrimental effects. First, it fueled the flames of the "Red Scare," the feeling of fear and paranoia about Communist infiltration. To make matters worse, the Soviet Union tested its first atomic bomb in 1949, years before most Americans' predictions. The Fuchs affair also destroyed the trust between the United States and Britain, severely undermining efforts by the British to reestablish wartime scientific exchanges of information. Moreover, it highlighted the role of scientists in matters of national security and demonstrated the need to keep them loyal. The arrest of Klaus Fuchs empowered members of Congress in their hunts for Communists in the federal government, lent credibility to "loyalty oaths" at major universities, and made Americans wary of a trusted ally, namely, Britain. It put scientists under close scrutiny; in ensuing years, Americans questioned the reliability and loyalty of leading scientists such as Edward Condon (1902–1974) and J. Robert Oppenheimer (1904–1967). Like them, Fuchs had been at the top level of science, with knowledge of closely guarded secrets.

See also Cold War; Crime Detection; Loyalty; Manhattan Project; Soviet Science

References

Holloway, David. *Stalin and the Bomb: The Soviet Union and Atomic Energy, 1939–1956.* New Haven, CT: Yale University Press, 1994.

Kramish, Arnold. *The Griffin: The Greatest Untold Espionage Story of World War II.* Boston: Houghton Mifflin, 1986.

Moss, Norman. *Klaus Fuchs: The Man Who Stole the Atom Bomb.* New York: St. Martin's Press, 1987.

Overy, Richard. *Why the Allies Won.* New York: W. W. Norton, 1995.

Richards, Pamela Spence. *Scientific Information in Wartime: The Allied-German Rivalry, 1939–1945.* Westport, CT: Greenwood Press, 1994.

Williams, Robert Chadwell. *Klaus Fuchs, Atom Spy.* Cambridge, MA: Harvard University Press, 1987.

Eugenics

Eugenics literally means "good breeding," and it was the science of racial improvement. It was popularized in the nineteenth century by the work of British biologist Francis Galton (1822–1911), who had demonstrated that the hereditary traits of certain peas were relatively stable when passed from one generation to the next, despite the appearances of variations. The average of the traits in the entire population remained the same. Galton and his disciples extended this conclusion to human populations as well, noting the occasional dissimilarity between parents and children. He claimed that heredity was based more on the totality of one's ancestors rather than one's parents. Each new child's heritable traits—for example, intelligence—would tend toward the average. Eugenicists of the twentieth century took this to mean that selective breeding might improve the average. They tried to use breeding either to improve racial characteristics or to maintain the integrity of racial groups.

The pervasive acceptance of the basic tenets of eugenics had far-reaching consequences in the early twentieth century. One was the abuse of a scientific idea to reinforce racial stereotypes and discrimination and to categorize entire ethnic groups hierarchically in order of superiority. Eugenics appeared to demonstrate that intermarriage could disturb and perhaps "pollute" racial stocks, leading inadvertently to what many—including U.S. president Theodore Roosevelt (1858–1919)—dubbed "race suicide." Such fears were fueled by reports of an increase in mental deficiency in both the United States and Britain, which reinforced, among the supposedly "well-bred" white population, militant desires to preserve national integrity by identifying it with racial strength or survival. For example, in the United States, eugenics provided a strong rationale for anti-Asian sentiment on the West Coast, leading to strict immigration laws during the Progressive era in the first decade of the century. Political Progressives believed race to be one of the country's most important nonrenewable resources.

British eugenicists founded the Eugenics Education Society in 1907, hoping to influence the development of a more rational society based on science's findings about heredity. It sought to find ways, through legislation and general influence on government, to encourage programs for positive and negative eugenics—that is, encouraging some elements of society to breed and others not to do so. In contrast to the U.S. experience, the alleged dividing line between the well-bred and the ill-bred was not always racial, but based more often on class distinctions (these, of course, were often combined, as in the case of discrimination against the Irish). Some of the eugenicists hoped to reduce social strife by decreasing, through eugenics laws, procreation among those most likely to join labor unions and other working-class groups.

In Germany, eugenics was known as racial hygiene. Before the Nazi period, German eugenics was similar to those of Britain and the United States, although it was more closely tied to the field of medicine. Although one of the principal eugenicists, Wilhelm Schallmayer (1857–1919), often critiqued the widespread linkage of eugenics with racism in society, most German eugenicists wished to use science to construct a healthier, more rational, and more efficient society, which typically meant paying close attention to racial purity. The tendency to adopt managerial outlooks toward race in society during the early twentieth century continued into the 1930s. Under the racist agenda of the Nazis, German eugenics focused on the importance of preserving the Aryan race, citing the need to keep it free of Jewish and Slavic influences. Adolf Hitler (1889–1945) believed that Jews in particular were sucking the vigor out of the Aryan race, and the Nazis passed the Nuremberg Laws in 1935 to redefine German citizenship along racial lines and to discriminate

Evolution

Evolution became a prominent scientific theory in the nineteenth century, but it remained controversial in the twentieth. The most influential evolutionary theories were proposed by French zoologist Jean-Baptiste Lamarck (1744–1829) in *Philosophie Zoologique* (1809) and English botanist Charles Darwin (1809–1882) in *On the Origin of Species by Means of Natural Selection* (1859). Each was controversial in the early twentieth century for scientific and philosophical reasons. Evolution in general was virulently attacked for religious reasons and also met with a serious critique from geneticists. Although Lamarckian evolution survived in countries such as the Soviet Union, and religious conservatives continued their hostility toward evolution in general, the scientific community in most Western countries accepted by the 1940s a synthesis of Darwinian evolution and genetics.

The most attractive evolutionary idea at the dawn of the twentieth century was Lamarck's vision of adaptive change. Lamarck had denied the existence of species, and instead argued that organisms continually adapt to their environments, causing changes in themselves that could be passed down to their offspring. This idea, the inheritance of acquired characteristics, had philosophical implications: The organism was the agent of change, improving its ability to live productively in any environment, and such improvement was permanent. Unfortunately, experiments made in the 1890s by August Weismann (1834–1914) had demonstrated that cutting off the tails of mice, generation after generation, did not lead to change in the length of tails in mice in any succeeding generation. This cast serious doubt on the inheritance of acquired characteristics. Thus in the first years of the twentieth century, evolutionists had to choose between two competing visions (Lamarckian and Darwinian) whose boundaries had been blurred for years. Despite the lack of evidence in its favor,

Francis Galton, founder of the eugenics movement. (Bettmann/Corbis)

strongly against Jews in society. After World War II began, this policy became one of outright genocide. Horrified reactions to the Holocaust, during which the Nazis executed and gassed some six million Jews, contributed to a dramatic decline in sympathy for eugenics after the war.

See also Biometry; Birth Control; Intelligence Testing; Nazi Science; Race; Social Progress

References
Kevles, Daniel J. *In the Name of Eugenics: Genetics and the Uses of Human Heredity*. New York: Alfred A. Knopf, 1985.
Pickens, Donald K. *Eugenics and the Progressives*. Nashville, TN: Vanderbilt University Press, 1968.
Searle, Geoffrey. *Eugenics and Politics in Britain, 1900–1914*. Leyden: Noordhoff International Publishing, 1976.
Weiss, Sheila Faith. *Race Hygiene and National Efficiency: The Eugenics of Wilhelm Schallmayer*. Berkeley: University of California Press, 1987.

Lamarckian evolution proved popular, largely because the alternative theory made the world appear to be a senseless, random, purposeless accident. In the Soviet Union, Lamarckian evolutionists such as Trofim Lysenko (1898–1976) were hostile to Weismann's conclusions. Science, Lysenko was arguing even in the late 1940s, was the enemy of chance. Humans should feel empowered to make changes to their environment, such as improving agriculture by stimulating designed evolution.

The leading Darwinists were biometricians who hoped to use large populations to demonstrate that natural selection could have a lasting effect. By the early twentieth century, Darwinian evolution was faced with the problem of blending. If a drop of black paint fell into a vat of white paint, the black would blend so thoroughly that no effect could be perceived. The same might be true of evolution; even if new traits were introduced into a population, how could that trait have a lasting effect? Karl Pearson (1857–1936) was the most active and influential of the biometricians trying to save Darwin from the effects of blending. Not only did Pearson believe that natural selection could change a population, he also believed in selective breeding to improve the general stock of the whole population. This movement was called eugenics and had been pioneered by Charles Darwin's cousin Francis Galton (1822–1911). Pearson promoted social reforms along these lines, hoping to have governments encourage the procreation of the most intelligent and productive groups in society.

A powerful threat to Darwinian evolution came from the development of genetics in the first decade of the century. The modern agreement between genetics and natural selection often obscures the strong mutual disaffection between evolutionists and geneticists in the 1910s. This conflict, in part, was owing to the personal animosity between leading geneticist William Bateson (1861–1926) and leading biometrician Walter F. R. Weldon (1860–1906). But genetics, by

allowing dominant and recessive genes to be transmitted by mathematical laws, without blending, had in fact saved Darwinism from its most serious critique. Still, one salient issue appeared to make Darwinism and genetics irreconcilable, namely, the problem of continuous versus discontinuous change. The Darwinians insisted that random change occurred continuously, whereas the geneticists proposed that changes occurred in bursts, through sudden mutation.

The evolutionary synthesis achieved a relatively harmonious combination of Darwinism with genetics. U.S. biologist Thomas Hunt Morgan's (1866–1945) experiments with the fruit fly, beginning in 1910, revealed small naturally occurring mutations that were inherited according to Mendelian laws. The new traits, however, only spread through the population by means of natural selection. Building on this work, British biologist Ronald A. Fisher (1890–1962), trained in biometrical methods, realized that Mendelian inheritance would allow traits to be preserved without blending, thus allowing a large number of traits to be present in a population without always showing themselves, in turn explaining the large degree of variability. Another British biologist, J. B. S. Haldane (1892–1964), demonstrated that the process could work very quickly in a population; he offered the famous example of the dark-colored moths that spread quickly in industrial, urban areas, where they flourished unseen by predators. Other researchers in the 1920s and 1930s such as Sewall Wright (1889–1988) and Theodosius Dobzhansky (1900–1975), both working in the United States, helped to found a new field of evolutionary science called population genetics. George Gaylord Simpson's (1902–1984) *Tempo and Mode in Evolution* (1944) persuasively argued that the evolutionary processes described by population geneticists could be reconciled with the fossil record, especially the aspects of Darwinism that emphasized branching and common ancestry rather than linear progressions of species. By mid-century, most scientists in Western Europe and North America accepted the evolutionary synthesis as it was developing in the hands of population geneticists.

Evolution was more than a scientific idea; it became a symbol for the conflict between science and religion. There is no mention of species evolving in the Bible, and in fact the Bible presents a different story altogether, that all creatures were created by God during the six days described in the book of Genesis. The idea of being descended from an ape (which was not the true position of the Darwinists, who held that men and apes were not related linearly but instead shared a common ancestor), seemed to contradict the word of God. Especially in the United States, Christian fundamentalists and other concerned religious-minded people disliked the fact that such a controversial theory was being taught to children in schools. The 1925 Butler Act, passed in Tennessee, prohibited the teaching of evolution in public schools. In response, the American Civil Liberties Union argued that the law ignored the principle of separating church and state and asked substitute teacher John T. Scopes (1900–1970) to break the law by teaching it. He agreed, thus provoking an arrest and a widely publicized trial, which Scopes lost. The trial drew attention to the vehement opposition in the southern United States to scientific ideas that threatened religious values. The Scopes trial also provoked a wave of anti-Southern ridicule, but also discouraged publishers from including evolution in secondary school textbooks.

See also Biometry; Determinism; Genetics; Haldane, John Burdon Sanderson; Kammerer, Paul; Missing Link; Peking Man; Religion; Scopes Trial; Simpson, George Gaylord; Wright, Sewall

References

Bowler, Peter J. *Evolution: The History of an Idea.* Berkeley: University of California Press, 1989.

Grene, Marjorie. *Dimensions of Darwinism: Themes and Counterthemes in Twentieth Century Evolutionary Theory.* Cambridge: Cambridge University Press, 1983.

Larson, Edward J. *Summer for the Gods: The Scopes Trial and America's Continuing Debate over Science and Religion.* New York: Basic Books, 1997.

Mayr, Ernst. *The Growth of Biological Thought: Diversity, Evolution, and Inheritance.* New York: Belknap Press, 1985.

Extraterrestrial Life

Extraterrestrial life refers to life originating from somewhere outside the earth. Before the twentieth century, the concept received a great deal of attention, particularly among philosophers and scientists who insisted on the plurality of worlds (that the earth is not unique, and there are many worlds in the universe). The iconoclast of the American Revolution, Thomas Paine (1737–1809), argued that the plurality of worlds necessitated the rejection of human-centered Christianity. Typically, however, this argument ran in the other direction: The story of creation in the book of Genesis, if taken literally, necessitated the rejection of a belief in life on other planets. At the end of the nineteenth century, the religious basis for discarding the notion of alien life proved very strong, particularly among those who already felt challenged by the evolutionary ideas of Charles Darwin (1809–1882). In the twentieth century, those concerned with extraterrestrial life were interested not only in life on other planets but also in the possibility that such life might have visited the earth.

Those seeking to find life on other planets often looked to the earth's neighbor, Mars. The wealthy U.S. astronomer Percival Lowell (1855–1916), for example, was convinced that the seemingly geometric lines discovered on the face of Mars were in fact canals built by intelligent beings, which he wrote about in *Mars and Its Canals* (1906). These lines appeared to connect great masses of water, littering the planet's surface with groups of straight lines that could not have been formed naturally. These "canals" had been identified in the late nineteenth century by the Italian Giovanni Schiaparelli (1835–1910), who, along with French pluralist Camille Flammarion (1842–1925), felt that studies of the canals would vindicate the old notion of the plurality of worlds. But investigations of the canals did quite the opposite; in 1912, analyses of the Martian atmosphere indicated that the lines were not as geometric as previously believed, and their previous appearance had been a result of poor resolution in telescopic lenses. The canals of Mars were simply optical illusions.

Although scientists abandoned the notion of canals on Mars, the logical conclusion about the plurality of worlds remained: If there were many stars, there must be many stellar systems and many planets like the earth capable of sustaining life. The theoretical possibility of life in other star systems remained plausible but impossible to demonstrate. In the 1930s, U.S. astronomer Henry Norris Russell (1877–1957) observed that the origin of the solar system was the most important problem in astronomy, because there still was no evidence to suggest the existence of planets outside of the sun's system. The lack of evidence buttressed scientific and religious claims denying plurality, making it possible to argue the earth's (and man's) uniqueness in the universe. Substantial evidence for the existence of planets in other stellar systems never arose in the first half of the twentieth century.

The most widely known suggestion of extraterrestrial life came from sightings of unidentified flying objects (UFOs). The year 1909 saw a wave of sightings of unidentified airships, in England, the United States, and even New Zealand. However, these airships resembled "zeppelins," a recent German innovation, and the fears were not necessarily of aliens, but of German invaders. In 1946, residents of Norway and Sweden saw "ghost rockets" above their cities and towns, which some believed were extraterrestrial. Others reported that the Americans or Soviets were testing rocket designs recovered from Germany during the war and that their countries were being used as a firing range. Unexplained UFO sightings in 1947 and 1948 near Roswell, New Mexico, provided one of the strongest bases for UFO enthusiasm in the United States, although these, too, could have been connected to military flight or rocket testing. In 1947, the United States Air Force began to collect reports of UFO sightings. It issued two reports in 1949 (the Sign Report and the Grudge Report), both of which attributed the sightings to ordinary phenomena, illusions, or hoaxes. Nevertheless, in subsequent decades, their unexplained nature and the fact that the Air Force kept a great deal of its UFO data secret sparked the curiosity of the public about the possibility of intelligent extraterrestrial life visiting the earth.

See also Astronomical Observatories; Lowell, Percival; Radio Astronomy; Religion; Science Fiction

References

Bartholomew, Robert E., and George S. Howard. *UFOs and Alien Contact: Two Centuries of Mystery.* Amherst, NY: Prometheus Books, 1998.

Boss, Alan. *Looking for Earth: The Race to Find New Solar Systems.* New York: John Wiley & Sons, 1998.

Dick, Steven J. *The Biological Universe: The Twentieth-Century Extraterrestrial Life Debate and the Limits of Science.* New York: Cambridge University Press, 1996.

Jacobs, David Michael. *The UFO Controversy in America.* Bloomington: Indiana University Press, 1975.

F

Federation of Atomic Scientists

The Federation of Atomic Scientists (the Federation), formed in 1945, was one of the first scientific organizations to embrace the new political responsibilities of scientists in the postwar world. After the development of the atomic bomb, the work of science became the centerpiece of U.S. national security strategy, and thus scientists assumed a new role of power, prestige, and—to the minds of some scientists—responsibility for the future of humanity. Some of the scientists who participated in the Manhattan Project believed that the new weapon had heralded a new age that radically would change global politics. The hallmark of that change would be an arms race between the United States and the Soviet Union. To prevent such a race, scientists such as Leo Szilard (1898–1964) and James Franck (1882–1964) had argued during the war against the use of the bomb on Japanese cities, and they had urged international control rather than a U.S. atomic monopoly. When the war ended, much of this effort was taken up by the Federation.

The political focus of the Federation was the control of atomic energy at home and abroad. In what some have dubbed the "scientists' movement," the Federation took the lead in putting the political issues of the bomb at the forefront of policy-minded scientists' concerns, edging out other important issues such as creating a national science foundation. They successfully lobbied in 1946, for example, to have Senator Brien R. McMahon's (1903–1952) atomic energy bill passed over a rival bill that would have given more power to the military. The Federation scientists believed that they had created a powerful weapon that threatened all of mankind, and that they should bear the responsibility for shaping atomic weapons and atomic energy policies.

The Federation first published *The Bulletin of Atomic Scientists* in December 1945, as an effort to educate other scientists about the growing relationship between them and national and international politics. The *Bulletin of Atomic Scientists* was supposed to reveal the dangerous ramifications of the "Pandora's box" of modern science. It did not limit itself to atomic matters, but instead gave attention to science's connections to other issues such as religion, ethics, and law. In an effort to broaden its focus and membership, the Federation soon changed its name to Federation of American Scientists. But the perils of the atomic age stood at the forefront of the *Bulletin of Atomic Scientists,* and it included a "clock of doom" on its cover. When the clock was turned closer to midnight, it reflected the editors' assessment of the dangers of world events. For example, when the Soviet Union tested its first atomic bomb in 1949, the clock moved closer to midnight.

The failure of scientists to negotiate international control of atomic energy with the Soviet Union sent the Federation into disarray. The 1946 Baruch Plan, named for U.S. delegate Bernard Baruch (1870–1965), had proposed both entrusting an international body with atomic secrets and destroying existing weapons. But the Soviets refused because the plan also insisted on inspections, which for the Soviets meant spying. They delayed agreement to the plan, and negotiations about inspections continued until 1949, when the first Soviet atomic bomb test was conducted. The Baruch Plan had failed, and the arms race continued. Some Federation scientists, on the pages of the *Bulletin of Atomic Scientists* and elsewhere, adopted new strategies to advocate arms control in the 1950s, while others worked within the military establishment to strengthen national security through weapons research.

See also Atomic Bomb; Cold War; Manhattan Project; Social Responsibility; Szilard, Leo

References

Gilpin, Robert. *American Scientists and Nuclear Weapons Policy*. Princeton, NJ: Princeton University Press, 1962.

Grodzins, Morton, and Eugene Rabinowitch. *The Atomic Age: Scientists in National and World Affairs*. New York: Simon & Schuster, 1963.

Kevles, Daniel J. *The Physicists: The History of a Scientific Community in Modern America*. Cambridge, MA: Harvard University Press, 1995.

Smith, Alice Kimball. *A Peril and a Hope: The Scientists Movement in America, 1945–1947*. Chicago: University of Chicago Press, 1965.

Fermi, Enrico

(b. Rome, Italy, 1901; d. Chicago, Illinois, 1954)

"The Italian navigator has landed in the New World," the physicist Arthur Compton (1892–1962) said on a long-distance call to a colleague, trying to talk in code. It was 1942.

At the time, Enrico Fermi was working on a secret project and had achieved what so far only had been theoretically possible: the first controlled nuclear chain reaction. Soon Fermi and his colleagues moved to Los Alamos, New Mexico, as part of the Manhattan Project, the effort by the United States to build the atomic bomb.

Fermi already was a world-renowned physicist by this time. He had received his doctorate from the University of Pisa in 1922, but studied also in Göttingen, Germany, and Leiden, Holland, in order to acquaint himself with the latest developments in physics (Italy was not a major center of activity at that time). Throughout the 1920s, Fermi published papers in theoretical physics and in 1926 developed the statistics, named for him, that describe the behavior of subatomic particles while taking into account Wolfgang Pauli's (1900–1958) 1925 exclusion principle, which prevents more than one electron from occupying the same quantum orbit in an atom. The following year, he became a professor of theoretical physics at the University of Rome, where he would remain for more than a decade, conducting research on the inner workings of the atom.

One fruitful area of research was in neutron bombardment. Frédéric Joliot (1900–1958) and Irène Joliot-Curie (1897–1956) discovered that bombarding stable nuclei with alpha particles in a laboratory resulted in the creation of radioactive isotopes. In other words, they discovered artificial radioactivity. Fermi was convinced that radioactive isotopes might be produced by bombarding existing elements with neutrons. Because they have neither a positive nor negative charge, neutrons would not be repelled and would have a better chance of reaching (and being captured by) the tiny nucleus, thus producing an isotope. Fermi systematically subjected more than sixty elements to such bombardment, and nearly forty of them exhibited the activity he had hypothesized. He also found, to his surprise,

Enrico Fermi seated at the control panel of a particle accelerator, the "world's most powerful atom smasher." (Library of Congress)

that bombarding uranium resulted in the manufacture of elements previously unknown to man. In the process of his experiments, Fermi discovered that collisions with hydrogen atoms slowed down some neutrons, and these "slow neutrons" became the focus of much of his subsequent research.

His work on neutrons and the discovery of new radioactive elements culminated in a Nobel Prize for Physics in 1938; Fermi took the travel opportunity (the Nobel Prize ceremonies were in Stockholm, Sweden) to make an exit from Fascist Italy. Although he had enjoyed the support of the Italian government, he had been offered an attractive position at Columbia University in the United States. In addition, racial laws in Benito Mus-

solini's (1883–1945) Italy appeared to threaten his wife, who was Jewish. He traveled directly from Sweden to New York, leaving his laboratory and colleagues behind.

After the 1938 discovery of nuclear fission (splitting the atom) by German scientists Otto Hahn (1879–1968) and Fritz Strassman (1902–1980), Fermi realized that some of his neutron experiments in Rome had resulted in fission without him recognizing it. Looking back on these experiments, Fermi realized that if fission resulted from the bombardment of elements with neutrons, and if the process of fission produced more neutrons (i.e., if one neutron produced fission in an atom but two neutrons were released after fission), the excess neutrons could produce further fission in nearby atoms. Theoretically, this could lead to a fission chain reaction.

After World War II began, the United States hired Enrico Fermi to work in the first phase of the secret atomic bomb project at the University of Chicago. Working closely with Leo Szilard (1898–1964), a Hungarian-born physicist, Fermi's team of scientists achieved the first controlled nuclear chain reaction, by designing a reactor (or "pile" as they called it) in a squash court underneath University of Chicago's Stagg Field. In a dramatic experiment in 1942, the first atomic pile went critical, demonstrating that a chain reaction, and possibly an atomic bomb, was possible. Fermi and others assumed that Adolf Hitler's (1889–1945) scientists were doing the same thing, although the highly effective method of slowing down neutrons by using graphite of very high purity was never adopted by the German scientists, and they did not achieve a chain reaction.

Without the work of Fermi and Szilard, the United States would not have built the atomic bomb during World War II. After this stunning success, which some hoped would not prove possible, Fermi moved to Los Alamos, New Mexico, to work as a consultant for designing and building the atomic bombs that ultimately would be dropped on Hiroshima and Nagasaki in 1945. He later served in atomic energy policy advising positions and even visited Italy after the war. In 1954 his health rapidly declined; he had developed incurable stomach cancer, and he died the same year.

See also Atomic Bomb; Atomic Structure; Fission; Physics

References

MacPherson, Malcolm C. *Time Bomb: Fermi, Heisenberg, and the Race for the Atomic Bomb.* New York: E. P. Dutton, 1986.

Segrè, Emilio. "Fermi, Enrico." In Charles Coulston Gillispie, ed., *Dictionary of Scientific Biography,* vol. IV. New York: Charles Scribner's Sons, 1971, 576–583.

———. *Enrico Fermi, Physicist.* Chicago: University of Chicago Press, 1970.

Fission

Few processes of nature have had so great an impact on the course of history in the twentieth century as nuclear fission. Its discovery led almost immediately to bomb projects in several countries. Some six years after scientists achieved fission in a laboratory, two cities in Japan were decimated, each by a single bomb exploiting the principles of atomic physics. Atomic energy then transformed the world, through peaceful uses such as power production and, more importantly, through nuclear weapons. These weapons became symbols of international power; their numbers and the increasing complexity of delivery systems became the hallmark of the arms race between the United States and the Soviet Union. They also promised to make the next war, should it ever occur, a worldwide holocaust.

The scientific background of nuclear fission begins with the discovery of radioactivity in 1896. Henri Becquerel (1852–1908) was investigating the properties of uranium, which he believed was one of several elements that phosphoresced for extended periods after being exposed to the sun. He soon discovered that his samples of uranium emitted some kind of radiation, detectable from exposed photographic plates, even without being exposed to the sun. The property was inherent to the material, not a reaction to sunlight. Radioactivity remained an unspectacular field (compared, for example, with the study of X-rays) until Marie Curie (1867–1934) and her husband, Pierre, began their studies of uranium ores in the last few years of the nineteenth century. They found two new elements, polonium (named for Poland, Marie's homeland) and radium. She coined the word *radioactive* to describe the property of giving off some kind of emission.

Studies of radioactivity increased rapidly in ensuing years, especially under the influence of New Zealander Ernest Rutherford (1871–1937). He and others began to recognize a number of different properties of ionizing radiation (meaning the radiation that produced charged particles as it passed through surrounding gas). Rutherford named the "alpha" and "beta" particles emitted. Alpha particles appeared strongly charged but could be stopped by a sheet of paper, whereas beta particles seemed weaker but more penetrating. Other kinds of radiation, such as gamma rays, were also discovered, but a full understanding of the particulate and electromagnetic forms of radiation would be slow in coming.

Rutherford recognized that the release of alpha and beta particles meant that a process was taking place that seemed less like science and more like the alchemy of centuries past. But whereas alchemists had tried to turn lead into gold, Rutherford and his colleagues came to realize that radioactive elements were in fact changing, over huge spans of time, into lead. He and Frederick Soddy (1877–1956) argued that radioactive atoms were not stable, and by ejecting an alpha or beta particle, the atom of one element "transmuted" into an atom of another element. This process continued in a chain of unstable elements—a decay chain—until finally a stable element resulted. Decay chains end with lead.

After World War I, scientists began to experiment with artificial disintegrations. They bombarded stable elements with alpha particles, resulting in the ejection of particles, thus using natural radioactivity to produce artificial disintegration. Most elements could not be bombarded in this way, because alpha particles were charged particles, and so were the target nuclei. Their electric charges tended to repel each other. But John Cockcroft (1897–1967) and Ernest Walton (1903–1995), researchers at Rutherford's Cavendish Laboratory at England's Cambridge University, built a machine to accelerate alpha particles artificially to achieve high enough energy to overcome this obstacle. Here scientists were creating alpha particle speeds not found in nature in order to create a nuclear disintegration also not found in nature. Soon physicists in many countries began to build particle "accelerators" to overcome

the barriers of nature, including U.S. physicist Ernest Lawrence's (1901–1958) famous cyclotron. Around the same time, in 1932, Rutherford's chief assistant, James Chadwick (1891–1974), discovered what Rutherford had predicted must exist: a particle roughly the same size as a proton but with no charge, called the neutron.

The discovery of the neutron opened up new doors for experiments with artificial disintegrations of elements. Neutrons were the ideal projectiles for experiments. Alpha particles (identified by this time as helium atoms) had an electric charge, but neutrons did not. Thus they had the mass needed to collide with atomic nuclei but no charge to deflect them away from it. The Italian physicist Enrico Fermi (1901–1954) began to bombard elements with neutrons, describing a number of nuclear reactions and even finding new elements. Determining the properties of the products from such bombardments was the basis of research that led to nuclear fission.

Nuclear "disintegration" previously described was not the same as nuclear fission. Disintegration described the decay of one element into another, caused by the emission of a particle. Emitting that particle is what makes the element "radioactive." Fission, however, is not merely the ejection of a particle. It requires that the atom itself split roughly in half. Research on nuclear reactions in the late 1930s was complex, mainly because of a number of confusing disintegrations taking place, because there is not a single decay chain to record. In addition, with minute quantities of radioactive material, scientists needed to mix in similar, nonradioactive elements to help track the processes.

In 1938, German scientists Otto Hahn (1879–1968) and Fritz Strassman (1902–1980) noted that their experiments with uranium were not resulting in disintegration toward radium, as expected. Instead, they appeared to have produced a new sort of radioactive barium, closer to the middle of the periodic table. In fact, barium was roughly half of radium's weight. They knew that, if uranium were to split, something like barium might be expected, but nothing bigger than an alpha particle had even been seen coming from an atom before. Hahn's longtime colleague Lise Meitner (1878–1968), having fled Nazi Germany by this time, provided the explanation. She and her nephew Otto Frisch (1904–1979) interpreted the results as a deformation of the uranium nucleus from collision with a neutron. That deformation led to a break. The atom had been split. The addition of a neutron to the uranium nucleus had created two atoms of barium.

Meitner and Frisch also calculated that the sum of the two new atoms' masses did not exactly equal that of the old atom and the neutron. Some mass was lost in the fission process, being converted into energy. The amount of released energy could be calculated with Albert Einstein's (1879–1955) famous formula, $E = mc^2$, or energy equals mass multiplied by a huge number (the speed of light, squared). This reaction produced millions of times more energy than the most energetic chemical reactions then known.

Nuclear fission had been discovered. For many, the ramifications were clear. Aside from being a fascinating scientific phenomenon, the energy released suggested that, if a fission chain reaction could be sustained, one could create a weapon of such power that an entire city might be destroyed in an instant. Fission became the science of atomic bombs.

See also Atomic Bomb; Fermi, Enrico; Hahn, Otto; Manhattan Project; Meitner, Lise; Physics; Radioactivity; Uranium

References

Badash, Lawrence. *Scientists and the Development of Nuclear Weapons: From Fission to the Limited Test Ban Treaty, 1939–1963*. Atlantic Highlands, NJ: Humanities Press, 1995.

Kragh, Helge. *Quantum Generations: A History of Physics in the Twentieth Century*. Princeton, NJ: Princeton University Press, 1999.

Williams, Robert C., and Philip Cantelon, eds. *The American Atom: A Documentary History of Nuclear Policies from the Discovery of Fission to the Present, 1939–1984*. Philadelphia: University of Pennsylvania Press, 1984.

Franck, James

(b. Hamburg, Germany, 1882; d. Göttingen, Germany, 1964)

James Franck was an important figure in the physics community in Germany prior to World War II, then later as the leader of a group of scientists in the United States who opposed using the atomic bomb against Japan. In the first decade of the century, Franck worked in Berlin, attending colloquia led by eminent physicists such as Max Planck (1858–1947) and Albert Einstein (1879–1955). He and his colleague Gustav Hertz (1887–1975) conducted studies of electron collisions. They found that, although collisions between electrons and noble gas atoms usually were elastic, without transfer of kinetic energy, some collisions were inelastic, leading to the transfer of energy to atoms. Their experiments showed that energy transfer occurred only when kinetic energy exceeded a certain level, and at that level the entire amount of energy up to that level was transferred to the atom. These experiments, conducted prior to World War I, showed that energy was transferred in discrete amounts, not continuously. They helped to demonstrate the "quantized" nature of energy, which had been postulated by Niels Bohr (1885–1962). They did not immediately see this "quantum" connection, nor did they recognize the fundamental significance of Bohr's work. But as Franck later said, they followed "many a false trail and roundabout path" before finding the direct path provided by Bohr's theory. The work of Franck and Hertz appeared to be the experimental proof of a fundamental concept in the new quantum physics. For this work, the two men shared the 1925 Nobel Prize in Physics.

Like many of his colleagues, Franck was pained to see the Nazis rise to power in Germany in 1933. Franck was Jewish, but because of his scientific renown he was able to keep his job while others left the country. But faced with the expectation that he would dismiss coworkers because of their ethnic background or political beliefs, Franck de-

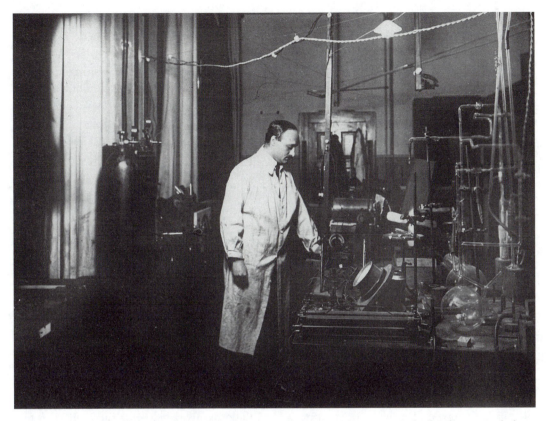

James Franck, the German physicist awarded the Nobel prize in 1925, in his laboratory. Franck is best known outside the realm of physics as the author of the Franck Report. This document was written under the veil of secrecy that covered the U.S. atomic bomb project. (Bettmann/Corbis)

cided to resign his professorship at the University of Göttingen. Soon he left Germany, first joining Niels Bohr in Copenhagen, and then accepting a professorship at Johns Hopkins University in the United States. During World War II, Franck joined many of his émigré colleagues in the secret project to build the atomic bomb.

Franck is best known outside the realm of physics as the author of the Franck Report. This document was written under the veil of secrecy that covered the U.S. atomic bomb project. Many of the scientists who worked on the bomb believed that they were doing so because Adolf Hitler (1889–1945) himself was pursuing such a weapon. After Germany surrendered, Franck and others drafted a report arguing against its use against Japan.

This new weapon would have ramifications beyond its military use, and the scientists urged the government to consider carefully the long-term political consequences of using the atomic bomb without warning. The writing of the Franck Report marked the beginning of a growing consciousness in the United States of the connections between science and social responsibility.

See also Manhattan Project; Physics; Quantum Theory; Social Responsibility; Szilard, Leo
References
Franck, James. "Transformations of Kinetic Energy of Free Electrons into Excitation Energy of Atoms Impacts" (Nobel Lecture, 11 December 1926). In *Nobel Lectures, Physics, 1922–1941.* Amsterdam: Elsevier, 1964–1970, 98–108.

Kuhn, H. G. "Franck, James." In Charles Coulston Gillispie, ed., *Dictionary of Scientific Biography,* vol. V. New York: Charles Scribner's Sons, 1972, 117–118.

Price, Matt. "The Roots of Dissent: The Chicago Met Lab and the Origins of the Franck Report." *Isis* 86:2 (1995): 222–244.

Freud, Sigmund

(b. Freiberg, Moravia [later Czechoslovakia], 1856; d. London, England, 1939)

Sigmund Freud was the founder of psychoanalysis and proved to be the most influential writer about the unconscious mind in the twentieth century. He received a medical degree from the University of Vienna in 1881 and took a position as a doctor in a hospital. He also set up a private practice to treat psychological disorders such as hysteria; from his patients came the evidence used for many of his theories about human psychology. Freud eventually believed that he was creating a new science, and he attempted to keep his work as "scientific" as possible, tying it to biology and physiology when he could. He even applied the law of conservation of energy, from physics, to mental processes. Yet he also determined that psychology required its own vocabulary, because its largely unexplored territory resisted simple identification with physical or biological processes. Freud turned to data that could not be quantified: dreams and fantasies. Subjecting such phenomena, which existed purely in the mind, to rigorous analysis formed the basis of psychoanalysis.

In the late nineteenth century, Freud worked to treat victims of hysteria, using techniques such as hypnosis. He and his colleague Josef Breuer (1842–1925) determined that many neuroses originated in traumatic experiences from early life that had somehow been forgotten. Through his clinical practice, Freud recorded what he came to call the workings of the "unconscious" mind. Part of the role of psychoanalysis, when treated as a therapy, was to bring the elements of the "unconscious" mind into consciousness, to better understand the conflicts that influence human thought and activity without the person knowing it. In 1900, he published *The Interpretation of Dreams,* which became a seminal work in the history of psychoanalysis. Not only did it include his ideas about the unconscious and the conscious, it also revealed Freud's tendency to view many psychological conflicts as rooted in sexuality. After establishing dream interpretation as the cornerstone of psychoanalysis, Freud published *Three Essays on the Theory of Sexuality* in 1905. Here he analyzed the development of the libido (sexual drive), drawing connections between its development and the formation of character traits. The libido, according to Freud, was the most important natural motivating force in life.

Freud's emphasis on sexuality alienated many of his colleagues, including Breuer, and initially his work was not received with enthusiasm. He described human instincts, of which there were many, in two general categories: life (Eros) instincts and death (Thanatos) instincts. The death instincts included destructive impulses and aggression, whereas the life instincts were oriented not only toward self-preservation but also toward erotic desire. By defining these categories broadly, Freud gave an unprecedented importance to sexuality, which seemed scandalous at the time. Gradually, however, his ideas became influential, and he embarked on a lecture tour in the United States, which culminated in his 1916 book, *Five Lectures on Psycho-Analysis.* Freud's work entered popular culture, with references to the unconscious mind—"Freudian slips" referred to misstatements that perhaps reflected unconscious desires, and the "Oedipus complex" came to describe father-son antagonism as competition for the love of the wife/mother.

In his 1923 book, *The Ego and the Id,* Freud extended these ideas further, constructing a

The founder of psychoanalysis, Sigmund Freud proved to be the most influential writer about the unconscious mind in the twentieth century. (Library of Congress)

theory of the mind based on the id, the ego, and the superego. The id represented instinct, the great sexual motivating force. The superego was based on external influences throughout life, and sought to control or limit the desires of the id. Both of them were unconscious. The ego was the conscious mind, representing the tension between instinct and control, the "self" that must satisfy the demands of each. The key concepts of Freudian psychoanalysis derived from the dynamic relationships among these three aspects of the mind. "Repression," for example, occurs when a strong instinctive desire comes into conflict with an even stronger value of the superego; to avoid traumatic conflict, the desire is pushed into the unconscious. A boy's erotic desire for his mother is the classical example of this; when the superego finds such thoughts detestable, it is repressed.

Freud's conception of psychoanalysis, despite its influence, sparked rival schools of thought and competing interpretations. In particular, two of his most noted followers, Alfred Adler (1870–1937) and Carl Jung (1875–1961), ultimately broke with him and developed their own interpretations of the meaning of the unconscious. All of them believed that they were laying the groundwork for a new science. Freud continued his work in Vienna well into the 1930s. However, the politics of Austria swayed toward Nazism in the late 1930s, culminating in annexation by Germany. Because he was a Jew, Freud decided to leave Austria and move to England. He died of cancer there in 1939.

See also Jung, Carl; Psychoanalysis; Psychology

References
Gay, Peter. *Freud: A Life for Our Time.* New York: W. W. Norton, 1988.
Jones, Ernest. *The Life and Work of Sigmund Freud, vol 2: Years of Maturity, 1901–1919.* New York: Basic Books, 1955.
Sulloway, Frank J. *Freud, Biologist of the Mind: Beyond the Psychoanalytic Legend.* New York: Basic Books, 1979.
Wollheim, Richard. *Sigmund Freud.* New York: Viking Press, 1971.

G

Game Theory

Game theory sprung from efforts by mathematicians to identify rational choices in different scenarios. In 1921, Émile Borel (1871–1956) of France published on *la theorie du jeu* (the theory of the game), in which he discussed aspects of bluffing in poker. Also in the 1920s, Hungarian mathematician John Von Neumann (1903–1957) took up the question of determining best strategies; his 1928 article, "Zur Theorie der Gesellschaftspiele" ("On the Theory of Party Games"), provided a proof of the minimax theorem, attracting great interest to game theory. Through Von Neumann's work, especially after he immigrated to the United States prior to World War II, game theory became a major field of study for mathematicians, and game theory has been applied widely to myriad aspects of society, such as economic theory and military strategy.

Von Neumann developed the minimax theorem to provide a rational choice in zero-sum games, in which the total amount to be won is static; in other words, a gain for one player is a loss for the other. Such games involve only two players. Von Neumann reasoned that, in situations where one player is at a slight disadvantage and must choose a rational action, one can identify an equilibrium choice that fits the self-interest of both players. Player A knows that Player B will always be able to minimize Player A's gains and maximize his own. Given the range of options that Player A has, he always will choose the one that maximizes what would be left *after* Player B acts to minimize it. This is the essence of Von Neumann's minimax theorem. Player A has to avoid the worst and salvage what he can; in a zero-sum game, this also means minimizing the benefits to one's opponent. The rational choice is determined by both players' self-interest. Take a familiar example: The best way to divide a piece of cake between two children is to ask one to cut the piece (Player A) and the other to choose (Player B). Player A would never cut himself a huge piece, although it might be tempting, because Player B would choose it for himself. In fact, Player A always will end up with the smaller portion, but by cutting the cake as evenly as possible, he will maximize his minimum, and also minimize his opponent's maximum.

Von Neumann felt that game theory could have applications in human interactions, particularly economics, where "utility" already was a crucial aspect of predicting what people and markets would do. He found a sympathetic listener in economist Oskar Morgenstern (1902–1976). They published their lengthy (some 600 pages) book, *Theory of*

Games and Economic Behavior, in 1944. The relationship between economic principles and game theory had never been explicitly outlined, and this book became a classic of game theory.

Some mathematicians believed that game theory could be developed further, and equilibrium solutions reached for many situations (or "games"). These ideas were pursued vigorously not only in university settings but also in institutions such as the RAND Corporation, with close ties to the military. The most challenging games would be those in which both players do not play in turns, but rather they must make their decisions simultaneously; such games grow more complex when they are not zero-sum, and the payoffs for each player are uneven. These scenarios require an analysis of the self-interest of each player and an evaluation of whether one's opponent will cooperate or cheat. Massachusetts Institute for Technology mathematician John Nash (1928–), a consultant at the RAND Corporation, determined that equilibrium solutions could be determined even for games that were not zero-sum and were played simultaneously. Nash found that "equilibrium points" could be identified in rational players' strategies given a range of divergent outcomes.

Game theory captivated mathematicians, politicians, and military leaders in the years following World War II. As the political confrontation with the Soviet Union grew more intense, U.S. theorists started to cast their views of global strategy in terms of game theory. Assuming that the United States and Soviet Unions were rational, what should each do? The simplified Cold War view divided the world into the "free world" and the "Communist world," each competing for the countries on the globe. Losses for one side automatically were gains for the other. In the parlance of game theory, this seemed like a zero-sum game, where the total amount to be won always remained divided between the two camps. In such a game, cooperation is not possible. Bertrand Russell (1872–1970), a mathematician, philosopher, and (occasionally) pacifist, observed in the early years of the Cold War that the United States might be better off declaring war early; it would be a disaster, but without consequences as horrible were the United States to wait until both countries had larger nuclear stockpiles. Preventive war appeared to be a natural application of the minimax theorem.

See also Cold War; Mathematics; Russell, Bertrand

References

Macrae, Norman. *John Von Neumann.* New York: Pantheon, 1992.

Nasar, Sylvia. *A Beautiful Mind: A Biography of John Forbes Nash, Jr., Winner of the Nobel Prize in Economics, 1994.* New York: Simon & Schuster, 1999.

Poundstone, William. *Prisoner's Dilemma: John Von Neumann, Game Theory, and the Puzzle of the Bomb.* New York: Anchor Books, 1992.

Weintraub, E. Roy, ed. *Toward a History of Game Theory.* Durham, NC: Duke University Press, 1992.

Gamow, George

(b. Odessa, Russia, 1904; d. Boulder, Colorado, 1968)

George Gamow began his career as a physicist in the Soviet Union, attracted to the new theoretical physics coming from Germany, such as Albert Einstein's (1879–1955) theories of relativity and others' work on quantum physics. He traveled to Göttingen, Germany, in 1928, where he made a name for himself through a peculiar theory of alpha particles. According to the prevailing view of the atom, and according to classical physics, it seemed impossible for an alpha particle to be able to escape the nucleus. Its positive charge seemed to exclude the possibility of it moving through the barrier surrounding the nucleus. And yet scientists had detected alpha decay, or the release of alpha particles. This presented a major contradiction between theory and observation. Gamow's accomplishment was in explaining this process through quantum mechanics, which describes the particle's motion as a complicated wave function and, more important, provided a mathematical demonstration of the possibility of movement through the barrier. In characteristic lighthearted imagery, Gamow described this as the alpha particle "tunneling through" the barrier, thus allowing for alpha decay.

Gamow's subsequent career spanned numerous leading laboratories in several countries: He worked with Danish physicist Niels Bohr (1885–1962) at the Copenhagen Institute for Theoretical Physics and with Ernest Rutherford (1871–1937) at the Cavendish Laboratory in Cambridge, England; later he worked briefly at the Pierre Curie Institute in Paris. He periodically returned to the Soviet Union, but in 1931 he was denied a visa to leave again. After two years as a professor of physics at the University of Leningrad, he managed to get permission to take his wife to attend the Solvay Conference (for physics) in Brussels, Belgium. Once safely away from his country of birth, he decided to use the conference as an opportunity to leave the Soviet Union permanently. In 1934 he became a professor of physics at George Washington University, in Washington, D.C., where he remained for over two decades.

Although Gamow continued to work briefly on nuclear theory (he published on beta decay in the mid-1930s), most of his subsequent efforts turned toward astronomy and cosmology. Gamow applied his knowledge of nuclear processes to the evolution and energy production of stars. He approved of the theory of the expanding universe espoused by Edwin Hubble (1889–1953), and the "big bang" idea formulated by Georges Lemaître (1894–1966). Gamow proposed the existence of a primordial state of the universe prior to the "big bang," which he named "ylem" and described as a mixture of protons, neutrons, electrons, and high-energy radiation. He believed that an explosion would have given birth to light elements in the first few moments after the "big bang," as temper-

George Gamow in his apartment at the University Club, revising material for a college textbook. (Bettmann / Corbis)

atures cooled enough to keep subatomic particles together. He showed his sense of humor when in 1948 he published his views with colleagues Ralph Alpher (1921–) and Hans Bethe (1906–). Although Bethe had nothing to do with the work, he let his name be used to allow it to become known as the Alpher-Bethe-Gamow article. This joke makes the work memorable, because it resembles the first three letters of the Greek alphabet: alpha, beta, gamma.

This lighthearted approach to science led Gamow to publish many popular works that tried to put difficult scientific ideas into terms readily understandable to a lay audience. He wrote nearly thirty books, most of them written to popularize science. For these efforts he was awarded the Kalinga Prize in 1956 by the United Nations Educational, Scientific, and Cultural Organization.

See also Big Bang; Physics; Quantum Mechanics; Solvay Conferences; Soviet Science

References:

Kragh, Helge. *Quantum Generations: A History of Physics in the Twentieth Century.* Princeton, NJ: Princeton University Press, 1999.

North, John. *The Norton History of Astronomy and Cosmology.* New York: W. W. Norton, 1995.

Stuewer, Roger H. "Gamow, George." In Charles Coulston Gillispie, ed., *Dictionary of Scientific Biography,* vol. V. New York: Charles Scribner's Sons, 1972, 271–273.

Genetics

The history of genetics in the twentieth century began with the rediscovery of Gregor Mendel's (1822–1884) laws of inheritance by researchers such as Hugo De Vries (1848–1935). Gregor Mendel lived and died in obscurity, in the nineteenth century. His experiments with garden peas demonstrated that some characteristics did not blend, but rather were transmitted whole from one generation to the next, even if they did not visibly manifest themselves. A short variety of pea, for example, could be bred with a tall variety and the result might be tall or short, but not a blend of the two. One of the characteristics would dominate the other, leading to all the offspring plants being tall, but both "tall" and "short" characteristics would be passed on. After crossbreeding the offspring plants, about three-fourths of the next offspring would be tall, and a fourth of them would be short. This 3:1 ratio became the cornerstone of Mendelian genetics, because it assigned a mathematical law to the inheritance of characteristics.

The immediate controversy about genetics arose from disagreements between geneticist William Bateson (1861–1926) and the biometricians Karl Pearson (1857–1936) and Walter F. R. Weldon (1860–1906). It was Bateson who coined the term *genetics* to describe the mathematical laws of inheritance, and the term *gene* came into currency over subsequent years to describe the carrier of characteristics, though the term often was defined vaguely. The main point of contention was variation in evolution: Was it continuous, as the biometricians would have it, or did it proceed in "jumps," or mutations, as the geneticists insisted? The methodology of the biometricians was statistical; they hoped to show that natural selection based on continuous variation could act on large populations. Geneticists denied the possibility. The intensity of the conflict was fueled by personal animosity between Bateson and Weldon, deriving in part from their institutional bases and their career ambitions, and their argument created a rift between the Darwinians and the Mendelians.

Despite the important work of the biometricians, the discontinuous nature of change was strengthened by genetics studies. After making Mendel's work widely known, De Vries proposed a theory about discontinuous change, noting that mutations could be responsible for producing changes in populations, and those mutations would then be subject to Mendelian laws of inheritance. Wilhelm Johannsen (1857–1927), who used the term *gene* in 1909, also conducted studies of hybridization. He had identified in 1903 what he called "pure lines," which were concealed by the range of perceived variability in any given hybrid. Pure lines did not change, but individual organisms might appear different because of environmental factors. This concept diminished the importance of perceived continuous variation, dismissing it as ephemeral. Most variation, according to Johannsen, was within the range of variability for any given pure line, and thus was not truly fundamental change. Only through mutation—the creation of a new pure line, with its own range of variation—could a true change occur. Johannsen's 1905 *Elements of Heredity* became an influential reference work strengthening Mendelism against the advocates of continuous change.

In the United States, genetics received a serious boost from the work of Thomas Hunt Morgan (1866–1945) and his colleagues in the "fly room" at Columbia University. The Carnegie Institution of Washington funded this work beginning in 1906, and the work

Thomas Hunt Morgan with a microscope in his laboratory, with diagrams of fruit flies on the wall. (Bettmann/Corbis)

continued when Morgan moved to the California Institute of Technology in 1928. Experiments with the fruit fly, or *Drosophila melanogaster,* demonstrated the transmission of characteristics with the expected Mendelian ratios, while also revealing mutations in the flies and the action of natural selection upon their survival. Morgan and his students, such as Alfred H. Sturtevant (1891–1970), became the leading figures in modern genetics in the 1910s and 1920s. Morgan's work was crucial in understanding the roles of chromosomes, microscopic entities that he believed were the carriers of genetic information. Morgan noted that these chromosomes were mobile and could segregate and pair off; the ramifications of this phenomenon became the focus of most genetics research until the 1930s.

The possibility of seeing natural selection at work on the fruit flies made Morgan's work the beginning of the reconciliation between Darwinism and genetics. With the waning influence of earlier antagonists such as Bateson, Pearson, and Weldon, new researchers bridged the gap between these two fields, finding ways to incorporate saltative (discontinuous) change into a population-scale conception of Darwinian natural selection. Theodosius Dobzhansky (1900–1975) joined Morgan's group in 1927, having come from the Soviet Union, and began work on bringing the two views together. One of his principal findings was that the large number of genes possessed by humans could provide an adequate explanation of adaptation to changing environments. His ideas culminated in the 1937 work, *Genetics and the Origin of Species.* Along with population geneticists Sewall Wright, Ronald A. Fisher (1890–1962), and J. B. S. Haldane (1892–1964), Dobzhansky helped to create a new synthesis based largely on the methods of two previous archenemies, namely, the biometricians and the geneticists, allowing reconciliation between Darwinian and Mendelian outlooks.

In the 1940s, work on genetics was not exclusively concerned with developmental issues. Even the traditional species of study was abandoned. In 1941, researchers led by George Beadle (1903–1989) turned away from fruit flies and explored the connections between genetics and biochemistry. Working with *Neurospora,* a mold, he produced biochemical mutants (through irradiation) and observed how the new mutants disrupted the expected chemical reactions. Beadle and colleagues Edward Tatum (1909–1975) and Joshua Lederberg (1925–) later won the 1958 Nobel Prize in Physiology or Medicine for their work, which showed how genes act as regulators in biochemical processes. The concept that the function of a gene is to direct the formation of a single enzyme, formulated by Beadle and Tatum, became known as the "one gene, one enzyme" principle. Other important discoveries followed. In 1944, Barbara McClintock (1902–1992) discovered "jumping genes," a name given the phenomenon of genes reconfiguring themselves on chromosomes (she won the 1983 Nobel Prize in Physiology or Medicine for this work). In 1943, X-ray diffraction techniques yielded the first visual representation of deoxyribonucleic acid (DNA). In 1944, Oswald Avery (1877–1955), Colin MacLeod (1909–1972), and Maclyn Mc-Carty (1911–) identified DNA as the agent controlling the nature of cells (rather than proteins, as previously believed). DNA thus became the key to understanding how biological specificity is transmitted from parent to offspring. They suggested that genes were in fact made from DNA. The 1953 identification of DNA's structure as a double helix by James Watson (1928–) and Francis Crick (1916–), along with their outlines of the pathways of genetic transmission, provided powerful tools for geneticists in the second half of the twentieth century.

See also Bateson, William, Biometry; Chromosomes; DNA; Evolution; Johannsen, Wilhelm; McClintock, Barbara; Morgan, Thomas Hunt; Mutation; Rediscovery of Mendel

References
Keller, Evelyn Fox. *The Century of the Gene.* Cambridge, MA: Harvard University Press, 2000.
Kohler, Robert E. *Lords of the Fly: Drosophila Genetics and the Experimental Life.* Chicago: University of Chicago Press, 1994.
Peters, James A., ed. *Classic Papers in Genetics.* New York: Prentice-Hall, 1959.
Provine, W. B. *The Origins of Theoretical Population Genetics.* Chicago: University of Chicago Press, 1971.
Sturtevant, A. H. *A History of Genetics.* New York: Harper & Row, 1965.

Geology

Geology, the study of the earth, was a branch of natural history. Its sources of information were rocks and the fossils found in them. The majority of the important controversies in geology were debated before the twentieth century, particularly debates about the development of the earth. The eighteenth-century concepts of Neptunism and Vulcanism, for example, each posited different visions of geological development, one based on peaceful sedimentation from a primordial ocean and the other based on violent change with its source in the earth's interior. By the end of the nineteenth century, most geologists accepted the view of Charles Lyell (1797–1885) that geological history was quite long, that the earth's changes still were in progress, and that such change was gradual, requiring no major cataclysms to produce meaningful change.

By 1900, geological time was somewhat established, but controversial. The geological epochs were recorded in strata, in the differing kinds of rocks layered on top of each other. These strata were most obvious in mountainous regions where portions had fallen away, exposing the strata like layers of a cake, or similarly in canyons where erosion provided very clear evidence of stratification over time. Although the periods—Quaternary, Tertiary, Cretaceous, etc.—had been defined already, the relative life spans of each had not. The only reliable source on any estimation of earth ages was William Thompson's (Lord Kelvin) (1824–1907) nineteenth-century calculation of the age of the entire earth, based on the recently formulated laws of thermodynamics. Because the earth released heat, it was in the process of cooling; its present age could be calculated based on the rate of heat flow. This proved controversial because Kelvin's estimate was very short—too short to fit with the long-term, gradualist views both of geological change and the evolution of species. The controversy did much to discredit geology, because geologists' own views contradicted the laws of physics. But the age of the earth and the age of particular rock strata were revised after the discovery of radioactivity, and radioactive dating was developed over the first decades of the twentieth century. The presence of radioactivity required revision of Kelvin's estimate, because radioactivity was a heat source that had not been known and had not entered into his calculations; thus the age

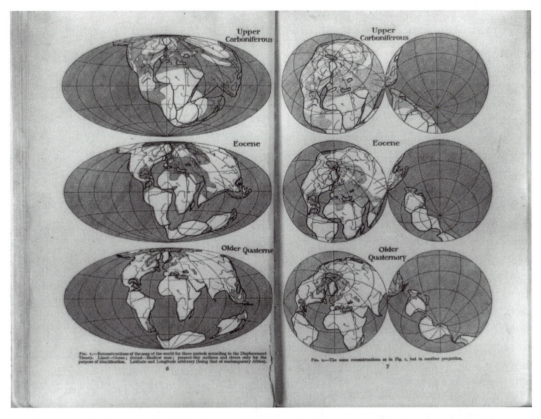

The drawings in Alfred Wegener's book, The Origin of Continents and Oceans, *tried to show the jigsaw fit of continents. (Library of Congress)*

of the earth could be extended dramatically. The discovery of radioactivity saved geological theories from the constraints placed on them by nineteenth-century physics.

Another controversy was the theory of continental drift, which was firmly rejected by most major geologists. Alfred Wegener (1880–1930) believed that the apparent jigsaw fit of South America and Africa was no coincidence and proposed in 1915 that there had once been a giant supercontinent called Pangaea (meaning "all-earth"). Most dismissed his views, largely because there was no known mechanism for moving such huge masses of land horizontally across the surface of the earth. Once again it was physics that stood in the way, and British physicist Harold Jeffreys (1891–1989) derided geologists who tried to jostle continents without due atten-

tion to the laws of physics. The exceptions to such skeptics were British geologist Arthur Holmes (1890–1965) and South African geologist Alexander Du Toit (1878–1948), both of whom argued in the 1930s that the objections to Wegener's ideas could be overcome. Holmes argued that if the heat flow of the earth's interior could be considered as convection rather than conduction, thus requiring the physical movement of hot masses, convection currents in the earth might provide a mechanism for continental motion. Du Toit, working in South Africa, noted the fossil similarities between those in his homeland and those in countries he visited in South America and was convinced that the two continents once had been connected. In 1937, he published a book entitled *Our Wandering Continents,* in which he proposed the prehistoric existence of two vast supercontinents (differing from Wegener, who proposed only one), called Gondwanaland and Laurasia. Du Toit's evidence failed to attract the support of leading geologists in Europe and North America.

Aside from these controversies (the age of the earth and continental drift), most of geology during the first half of the twentieth century was descriptive and local, entailing few all-encompassing theoretical contributions. In the United States, for example, the Geological Society of America had been founded in 1888, and it continued to support descriptive work for the benefit of both science and society, usually for business interests. One of the strongest fields of endeavor was economic geology, which was oriented toward surveys of minerals and petroleum reservoirs. Such work was well funded and pervasive, especially in the United States, where nineteenth-century geological surveys had only begun to assess the vast natural resources available to Americans. Some of the research used new geophysical methods, such as seismology, to penetrate the interior of the earth.

Many of the fundamental changes in geology occurred after World War II, when geophysical techniques were combined with the relatively new field of marine geology to harvest interesting and often baffling information about the ocean floor. MIDPAC, one of the first deep-sea expeditions to emphasize marine geology, was launched by the Scripps Institution of Oceanography (in California) in 1950. The scientists were surprised to find that the sea floor was younger than previously thought, making it geologically distinct from the continents. A major reevaluation of the nature and origins of the sea floor then began, which would in the next two decades revive of the idea that continents might be mobile.

See also Age of the Earth; Carbon Dating; Continental Drift; Geophysics; Jeffreys, Harold; Seismology; Oceanography; Patronage; Wegener, Alfred

References
Burchfield, Joe D. *Lord Kelvin and the Age of the Earth.* New York: Science History Publications, 1975.
Hallam, A. *Great Geological Controversies.* New York: Oxford University Press, 1990.
Menard, Henry W. *The Ocean of Truth: A Personal History of Global Tectonics.* Princeton, NJ: Princeton University Press, 1986.
Oldroyd, David R. *Thinking about the Earth: A History of Ideas in Geology.* Cambridge, MA: Harvard University Press, 1996.

Geophysics

Geophysics, the physics of the earth, explores processes in the earth and the forces that cause them. In the early twentieth century, studies in geophysics typically concentrated in three areas: seismology, magnetism, and gravity. Often the scientific efforts were devoted to measurement rather than comprehensive study, and such activities more aptly were called geodesy, which combined measurement of the earth with applied mathematics. But the two went hand in hand, and the end goal was twofold: better understanding of the earth's interior, and better use of science to exploit resources.

Seismology initially was strongest in Germany. Seismologists used the speed and triangulation of sound waves to understand the composition of the earth. At the University of Göttingen, Emil Wiechert (1861–1928) had proposed in the 1890s that the earth possessed an iron core. German imperialism helped to create a seismological network, connecting Germany to Samoa and parts of China; the Germans also coordinated their work with U.S. stations. One of Wiechert's students, Beno Gutenberg (1889–1960), in 1914 developed a three-part model of the earth, with two shells surrounding a core. Wiechert strove to expand the study of seismology by creating a worldwide network of seismology stations, as well as creating the Geophysical Institute at the University of Göttingen. Seismological stations were built in other countries as well, including Japan, Russia, and the United States; in the last, Jesuit scholars played the leading role in constructing a useful seismic network.

Another important pursuit in geophysics was the investigation of terrestrial magnetism. British physicist Arthur Schuster (1851–1934) attempted to evaluate its cause in the first decades of the century, believing that the answer lay in understanding the differences between the earth's geographic poles and magnetic poles. He sought to identify the relationship by simulating in a laboratory the pressures at work in the earth, but he arrived at no conclusions. In 1919, Joseph Larmor (1857–1946) proposed that the sun's magnetism could result from convection deep within it, causing it to act like a giant dynamo, generating electric currents (and thus an electromagnetic field). Applying this theory to the earth generated pro-dynamo and anti-dynamo theories in the 1920s. The anti-dynamo theorists carried the day until 1939, when German émigré (then in the United States) Walter Elsasser (1904–1991) began to suggest persuasively that the earth's core could indeed produce a dynamo effect.

Beginning in the 1920s, scientists increasingly used geophysical methods to exploit the earth's resources, not just for mining purposes but also in search for sources of oil. Not only were seismic techniques used, but also a simple pendulum apparatus could measure gravitation and thus gain further understanding about the character of subsurface rocks and potential reservoirs of petroleum. Several types of gravimeters were used in the 1930s, including ones based on gas pressure; their technology developed quickly, spurred by the financial opportunities of oil prospecting. Dutch geophysicist Felix Andries Vening Meinesz (1887–1966) even used a pendulum gravimeter aboard a submarine on an expedition in 1928 backed by the U.S. Navy, which hoped to avoid reliance on oil imports in time of war.

Training in geophysics was a somewhat haphazard process; only after World War II would geophysics itself become a common independent field of study. Those interested in the topic typically had been educated in other fields, such as physics or geology; particular emphasis depended on the traditions of the national setting. German geophysicists focused on physics and magnetism, for example, while the British tended to focus on earth structure or measurements of the earth's surface—geodesy. In North America, the approach was largely seismological, owing especially to the number of earthquakes in the vicinity of the California Institute of Technology in Pasadena and the University of California in Berkeley. Often geophysical research was relegated to geology departments, where it was part of the "petroleum geology" curriculum. Despite growing appreciation for geophysics, by the 1930s few universities offered broad training in it, except in Germany; in fact, most of the leaders in geophysics in the 1940s and later could trace their intellectual lineage back to a major German university.

Organizations to facilitate collaboration among scientists in disparate geographic regions were created in the early twentieth century. Some of these were based on imperial relationships, as in the case of Germany's seismological network. Others were based on the idea of international cooperation. The International Association of Seismology was founded in 1899, and its first international conference was held two years later. After World War I, the International Union of Geodesy and Geophysics was established (1919) to promote cooperation and forge intellectual ties across political boundaries. In 1932–1933, scientists organized the second International Polar Year in order to carry out simultaneous investigations of geophysical studies (and other kinds of scientific work) at the poles. In 1950, another such venture was conceptualized by U.S. and British scientists as a way to use new technology developed during World War II, and to promote international cooperation. This project, called the International Geophysical Year, would in 1957–1958 encompass a broad range of geophysical observations throughout the world and include more than sixty nations.

See also Colonialism; Continental Drift; Geology; Gutenberg, Beno; Mohorovičić, Andrija; Patronage; Richter Scale; Seismology

References
Doel, Ronald E. "Geophysics in Universities." In Gregory A. Good, ed., Sciences of the Earth: An Encyclopedia of Events, People, and Phenomena. New York: Garland Publishing, 1998, 380–383.
Oreskes, Naomi. "Weighing the Earth from a Submarine: The Gravity Measuring Cruise of the U.S.S. S–21." In Gregory A. Good, ed., The Earth, the Heavens, and the Carnegie Institution of Washington. Washington, DC: American Geophysical Union, 1994, 53–68.
Parkinson, W. Dudley. "Geomagnetism: Theories since 1900." In Gregory A. Good, ed., Sciences of the Earth: An Encyclopedia of Events, People, and Phenomena. New York: Garland Publishing, 1998, 357–365.
Pyenson, Lewis. Cultural Imperialism and Exact Sciences: German Expansion Overseas, 1900–1930. New York: Peter Lang, 1985.

Gödel, Kurt

(b. Brünn, Austria-Hungary [later Brno, Czech Republic],1906; d. Princeton, New Jersey, 1978)

For contributions in the fields of logic and mathematics, few thinkers of the twentieth century could rival Kurt Gödel. He received his doctorate from the University of Vienna in 1929, studying under Hans Hahn (1879–1934). The next year he proposed a theorem that challenged the prevailing notion that all mathematics could be reduced to axioms. His ideas informed the development of not only mathematics and logic, but also computers and artificial intelligence.

At the Second Conference on Epistemol-ogy of the Exact Sciences, held in Königsberg, Germany, in 1930, Gödel first publicly announced his "incompleteness" theorem. It would soon make him well known beyond the confines of Vienna. The theorem held that any axiomatic system must contain propositions that cannot be proven (or refuted) using the rules of the system. This held true not only for complex mathematics, but also for the most basic whole number arithmetic. Mathematicians had assumed that the elusive logical proofs of mathematical relations within any such system could eventually be found, using the rules of the system itself. Leading mathematician David Hilbert (1862–1943) advocated "formalism," a methodology that avoids reference to implications or meanings outside of a given system. But Gödel's theorem indicated the opposite; for him, logic forbade the possibility of proving the truth or falsity of all statements using only the terms of the system itself.

Gödel's paper on incompleteness was published in 1931, and the results were accepted by almost everyone. But part of the reason for this was that few understood its implications. There were notably exceptions, such as John Von Neumann (1903–1957), who pulled Gödel aside during the initial conference to discuss the ramifications for mathematical proofs. Gödel's challenge to Hilbert's formalist program set Von Neumann thinking about the nature of rationality, and Gödel was a major influence on his own work. But only slowly did other mathematicians work out the implications and ask the question: Does this mean that the outstanding problems of mathematics, which have plagued the best minds for centuries, are in fact not solvable? Or was Gödel's theorem simply a logician's trick posing no serious problems for the mathematician? At the very least, Gödel's work seemed to indicate that mathematics could not be reduced to a set of fixed axioms. Given any set of axioms, statements could

Mathematician Kurt Gödel (second from right) was the co-recipient with Julian Schwinger (right) of the first Albert Einstein Award for Achievement in the Natural Sciences. (Bettmann / Corbis)

be made that the axioms neither prove nor disprove. One controversial implication of Gödel's work was in the area of computers. Because computers must be programmed with a set of axioms, they cannot recognize some truths that are readily understood by human beings. Therefore, some have concluded, artificial intelligence always shall have severe inherent limitations.

Dissatisfied with the Nazi regime, Gödel left his home and traveled to the United States in 1940 (by way of the Soviet Union and Japan). In 1953, he became a member of the Institute for Advanced Study at Princeton University, joining such scientific luminaries

as Albert Einstein (1879–1955). He later became increasingly paranoid, convinced that someone wanted to poison him. He was afraid to eat and eventually he died of malnourishment at the unhealthy weight of some sixty pounds.

See also Computers; Mathematics; Philosophy of Science

References

Dawson, John W., Jr., *Logical Dilemmas: The Life and Work of Kurt Gödel.* Wellesley, MA: A. K. Peters, 1997.

Casti, John L., and Werner DePauli, *Gödel: A Life of Logic.* New York: Perseus, 2001.

Great Depression

In some ways, the 1930s were the golden years of science, particularly in physics. The discovery of neutrons, artificial radioactivity, and a host of other findings about the atom led to major new avenues of research, several Nobel Prizes, and ultimately the atomic bomb. But the economic hardship of the Great Depression in the 1930s also transformed the nature of scientific activity and its place in society.

Although individuals could not necessarily pursue science because of economic and political difficulties, science itself continued to inform major events. The Depression made scientific careers, like most jobs, difficult to find and pursue. Ernest Rutherford's (1871–1937) belief that scientists in Britain could still do a great deal of physics with wax and string was taken rather seriously during these lean years. Despite the economic crisis, physicists had plenty of research questions to answer, grappling with the theoretical contributions of the 1920s made largely by German scientists. Still, about 80 percent of college graduates in Germany who received technical degrees in 1932 had to find employment outside of the engineering fields for which they had been educated. The rise of Nazism, itself a product of the economic hardship of the interwar years, changed the character of science in Germany considerably, as scores of Jews were fired from their posts and fled the country. The Nazis themselves tried to reaffirm the importance of science and technology (particularly concepts such as Social Darwinism) as a means to lift Germany out of the economic and social morass in which they believed it had fallen.

In the United States, the Great Depression reshaped patronage practices and forced a reevaluation not only of scientific prestige but also of the allegedly progressive trajectory of a "scientific" society. The U.S. government in the late 1920s forged links between itself and private industries to cosponsor scientific research for the stimulation of enterprise and the national interest.

Under President Herbert Hoover (1874–1964), the Department of Commerce became a wide-ranging quasi-scientific organization. But state-supported scientific activities faltered in the early years of the Depression, cut out of the national budget by Franklin Roosevelt's (1882–1945) New Dealers. Congress cut the budgets of federal scientific agencies by some 12.5 percent on average; the National Bureau of Standards alone lost 26 percent of its budget between 1931 and 1932. The following year, the Bureau of Standards fired nearly half of its technical staff. The story was much the same in industry. About half of the research personnel at General Electric and AT&T were laid off. Patronage for science from philanthropic bodies such as the Rockefeller Foundation also suffered. Donations dropped off, and investment incomes diminished drastically. University salaries were cut, and research positions disappeared.

Along with the general hardship came a reassessment of the value of science. The association between scientific research and productivity was not questioned; but overproduction appeared to be part of the problem. For government programs, Roosevelt concentrated on those for relief, recovery, and reform. Scientists tried to convince government patrons that science could provide much-needed jobs, but they were not particularly successful. Part of the reason such arguments fell on deaf ears was the growing concern about technocracy, or rule by a scientific and technological elite. Confidence that science and progressive society went hand in hand was challenged as early as the Great War, when chemical weapons provided chilling evidence that science does not always serve the best interests of humanity. But if the Great War produced science skeptics, the hard times of the Great Depression evoked outright hostility to scientists. They found themselves confronting a general revolt against science. The application of scientific ideas, from "talkies" at the cinema to any number of mechanical inventions in the workplace, appeared to be putting people out of work. Increasing productivity, long promised by science advocates, was the last thing the world needed to pull itself out of an economic disaster in which supply far exceeded demand. Society had proved incapable of keeping up with scientific progress, and the viability of the capitalist system itself seemed at stake.

Some scientists tried to restore faith in science by revitalizing the idealism of the 1920s. One way was through world's fairs, which showcased the inventions and innovations of science of the past century. By reestablishing a sense of wonder and faith in science, scientists hoped to remind Americans of how closely intertwined U.S. values were with scientific values. Roosevelt's secretary of agriculture, geneticist Henry Wallace (1888–1965), felt that scientists bore a serious responsibility for some of the problems of the Depression. In 1933, he convinced the president to create the Science Advisory Board (SAB) to come up with the means to put science to work to solve the country's economic woes. Chaired by physicist Karl Compton (1887–1954), the SAB received no funding and lasted only two years. Compton tried to use the SAB to create a new body to fund the best science in the country, ultimately (so he reasoned) to help in such issues as land and mineral policies. Compton asked the president for $75 million, but the president insisted that 90 percent of the funds should go to people taken from the country's relief rolls. Such a requirement would hardly produce the best science, in Compton's view; the issue turned into one of elitism in science, and the idea collapsed. Although Compton's attitudes resonated among conservatives, New Dealers were more interested in programs to provide jobs and public works rather than provide money for already-employed scientists to do better research.

Increasingly, support for science needed to be tied to specific projects that connected clearly to the public's welfare. In the later years of the Great Depression, scientists were more successful in gaining federal money by tying their work to national interests. They did so by connecting their expertise to land use, soil conservation, the eradication and treatment of disease, and engineering projects. The Rockefeller Foundation tried to economize by concentrating its interests in biology. The director of its natural science division, Warren Weaver (1894–1978), reasoned that the steady breakthroughs in physics were unlikely to be hastened by Rockefeller money, but that biology needed a push. In addition, that field seemed well positioned to make genuine contributions to the welfare of mankind. More attention needed to be paid to the living parts of nature, not merely the inanimate forces of physics. This reorientation provided a boost for biological research, such as that by geneticist Thomas Hunt Morgan (1866–1945), but increased the strain already placed on the physical sciences. During the Great Depression, science for its own sake, an ideal of the 1920s, seemed to be an elitist dream with potentially disastrous consequences for society.

See also Elitism; Industry; Patronage; Social Progress; Technocracy

References

Dupree, A. Hunter. *Science in the Federal Government: A History of Policies and Activities.* Cambridge, MA: Harvard University Press, 1957.

Herf, Jeffrey. *Reactionary Modernism: Technology, Culture and Politics in Weimar and the Third Reich.* New York: Cambridge University Press, 1984.

Kargon, Robert, and Elizabeth Hodes. "Karl Compton, Isaiah Bowman, and the Politics of Science in the Great Depression." *Isis* 76:3 (1985): 300–318.

Kevles, Daniel J. *The Physicists: The History of a Scientific Community in Modern America.* Cambridge, MA: Harvard University Press, 1995.

Kohler, Robert E. *Partners in Science: Foundations and Natural Scientists, 1900–1945.* Chicago: University of Chicago Press, 1991.

Rydell, Robert W. "The Fan Dance of Science: American World's Fairs in the Great Depression." *Isis* 76:4 (1985): 525–542.

Gutenberg, Beno

(b. Darmstadt, Germany, 1889; d.
Pasadena, California, 1960)

Beno Gutenberg was the leading scientist in the relatively new field of seismology in the first half of the twentieth century. His education in this field began in Germany, where Emil Wiechert (1861–1928) had developed a course on observing geophysical phenomena with instruments. Although Gutenberg's interest in weather forecasting led him to this course, in short order he had learned everything currently known about seismology. He realized that this was a new field ripe for exploration and new discoveries. Gutenberg began research on microseisms for his Ph.D. thesis and published his first paper on them in 1910. Microseisms are disturbances in the earth's crust, but are always present and are too small in intensity to be called earthquakes. Gutenberg's work, however, showed that microseisms can be correlated to other phenomena such as oceanic waves or storms.

Gutenberg recognized that the amplitude of seismic waves could provide a powerful tool for determining the structure of the earth. One of his lasting contributions was his calculation of the existence of the earth's core. Gutenberg's mentor, Wiechert, had proposed in 1897 that the earth had an iron core, beginning at a depth of about 3,900 km. Assuming that the core could be identified by its effects on seismic waves, Gutenberg calculated the travel times of waves both reflected and refracted at the surface of the core. He established its depth at 2,900 km in 1914. Fast becoming one of the leading the seismologists in Europe, he began work at the International Association of Seismology in Strasbourg.

His work was disrupted by World War I, and with the loss of Alsace and Lorraine to France, he was unable to regain his position in Strasbourg in subsequent years. He ran his family's soap factory for some time while working on geophysical problems from home, unable to find a satisfactory position. But after accepting a professorship at the California Institute of Technology in 1930, Gutenberg expanded the field of seismology using the institute's ample resources and variety of talented scientists, among them Hugo Benioff (1899–1968) and Charles F. Richter (1900–1985). And of course, southern California's active seismic environment made it seem a natural home for Gutenberg.

With Richter, Gutenberg improved methods for recording earthquakes and identifying the location of epicenters. He also was convinced that there were major differences between the structure of the continents and that of the oceans, and was sympathetic to ideas of continental drift long before such ideas were current among most U.S. geologists and geophysicists. Working together on a series of papers on seismic waves, Gutenberg and Richter provided "a bible" for observational seismology, as one of Gutenberg's colleagues later wrote, emphasizing the foundation work they accomplished for modern seismology.

See also Geophysics; Richter Scale; Seismology

References
Knopoff, Leon. "Beno Gutenberg." In *Biographical Memoirs, National Academy of Sciences,* vol. 76. Washington, DC: National Academy Press, 1998, 3–35.
Shor, George G., Jr., and Elizabeth Noble Shor. "Gutenberg, Beno." In Charles Coulston Gillispie, ed., *Dictionary of Scientific Biography,* vol. V. New York: Charles Scribner's Sons, 1972, 596–597.

H

Haber, Fritz

(b. Breslau, Germany [later Wrocław, Poland], 1868; d. Basel, Switzerland, 1934)

Fritz Haber was both a Nobel Prize–winning scientist and a fervent patriot. He also became notorious for putting his extraordinary abilities at the service of Germany during World War I, helping it to develop chemical weapons. Eventually he was betrayed by the same country that he had fought so hard to strengthen through science.

Haber was most noted for his work in physical chemistry, a field in which he had little formal training. He pursued work that he thought would bestow practical benefits, rather than being purely of scientific interest. One such effort was combining nitrogen and hydrogen to create ammonia. His success with the "ammonia synthesis" drove Haber's subsequent research, which aimed to obtain, from widely available substances, materials of commercial or military value. By 1912 Haber firmly established himself as the leader in this field, becoming director of the Kaiser Wilhelm Institute for Physical Chemistry and Electrochemistry, in Dahlem (outside Berlin).

When war broke out in 1914, Haber put his skills to use for his country. He hoped to use chemical processes to supply the kaiser's armies with its necessities, such as a substitute for toluene as antifreeze in motor fuel. But when the War Ministry consulted scientists at the Kaiser Wilhelm Institute about methods to flush Allied armies out of their well-protected trenches, to force them to engage in open warfare, Haber and his colleagues, including Walther Nernst (1864–1941), took a fateful step toward the first development and use of chemical weapons in a time of war. These early investigations led to some unsuccessful experiments before the end of 1914, when a laboratory explosion killed physical chemist Otto Sackur (1880–1914).

Initially, scientists and military leaders envisioned irritant gases designed to move men away from entrenched positions; later, such gases were designed more explicitly to kill. By January 1915, Haber had developed chlorine gas as a weapon. Three months later, the weapon was ready for a wartime test. On April 11, German forces released chlorine gas along a 3.5-mile front near the town of Ieper, Belgium. Chemical weapons, today regarded as weapons of mass destruction, were soon developed by both sides in the conflict. Germany appointed Haber chief of the Chemical Warfare Service. His practical brand of science

German chemist Fritz Haber pioneered the techniques of chemical warfare and won the Nobel Prize in Chemistry for the synthesis of ammonia. (Hulton-Deutsch Collection / Corbis)

supplied the country not only with gas weapons, but also with a number of useful nitrogen compounds, from fertilizers to explosives.

Haber's work, deplored by many scientists outside Germany, became even more controversial when Haber was awarded the Nobel Prize in Chemistry in 1919. Although the war had ended, the choice of Haber was denounced by French, British, and U.S. scientists. By giving himself fully to his own country during the war and especially for his role in developing chemical weapons, Haber had to endure the remonstrations and rejection of his colleagues abroad. This did not stop him, however, from putting science to work for the good of Germany. During the interwar years, when Germany was saddled with reparations payments to former ene-

mies, Haber was struck with a fantastic plan: Perhaps chemists could help make these payments by extracting gold from seawater. Basing his calculations on some old data, Haber's idea led to several oceanic excursions and some improved data but not a profitable means to extract gold from the sea.

Throughout the 1920s, Haber remained very influential in Germany and became involved in international scientific organizations. Yet in the early 1930s, when the Nazis rose to power, the scientific community in Germany began to change. Haber was Jewish. Despite his devotion to Germany in war and peace, he soon found himself an outsider in an increasingly anti-Semitic society. Haber joined many eminent German scientists in resigning their posts in 1933 under pressure from the Nazis. He moved briefly to Cambridge, England, and then decided to take a leading position at a research institute in Israel. He died en route.

See also Chemical Warfare; Kaiser Wilhelm Society; Nazi Science; Nobel Prize; World War I
Reference
Goran, Morris. "Haber, Fritz." In Charles Coulston Gillispie, ed., Dictionary of Scientific Biography, vol. V. New York: Charles Scribner's Sons, 1972, 620–623.
———. The Story of Fritz Haber. Norman: University of Oklahoma Press, 1967.
Szöllösi-Janze, Margit. Fritz Haber, 1868–1934: Eine Biographie. Munich: Verlag C. H. Beck, 1998.

Haeckel, Ernst

(b. Potsdam, Germany, 1834; d. Jena, Germany, 1919)

After Thomas H. Huxley (1825–1895), Haeckel was the best-known promoter and popularizer of Charles Darwin's (1809–1882) theory of evolution. He embraced controversy and was openly hostile to religion. Originally trained as a physician, Haeckel left his work to become an apostle of biological evolution. He became a zoologist, and took a position at the University of Jena in 1865. Much of what was known about embryology during the first decades of the twentieth century derived from Haeckel's view that evolution could be proved simply by examining the embryological parallels.

In the nineteenth century, Karl Von Baer (1792–1876) proposed that developing embryos of higher animals (such as humans) passed through stages that essentially were similar to the stages of development in lower animals. Haeckel took this principle to describe a "biogenetic law" that affirmed a parallel between the embryological development of an individual with the development—or evolution—of species. Haeckel stated it as "ontogeny recapitulates phylogeny," which is not exactly what Von Baer had intended. The phrase, however, caught on. Haeckel was a great promoter of scientific ideas, particularly in the biological sciences. In addition to *ontogeny* and *phylogeny*, Haeckel coined the term *ecology*. Haeckel's recapitulation theory was his most well known.

Haeckel gained further notoriety for his involvement in the monist movement. The essence of monism was the belief that there was one spirit in all things, and that the whole of the world is derived from one fundamental law. For Haeckel, only science could provide that law. U.S. biologist David Starr Jordan (1851–1931) criticized the creed in 1895: "It is an outgrowth from Haeckel's personality, not from his researches" (Jordan 1895, 608). The creed appeared in English in 1901, in Haeckel's book *The Riddle of the Universe*. He emphasized the unity of nature, living and nonliving. In 1906, Haeckel founded the Monistenbund (Monist Alliance) to spread his idea, which had far-reaching philosophical implications, such as the rejection of a supernatural God and the soul. If God existed, he was one with nature.

As historian Daniel Gasman has noted, Haeckel can be seen as the founder of ideas that fueled racial antagonism and informed the National Socialist movement in Weimar Germany. He himself was a fervent nationalist and anti-Semite; his claims of universality for scientific laws might have played an important role in forging the links between "scientific" racial theories and plain racism at the turn of the century and beyond. The problem with monism was that, if there was only one fundamental law, then Haeckel's passion—evolutionary biology—could be applied to every aspect of life. Haeckel's embrace of social Darwinism lent it the scientific credibility it might have lacked without the endorsement of so prominent a man of science. Perhaps Haeckel's most significant contribution was in promoting the attitude of scientism: All things should be understood in the same terms as science. Such attitudes proved to be powerful tools in the hands of racists, nationalists, and social Darwinists.

See also Ecology; Evolution; Philosophy of Science; Race; Scientism

References

Gasman, Daniel. *Haeckel's Monism and the Birth of Fascist Ideology.* New York: Peter Lang, 1998.

Haeckel, Ernst. *The Riddle of the Universe: At the Close of the Nineteenth Century.* New York: Prometheus Books, 1992.

Jordan, David Starr. "Haeckel's Monism." *Science, New Series* 1:22 (31 May 1895): 608–610.

Hahn, Otto

(b. Frankfurt, Germany, 1879; d. Göttingen, Germany, 1968)

The German chemist Otto Hahn was one of several key figures in the history of radioactivity. He bears the fame, and the responsibility, for the discovery of nuclear fission—that is, splitting the atom. His entry into the field in 1904, at Sir William Ramsay's (1852–1916) laboratory at University College, London, was a surprising portent of things to come. Despite almost no background in radioactivity, his first project led to the discovery of a new radioelement, radiothorium (a radioactive isotope of thorium). In 1905 he traveled to Montreal to work with the leading scientist in the field, Ernest Rutherford (1871–1937), before taking up residence in Berlin in 1906.

Hahn's activities in Berlin included a productive collaboration with physicist Lise Meitner (1878–1968) that would continue for the next three decades. He took leave of radioactivity briefly during World War I, when he joined Fritz Haber's (1868–1934) chemical warfare group. He was actively involved in researching, developing, and implementing such weapons during the war. But he never left his own field of interest completely. He and Meitner discovered and named protactinium in 1917. This fruitful field became less so in the 1920s, when nearly all naturally occurring radioactive elements were identified by various researchers.

Fortunately for Hahn, radiochemistry (or, as it became known, nuclear chemistry) was stirred up again by several developments in the early 1930s. In England, James Chadwick (1891–1974) discovered the neutron; in France, the Joliot-Curies discovered artificial radioactivity; in Italy, Fermi (1901–1954) began producing radioactive material from bombardment by neutrons. Hahn and Meitner set out to identify the new products and their radioactive decay patterns. Soon Hahn's direct work with Meitner ended when, because she was Jewish, she fled Nazi Germany in 1938.

Hahn and Fritz Strassman (1902–1980) continued the experiments and found that bombarding uranium with neutrons seemed to result in a radioactive form of barium. But finding barium seemed highly unlikely, since no one yet had witnessed a reaction emitting anything heavier than alpha particles. Barium was in the middle of the periodic table! The chemist Hahn wrote to his physicist colleague Meitner, now in Sweden, about the unexpected results. It was Meitner, working with her nephew Otto Frisch (1904–1979), who in 1939 determined that the results could be explained by the "splitting" of the uranium nucleus. They named this process fission, and determined that the mass of two barium atoms was slightly less than the mass of a uranium atom and a neutron. Thus, they reasoned, some energy would be released as well. Hahn's Nobel Prize, awarded in 1944 for his discovery of fission, might well have been shared with Meitner.

News of Hahn and Strassman's achievement spread rapidly in the international physics community. Danish physicist Niels Bohr (1885–1962) brought the news to U.S. physicists, who immediately began to work out the ramifications. The release of energy, realized scientists such as Enrico Fermi (then at Columbia University), would be enormous if a fission chain reaction could be sustained. Although the Americans were not the only ones to realize this, they became the first to apply Hahn's discovery by building an atomic bomb.

Hahn was arrested by Allied troops in 1945 and sent to England, where he and several colleagues were interned and put under surveillance for over six months. His conversations with other physicists such as Werner Heisenberg (1901–1976) were taped, and the transcriptions helped the United States and Britain understand the strengths and weaknesses of wartime German nuclear physics. Only then did he learn that his discovery of fission had led to the use of atomic weapons against the Japanese cities Hiroshima and Nagasaki. After the war, he became president of the Kaiser Wilhelm Society (soon renamed the Max Planck Society), and in the 1950s he publicly warned against the dangers of atomic energy.

See also Atomic Bomb; Fission; Heisenberg, Werner; Meitner, Lise; Nazi Science; Physics

References

Badash, Lawrence. "Hahn, Otto." In Charles Coulston Gillispie, ed., *Dictionary of Scientific Biography,* vol. VI. New York: Charles Scribner's Sons, 1972, 14–17.

———. *Scientists and the Development of Nuclear Weapons: From Fission to the Limited Test Ban Treaty, 1939–1963.* Atlantic Highlands, NJ: Humanities Press, 1995.

Shea, William R., ed. *Otto Hahn and the Rise of Nuclear Physics.* Dordrecht: D. Reidel, 1983.

Haldane, John Burdon Sanderson

(b. Oxford, England, 1892; d. Bhubaneswar, Orissa, India, 1964)

J. B. S. Haldane was a rare individual whose interests crossed many disciplines. His works, including a science book for children, were both technical and popular. As a young man, he fought in World War I, both on the Western Front and in the Middle East, and was wounded in both theaters. He helped his father, physiologist John Scott Haldane (1860–1936), to develop gas masks to protect against chemical weapon attacks by the Germans. When the war ended, he began his scientific career by teaching physiology at Oxford.

His work on human physiology made Haldane seem like a mad scientist. He consumed solutions of ammonium chloride to study the effects of hydrochloric acid in blood; he exercised to exhaustion to study the relation-

ship between carbon dioxide and muscles; and he conducted other experiments on himself and his laboratory colleagues. He studied enzymes in the 1920s and showed that they obey the laws of thermodynamics, and by 1930 he had produced a book, *Enzymes,* to provide an overview of the field.

But Haldane is most well known for his contributions to genetics. He conducted important work on gene linkage, and in 1922 he formulated Haldane's Law. The law attempted to explain cases where crossing two animal species resulted in a population of offspring in which one sex was absent, sterile, or rare. In such cases, according to the law, that sex must be the heterogamic one (in other words, that sex has two different sex chromosomes, rather than two of the same sex chromosomes).

Haldane was one of three key individuals who, through population genetics, brought about the evolutionary synthesis that reconciled Darwinism with some of the outstanding problems in biology—the other two were Ronald A. Fisher (1890–1962) and Sewall Wright (1889–1988). Like many others in the field, Haldane regarded Charles Darwin's (1809–1882) theory of evolution by natural selection as flawed. Darwin's view of continuous variation and adaptation through natural selection had serious problems. For one, how could a favorable trait survive if it blended with other traits? In a huge population, even the most favorable trait could not survive subsequent generations of breeding. But natural selection had been resurrected when scientists such as Wilhelm Johannsen (1857–1927) demonstrated that it could fit into Mendelian genetics, which described characteristics as individual and particulate, not subject to mixing. Genetics gave individual traits a kind of durability that Darwin had not articulated.

Haldane insisted that natural selection was the mechanism for evolutionary change in large populations following Mendelian laws of inheritance. He claimed that those genotypes (the totality of genes) that produced phenotypes (outward appearance) that were well adapted to their environment would tend to survive. The effects of natural selection could be quick and dramatic. The famous example of Haldane's sped-up natural selection was the dark-colored moth (now known as *Biston betularia*), which had first been noted in British industrial towns in 1848. Its color, like the soot in the dirty cities, offered protection from predators. By 1900, these dark moths almost completely replaced the previous gray version. Such a rapid pace pointed to natural selection on a massive scale, where the dark moths reproduced twice as fast as their competitors in such a friendly environment. These ideas he set forth in his 1932 book *The Causes of Evolution*.

Politically, Haldane veered to the left. He joined the Communist Party after the outbreak of civil war in Spain in 1936. He contributed more than 300 articles on popular science to the Communist *Daily Worker*. He also joined its editorial board. More surprising for a geneticist, he even (for a time) supported the work of Trofim Lysenko (1898–1976), the notorious anti-Mendelian agricultural scientist in the Soviet Union. Such left-wing politics did not disallow Haldane from working for his own country during World War II. He helped the British Admiralty conduct research on human physiology aboard submarines, such as the problems of escaping them and the operation of miniature submarines.

Haldane openly condemned his country's action during the 1956 Suez crisis, when both Britain and France occupied the canal against the wishes of Egypt. In 1957 he decided to leave England for good and settle in India. In India there were good facilities for research, so he had a justification for leaving. Once there, he established a genetics and biometry laboratory. He remained there until his death from cancer in 1964. He kept his sense of humor, writing a poem for the *New Statesman* about his cancer, including the lines "My final word before I'm done/Is 'Cancer can be rather fun.'"

See also Cancer; Chemical Warfare; Evolution; Genetics

References

Clark, Ronald W. "Haldane, John Burdon Sanderson." In Charles Coulston Gillispie, ed., *Dictionary of Scientific Biography*, vol. VI. New York: Charles Scribner's Sons, 1972, 21–23.

Mayr, Ernst, and William B. Provine, eds. *The Evolutionary Synthesis: Perspectives on the Unification of Biology*. Cambridge, MA: Harvard University Press, 1980.

Hale, George Ellery

(b. Chicago, Illinois, 1868; d. Pasadena, California, 1938)

Aside from his scientific accomplishments in astronomy, George Ellery Hale was best known as an organizer of U.S. science. He was largely responsible for the founding of three of the world's great observatories— Yerkes, Mount Wilson, and Palomar. In addition, he played a leading role in reforming the National Academy of Sciences and in creating the National Research Council, the International Research Council, and the California Institute of Technology. Hale was plagued with illness throughout much of his life. As child, he had intestinal ailments and typhoid, and as an adult he suffered three major breakdowns from what doctors called "brain congestion." This condition stemmed, they said, from his intensity and inability to relax.

Such intensity served Hale's scientific work and the scientific community very well. His early years in astronomy were devoted to understanding the physical properties of stars, rather than the distribution and motions of stars, which occupied the attention of most other researchers. While attending the Massachusetts Institute of Technology in the 1880s, he volunteered at the Harvard College Observatory. There he conceived of a new instrument, the spectroheliograph, which would allow him to photograph solar prominences in full daylight. At the time, the sun was the only star for which meaningful observations were possible, and Hale was obsessed with it. His thesis at MIT described his results, providing the basis for later solar observational astronomy. Hale never earned a doctorate, although later many honorary ones were bestowed upon him.

After Hale finished his university degree, he got married and honeymooned in California, where he visited the Lick Observatory near San Jose. Scientists there were using a thirty-six-inch telescope to observe planetary nebulae. Upon his return, Hale convinced his father to buy him a telescope of his own, a twelve-inch refractor set up in 1891 behind their house, in what became known as the Kenwood Observatory. This was the first of many successes persuading wealthy philanthropists to contribute funds for scientific purposes. The following year, Hale became an associate professor of astrophysics at the University of Chicago.

At Chicago, Hale persuaded businessman Charles T. Yerkes (1837–1905) to pay for a telescope that would surpass all others in focusing power, allowing scientists to see further into space. The Yerkes Observatory was dedicated in 1897, where Hale continued his research on solar properties and attracted astronomers from all over the world to work in the only observatory having a telescope with a forty-inch lens. During these years, Hale cofounded the journal *Astronomy and Astro-Physics* and also the *Astrophysical Journal*.

Hale never rested on his laurels and constantly was trying to secure a means to improve astronomical instruments. In 1896, his father agreed to pay for the lens of a sixty-inch reflecting telescope, on the condition that the University of Chicago would pay to mount it. The university did not fulfill its end of the bargain, and Hale began to look elsewhere. In 1902 he received a large sum—$150,000—to found the Mount Wilson Solar Observatory in the mountains above Pasadena, California. This location proved to

be a fruitful one for astronomy. Here the first photograph of a sunspot spectrum was taken in 1905. Hale and his colleagues determined from the spectrum that sunspots were cooler than the surrounding surface of the sun. In 1908, Hale showed the presence of magnetic fields in sunspots, the first observation of such a field outside the earth.

Although Hale's work on sunspots was fundamental, his influence in subsequent years was in supporting science through better facilities and more effective organization. The sixty-inch reflecting telescope became available at Mount Wilson in 1908, making it possible to study, in depth, stars other than the sun. But Hale was not satisfied and made plans to acquire a 100-inch lens through the philanthropy of local businessman John D. Hooker. This telescope saw its first use in 1917. With it, Edwin Hubble (1889–1953) was able to settle a longstanding dispute about the nature of nebulae, establishing them as separate star systems, or "galaxies." The 100-inch telescope, under pioneers such as Hubble, initiated a new phase in understanding the nature of the universe and the vast distances between galaxies.

In the first decade of the century, Hale became a trustee of Throop Polytechnic Institute, a small school in Pasadena. Hale was instrumental in transforming it into an MIT for the West Coast, but focusing as much on fundamental science as on technology. The school, soon known as the California Institute of Technology, became part of Hale's vision to see Pasadena as a center for science and culture. The institute evolved over the next two decades to become a major research and educational facility. In 1928, it received $6 million from the Rockefeller Foundation to build a 200-inch telescope, which gave rise to a new observatory on Palomar Mountain. The two southern California observatories, Mount Wilson and Palomar, worked cooperatively; they were renamed the Hale Observatories in 1969.

Hale's activities went far beyond astronomy. At the turn of the twentieth century, Hale wanted to transform the National Academy of Science from a mutual admiration society to a useful body with international prestige and influence. One of his methods was to encourage the membership of younger scientists with productive years ahead of them rather than behind them. He also encouraged the academy to participate in international ventures, although these faltered during World War I. The war spurred Hale to help create the National Research Council (NRC), a component of the academy dedicated to increasing knowledge for the benefit of national defense and public welfare. Hale wanted to use the NRC to consolidate relationships between scientists and sources of patronage, such as industry or philanthropic organizations. He believed this was an essential step if the United States expected to compete with the scientific powerhouses of Europe, especially Germany.

When World War I ended, Hale's previous sentiments of internationalism were revised considerably. Like many other scientists, he held German scientists responsible for their country's misdeeds in the war. He proposed the International Research Council as a rival to the existing International Association of Academies. The latter included Germany and its wartime allies, whereas the former excluded them. It was even conceived as a federation of national research councils, bodies devoted as much to national security as to science. The council was inaugurated in 1919, and it kept Germany out for years. In 1931, it took the name of International Council of Scientific Unions, and in 1932 Hale became its president. Hale's gifts as an organizer served his country and its scientific community well, despite his failure to keep science above international politics.

See also Astronomical Observatories;
  International Research Council; National
  Academy of Sciences; Nationalism; Patronage;
  World War I

References

Kevles, Daniel J. *The Physicists: The History of a Scientific Community in Modern America.* Cambridge, MA: Harvard University Press, 1995.

Wright, Helen. *Explorer of the Universe: A Biography of George Ellery Hale.* New York: Dutton, 1966.

———. "Hale, George Ellery." In Charles Coulston Gillispie, ed., *Dictionary of Scientific Biography,* vol. VI. New York: Charles Scribner's Sons, 1972, 26–34.

Heisenberg, Werner

(b. Würzburg, Germany, 1901; d. Munich, Germany, 1976)

Werner Heisenberg was the founder of quantum mechanics and the creator of the uncertainty principle, cornerstones of modern theoretical physics. His professional life proved controversial; not only was he the father of quantum mechanics, but he also would have been (had he succeeded) the father of the Nazi atomic bomb. He took his doctoral degree from the University of Munich in 1923, just two years before publishing the work that would elevate him to celebrity and win him the Nobel Prize.

Heisenberg was part of an extraordinary community of physicists attacking problems brought about by Max Planck's (1858–1947) quantum theory and new models of atomic structure. Heisenberg took part in this community in various locales, in institutions led by leading theorists: Arnold Sommerfeld (1868–1951) in Munich, Max Born (1882–1970) in Göttingen, and Niels Bohr (1885–1962) in Copenhagen. By the 1920s, Heisenberg's efforts were focused on the dynamics of physics. Quantum physics (particularly Niels Bohr's quantum model of the atom) and new ideas about radiation had necessitated a replacement of classical mechanics, but its details were elusive. By 1925, the lack of a workable theory of mechanics posed a serious crisis to physicists.

Werner Heisenberg provided his new interpretation of mechanics in 1925. It appeared to be more elegant mathematically than empirically, because it was not easily visualized. Still, mathematicians found it familiar, as it could be expressed in terms of matrix calculus; the variables of quantum mechanics could be understood as matrices. Thus Heisenberg's version of quantum mechanics is often dubbed "matrix mechanics." Another version of quantum mechanics, building on Heisenberg's work, was developed also in 1925 by Paul Dirac (1902–1984), using algebraic methods. Many physicists were repelled by the mathematical emphasis, particularly Erwin Schrödinger (1887–1961), who soon developed a version of mechanics centered upon the behaviors of waves. His version, called "wave mechanics," was easier to conceptualize. It was later found essentially to be equivalent to Heisenberg's matrix mechanics. Heisenberg received the Nobel Prize in Physics in 1932 for his creation of quantum mechanics.

Heisenberg was most well known for developing the principle of uncertainty in 1927. The complex statistical calculations of quantum mechanics provoked serious questions about the nature of reality: Statistics were useful tools for a mathematician, but a physicist was supposed to be concerned with real relationships between particles. These relationships had yet to be described. Already Heisenberg, in his discussions with Wolfgang Pauli (1900–1958), had noted that it was meaningless to talk of particles with fixed velocities. He could conceive of no experiment, even theoretically, that could identify the position of an electron precisely. Attempts at more specificity simply could not correspond to reality and were thus "meaningless"; more vague estimations would have to be acceptable in making calculations. Heisenberg elevated this to a fundamental principle of quantum mechanics, that it is impossible to determine coordinates precisely at the quantum level. For example, one could not know simultaneously both the position and velocity of an electron. Given a

certain position, one had to accept a range of possibilities (or a probability statement) for the velocity, and vice versa. At the quantum scale, greater certainty about one variable meant greater uncertainty about others.

The uncertainty principle's impact on the scientific community had much to do with its philosophical implications. Also called the indeterminacy principle, it dealt a serious blow against determinism and opened in physics a similar philosophical debate that had characterized biology since Charles Darwin's (1809–1882) insistence on random, purposeless variation in evolution. The uncertainty principle appeared to abandon traditional notions of cause and effect, noting that some things simply cannot be known to any observer; they must be understood instead in terms of probabilities and statistics. The present is unknowable, and thus knowledge of future effects must always consist of a range of possibilities. This profound conclusion troubled many of the leading physicists, particularly the philosophically minded ones. It was in response to the uncertainty principle, for example, that Albert Einstein (1879–1955) made his famous statement that God does not play dice. Others embraced Heisenberg's uncertainty principle. Niels Bohr, for example, placed it alongside his complementarity principle, which proclaimed the mutual validity of seemingly opposed systems of understanding physical reality (such as the wave and particle theories of matter). The lack of determinism inherent in the uncertainty principle became a key feature of the Copenhagen interpretation of quantum mechanics (Bohr lived in that city).

Quantum mechanics and the uncertainty principle made Heisenberg a celebrated figure worldwide. But the community of physics in which his ideas had been born was falling apart with the rise of Nazism in Germany. Heisenberg found himself the target of virulent critiques; in an increasingly racist Germany during the 1930s, he was dubbed a "white Jew" who cavorted with ethnic Jews.

Theoretical physics, his domain, was criticized by leading experimental physicists as the province of Jews such as Einstein. Despite this, Heisenberg chose to stay in Nazi Germany, although he was offered jobs elsewhere. When Einstein emigrated, he became Germany's foremost physicist.

In the late 1930s, Werner Heisenberg became the leader of Nazi Germany's atomic bomb project. Heisenberg was captured toward the end of the war during the U.S. mission *Alsos,* which was designed to learn about the progress on the project. His conversations with several other high-ranking scientists, notably Otto Hahn (1879–1968) and Max Von Laue (1879–1960), were recorded secretly at Farm Hall, a manor in England. He appeared surprised to hear about the bombings of Hiroshima and Nagasaki. A member of the *Alsos* mission, Samuel Goudsmit (1902–1978), concluded that Heisenberg never understood the details of how the bomb should work. Heisenberg denied this, and historians still debate what role Heisenberg played in the failure of the Nazi bomb project. He claimed that he did not want it to be built, whereas critics tend to say that the Germans, despite their best efforts, simply did not get as far as the Americans did. The major hurdle of creating a fission chain reaction was accomplished in the United States in 1942 by using extraordinarily pure graphite as a moderator. The Germans, like the Americans (initially), had assumed that the use of graphite was not practical and that heavy water must be used. This technical point was crucial, because the chain reaction was a critical test to prove or disprove the possibility of an atomic bomb. Whether by failure or by design, Heisenberg's group never accomplished it.

After the war, Heisenberg returned to Germany to help rebuild the physics community there. He became part of the Institute of Physics in Göttingen, renamed in 1948 the Max Planck Institute for Physics. It was later moved to Munich. He traveled widely and lectured in Britain and the United States. In the 1950s, his research turned to plasma physics; he continued close involvement in nuclear physics, occasionally serving in a policy advising role for his government.

See also Atomic Bomb; Bohr, Niels; Determinism; Nazi Science; Philosophy of Science; Quantum Mechanics; Social Responsibility; Uncertainty Principle

References
Cassidy, David C. *Uncertainty: The Life and Science of Werner Heisenberg.* New York: W. H. Freeman, 1992.
Heisenberg, Elisabeth. *Inner Exile: Recollections of a Life with Werner Heisenberg.* Cambridge: Birkhäuser, 1984.
Kragh, Helge. *Quantum Generations: A History of Physics in the Twentieth Century.* Princeton, NJ: Princeton University Press, 1999.
Powers, Thomas. *Heisenberg's War: The Secret History of the German Bomb.* New York: Knopf, 1993.
Price, William C., and Seymour S. Chissick, eds. *The Uncertainty Principle and Foundations of Quantum Mechanics: A Fifty Years' Survey.* New York: Wiley-Interscience, 1977.

Hertzsprung, Ejnar

(b. Frederiksberg, Denmark, 1873; d. Roskilde, Denmark, 1967)

Ejnar Hertzsprung should inspire all university students who find themselves unable to choose a major field of study. Although he is known for his contributions to astronomy, he started out studying chemical engineering and even worked as a chemist for a while after taking his university degree. In 1901 he moved to Leipzig, Germany, to study photochemistry under Wilhelm Ostwald (1853–1932), until finally a year later he returned to his native Denmark in order to pursue astronomy.

Work on astronomy around the turn of the century focused on the distribution of stars in the universe and their motions. The physical nature of stars was relatively unknown and little studied. Most of what was known came from the late-nineteenth-century work of Angelo Secchi (1818–1878) and

William Huggins (1824–1910), who brought photography into wide astronomical use by their photographic plates of stellar spectra. These two men had identified several different kinds of stars, classified by the placement of absorption lines visible in the spectra from the stars' light. The last decade of the nineteenth century saw the publication of a series of star catalogues using various classification schemes based on stellar spectra. The result was a sequence of star "types," labeled O, B, A, F, G, K, M, R, N, S, each one less hot than the next.

Hertzsprung once said that he came to astronomy through his interest in blackbody radiation, a question being studied by a number of leading physicists, including Max Planck (1858–1947). A blackbody is a substance that, when heated, radiates all frequencies of light. Hertzsprung became interested in studying the light from stars, and in 1905 and 1907 published two major papers on the spectra of stars. By correlating the existence of sharp and deep spectral lines with greater luminosity, he found a means of measuring luminosity more precisely by examining spectra. Later, when luminosity came into use as a means of calculating distances between stars, Hertzsprung's discovery gave astronomers a formidable tool for calculating such distances with greater precision.

In addition to calculating distance, Hertzsprung differentiated stars in terms of size and temperature. He plotted the stars of the Pleiades cluster on an x-y axis according to temperature, or star type (on the x-axis) and size (on the y-axis). The diagram shows that most stars exist somewhere along a (roughly) diagonal line showing a clear correlation between large size and high temperature. These stars were the "main sequence" stars, which constituted one of two star varieties. In addition to these were the few red giant and white dwarf stars, which deviated from the main sequence and appeared to be either very large and relatively cool (red) or very small and relatively hot (white).

Hertzsprung's ideas were not known in the United States, where Henry Norris Russell (1877–1957) was developing similar ideas. In 1913, Russell presented a diagram showing the correlation, which became the fundamental basis for all future research on stellar evolution. It now is called the Hertzsprung-Russell diagram. Both scientists believed that the diagram showed stars at different points in stellar evolution, but they differed when interpreting the process. Russell felt that stars began as red (cool) giants, then condensed and heated to form blue (hot) stars, then cooled slowly without changing much in size. Hertzsprung thought the diagram revealed two separate evolutionary paths. Neither interpretation lasted in its entirety, but their work stimulated astronomers to think more about how stars are born, how they live out their lives, and how they die.

Although Hertzsprung is best known for his contributions to stellar evolution, he made at least one more fundamental contribution, related to Cepheid variables, or stars of fluctuating brightness. Henrietta Swan Leavitt (1868–1921), working at the Harvard College Observatory, discovered a correlation between the apparent magnitude of these variable stars in the Small Magellanic Cloud and their periodicity. Simply put, she saw that brighter stars had longer periods. Since these stars were roughly the same distance away, they could be used as a measuring stick for stars elsewhere. If one could find a pulsating star of the same brightness but with differing periodicity, one could judge relative distances. Hertzsprung aided in this in 1913 by trying to determine the distance to the Small Magellanic Cloud; his results led to scientific honors but were later revised considerably, as he placed the cloud about five times too distant. Hertzsprung continued research on variable stars and stellar spectra for the next several decades, even after his retirement in 1944.

See also Astronomical Observatories;
　Astrophysics; Chandrasekhar, Subrahmanyan;
　Eddington, Arthur Stanley

References

Herrmann, Dieter B. *The History of Astronomy from Herschel to Hertzsprung.* New York: Cambridge University Press, 1984.

North, John. *The Norton History of Astronomy and Cosmology.* New York: W. W. Norton, 1995.

Strand, K. A. "Hertzsprung, Ejnar." In Charles Coulston Gillispie, ed., *Dictionary of Scientific Biography,* vol. VI. New York: Charles Scribner's Sons, 1972 350–353.

Hiroshima and Nagasaki

Atomic bombs have been used twice in war. The United States used an atomic bomb against the Japanese city of Hiroshima on 6 August 1945. Three days later, on August 9, it dropped a bomb on Nagasaki. These cities were chosen for their strategic significance—Kyoto initially was a target but was removed from the list because of its status as an ancient capital and major cultural center. The bombs resulted from a large-scale effort of scientists to harness the energy of the atom, using the knowledge gained from half a century of research on nuclear physics. The two cities were devastated; more than 200,000 people were killed or wounded in the blasts, and more would die later of radiation sickness. The names *Hiroshima* and *Nagasaki* have become synonymous with the dawn of the atomic age.

The scientists, particularly Albert Einstein (1879–1955), who pressed President Franklin Roosevelt (1882–1945) to pursue the bomb, had done so in fear that the Nazis would develop one first. In the hands of Adolf Hitler (1889–1945), they reasoned, an atomic bomb would seal Europe's fate and leave it under the dominance of the Third Reich. But as the war came to an end, they realized how far short the Germans had fallen in their efforts. Many argued that the weapon, designed for use against Germany, should not be used against Japan; one scientist, Joseph Rotblat (1908–), even left the project once the German threat seemed under control. Most of them, however, continued their work and developed weapons to be used against Japan.

After meeting with Allied leaders at Potsdam in July 1945, President Harry S Truman (1884–1972) issued a declaration, demanding the surrender of Japan and promising "prompt and utter destruction" in the event that Japanese leaders refused. Some scientists in Chicago and Los Alamos who had been active in reactor design and bomb development presented him with alternatives to bombing Japanese cities; one suggestion was to provide a demonstration to the Japanese of the bomb's effectiveness. Such alternatives were rejected, and after a period in which Americans believed Japan had no intention of surrendering, Truman decided to use the weapon. On 6 August 1945, a little after eight o'clock in the morning, the U.S. B-29 bomber *Enola Gay* dropped an atomic bomb on Hiroshima. The explosion was estimated at about 20 kilotons (revised later to about 12.5 kilotons, equivalent to about 12,500 tons of dynamite). The cloud from the blast rose to 40,000 feet and formed what universally has been referred to as a "mushroom cloud."

Hiroshima was devastated by the bomb. About 80,000 people were killed, with roughly the same number wounded. Buildings were leveled by the initial pressure wave. Fires in the city reached massive proportions, creating firestorms that sucked in outside air and produced winds of some thirty or forty miles per hour. With Hiroshima's infrastructure destroyed, doctors, police, and firemen were severely limited in what they could do. Truman issued a statement promising further destruction if the Japanese refused to surrender. Meanwhile, the Soviet Union informed Japan that their countries also would be at war beginning August 9. When that day came, another U.S. B-29 bomber, *Bock's Car,* dropped an atomic bomb on the city of Nagasaki. The results were similar to those in Hiroshima, although

The wrecked framework of the Museum of Science and Industry as it appeared shortly after the blast. (Bettmann / Corbis)

in this case there was no firestorm, owing to the topography of the city. Because the blast was confined largely to an industrial valley, the death toll was about half that of Hiroshima, around 40,000.

Japan's emperor, Hirohito, insisted to his cabinet that the war should end; in doing so, he broke with tradition by getting involved in decisions of state. On August 10, Japan sent word to the United States that it would surrender on the condition that the emperor still would be allowed to retain his position. When the United States responded that the emperor would be subjected to the rule of U.S. occupation forces, the Japanese cabinet refused to surrender. Again, Hirohito insisted that Japan stop fighting, and on August 14, Japan accepted the U.S. terms.

Part of the reason that Truman did not want to demonstrate the weapon beforehand was that it might not work, which would prove embarrassing, and the United States did not want to waste the bomb. Although hoping that the Japanese would believe otherwise, the Americans actually did not have a large supply of atomic bombs in August 1945. The Hiroshima bomb, called "Little Boy," was fabricated from enriched uranium, whereas the Nagasaki weapon, "Fat Man," was made from plutonium. Neither one of these weapons was simple or quick to manufacture. The technical details of fission chain reactions were well known in theory and had been achieved in practice in Chicago in 1942. The detonation designs of the bombs, either through high-velocity impact ("gun-type") or implosion, had been created at Los Alamos, and they could be replicated. But the material for the weapons did not come from ordinary uranium, or U-238; instead, top-secret installations were required to separate U-238 from U-235 ("enriched" uranium, needed for bombs, contains a higher proportion of U-235 than normal), or to create the element plutonium, with an atomic weight of 239. This process was long and costly. Despite the massive destruction caused by the first two bombs, in one sense they were a bluff: The United States simply did not have enough material to continue its atomic bombardment indefinitely.

Considerable historical controversy surrounds Truman's decision to use atomic weapons against Hiroshima and Nagasaki. Conventional wisdom holds that the atomic bombs were not "war-winning" weapons, but perhaps—as Truman and his advisors reasoned—they would shorten the war and thus prevent an invasion of Japan's home islands, which would cost U.S. lives. But there is also evidence that suggests that in August 1945 the United States missed opportunities to end the war because it was concerned with strengthening its future position vis-à-vis the Soviet Union. The development of the atomic bomb had been a cooperative effort, including the United States and Britain but excluding the Soviet Union. Cases of espionage in later years would show that the Soviets knew much more about the project during the war than the Americans realized. But in deciding to drop the bombs on Hiroshima and Nagasaki, perhaps Truman and his advisors were sending a message to the Soviet Union, in what some historians have called the first round of "atomic diplomacy" in the Cold War.

See also Atomic Bomb; Cancer; Cold War; Manhattan Project; Radiation Protection; Uranium; World War II

References

Alperovitz, Gar. *The Decision to Use the Atomic Bomb, and the Architecture of an American Myth.* New York: Alfred A. Knopf, 1995.

Badash, Lawrence. *Scientists and the Development of Nuclear Weapons: From Fission to the Limited Test Ban Treaty, 1939–1963.* Atlantic Highlands, NJ: Humanities Press, 1995.

Hersey, John. *Hiroshima.* New York: Vintage, 1989.

Maddox, Robert James. *Weapons for Victory: The Hiroshima Decision Fifty Years Later.* Columbia: University of Missouri Press, 1995.

Hormones

Hormones are substances that are secreted internally by endocrine glands, to regulate the activities of organs and tissues. They control a host of bodily functions, such as growth, metabolism, and reproduction. In the late nineteenth century, the medical use of biological secretions (such as crushed testicles) was called organotherapy. In 1905, British physiologist Ernest Starling (1866–1927) introduced the term *hormone* to describe these chemical messengers that regulated organs. With hormones, physiological processes were studied and manipulated.

One of the first successful uses of hormones as therapy was insulin. Canadian researcher Frederick Banting (1891–1941) knew that diabetes was said to result from the inability of the pancreas to secrete insulin (a hormone) for the regulation of the body's sugars. He sought to use other animals' insulin as a therapy for humans. Despite other scientists' failure to do so, in 1921 Banting and John J. R. MacLeod (1876–1935) extracted insulin from a dog's pancreas, and that insulin was used successfully to treat diabetes in humans. The two men shortly (in 1923) shared the Nobel Prize in Physiology or Medicine for their work.

The Dutch pharmaceutical company Organon, founded in 1923, was the world's major producer of hormone drugs. After Banting and MacLeod isolated it, Organon began to manufacture and sell insulin while conducting its own research. In 1931, it founded its own research journal, *Het Hormoon* (The Hormone). In 1925, its re-

One of the first successful uses of hormones as therapy was insulin. Canadian researcher Frederick Banting (pictured here) knew that diabetes was said to result from the inability of the pancreas to secrete insulin (a hormone) for the regulation of the body's sugars. (National Library of Medicine)

searchers isolated the hormone estrogen, and in 1935, one of its founders, Ernst Laqueur (1880–1947), isolated testosterone. The cooperative arrangement between researchers and commercial enterprise maintained Organon as the world's leading hormone company until World War II.

In humans, the study of sex hormones sparked a scientific and cultural reevaluation of what it meant to be male and female. The field of sex endocrinology posited the concept of male and female sex hormones, which made these secretions out to be the defining characteristics of gender differences. Femininity and masculinity were encapsulated, according to this view, in chemicals unique to each. Nelly Oudshoorn has argued that this view of hormones allowed popular Western culture to ascribe to hormonal actions many of the perceived behavioral differences between men and women. Even more important, it often led to a stigmatization of women (although ironically not men) as being completely controlled by hormones.

During the 1920s and 1930s, hormone therapy typically was practiced on women. Pharmaceutical companies in the 1930s and 1940s pursued drugs to help regulate menstruation and menopause, with comparatively little attention to male concerns. Ernst Laqueur, a researcher at the Pharmaco-Therapeutic Laboratory at the University of Amsterdam (and, as mentioned, a founder of Organon), found that a hormone thought to be unique to female horses turned up in the urine of males. Laqueur's work began a long critical reexamination of the assumption of stable, vastly different chemical differences between males and females, not just in horses, but in humans as well. Despite Laqueur's discovery, the cultural representation of gender-defining male and female hormones did not disappear.

Research on hormonal action in plant physiology sparked interest in growth-promoting hormones and also led to the first powerful, selective herbicides of the 1940s. Focusing on growth control, scientists reasoned that one must understand normal growth in order to inhibit it. Hormones seemed to hold the key to controlling and manipulating growth. Investigations of the role of auxin, identified in 1926 by Dutch scientists as the hormone permitting stem elongation, were undertaken by U.S. scientists in the 1930s. At institutions such as the Boyce Thompson Institute, researchers tried to use hormones to promote growth and thus address the hunger problems of the Great Depression. Plant hormone research was stepped up intensively during World War II, with the same purpose in mind, under the auspices of the Department of Agriculture. But elsewhere in government, top-secret committees discussed using hormones as a weapon to damage crops sorely needed by the enemy and to clear away the foliage concealing enemy fortifications. By war's end, such projects became an official part of the Chemical Warfare Service.

See also Chemical Warfare; Endocrinology; Industry; Medicine; World War II

References

Bliss, Michael. *The Discovery of Insulin.* Chicago: University of Chicago Press, 1982.

Kendall, Edward C. *Cortisone: Memoirs of a Hormone Hunter.* New York: Charles Scribner's Sons, 1971.

Oudshoorn, Nelly. *Beyond the Natural Body: An Archaeology of Sex Hormones.* London: Routledge, 1994.

Rasmussen, Nicolas. "Plant Hormones in War and Peace: Science, Industry, and Government in the Development of Herbicides in 1940s America." *Isis* 92 (2001), 291–316.

Hubble, Edwin

(b. Marshfield, Missouri, 1889; d. San Marino, California, 1953)

When Edwin Hubble was a young man at the University of Chicago, a sports promoter saw him box and offered to train him to fight the heavyweight champion. Instead, the athletic Hubble became a Rhodes scholar and studied at Oxford. While abroad, he boxed in an exhibition match with French champion Georges Carpentier. He returned to the United States and finished his studies in 1912. Although he had admired his science professors at Chicago, especially physicist Robert Millikan (1868–1953) and astronomer George Ellery Hale (1868–1938), he took a degree in jurisprudence and became a lawyer. A couple of years later, he abandoned this profession and became a graduate student at the University of Chicago's Yerkes Observatory, finishing his Ph.D. in 1917. Although offered a job at the Mount Wilson Observatory in California, Hubble joined the army to fight in World War I.

Hubble joined the researchers at Mount Wilson in 1919, after the war. He began his work using a 60-inch telescope, but soon the Hooker 100-inch telescope was ready for the scientists to use. Hubble's contributions would not have been possible without this powerful telescope. He turned his research to exploit its capabilities, and began to examine the most distant observable objects. His first major discovery, in 1923, was in identi-

fying a Cepheid variable star in the great nebula in Andromeda. Cepheid variables had
been demonstrated a decade earlier by Henrietta Swan Leavitt (1868–1921) as useful for
determining distances. Now, the Cepheid
could act as a marker to aid Hubble to determine the distance to the nebula.

Hubble's determination of distance to the
Andromeda nebula was crucial because of an
outstanding argument between astronomers
at the time. A famous debate between Harlow Shapley (1885–1972) and Heber D.
Curtis (1872–1942) in 1920 had centered on
whether nebulae—the cloud-like bodies in
space—were in reality star systems of their
own and, more important, whether they
were part of our own system. Curtis took the
position that the nebulae were indeed star
systems separate from our own galaxy and
could be regarded as "island universes." Shapley took the opposing view, that the nebulae
belonged to a single system shared by all. By
1924, Hubble had found thirty-six variable
stars in the Andromeda nebula, and he calculated a distance of approximately 900,000
light-years. But the maximum estimate of the
Milky Way's diameter was accepted at about
100,000 light-years. Hubble's results indicated that the nebula was very distant and
well beyond the reaches of the Milky Way,
our own galaxy. His results were announced
at a meeting of the American Astronomical
Society in December 1924. Curtis was vindicated, and Shapley's view of a single galaxy
was discredited. Shapley received a letter
from Hubble with the results, and Shapley
said that the letter destroyed his universe.

Hubble was convinced that galaxies were
the basic structural units of the entire universe. He was the first, in 1925, to create a
significant classification system for them.
Most galaxies, he said, rotate around a central nucleus either in a spiral or elliptical
shape. Hubble identified a number of other
elements used to classify the various observable galaxies. He devoted the latter half of
the 1920s to determining distances to other
galaxies, and by 1929 he had obtained dis-

tances for eighteen isolated galaxies and for
four within the Virgo cluster.

Also in 1929, Hubble made a more remarkable discovery that had ramifications for
both astronomy and cosmology. In 1912, scientists first measured radial velocity of stars
(radial velocity is the rate at which the distance between object and observer is changing) by observing the displacement of spectral
lines from starlight. The displacement was attributed to the Doppler effect. This effect
usually is understood in relation to sound, like
the ambulance siren changing tones as it
comes closer or moves farther away, as sound
waves are compressed or elongated. It can
also be applied to light, in the case of astronomy. Hubble noted that radial velocities increased with distance, and he calculated a
ratio of that change, establishing the proportionality of radial velocity to distance (from
the sun). Hubble and his colleague Milton L.
Humason (1891–1972) examined spectra
from the most distant observable stars, using
a new type of fast lens to take photographs of
faint spectra. Through this work, Hubble
showed that the previously stated proportionality of radial velocity to distance could be extended to bodies at a distance of more than
100 light-years. This proportionality became
known as Hubble's Law.

The spectral lines indicating radial velocity
were displaced, or "shifted," always toward
the red end of the spectrum, never to the violet end. The more distant stars exhibited a
greater "red shift" than the closer stars, indicating greater radial velocity. The conclusion
was staggering for cosmologists, who knew
that red shifts indicated objects moving away
from each other, whereas violet shifts (had
they existed) would indicate objects moving
closer. Because only red shifts were observed, all galaxies must be moving away
from each other. The universe must be expanding. Such a conclusion, based on empirical evidence from Mount Wilson Observatory, stimulated renewed interest in
cosmological theories, such as the "fireworks" or "Big Bang" theory of Georges

Lemaître (1894–1966). Hubble's discovery, which changed scientists' conception of the universe, has been compared with those of Copernicus and Galileo.

Hubble himself was not convinced by this interpretation and was suspicious of red shift. He preferred to use the term *red shift* to *velocity of recession,* because the former was neutral, based on observation, whereas the latter was an interpretation. By the late 1930s, Hubble began to reject the Doppler interpretation altogether, although cosmologists in turn rejected his interpretations of his own data and stuck to the expanding universe.

During World War II, Hubble left astronomy and became the chief of ballistics and director of the Supersonic Wind Tunnel Laboratory at Maryland's Aberdeen Proving Ground. After the war, he was instrumental in erecting the Hale 200-inch telescope at Palomar Observatory to penetrate even further into space.

> *See also* Astronomical Observatories;
>   Astrophysics; Big Bang; Cosmology; Leavitt,
>   Henrietta Swan; Shapley, Harlow
> *References*
> Christianson, Gale E. *Edwin Hubble: Mariner of the*
>   *Nebulae.* New York: Farrar, Strauss, and
>   Giroux, 1995.
> North, John. *The Norton History of Astronomy and*
>   *Cosmology.* New York: W. W. Norton, 1995.
> Trimble, Virginia. "The 1920 Shapley-Curtis
>   Discussion: Background, Issues, and
>   Aftermath." *Publications of the Astronomical*
>   *Society of the Pacific* 107 (December 1995):
>   1133–1144.
> Whitrow, G. J. "Hubble, Edwin Powell." In
>   Charles Coulston Gillispie, ed., *Dictionary of*
>   *Scientific Biography,* vol. VI. New York:
>   Charles Scribner's Sons, 1972, 528–533.

Human Experimentation

Human experimentation, particularly when involuntary, raised fundamental questions of medical ethics that date to the ancient Greeks, when physicians adopted the Hippocratic Oath, swearing to do no harm to one's patients. The ethical problem was debated in the United States at the turn of the twentieth century, owing to questionable practices in previous years. The American Antivivisection Society (established in 1883) worked against experimentation on animals, and between 1896 and 1900 the U.S. Congress conducted hearings to restrict it. Vivisection is like dissection, but involving a living body. Such efforts to establish strict regulations failed, while bringing to light attitudes condoning even human experimentation in mainstream groups such as the American Medical Association. In subsequent years, leading researchers and physicians condoned treating patients with experimental drugs and techniques. This led to controversy in 1911 when scientists at the Rockefeller Institute inoculated patients with an inactive form of syphilis in order to develop a diagnostic tool. Also controversial was the fact that researchers could gain access to patients in mental institutions, to orphans, and to convicts, in order to conduct invasive experiments without consent. In 1916, a leading U.S. physician, Walter Cannon (1871–1945), editorialized in the *Journal of the American Medical Association* that researchers needed to be more attuned to the rights of their own patients.

Despite such cautions, human experimentation continued. The U.S. Public Health Service, for example, conducted an experiment to determine whether syphilis affected blacks differently than whites. Beginning in 1932, four hundred syphilitic black men in the area of Tuskegee, Alabama, became part of a long-term observation of the disease. The patients were told that they had "bad blood" and were given pills, but these were really placebos made of aspirin. None of the patients were aware that they had syphilis, and no attempts were made to inform the patients or to administer any known medication for the disease. Even after the development of penicillin to effectively treat syphilis during World War II, these patients received no help, to preserve the sanctity of the secret, long-term observation. The study continued into the 1970s.

The ruined foundations of the hospital at the Buchenwald concentration camp decay next to a standing camp building. Doctors at the camp carried out varying kinds of experimentation on human subjects, including exposure to dangerous diseases and experimental vaccines. (Ira Nowinski/Corbis)

The most infamous cases of human experimentation were carried out by German scientists during World War II. Under the Nazi regime, Jews, Gypsies, and people with mental retardation were considered legitimate subjects for scientific research. They and other supposed "racial pollutants" were deemed a threat to the survival of the Aryan race. "Scientific" killing had the added benefit of contributing to knowledge. The Kaiser Wilhelm Institute for Brain Research, for example, was eager to dissect the brains of those killed who had suffered from cognitive disabilities. At the death camp in Auschwitz, Poland, physician Josef Mengele (1911–1978?) conducted experiments (including killings and dissections) on twins, dwarves, and Jews with genetic abnormalities, all as part of his own intellectual research interests. Mengele shipped body parts back to his colleagues in German universities for analysis.

These activities were not carried out merely by uneducated brutes in the Nazi Party; Germany's leading scientists and physicians participated in and encouraged them.

When Japan installed the puppet state of Manchukuo (in Manchuria) in the 1930s, the government sent bacteriologist Shiro Ishii (1883–1959) there to set up a biological warfare unit. After World War II began, Ishii's Unit 731 tested defensive weapons (such as vaccines) and offensive weapons (disease-causing pathogens) on human subjects, the vast majority of them Chinese. Unit 731 tested the effects of plague, cholera, anthrax, typhus, and other diseases in three ways: individual human exposure in a laboratory, open-air exposure through experimental delivery systems (bombs), and field tests during which both civilian and military personnel were exposed. Experiments on disease dissemination required unorthodox explorations, such as forcing victims to drink copious amounts of infected milk. Some of these experiments were reported in published papers, but referring to the human subjects as "Manchurian monkeys." When the war ended, the United States helped Japan to conceal these events. Because the Japanese had operated under no legal or ethical constraints, the data they possessed could not have been attained by U.S. researchers. Consequently, the United States decided not to prosecute the Japanese scientists for war crimes; in return, the scientists provided Americans with details of the experiments, which were integrated into the U.S. biological warfare program.

During the war, the United States perceived that atomic weapons would transform the nature of warfare. At war's end, U.S. officials knew that this would require considerable research on nuclear energy and the biological effects of radiation. Scientists under the auspices of the Army and later the Atomic Energy Commission conducted human experiments to test the effects of exposure to plutonium or other radioactive substances. In 1945 and 1946, for example, patients at a civilian hospital in Rochester, New York, were injected with plutonium without their informed consent, at five times the levels that scientists believed a body could sustain without harm. Other experiments entailed giving pregnant women radioactive elixirs while telling them that they had nutritional value, and putting radioactive iron or calcium into children's cereal at a boys' school in Massachusetts. Such experiments continued into the second half of the twentieth century and were not acknowledged by the U.S. government until the 1990s.

See also Atomic Energy Commission; Cold War; Medicine; Nazi Science; Public Health; Radiation Protection; Venereal Disease; World War II

References

Harris, Sheldon H. *Factories of Death: Japanese Biological Warfare, 1932–1945, and the American Cover-Up.* New York: Routledge, 2002.

Jones, James H. *Bad Blood: The Tuskegee Syphilis Experiment.* New York: Free Press, 1981.

Lederer, Susan E. *Subjected to Science: Human Experimentation in America before the Second World War.* Baltimore, MD: Johns Hopkins University Press, 1995.

Müller-Hill, Beno. *Murderous Science: Elimination by Scientific Selection of Jews, Gypsies, and Others, Germany 1933–1945.* Oxford: Oxford University Press, 1988.

Welsome, Eileen. *The Plutonium Files: America's Secret Medical Experiments in the Cold War.* New York: Delta, 1999.

I

Industry

Because of widely held beliefs in close connections between technological innovation and economic growth, science and industry maintained a tight relationship in the early twentieth century. Some reformers hoped to shape a progressive society on scientific, rational principles, and industry often appeared to be an ideal realm in which to promote such progress. Efforts to reform scientific and educational institutions in the nineteenth century convinced leading intellectuals that science should be put to work for overall economic well-being. Some institutions explicitly conformed to that mission. For example, Britain's Imperial College of Science and Technology was founded in 1907 to educate manufacturers and entrepreneurs in the sciences in the hope of providing both industry and government with scientifically minded leaders.

Scientific work contributed to industry in numerous fields, such as chemistry, medicine, and electricity. The German dye industry, for example, established industrial research laboratories in the late nineteenth century. Although the dye industry was first developed in England, German chemists at companies such as Bayer developed not only dyes but also a host of organic chemical compounds. The work of industrial scientists could be applied in other realms, such as the pharmaceutical industry. Industry was put to work for the German government during World War I. In later years, the marriage of science and industry established the backbone of industrial giants, such as I. G. Farben, that relied on research laboratories to design, test, and help manufacture chemicals.

Often governments took an active role in trying to promote industry through science. In the Soviet Union, for example, Stalinist reformers sought to accelerate industrialization through rational, state-planned management. Centralized effort, many felt, could speed the dissemination of technology and scientific ideas throughout the country, and government control could break down existing hierarchies of power in the factories, allowing innovation by younger men. Scientists and engineers were not united in this effort, however, and many believed the policies were fundamentally flawed.

During the 1920s, the perceived connections between science and industry sparked efforts to have major corporations fund scientific research. In the United States, the National Academy of Sciences launched a scheme to have businesses contribute money for academic science. The funds were supposed to be managed by a board of trustees made up of leading scientists and industrialists. But the project, called the National Research Fund, smacked of philanthropy and few contributed except a few giants of industry, largely because there could be no secrecy in such a scheme—no business edge for anyone who came up with a new idea. After 1929, amidst the years of the Great Depression, joint scientific efforts were abandoned.

Individual companies continued to fund research, hoping to tap into new markets. This was occasionally successful, as in the case of the pharmaceutical industry; the companies required expertise, and the scientists needed funding for their research. Companies such as Merck and Eli Lilly, for example, became leading producers of new drugs to treat diseases, using recent innovations from scientists. The isolation of insulin was the most widely celebrated of these; its mass production allowed the widespread treatment of diabetes, and the scientists—Frederick Banting (1891–1941) and John Macleod—were recognized with the Nobel Prize in Physiology or Medicine in 1923. It came about through a collaborative arrangement between Eli Lilly and Company and the University of Toronto.

Outside of the pharmaceutical industry, successful industrial laboratories were operated by companies such as General Electric and AT&T. In part, laboratories were maintained in order to legitimize these companies' controversial monopolistic control of the marketplace. At the turn of the century, huge industrial combinations were met with skepticism by the U.S. government, and industrial laboratories became crucial for helping large corporations establish legitimate patent claims for their domineering positions in the marketplace. Although developing new technology was a concern for these companies, they were far more concerned with developing new patents to outmaneuver competitors. Because of this, industrial research was geared toward design and innovation, despite being managed by leading men

of science such as Frank Jewett (1879–1949) of Bell Laboratories and Irving Langmuir (1881–1957) of General Electric. Langmuir's research on surface chemistry led to a Nobel Prize, as did that of Clinton Davisson (1881–1958), whose work at Bell Laboratories helped provide experimental evidence for wave mechanics. Industrial research continued strongly after World War II. Nonetheless, industry's importance in financing science became secondary to that of the federal government, which became an active promoter and financial supporter of basic research in the late 1940s and beyond.

See also Davisson, Clinton; Great Depression; Hormones; Medicine; Patronage; Penicillin

References

Aftalion, Fred. *A History of the International Chemical Industry.* Philadelphia: University of Pennsylvania Press, 1991.

Beer, John Joseph. *The Emergence of the German Dye Industry to 1925.* Urbana: University of Illinois Press, 1959.

Hall, A. Rupert. *Science for Industry: A Short History of the Imperial College of Science and Technology and Its Antecedents.* London: Imperial College, 1982.

Haynes, William. *American Chemical Industry, vol. 5: The Decade of New Products, 1930–1939.* New York: Van Nostrand, 1954.

Kevles, Daniel J. *The Physicists: The History of a Scientific Community in Modern America.* Cambridge, MA: Harvard University Press, 1995.

Reich, Leonard S. *The Making of American Industrial Research: Science and Business at GE and Bell, 1876–1926.* New York: Cambridge University Press, 1985.

Shearer, David R. *Industry, State, and Society in Stalin's Russia, 1926–1934.* Ithaca, NY: Cornell University Press, 1996.

Swann, John P. *Academic Scientists and the Pharmaceutical Industry: Cooperative Research in Twentieth-Century America.* Baltimore, MD: Johns Hopkins University Press, 1988.

Intelligence Testing

Modern intelligence testing owes its origin to Alfred Binet (1857–1911), a psychologist at the Sorbonne in Paris. In the 1890s, he was part of the school of French craniometers (skull measurers) who believed in a relationship between skull size and intelligence. Binet ultimately found his own data unconvincing, and he turned to a psychological approach in 1904. His motivation came from the French government, which wanted him to devise a way to identify children with special needs, whose inclusion in normal classrooms had been unproductive. Binet tried to construct a test that assessed basic reasoning through a series of diverse problems and activities. From a large variety of small tests, he arrived at an overall score; he published the scale initially in 1905. He assigned an age level to each kind of problem, and children taking the test were assigned a "mental age" at the conclusion of the test. In 1912, after Binet's death, psychologists began to use the test by dividing mental age by actual age to arrive at the intelligence quotient, or IQ for short. IQ was the first widely used, quantifiable measure of intelligence.

Binet warned against using his test as anything more than a practical device; it should not be used, he said, to prop up a theory of intelligence. He made no claim of intelligence as innate. He wanted to use the tests to identify children in need of special attention, not to label them as imbeciles incapable of improvement. But U.S. psychologists, notably Henry Herbert Goddard (1866–1957) and Lewis M. Terman (1877–1956), expanded Binet's intelligence tests to provide for widespread testing of the general population. Their work turned Binet's practical, crude tool into a widely used measure for psychological study. Terman's position at Stanford University gave the new test its lasting name, the Stanford-Binet IQ test. Terman's innovations to the test, first published in 1916, were designed to provide a score of 100 for average people at each age level. Intelligent people scored above average, while the more feebleminded scored far below. The Stanford-Binet IQ test was not limited to children; in

Alfred Binet (1857-1911), French psychologist who originated the intelligence test. (Bettmann/Corbis)

fact, it was administered to soldiers during World War I, highlighting serious differences in intelligence between young men from northern states and those from southern states, and indicating low scores for recent immigrants and African Americans. These results fueled a wave of racism and nativism in the United States that helped pass the Immigration Restriction Act of 1924. Most intelligence tests were modeled on the Stanford-Binet one and were designed to agree with its results; psychologists and laymen alike looked to Terman's 1919 book, *The Measurement of Intelligence,* for guidance on intelligence testing.

After quantifying intelligence, psychologists sought to prove that it was a permanent, static, and heritable feature of humans. Henry Herbert Goddard brought Binet's intelligence test to the United States and transformed it into a tool for classification and the study of *morons,* a term he coined. He urged the spread of intelligence testing beyond the confines of psychology, into the military, the medical profession, and at immigration ports. Goddard even argued, successfully in some cases, that people with low IQ scores ought not to be held criminally responsible for their deeds, because they were imbeciles. Under Goddard, the pupils at New Jersey's Training School for Feeble-Minded Girls and Boys became the subjects of experiments on mental retardation. In 1912 he published *The Kallikak Family,* in which he tried to demonstrate the permanence of degenerate minds from one generation to the next. This book was instantly hailed as strong support for the eugenics movement, which emphasized good breeding, and it was published in 1933 by the Nazis as scientific proof for their propaganda against "inferior" peoples. In the late 1920s, Goddard recanted his views that being a moron was not curable by education and that morons ought to be institutionalized. In the late 1930s, Terman also moderated his views, noting the strong environmental influences on individual intelligence.

The consequences of believing in innate, static, and inheritable levels of intelligence were vast. Both Goddard and Terman advocated social action against children with low scores and against groups with supposedly innate, genetic defects in intelligence. Terman suggested that feebleminded people would tend toward antisocial or immoral behavior, such as prostitution in the case of women. He also noted that intelligence testing could predetermine one's access to better jobs, which should be closed off from children who exhibited below-average IQs. Belief in permanent, inheritable, and thus inevitable intelligence levels provided a justification for existing social classes and obviated the need to address them through regulation or education reform. It also rewarded those whose socioeconomic position fostered better education, while calling it "natural." If intelligence is innate and unchanging, then intelligence testing would provide the ultimate means of a true meritocracy. But if it is not, and intelligence is formed through genetic and environmental influences, then schools relying solely on intelligence testing simply acted as "sorters" to validate and encourage preexisting inequities between social groups.

See also Eugenics; Mental Retardation; Race
References

Chapman, Paul Davis. *Schools as Sorters: Lewis M. Terman, Applied Psychology, and the Intelligence Testing Movement, 1890–1930.* New York: New York University Press, 1988.

Gould, Stephen Jay. *The Mismeasure of Man.* New York: W. W. Norton, 1996.

Minton, Henry L. *Lewis M. Terman: Pioneer in Psychological Testing.* New York: New York University Press, 1988.

Zenderland, Leila. *Measuring Minds: Henry Herbert Goddard and the Origins of American Intelligence Testing.* New York: Cambridge University Press, 1998.

International Cooperation

The notion that science is universal—the same laws apply everywhere—often contributed to the belief that science inherently was international. Although scientists often worked in the service of their respective countries, helping them compete for prestige or technological dominance, scientists also claimed to belong to a community of science that transcended national borders. There were undoubtedly many challenges to such notions: The efforts of Allied scientists to exclude Germans from international scientific organizations after World War I was the most obvious example, but there were a host of others. Nazis asserted that certain ethnic groups (i.e., Jews) were more inclined toward fantastical, theoretical physics; Soviets claimed to eschew idealistic theories and tried to develop a Marxist, proletarian science.

In the first decades of the twentieth century, international cooperation existed informally, in the form of international study and international conferences. Many U.S. scientists, for example, studied in leading European institutions in the first decades of the century. In physics, the Solvay Conferences were an early indication of the usefulness of an international forum to discuss scientific ideas, especially in times of great scientific controversy. For example, such conferences provided the Swiss-born German Albert Einstein (1879–1955) and the Dane Niels Bohr (1885–1962) to debate their views about quantum mechanics in the presence of colleagues from outside their own institutions. After World War I, contacts were more difficult and relations were strained. But after Germany became a member of the League of Nations in 1926, attitudes toward German scientists relaxed. The same year, they were permitted to join the International Research Council (IRC), which previously had banned them. The IRC soon tried to refashion its image and to take up the mantle of a truly international scientific body that should rise above politics. In 1931, it renamed itself the International Council of Scientific Unions (ICSU). It was composed of various unions representing scientific disciplines; its members were supposed to represent science, not countries.

International cooperation often was most successful in the earth sciences, whose laboratory was itself international in scope. One avenue for international cooperation that proved successful was in oceanography. Beginning in 1902, scientists in several European countries participated in the International Council (later called the International Council for the Exploration of the Sea), a body directed at finding ways to manage plaice, cod, herring, and other fish. The International Council became a lasting body that mixed scientific and economic interests on an international level. Other efforts included the Pacific Science Association (PSA), which was founded in 1920 after the Pan-Pacific Science Congress in Honolulu, Hawaii. The PSA initiated periodic congresses aimed to discuss research of interest to scientists (largely oceanographers, marine biologists, and geologists) of the entire region. Despite their successes, both the International Council and the Pacific Science Association were severely disrupted by World War II (and also World War I, in the case of the International Council), but both managed to survive.

Science in the Polar Regions also sparked international cooperation. Although both the Arctic and Antarctic were noted more for national competition (getting to the poles first; traversing the southern continent; exploring inlets; staking territorial claims), the relatively unknown conditions gave rise to the Second International Polar Year, in 1932. Fifty years after the first "year," the Second IPY was designed to provide a better portrait of meteorological conditions by having scientists of various countries take the same kinds of data on specified days. Often seen as benchmark of success in international cooperation, the Second IPY inspired a third installment that was far more ambitious: the International Geophysical Year, in 1957–1958. First conceived in 1950, scientists in the United States and Britain hoped that international cooperation after World War II would give scientists "snapshots" of the earth.

See also Conservation; Eddington, Arthur Stanley; International Research Council; Oceanography; World War I

References

Greenaway, Frank. *Science International: A History of the International Council of Scientific Unions.* New York: Cambridge University Press, 1996.

Rozwadowski, Helen. *The Sea Knows No Boundaries: A Century of Marine Science under ICES.* Seattle: University of Washington Press, 2002.

Schröder-Gudehus, Brigitte. "Challenge to Transnational Loyalties: International Scientific Organizations after the First World War." *Science Studies* 3:2 (1973): 93–118.

International Research Council

The International Research Council (IRC) was founded in 1919 to coordinate international science among those countries that had been victorious or neutral in World War I. It was composed of international unions, each devoted to a particular field of science. Although it was designed to promote international cooperation, it explicitly barred participation by scientists from the Central Powers. In particular, the organizers of the IRC, notably U.S. astronomer George Ellery Hale (1868–1938), hoped to prevent Germany from coming to dominate any field of endeavor, including science.

The need for a body to coordinate international cooperation in science was evident when the International Association of Academies (IAA) dissolved during the war. The IAA was founded at the turn of the twentieth century for exactly that purpose, facilitating communication among various national academies of science. It had done little to accomplish that goal, although it comprised over twenty academies by the time the war broke out. Although some, especially Hale, initially tried to preserve the IAA during the war, the severity of animosities among European scientists made personal relationships (on which scientific cooperation depended) seem unlikely after the war. The casualties of war were high, and many leading scientists had sons and nephews who had been killed; the prospect of sitting across a meeting table with Germans was anathema to French and British scientists, despite the alleged internationalism of science. When the United States entered the war on the side of the Allies, Hale's own sentiments became less inclusive, and he and others turned their attention toward creating a new body.

The IRC was modeled largely on what Hale had created during the war in the United States, the National Research Council. The latter was supposed to coordinate research and promote activities that might serve the national interest, whether technical, strategic, or industrial. The IRC could do the same thing, ensuring the dominance of the former Allied countries in scientific activities. Yet the IRC was different because it wished to include neutral countries rather than "lose" them to a different international organization designed by the Germans; the goal was not merely to strengthen Allied science, but rather to isolate and exclude scientists from Germany. The motivation behind the IRC obviously was not in keeping with scientific internationalism, but it certainly was in keeping with leading Allied scientists' sense of morality and justice. In their view, German scientists had supported German militarism; just as politicians wanted to punish Germany with reparations, scientists wanted to prevent them from participation in international science.

Some scientists, typically those from countries that had been neutral during the war, objected to the exclusion of Germany. Notable scientists from Allied countries also objected, but the powerful role played by national scientific organizations such as the National Research Council (United States) and the Royal Society (Great Britain) helped to silence these views. Critics of exclusion noted that scientists were simply trying to carry on the war in science after the war between soldiers had ended. Nevertheless, on 28 July 1919, a month after the Treaty of Versailles ended the war, the International Research Council came into being and most of the invited neutral countries joined it. The clause that barred Germany was finally repealed in 1926, around the same time that Germany was allowed to join the League of Nations.

Given the widespread belief in science's inherent internationalism (i.e., that science is the common language of mankind and is not subject to national prejudices), the IRC stood as testimony that scientists were neither above politics nor above sentiments of chauvinism toward researchers from other countries; to the contrary, scientists in the case of the IRC simply followed public opinion. The IRC was renamed the International Council of Scientific Unions (ICSU) in 1931, and that body continues to exist. ICSU developed a reputation for its internationalism and its nonpolitical character; it celebrated its fifty-year anniversary in 1981, not 1969, clear testimony that it felt no need to call attention to its embarrassing political beginnings.

See also Hale, George Ellery; International Cooperation; Nationalism; World War I

References

Cock, A. G. "Chauvinism and Internationalism in Science: The International Research Council, 1919–1926." *Notes and Records of the Royal Society of London* 37:2 (1983): 249–288.

Forman, Paul. "Scientific Internationalism and the Weimar Physicists: The Ideology and Its Manipulation in Germany after World War I." *Isis* 64:2 (1973): 150–180.

Kevles, Daniel J. "'Into Hostile Political Camps': The Reorganization of International Science in World War I." *Isis* 62: 1 (1971): 47–60.

Schröder-Gudehus, Brigitte. "Challenge to Transnational Loyalties: International Scientific Organizations after the First World War." *Science Studies* 3:2 (1973): 93–118.

J

Jeffreys, Harold

(b. Fatfield, England, 1891; d. Cambridge, England, 1989)

Harold Jeffreys was an influential mathematician and geophysicist who made contributions to understanding the earth's structure and the origins of the planets. Jeffreys graduated from Armstrong College in 1910 with a distinction in mathematics before going to St. John's College, Cambridge University. He became a fellow there in 1914 and, after working at the Cavendish Laboratory during World War I, he taught mathematics at Cambridge until 1932. In subsequent years he also taught geophysics and astronomy, becoming a leading authority in several areas of the earth and planetary sciences.

Direct evidence of the interior of the earth was not available then or now; the deepest borings made in the 1920s, when Jeffreys solidified his ideas, were at best little more than 2 kilometers. For this reason scientists like Jeffreys turned to other kinds of evidence to make inferences about the physics of the earth's interior at depths of more than 6,000 kilometers. Scientists relied on the laws of gravitation, heat conduction, radioactive decay, and other methods to estimate the constitution of the earth. From seismological studies, Jeffreys made the controversial claim that the earth's core must be liquid. He believed that this necessitated a change from the widely accepted "planetesimal hypothesis" of planet formation, which assumed the accretion of already-cooled solid chunks. Jeffreys suggested that planets formed while they were still liquid, and their slow cooling left liquid cores in their interiors.

Jeffreys took an interest in the motion and structure of the bodies of the solar system during the war years. In the 1920s, he developed his own view that revised the existing planetesimal hypothesis of U.S. scientist Thomas Crowder Chamberlin (1843–1928). The planetesimal hypothesis suggested that planets were formed by accretion of "planetesimals" ejected from the sun. This concept had been favored by geologists, but astronomers paid little attention to it; Jeffreys sought to rectify the situation with a theory that appealed to both groups. The tidal hypothesis, as it was known, was concerned with the rupture of fluid masses, the means by which bits would be ejected from the sun. Jeffreys and colleague James Jeans (1877–1946) proposed that a passing star could exert a gravitational influence much like the moon exerts on the tides on earth; if the tidal influences were strong enough, the rudimentary forms of planets would be torn from the sun.

Jeffreys set forth his ideas about the earth's structure in *The Earth* (1924), which went into several editions and became the most frequently cited book on the subject for the next three decades. It described the earth as a liquid core surrounded by a solid mantle, topped by a thin crust. He explained the formation of mountains by claiming that the earth contracted. In later years, because of the popularity of the text, Jeffreys became a principal authority whose views plagued the efforts of scientists who believed in continental drift.

Jeffreys was also known for his contributions to the field of statistics, and he published extensively on this subject in the 1930s and 1940s. His *Theory of Probability* (1939) synthesized his own work on Bayesian statistics and tried to rectify a controversy with statistician (and geneticist) Ronald A. Fisher (1890–1962), who had developed what he considered a superior approach to the Bayesian one. In 1946, Jeffreys developed what became known as Jeffreys prior in statistics. He became a fellow of the Royal Society in 1925 and was knighted in 1953.

See also Age of the Earth; Continental Drift; Cosmology; Earth Structure; Geophysics; Origin of Life; Wegener, Alfred

References

Brush, Stephen G. *Nebulous Earth: The Origin of the Solar System and the Core of the Earth from Laplace to Jeffreys.* New York: Cambridge University Press, 1996.

Jeffreys, Harold. *The Earth: Its Origin, History, and Physical Constitution.* Cambridge: Cambridge University Press, 1924.

Swirles, Bertha. "Harold Jeffreys from 1891 to 1940." *Notes and Records of the Royal Society of London* 46:2 (1992): 301–308.

Johannsen, Wilhelm

(b. Copenhagen, Denmark, 1857; d. Copenhagen, 1927)

Many European biologists of the twentieth century owed their understanding of heredity to the work of Wilhelm Johannsen. The 1909 German version of his *Elements of Heredity* contained the fundamental concepts, many created by Johannsen, of the new science of genetics.

Although genetics was born from the rediscovery of Gregor Mendel's (1822–1884) work around 1900, Johannsen already had conceptualized many of his ideas by that time. He had begun a career as a pharmacist, only to abandon this profession in favor of botany and chemistry. Initially, his research focused on the ripening, dormancy, and germination of various plants, and he was most attracted to areas of study that could be quantified and understood with mathematical precision. This led him to admire the biometricians of the late nineteenth century who used statistics to understand evolutionary change. He began to study variability in heredity in the 1890s, influenced not only by Charles Darwin's (1809–1882) evolutionary writings but also by those of Francis Galton (1822–1911), whose work carried the merit (to Johannsen's mind) of resulting from quantitative, statistical methods.

Despite his appreciation for Galton's work, Johannsen went a long way toward discrediting his conclusions about heredity. Galton, the leading figure in nineteenth century eugenics, had argued that any given characteristic of offspring will "regress" to the average of their parents. Populations as a whole regress toward mediocrity; organisms with less desirable traits tend to improve upon breeding, whereas organisms with more desirable traits tend to produce organisms inferior to themselves. His views defied Darwin's theory of evolution, which posited natural selection as the mechanism for change in a population. Galton seemed to indicate that no permanent change occurred from one generation to the next.

Johannsen pointed out some deficiencies in Galton's law of regression. Galton's experiments with self-fertilizing sweet pea plants showed that, despite efforts to simulate natural selection, each new population of offspring "regressed" to reflect the average (of some chosen characteristic) of the previous generation before selection. Galton thus rejected Darwinian natural selection. Johannsen experimented with princess beans, and chose weight as the character to watch. He found that the beans produced by self-fertilization appeared to exhibit a range of variability in weight. But taken as a whole, both these beans and their offspring were about the same average weight. While individuals may have seemed different, statistically the bean population did not change from one generation to the next. Even taking the lightest beans and self-fertilizing them (in effect, simulating natural selection) did not result in only light beans in the subsequent generation. Instead, the next generation—again, taken as a whole, averaging the weight of the whole population—was the same as the generation prior to the artificial selection. But Johannsen's interpretation was somewhat different from Galton's. They agreed that regression was taking place and that natural selection never was a factor because no heritable change ever took place. But Johannsen believed that this population of self-fertilizing beans was a "pure line," whose offspring would always have identical heritable characteristics as the parents, despite apparent variability. There were many such "pure lines" in nature. The choice to be made by Darwinian natural selection, according to Johannsen, was between competing pure lines, not minute environmental variations within the "pure line" population.

But if Johannsen, like Galton, rejected that continuous variation within a population produced choices for natural selection, how did evolution through natural selection occur? Johannsen's criterion was simple: A trait must be heritable in order to constitute a new "pure line." To provide such new, permanent traits, Johannsen turned to the work of the Dutch botanist Hugo De Vries (1848–1935), who in 1900 proposed muta-

tion as a form of discontinuous variability in a population. The concept of mutation, which could create a new pure line, would preserve Darwinian evolution within the new science of genetics.

De Vries had helped resurrect the ideas published more than three decades earlier by Gregor Mendel. Mendel's mathematical description of hybridization established the idea of dominant and recessive characteristics. Mendel's work transformed Johannsen's, providing evidence that led him in 1903 to discard Galton's interpretation of regression and replace it with his own conception of pure lines whose interactions were governed by Mendelian laws.

Johannsen's work solidified the concept of the gene as a stable entity that could be passed on to the next generation without environmental interference. He also added some of the basic vocabulary of genetics. He coined the word *gene* to describe the unit of heredity. He also created a way to distinguish the general characteristics of pure lines from the appearance of variability in individuals within the pure line. The *genotype* defined the pure line, whereas the *phenotype* described outward appearance. The two types often differed because the phenotype was subject to environmental influences, but only the genotype described heritable characteristics. Johannsen outlined these concepts in *The Elements of Heredity,* published in 1905 in Danish and rewritten in 1909 in German.

See also Bateson, William; Evolution; Genetics; Mutation; Rediscovery of Mendel
References
Bowler, Peter J. *The Mendelian Revolution: The Emergence of Hereditarian Concepts in Modern Science and Society.* Baltimore, MD: Johns Hopkins University Press, 1989.
Dunn, L. C. "Johannsen, Wilhelm Ludvig." In Charles Coulston Gillispie, ed., *Dictionary of Scientific Biography,* vol. VII. New York: Charles Scribner's Sons, 1973, 113–115.

Joliot, Frédéric, and Irène Joliot-Curie

(b. Paris, France, 1900; d. Paris, 1958)

(Irène Joliot-Curie b. Paris, 1897; d. Paris, 1956)

Before beginning her own scientific life, Irène Curie's fame came principally from her mother and father, Marie Curie (1867–1934) and Pierre Curie (1859–1906), whose research in radioactivity provided the early foundations of the field around the turn of the century. Having such illustrious parents worked to her advantage, and after World War I she studied at the Sorbonne, taking a *licence* in physics and mathematics. She had worked with her mother during the war, setting up radiographic equipment for medical use at the front. Her research, following in her parents' footsteps, was in radioactivity. She received a doctorate in 1925 and married her most important collaborator, Frédéric Joliot, in 1926.

Unlike Irène, Frédéric did not come from a famous French family. His father was a merchant who had taken part in the Paris Commune after the Franco-Prussian War, a fact that compelled him to flee France for a while to escape persecution. Frédéric was the youngest of six, and he studied at the École Supérieure de Physique et de Chimie Industrielle to become an engineer. But under the influence of physicist Paul Langevin (1872–1946), he altered his plans and began to work for Marie Curie at the Radium Institute in 1925. He fell in love with Irène and married her a year later. They combined their names, and both took the surname Joliot-Curie. He soon earned his *licence* and, later, his doctorate in 1930. The next year the couple began their fruitful collaborations.

Joliot's training as an engineer helped him to construct an improved version of C. T. R. Wilson's cloud chamber. It enabled an observer to see and photograph the paths of electrically charged particles as they passed through a gas. The chamber became his favorite tool, and it proved useful in a number of projects on subatomic particles. Their research led James Chadwick (1891–1974), to his fortune and not theirs, to discover the neutron, and they also missed discovering the positron (an electron with a positive charge), discovered by Carl Anderson (1905–1991). Both were discovered in 1932, the so-called annus mirabilis (Latin for "year of miracles") of physics. But the Joliot-Curies were not too late for another important discovery: artificial radioactivity.

In 1934, they observed that by bombarding aluminum with alpha rays, they produced radioactive atoms with a half-life of just more than three minutes. This radioactivity seemed much like the beta activity so often observed by physicists, but the electrons were positive (they were in fact positrons). The effect also seemed to show an ejection of neutrons. This meant that an aluminum atom was bombarded with an alpha particle, resulting in the ejection of a neutron and the creation of a short-lived radioactive isotope of phosphorus. This then decayed into a stable isotope of silicon. They conducted a similar experiment with boron, bombarding it with alpha rays to produce a radioactive isotope of nitrogen that had a half-life of more than ten minutes. They soon announced that they had created artificially radioactive elements from known stable elements. The discovery encouraged a flurry of work on artificial radioactivity, including Enrico Fermi's (1901–1954) efforts to do so by neutron bombardment.

The Joliot-Curies received the Nobel Prize in Chemistry in 1935. Ironically, in their Nobel lectures, Irène the chemist recounted the discovery of the positive beta decay, a physical property, whereas Frédéric the physicist explained the chemical identification of the radioisotopes. Although Marie Curie lived to see these fundamental results in the field she helped establish three decades earlier, she died before her daughter and son-in-law received the prize.

Both Frédéric and Irène received honors and important positions after their Nobel Prize. Irène served as France's secretary of

Frédéric Joliot and Irène Joliot-Curie shared the Nobel Prize in 1935. (Bettmann/Corbis)

state for a few months in 1936 under Léon Blum's (1872–1950) Popular Front. She then accepted a professorship at the Sorbonne. Frédéric became a professor at the Collège de France in 1937. After the discovery of fission by Otto Hahn (1879–1968) in 1939, Joliot was the one to furnish proof of the great kinetic energy of fission fragments. He also noted that fission was accompanied by the ejection of more neutrons than was necessary to create the reaction. This meant that it might be possible to create a fission chain reaction and, because of the great energy involved, even a bomb.

Joliot frustrated many other scientists because he refused to go along with Hungarian-born physicist Leo Szilard's (1898–1964) idea not to publish work on fission. Szilard, who had immigrated to the United States, argued that the Nazis were trying to develop a bomb, and thus the latest work should not be made available to them. But Joliot continued his work and published it. He ordered uranium oxide from the Belgian Congo and a large quantity of heavy water from Norway, to be used as a moderator to absorb neutrons in an experimental chain reaction. When Germany started World War II, Joliot sent some of his materials to England. Under Nazi rule, he diverted his research from atomic energy to other questions of physics. He also secretly joined the Communist Party and helped in the resistance against Germany. Irène escaped with their children to Switzerland.

After the war, Joliot helped convince General Charles De Gaulle (1890–1970) of the need for a French atomic energy commission. The Commissariat à l'Energie Atomique (CEA) was created in 1945 with Joliot at its head. Irène Joliot-Curie joined him there as one of the directors of research, while she also took up the directorship of Paris's Radium Institute. Under Joliot's leadership, France built its first atomic reactor in 1948, and soon it built a new nuclear research center at Saclay. But his activities with the Communist Party, and his declaration that he would never support a war against the Soviet Union, annoyed the French government. He was removed from his position of leadership in 1950. He spent the 1950s doing research, teaching, and agitating for peace groups.

Irène was like her mother in another way. Both worked for years with dangerous materials, exposing themselves to large quantities of X-rays and gamma rays. Both mother and daughter died of leukemia, a tragic side effect of a new and mysterious property of nature. When Irène died in 1956, Frédéric took over her position as director of the Radium Institute. He died after an operation in 1958 and was given a state funeral.

See also Atomic Structure; Curie, Marie;
     Loyalty; Physics; Radioactivity
References
Goldsmith, Maurice. Frédéric Joliot-Curie: A
     Biography. Atlantic Highlands, NJ: Humanities
     Press, 1976.
Perrin, Francis. "Joliot-Curie, Irène." In Charles
     Coulston Gillispie, ed., Dictionary of Scientific
     Biography, vol. VII. New York: Charles
     Scribner's Sons, 1973, 157–159.
————."Joliot, Frédéric." In Charles Coulston
     Gillispie, ed., Dictionary of Scientific Biography,
     vol. VII. New York: Charles Scribner's Sons,
     1973, 151–157.
Weart, Spencer R. Scientists in Power. Cambridge,
     MA: Harvard University Press, 1979.

Jung, Carl

(b. Kesswil, Switzerland, 1875; d. Küsnacht, Switzerland, 1961)

Carl Jung began an autobiographical sketch of himself by recounting his earliest memories and dreams, which he called "islands of memories afloat in a sea of vagueness." Like Sigmund Freud (1856–1939), Jung placed a great deal of importance on early experiences in shaping human personality, for discerning the causes of one's fears, appetites, and motivations. But whereas Freud emphasized the primacy of sexuality in unconscious motivation, Jung looked elsewhere. He was convinced of the existence of deep structures in the human psyche that rested in the unconscious. This went beyond the formation of personality that Freud had studied and popularized. He set forth these ideas in The Psychology of the Unconscious in 1913. For Jung, these structures were historical, shared in unconscious collective memory, expressed in myths and in symbolic images he called "archetypes." They found expression in individuals.

Jung abandoned psychoanalysis altogether in 1914 in favor of his own creation, analytical psychology. Because of his focus on the collective unconscious, Jung began a wide-ranging study of the history and myths of cultures all over the world. Here he hoped to identify elements of the psyche's historical structure, which lay beneath the personal level to which psychoanalysis confined itself. He also developed a system to classify personalities at the conscious level, organized by types. Attitude types included introverts and extroverts; function types were thinking, feeling, sensation, and intuition. These types existed in combination, constituting the ego in each individual. Jung called the process of interaction between the collective unconscious and the conscious ego "individuation."

Jung developed an international reputation, as well as an extensive coterie of followers that helped to expand the understanding of collective symbols throughout the world. One controversial, yet seemingly prescient, conclusion he drew was about the German people. Jung found the archetypal themes in his German patients ominous. Toward the end of World War I, he predicted that the threatening themes in the German collective unconscious pointed toward an-

Carl Jung, pictured here, eventually broke from his mentor, Freud, partly because of differences over the importance of sexuality. (Bettmann/Corbis)

other episode of Germany endangering Europe. In these years he engaged in comparative studies, traveling to the United States, throughout Europe, and to Africa to study tribes in Kenya and Uganda.

The rise of Nazism seemed to lend credibility to Jung's study of archetypes, given that he had predicted a further menace arising from Germany. Some accused him of being a Nazi sympathizer, which seemed born out by his presidency of the International General Medical Society for Psychotherapy, which had Nazi connections. One reason such accusations seemed justified was Jung's unwillingness to see Adolf Hitler (1889–1945) as an aberration from German culture. He saw Hitler as the mouthpiece of Germany's collective unconscious, not as an individual man manipulating the populace. He later claimed that his views stemmed not from sympathy toward Nazism, but rather from the need to acknowledge that the views

and aims of the Nazis were products of a general trend rather than the influence of one person.

World War II ended Jung's travels, but not his writing. He developed his famous theory of synchronicity, or "meaningful coincidences," which abandoned any emphasis on causality in favor of understanding the meaning of random occurrences. Jung pointed once again to the collective unconscious, which could bring forth events and images simultaneously with no apparent causal connection between them. Like all of his ideas, this one was controversial. His earlier work, on archetypes and individuation, has proven more lasting, as have his distinctions between extrovert and introvert attitude types. Jung's conflict with Freud was a feud that, if nothing else, provoked fruitful debate for the development of psychology in the twentieth century.

See also Freud, Sigmund; Psychoanalysis; Psychology

References
Fordham, Michael. "Jung, Carl Gustav." In Charles Coulston Gillispie, ed., *Dictionary of Scientific Biography,* vol. VII. New York: Charles Scribner's Sons, 1973, 189–193.
Jung, Carl Gustav. *Memories, Dreams, Reflections.* New York: Vintage Books, 1989.
Karier, Clarence J. *Scientists of the Mind: Intellectual Founders of Modern Psychology.* Urbana: University of Illinois Press, 1986.

Just, Ernest Everett

(b. Charleston, South Carolina, 1883; d. Washington, D.C., 1941)

Ernest Everett Just was an African American marine biologist and embryologist. Born and raised in South Carolina, Just attended college at the predominantly white Dartmouth College, where he excelled in biology and history and took his degree in 1907. He received his doctorate in 1916 from the University of Chicago, where he studied under Frank Lillie (1870–1947). He became a professor of zoology at Howard University, a historically black college. In 1915, Just won the first Spingarn award from the National Asso-

Ernest Everett Just felt that racial prejudice limited his role as a scientist in the United States and moved to Europe. (Moorland-Spingarn Research Center Library)

ciation for the Advancement of Colored People, for contributions in the service of his race. These laudable achievements characterized him as a black man with honors relative to other African Americans. Just's goal, never truly attained, was to be recognized as the distinguished scientist he was, not simply to be applauded as a distinguished black scientist.

The challenges of Just's scientific career underscored the racial prejudices of U.S. society. Prejudice barred Just from the research-oriented universities where he craved to be. To distinguish himself as leading scientist, he needed a position, research support, and graduate students. Howard University was a safe appointment, because it retained his identity as a scientist during a time when white research universities would not have considered him. But at Howard, Just was en-

couraged to concentrate on teaching, not research. To skirt around such problems, Just applied for funding from numerous sources to support his research, but these successes were often problematic. Some grant-providing organizations insisted that he show contributions to the advancement of his race, whereas often Just was simply interested in contributing to marine biology.

To stay connected to the research community, Just spent most of his summers in the 1910s and 1920s in Woods Hole, Massachusetts, at the Marine Biological Laboratory (MBL). MBL provided the research opportunities that he lacked at Howard University, allowing him to pursue the research that made him a leading marine biologist. His research focused primarily on the fertilization of marine invertebrates, and he became a member of the editorial board of the journal *Physiological Zoology*. But Just was dissatisfied by what he felt was ill treatment of him by the white scientific community. Aside from MBL, no other white research institution would tolerate him, and he rarely succeeded in getting funds for research. His mentor, Frank Lillie, made little use of his reputation to help Just gain financial support and urged him to stay at Howard.

Embittered, Just gave up on the U.S. scientific community and traveled instead to Europe, where he lived in the 1930s. Ironically (given the political futures of these countries), he turned to Italy and Germany; he worked at the Naples Zoological Station and the Kaiser Wilhelm Institute for Biology in Berlin. In Europe, Just further developed his theoretical ideas of fertilization and development, emphasizing the role of the ectoplasm (the outer portion of the cell). Just argued that the ectoplasm's behavior was one of the prime factors in differentiation during development and should be considered by biologists as more than just a membrane covering the cell. This was the main theme of his major work, *The Biology of the*

Cell Surface, published in 1939. In the same year, he also published *Basic Methods for Experiments in Eggs of Marine Animals.* Disillusioned with U.S. society, Just left everything behind when he moved to Europe, including his family. He remarried a German woman and settled in France, where he worked at the Sorbonne.

When war broke out, Just was interned briefly by the Nazis. He soon moved back to the United States and rejoined the faculty at Howard University. He died shortly thereafter. Frank Lillie wrote an obituary and drew attention to Just's frustrations at Howard University, noting that he was "condemned by race to remain attached to a Negro institution unfitted by means and tradition to give full opportunity to ambitions such as his."

See also Embryology; Marine Biology; Race

References

Lillie, Frank R. "Obituary: Ernest Everett Just." *Science,* New Series, 95:2453 (2 January 1942): 10–11.

Manning, Kenneth R. *Black Apollo of Science: The Life of Ernest Everett Just.* New York: Oxford, 1983.

K

Kaiser Wilhelm Society

The Kaiser Wilhelm Society was a scientific body that served as the umbrella organization for dozens of research institutes in Germany in the first half of the century. Beginning in 1905, several leading chemists in Germany hoped to found an organization that would connect science to national interests. Emil Fischer (1852–1919), Wilhelm Ostwald (1853–1932), and Walther Nernst (1864–1941) lobbied to create such a body devoted to chemical research, financed jointly by the chemical industry and the imperial government itself; the body would be both academic and practical. In their view, this would help to facilitate the process of imperial Germany's modernization, tying its fortunes closely to scientific advancement. Initially rejected, the project was modified to be more inclusive (not just chemistry), and it became the Kaiser Wilhelm Society for the Promotion of Science, or KWG (Kaiser-Wilhelm-Gesellschaft zur Förderung der Wissenschaften), named for the reigning emperor of Germany.

The main effort of the KWG was in establishing and managing research institutes. It financed the first two institutes with the help of the Verein Chemische Reichsanstalt and a wealthy industrialist named Leopold Koppel (1854–1933). They were both founded in 1911, devoted to chemistry and physical chemistry. The latter, formally called the Kaiser-Wilhelm Institute for Physical Chemistry and Electrochemistry, reflected a strong desire to link physics and chemistry research to areas of broad industrial interest. Each institute was controlled by a council composed of donors and government officials, along with a scientific board and the director of the institute. The first two institutes were located on a portion of the Prussian Royal Estate of Dahlem, today a suburb of Berlin. The opening ceremonies for the first two institutes, held in 1912, were attended by eminent scientists and Kaiser Wilhelm II (1859–1941). By 1914, two more institutes had been added: the Institute for Experimental Therapy (also in Dahlem) and the Institute for Coal Research (in Mülheim). More than thirty other institutes for science were created as part of the KWG system in subsequent years.

The goal of the KWG was shaped considerably by World War I, which began shortly after it was founded. Fritz Haber (1868–1934) became the first director of the Institute for Physical Chemistry, and he soon turned the focus of research at the institute to military matters. In particular, the scientists worked on improving explosives, resulting in a serious accident in 1914 during which an explosion killed one of the physicists, Otto Sackur (1880–1914). It was during this period that scientists began to develop chemical weapons as a way to break the stalemate of trench warfare. The institutes were engaged in this kind of work, either developing chlorine and mustard gas or devising ways to protect soldiers against enemy attack through special respiratory equipment (gas masks). The scale of this work turned the KWG into a large organization.

After the war, scientists at the institutes turned their attention to basic research or efforts to help their country survive a difficult period marked by harsh indemnity payments and generally poor economic conditions. Haber, for example, launched a project to study the possibility of extracting gold from seawater to help make the payments. The feasibility of such a scheme was based on early experimental evidence that turned out to be grossly overestimated. Meanwhile, specialized departments within the KWG system proliferated, including one for atomic physics led by James Franck (1882–1964). His work at the Institute for Physical Chemistry led to a Nobel Prize in 1925. His and others' work helped the institute become a favored destination for researchers throughout Europe and North America, seeking training and collaboration.

Although the institutes were at the center of science in the 1920s, especially the transformation of theoretical and experimental physics, this came to an end with the rise of fascism and the coming of Nazis into power in Germany. In 1933, Haber was ordered to fire racially inferior (i.e., Jewish) employees; Haber, himself a Jew, was exempt from this because of his past service and his eminence in Germany. But rather than carry out the orders to remove other Jews from their posts, Haber resigned. Other department heads also resigned; in general, many of the leading scientists in Germany emigrated.

Physical chemist Otto Hahn (1879–1968) initially took over Haber's position as director, but soon he was replaced by scientists whose views were more in line with those of the government.

Throughout this period, the president of the Kaiser Wilhelm Society, physicist Max Planck (1858–1947), attempted to help German science survive amidst the demands of the Nazi regime. The KWG depended on the Third Reich for support and Planck's goal was to salvage what he could of German science while trying not to offend the Nazis. Planck, the originator of quantum theory, was more than seventy years old; he used his position as a venerable elder physicist to protect Jewish colleagues while coddling the Nazis. This meant accepting Nazi agents throughout the "scientific" society and losing battles over leadership positions. Ultimately, other scientists more loyal to the Nazis decried the KWG as a society not for science, but for the social legitimization of Jews. Planck had to fire his colleagues and was himself forced out.

The KWG suffered deeply from the effects of Nazi rule. During World War II, its institutes were directed toward war research, as they had been during World War I. After Germany's defeat in 1945, science underwent a major reorganization under foreign occupation forces. The names of old institutions were changed. In 1948, the Kaiser Wilhelm Society became the Max Planck Society, a change linking it to a respected scientist rather than to an emperor whose name seemed (to the occupation forces) synonymous with militarism.

See also Chemical Warfare; Haber, Fritz; Nazi Science; Planck, Max; World War I; World War II

References
Goran, Morris. *The Story of Fritz Haber*. Norman: University of Oklahoma Press, 1967.
Heilbron, J. L. *The Dilemmas of an Upright Man: Max Planck and the Fortunes of German Science*. Cambridge, MA: Harvard University Press, 2000.

Johnson, Jeffrey Allan. *The Kaiser's Chemists: Science and Modernization in Imperial Germany.* Chapel Hill: University of North Carolina Press, 1990.

Macrakis, Kristie. *Surviving the Swastika: Scientific Research in Nazi Germany.* New York: Oxford University Press, 1993.

Kammerer, Paul

(b. Vienna, Austria, 1880; d. Austria, 1926)

Paul Kammerer was a Viennese biologist who defended the Lamarckian interpretation of evolution during a period when it vied for dominance against Darwinian natural selection and Mendelian genetics. He was best known for his role in a scandal involving his studies of midwife toads. Kammerer did not shy from drawing out the broad meaning of his scientific work, which seemed to indicate that man could actively shape the destiny of his own species. This elicited the suspicions of his peers and led ultimately to his downfall and suicide.

Kammerer claimed to have demonstrated, in the laboratory, that characteristics acquired in the lifetimes of various animals could be inherited. The midwife toad, *Alytes obstetricans,* was a land variety of toad that lacked a particular pad used by water varieties of toads during mating. Kammerer claimed to have bred midwife toads in water and, after a few generations, they developed the pad and a dark spot much like those of water toads. In addition, he claimed that the acquired trait was inherited from one generation to the next. The concept of the inheritance of acquired characteristics was fundamental to the Lamarckian interpretation of evolution, but it was not consistent with Darwinism. Jean-Baptiste Lamarck (1744–1829) had noted that, as in the case of giraffes' necks growing long because of continued exertion, species developed through the action of the organism. Charles Darwin (1809–1882) accepted change only as a result of random variation prior to the organism's devel-

Paul Kammerer was a biologist who claimed to demonstrate Lamarkian inheritance in the amphibians he bred. He committed suicide amid allegations of fraud. (Bettmann/Corbis)

opment. Darwinism rejected both design in nature and the inheritance of acquired characteristics.

The findings of Kammerer's experiments, performed prior to World War I, seemed to be a triumph for Lamarckian evolution. It was also a triumph for philosophers and thinkers of all stripes who were happy to see that nature allowed for progressive, directed evolution. It lent credence to the view that society itself could be engineered for the better. It became a weapon in the arsenal of eugenicists who wanted, through good breeding, to purify the racial composition of their own citizenry. The only problem was that Kammerer's results could not be replicated in other laboratories. After the war, he toured with preserved specimens, and in 1923 he lectured about Lamarckian inheritance in the United States and Britain. Skeptical, British geneticist William Bateson (1861–1926) insisted that Kammerer allow other scientists to have a closer look at his specimens. After a few years of reluctance, he allowed U.S. herpetologist Gladwyn K. Noble (1894–1940) to view a specimen; Noble found no pad at all, but he saw that the dark spot had been colored by an ink injection. He reported these results, to the dismay of the scientific world. Less than two months later, in 1926, Kammerer shot himself in the mountains of Austria.

Kammerer, a socialist, became a hero in the Soviet Union, where he was much respected (he had been offered a position there) for his defense of Lamarckian evolution. Biologists in that country rejected both Mendelian genetics and Darwinian natural selection; Kammerer's Lamarckian outlook was an attractive alternative. In his 1972 book, *The Case of the Midwife Toad,* author Arthur Koestler attempted to revive Kammerer's reputation. He observed that someone besides Kammerer probably tampered with the specimen, either to discredit him or to help him by making the marks more prominent. Many factors contributed to his suicide in addition to his scientific setback, including a complicated love affair. Still, the revelation of a scientific fraud, coupled with his suicide, did little to strengthen the position of Lamarckian evolution vis-à-vis Darwinian evolution.

See also Bateson, William; Evolution; Genetics; Lysenko, Trofim

References

Bowler, Peter J. *Evolution: The History of an Idea.* Berkeley: University of California Press, 1989.

Gould, Stephen Jay. "Zealous Advocates." *Science,* New Series 176:4035 (12 May 1972): 623–625.

Koestler, Arthur. *The Case of the Midwife Toad.* New York: Random House, 1972.

Kapteyn, Jacobus

(b. Barneveld, Netherlands, 1851; d. Amsterdam, Netherlands, 1922)

Dutch astronomer Jacobus Kapteyn was a leading figure not only in astronomy, but also in international cooperation. Kapteyln was based at the University of Groningen, and his interests lay in the distribution and motions of stars. But at the end of the nineteenth century, data on stars was poor; few had taken the time to provide the basic modern measurements. Kapteyn, in addition to compiling a major catalogue of his own toward the end of the nineteenth century, became a major impetus behind twentieth-century cooperative studies throughout the world to provide the astronomical community with the information needed to improve knowledge of the universe.

Unlike many other centers of astronomical study, the University of Groningen did not have a large telescope of its own, or even a useful observatory. Kapteyn conducted his work in a couple of rooms in the physiological laboratory. In place of a telescope, Kapteyn studied photographic plates taken from data collected at the Cape Observatory between 1885 and 1890. He used a theodolite, placed away from the photographic plates at the same distance as the focal length of the original telescope. In this way, an apparatus typically used for land surveying was used to measure the coordinates of stars. He also measured the apparent magnitudes of the stars. To help with some of the menial labor, a local prison put some of its convicts at Kapteyn's disposal. The result was a catalogue of 454,875 stars, published between 1896 and 1900.

In 1904, Kapteyn announced a discovery of two "star streams" while at a congress in St. Louis, Missouri. This stemmed from his investigations of stellar distances. Of the two methods for calculating distance, measuring either proper motion or apparent magnitude, Kapteyn preferred the former because magnitudes depended on the properties of the stars and thus involved too many variables. Proper motions, the movement of individual stars relative to each other, had been discovered in the eighteenth century by Edmond Halley (1656–1742)—stars previously had been believed to be fixed in place. Kapteyn investigated the distribution of stars according to their velocities. Like others, he assumed that stars moved about randomly, like gas molecules, with no rhyme or reason to the direction. But soon Kapteyn came to the conclusion that stars seemed to prefer some directions. He identified two separate, though intermingled, groups of stars having different mean motions in relation to the sun. The existence of two "star streams" astonished the astronomical community, suggesting some kind of order in the apparent randomness of the universe. In particular, Kapteyn's discovery prompted some of the influential work by Karl Schwarzchild (1873–1916) on stellar motion and distribution.

Kapteyn used his influence to coordinate an international plan of action to advance the science of astronomy. He realized that position and apparent brightness were known for fewer than a million stars, whereas proper motions were known for a few thousand, and trigonometric parallaxes were known for fewer than a hundred. After discussions with colleagues in various countries, he initiated a plan to have astronomers all over the world concentrate work on some two hundred stellar areas, taking the same kinds of data. Kapteyn's plan became a hallmark of international cooperation. It evolved under the guidance of a committee of prominent astronomers (including Kapteyn himself) and became a crucial component of the International Astronomical Union. Kapteyn was a strong supporter of scientific internationalism, believing that it was each scientist's duty to rise above political animosities. Although many shared his view, World War I hurt this vision badly, in all scientific disciplines. Kapteyn was saddened to see Germany excluded from international science after the

war, and he even resigned his membership to his homeland's Royal Netherlands Academy of Sciences and Letters because of its complicity in such actions.

Kapteyn capped his career with a model of the galaxy, published in the year of his death in 1922. He hoped to develop a dynamical theory, incorporating new knowledge in density distribution and motions. He represented the Milky Way as resembling a squashed sphere, rather than a rugby ball or U.S. football. The system revolved around an axis, about one-fifth the length of the diameter from one edge of the Milky Way to the other side. He estimated the thinning out of stars with distance from the axis, and he believed the sun to be fairly near the center. It was a major effort to provide a systemic understanding of stars, but rival models soon replaced many of his views.

See also Astronomical Observatories; Astrophysics; Pickering's Harem

References
Blaauw, A. "Kapteyn, Jacobus Cornelius." In Charles Coulston Gillispie, ed., *Dictionary of Scientific Biography,* vol. VII. New York: Charles Scribner's Sons, 1973, 235–240.
North, John. *The Norton History of Astronomy and Cosmology.* New York: W. W. Norton, 1995.

Koch, Robert

(b. Clausthal, Oberharz, Germany, 1843; d. Baden-Baden, Germany, 1910)

Most of Robert Koch's life and work belong firmly to the nineteenth century, but he also had considerable influence on the twentieth century. Many of the fundamental principles of bacteriology were developed or modified by him and his colleagues in Berlin, and he became an international celebrity for his isolation of the agents of anthrax, tuberculosis, and cholera, all scourges of the nineteenth century. Koch was directly responsible for shaping public health in the early twentieth century, based on knowledge of microbes as the origin of disease. His work led not only to scientific discoveries, but also to a public transformation in conceptions of personal hygiene as a way to prevent sickness and disease.

Koch's achievement in the late nineteenth century was in identifying the specific bacterium that caused a certain disease. In 1882, he demonstrated the role of the tubercle bacillus in causing all forms of tuberculosis. He developed a method of identification by isolating a microorganism in tissue, then injecting it into a healthy individual; if the identical disease resulted, the proper agent had been identified. By the early 1900s, scientists had used similar techniques to identify a host of diseases from the bubonic plague to syphilis.

Koch posited the novel hypothesis that some diseases, such as diphtheria and typhoid, could be carried by healthy individuals—showing no symptoms—and passed on to others. He also had shown that the tubercle bacillus was present in sputum; the implication was that there was a widespread ignorance about the contagious effects of spitting, coughing, and even breathing. Thus at the dawn of the twentieth century, public health officials were confronted with the severe challenges of tracking, and halting, infections across populations. What were the most likely intermediaries for bacteria? This problem sparked intensive studies of the channels of disease contagion.

When typhoid fever broke out in the Ruhr region of Germany in 1901, Koch was recruited to help find ways of containing it. Koch emphasized the need for sanitary water supplies and sewage disposal as well as the need to avoid contact with those already infected. He began investigations of the spread of the disease, which resulted in the creation of new laboratories and health officials trained in bacteriology. His efforts helped to halt the epidemic and reduce the death toll.

Koch traveled extensively, including several years in the first decade of the twentieth century in equatorial Africa, investigating local diseases. He experimented on monkeys

Robert Koch, the German bacteriologist whose work was influential in public health measures. (Bettmann/Corbis)

in order to isolate and study the bacteria of Coast and Texas fevers, sleeping sickness, and African relapsing fever. His work on tuberculosis, however, was the most widely praised, and he received the Nobel Prize for it in 1905. He returned to Africa in 1906 as leader of the German Sleeping Sickness Commission and conducted exhaustive studies of that disease over the next couple of years. He returned to Berlin in 1907 and spent the remainder of his life improving methods of tuberculosis control. His work in bacteriology provided not only a basis for scientific study of microorganisms, but also a transformation of public understanding of disease; his work demonstrated the need for rigorous public health measures.

See also Ehrlich, Paul; Medicine; Microbiology; Public Health

References

Brock, Thomas D. *Robert Koch: A Life in Medicine and Bacteriology.* Madison, WI: Science Tech Publishers, 1988.

Dolman, Claude E. "Koch, Heinrich Hermann Robert." In Charles Coulston Gillispie, ed., *Dictionary of Scientific Biography,* vol. VII. New York: Charles Scribner's Sons, 1973, 420–435.

Tomes, Nancy. *The Gospel of Germs: Men, Women, and the Microbe in American Life.* Cambridge, MA: Harvard University Press, 1998.

Kurchatov, Igor

(b. Sim, Ufimskaya guberniya [later Ufimskaya oblast], Russia, 1903; d. Moscow, USSR, 1960)

Kurchatov was the father of the Soviet atomic bomb. In 1934, he received his doctorate in physics and mathematics and spent the next several years studying the recently discovered neutron. Like scientists in Germany, Britain, and the United States, he became interested in nuclear fission and the intriguing possibility of nuclear chain reactions. The center of Soviet physics was Leningrad; it was home to the Leningrad Physical Technical Institute, led by Abram Ioffe (1880–1960), and the Radium Institute, led by V. G. Khlopin (1890–1950). Kurchatov joined Ioffe's group and was fast becoming a leader in nuclear studies. He and his colleagues published on nuclear research even after this field slipped under a veil of secrecy in other countries.

All this changed when Germany declared war on the Soviet Union in 1941. Scientists abandoned their institutes for safer locales in the east. Kurchatov left his nuclear work and put himself to more immediate use by helping the Black Sea fleet develop methods of protecting ships from magnetic mines. During the war some scientists, especially Georgii Flerov (1913–1990), urged Kurchatov and other leading scientists to push the government harder to support nuclear research, reasoning that an atomic bomb was going to be developed elsewhere while the Soviets were left behind. Flerov even wrote to Joseph Stalin (1879–1953) himself, in a letter perhaps analogous to the one Albert Einstein (1879–1955) wrote to Franklin Roosevelt (1882–1945) in the United States.

The Soviet decision to build an atomic bomb was made around the time of the battle of Stalingrad. Until this decisive turn of events in the war, Stalin had deemed scientific work on uranium as superfluous to the immediate concerns of the war and as a waste of resources. But after a long struggle at Stalingrad, Soviet forces surrounded the German invaders and compelled them to surrender in February 1943. The Soviet counteroffensive was codenamed "Uran," which some scholars argue is a reference to the fact that Stalin decided at this time to start up work again on the "uranium problem."

Although Soviet espionage had been successful in tracking atomic research, especially the British work, most Soviet scientists were not aware of it. By early 1943, Kurchatov still was not convinced that an atomic bomb was possible and had no idea how long such a project would take. Shortly after the battle of Stalingrad, the government turned over the intelligence materials to him, revealing some of the work already accomplished in Britain and the United States. Kurchatov's doubts disappeared.

Physicists Iulii Khariton (1904–1996) and Iakov Zel'dovich (1914–1987) had conducted research on nuclear chain reactions, hoping to use uranium and heavy water as moderator, but had concluded that such efforts were hopeless. The British data showed otherwise. In addition, the data indicated a promising route to the bomb by creating the artificial element plutonium. There is no doubt that espionage played a central role in convincing Soviet physicists, and thus the Soviet government, to begin work in earnest on an atomic bomb, and it is likely that the project was intended as a postwar weapon, not as a decisive weapon against Germany in the current war. Kurchatov became the scientific director of the project, and he continued in this role as leader of nuclear projects until his death in 1960.

When the war ended, the pressure on Kurchatov and others to produce an atomic bomb was immense. The confrontation between the Soviet Union and the United States intensified with each year, while the Americans enjoyed the diplomatic leverage of possessing a monopoly on atomic bombs. Buoyed by secrets provided about the U.S. project by Klaus Fuchs (1911–1988), who was later convicted and imprisoned for espionage, and the availability of uranium from mines in Germany and Czechoslovakia, the project under Kurchatov advanced faster than most expected. Kurchatov worked well with colleagues at all levels of the chain of command. He was known affectionately as "the Beard" (because of his distinctive long beard) and less affectionately as "Prince Igor" to the scientists who worked under him. He also had the skills needed to work under the Lavrentii Beria (1899–1953), notorious for his leadership of the secret police and his role in other internal affairs, who administered the atomic bomb project after 1945.

Kurchatov's team conducted the first Soviet test of an atomic device on 29 August 1949, a mere four years after the United States dropped atomic bombs on Hiroshima and Nagasaki. Kurchatov and his colleagues were showered with honors by the government; they had broken the atomic monopoly and provided Stalin with an atomic weapon. Kurchatov's work continued, including the development of the hydrogen (fusion) bomb in 1953. When he died, his ashes were laid to rest in the Kremlin.

See also Atomic Bomb; Cold War; Espionage; Soviet Science; World War II
References
Dorfman, J. G. "Kurchatov, Igor Vasilievich." In Charles Coulston Gillispie, ed., *Dictionary of Scientific Biography,* vol. VII. New York: Charles Scribner's Sons, 1973, 526–527.
Holloway, David. *Stalin and the Bomb: The Soviet Union and Atomic Energy, 1939–1956.* New Haven, CT: Yale University Press, 1994.
Josephson, Paul R. *Red Atom: Russia's Nuclear Power Program from Stalin to Today.* New York: W. H. Freeman and Company, 2000.

L

Lawrence, Ernest

(b. Canton, South Dakota, 1901; d. Palo Alto, California, 1958)

Ernest Lawrence, a U.S. physicist, invented the cyclotron in the 1930s and was a leading figure in high-energy particle physics. He attended the University of South Dakota, where he took a degree in chemistry in 1922. He received his doctorate in physics from Yale University in 1925. Three years later, he took up a position at the University of California–Berkeley; over the next two decades, Lawrence transformed that institution into the world's focal point in high-energy physics and the location of the most advanced particle accelerators.

In 1929, Lawrence conceived of a particle accelerator that would, by electromagnetic force, keep the particles rotating in a spiral before finally being ejected. By 1932, Lawrence and one of his graduate students, M. Stanley Livingston (1905–1986), had invented a device that produced 80,000-volt protons. Particle acceleration seemed to be the most effective way to conduct physical experiments on atoms; particles needed a lot of energy to overcome the repulsion of other atomic nuclei in order to collide with them. *Cyclotron* was the name used by local researchers to describe Lawrence's accelera-

tors, and the appellation stuck. Cyclotrons became effective "atom smashers," capable of creating artificial elements and artificial radioactivity. Other laboratories built them, or requested samples from the experiments conducted at Berkeley. Although the Nobel Prize in Physics was not typically awarded for technological inventions, in this case the Nobel Committee made an exception and awarded the prize to Lawrence in 1939.

Lawrence's name is often associated with *Big Science,* a phrase used to describe the kind of scientific inquiry he helped to create: large teams of researchers, well-funded laboratories, and expensive, complex equipment. The Berkeley group did not accomplish a great deal of pure physics in the 1930s, mainly because of Lawrence's focus on building newer and better particle accelerators. The results of these efforts were used by others. But with the development of more efficient and faster cyclotrons at Berkeley, Lawrence's ability to attract researchers and funding was all the more impressive because it was accomplished during times of severe financial hardship, the Great Depression. Because of his experience in the 1930s, he became an effective manager in the early stages of the U.S. atomic bomb project, the largest scientific project ever undertaken in the

United States, in terms of scientific and material resources. He led the effort to "enrich" uranium through the process of electromagnetic separation of U-238 and U-235.

In subsequent years, Lawrence continued to occupy a leading place in all nuclear matters. A political conservative, he found himself estranged from former colleagues such as J. Robert Oppenheimer (1904–1967), who had directed the scientific team at Los Alamos during the war (Lawrence had backed him for this post). His views during the early Cold War diverged sharply from those of other scientists who sought to limit the role of atomic weapons. He and Oppenheimer had become friends at Berkeley in 1929, and they had complemented each other there during the 1930s, Oppenheimer as a theoretician and Lawrence as an experimentalist; their relationship soured toward the end of the war and afterward. They disagreed particularly in 1949, on the question of whether the United States should develop a "super" weapon, a hydrogen bomb. Such weapons, based on principles of atomic fusion, would be a thousand times more powerful than the fission weapons dropped on Hiroshima and Nagasaki. Lawrence favored it, criticizing Oppenheimer's (and others') outspoken opposition to it. Lawrence believed that it should be developed to counter the Soviet threat, whereas Oppenheimer saw it as a weapon of genocide. Lawrence's view prevailed, and the United States developed thermonuclear weapons.

The success of high-energy particle physics at Berkeley owed a great deal to Lawrence's personal energy. He expected a great deal from his colleagues and students, and exerted a great deal of pressure on them. He evidently exerted similar pressures on himself: Lawrence suffered during his life from ulcerative colitis, a condition exacerbated by stress. Ultimately he died of it.

See also Artificial Elements; Atomic Bomb; Cyclotron; Manhattan Project; Physics

References
Childs, Herbert. *An American Genius: The Life of Ernest Orlando Lawrence, Father of the Cyclotron.* New York: Dutton, 1968.
Davis, Nuel Pharr. *Lawrence and Oppenheimer.* New York: Simon & Schuster, 1968.
Heilbron, John L., and Robert W. Seidel. *Lawrence and His Laboratory: A History of the Lawrence Berkeley Laboratory,* vol. 1. Berkeley: University of California Press, 1989.
Herken, Gregg. *Brotherhood of the Bomb: The Tangled Lives and Loyalties of Robert Oppenheimer, Ernest Lawrence, and Edward Teller.* New York: Henry and Holt, 2000.

Leakey, Louis

(b. Kabete Mission, near Nairobi, Kenya, 1903; d. London, England, 1972)

Louis Leakey was the dominant voice from the field of anthropology, who claimed that human beings originated in Africa. Most anthropologists of his day were convinced that Asia was the birthplace of man, because the oldest known fossils had been found in China and Java. As a paleoanthropologist, Leakey sought new fossil evidence to support his view that Africa held remains of human ancestors far more ancient. He was born and raised in Kenya, speaking not only English but also the language of the Kikuyu tribe. He was initiated into the tribe at age thirteen, and he later wrote a book about Kikuyu culture. He studied at Cambridge University, taking degrees in anthropology and archaeology in 1926. He met Mary Nicol (1913–1996), of London, in the 1930s while married to his first wife. By 1936 he was divorced, and he married Mary; they now were not only a couple but also scientific collaborators.

Leakey's notoriety came from his long experience in the field, searching for fossil remains in Africa. Owing to a head injury he received while playing rugby, Leakey took some time off from his studies at Cambridge and instead made a trip to Tanzania in 1924 as part of a paleontological expedition. Later, in the 1930s, he came to focus on Tanzania's

Olduvai Gorge. Leakey's first book, *The Stone Age Cultures of Kenya Colony* (1931), detailed some of his important findings of the late 1920s. One of these was the discovery of specific tools known to be in use in other parts of the world, which demonstrated a level of African social sophistication on par with other areas. In the early 1930s, he found a number of fossils that he believed to be ancient, and he insisted that a certain skull of a man in the Olduvai Gorge was among the oldest then known. Amidst scientific controversy, he had to retract his statements. Leakey continued excavations at Olduvai in the hopes of finding the most ancient human ancestors, but his reputation suffered in the mid-1930s because some of his claims turned out to be erroneous and his field methods were criticized. During this period he wrote his 1936 autobiography, *White African.*

Leakey came to international fame in the late 1940s, because of excavations made on Rusinga Island in Lake Victoria. In 1948, he and Mary Leakey discovered a skull that he believed to be 20 million years old. He named it *Proconsul africanus;* although some suspected it was a progenitor of man and ape, scientists soon discarded the notion that it was a direct ancestor of either. The discovery led to fame for the Leakeys and for increased financial support, particularly by the National Geographic Society. Leakey's quest for a hominid fossil to prove man's origins in Africa bore fruit in 1959, after some thirty years of searching. Mary Leakey found a skull in deposits accompanied by stone tools. Aged some 1.75 million years, Leakey claimed this hominid, *Zinjanthropus boisei,* as a human ancestor. Called "Zinj" for short, it was later determined to be *Australopithecus.* In later years, Leakey and his other collaborators identified more fossils as distant ancestors of humans.

Occasionally Leakey was viewed as a showman or adventurer rather than a scientist, more concerned with proving his theory about man's origins than with scientific evaluation of evidence. Nevertheless, one of Leakey's most valuable attributes to science was his attention to detail and context. When he and his team uncovered fossils, they recorded a great deal about the geological, paleontological, and archaeological context. Much of this was owing to the archaeological efforts of Mary Leakey. Neither Louis Leakey nor Mary, nor their son Richard—all prominent paleontologists—took doctoral degrees. Louis Leakey prided himself on being an outsider, working beyond the confines of the academic world. Yet he was loved by the public; he gave numerous lectures, and periodicals such as *National Geographic* made Leakey's work known throughout the English-speaking world. Leakey died of a heart attack while on his way to a speaking engagement in London.

See also Anthropology; Missing Link
References
Cole, Sonia. *Leakey's Luck: The Life of Louis Seymour Bazett Leakey, 1903–1972.* New York: Harcourt Brace Jovanovich, 1975.
Isaac, Glyn L., and Elizabeth R. McCown, eds., *Human Origins: Louis Leakey and the East African Evidence.* Menlo Park, CA: Benjamin, 1976.
Leakey, L. S. B. *By the Evidence: Memoirs, 1932–1951.* New York: Harcourt Brace Jovanovich, 1974.
Morell, Virginia. *Ancestral Passions: The Leakey Family and the Quest for Humankind's Beginnings.* New York: Simon & Schuster, 1995.

Leavitt, Henrietta Swan

(b. Lancaster, Massachusetts, 1868; d. Cambridge, Massachusetts, 1921)

Henrietta Swan Leavitt discovered a relationship between brightness and periodicity in variable stars, enabling some of the most far-reaching theoretical changes in astronomy and cosmology in the twentieth century. Leavitt attended what became Radcliffe College; after Leavitt graduated in 1892, an illness left her almost totally deaf. In 1895, she took a position at Harvard College Observatory, where she worked under

Henrietta Swan Leavitt was one of several women working in "Pickering's Harem" who made fundamental contributions to astronomy. (Photo courtesy Margaret Harwood, AIP Emilio Segrè Visual Archives, Shapley Collection)

the observatory's director, astronomer Edward Pickering (1846–1919). She became a permanent staff member in 1902.

Leavitt's work, in part, was to make a survey of Cepheid variable stars in the Magellanic Cloud, a cluster of distant stars. She developed expertise in photometric astronomy, taking measurements from photographic plates acquired by exposure under the observatory's telescopes. These plates were very useful because they were more sensitive than the human eye. Pickering was in the process of amassing a collection of such plates, not only from the Harvard College Observatory but also from other observatories throughout the world. Leavitt took a special interest in the plates of stars from the Magellanic Cloud, and she soon discovered the abundance of variable stars in them, called Cepheids. These stars appeared to "blink" over time, as their brightness changed in a cyclic fashion. In 1908, Leavitt noted that the brightest of these stars also had the longest period of variability.

In 1912, Leavitt extended this analysis to include more stars and proposed a relationship between the apparent brightness (or luminosity) of the Cepheid variables and their periodicity (the time between "blinks"). She and others realized that one needed only to calculate the distance to these Cepheids, which almost certainly were roughly the same distance from the earth, to have a useful yardstick for measuring other distances. For example, if a researcher found two stars of the same apparent luminosity but with differing periods, he or she could estimate the difference in distance based on the period. Unfortunately, Leavitt herself was not in a position to mount a research program to do this. In 1913, Danish astronomer Ejnar Hertzsprung

(1873–1967) determined the distance to some Cepheid variables; using Leavitt's method, the distances to other Cepheids then could be determined. The Cepheid variables, using Leavitt's discovery of the luminosity-period relationship, became powerful tools for astronomers and cosmologists in subsequent years. Among other things, in the 1920s, they led Harlow Shapley (1885–1972) to extend estimates of the size of the Milky Way by ten times, and they led Edwin Hubble (1889–1953) to demonstrate that some nebulae were far too distant to be considered part of the Milky Way. The latter discovery helped to establish the galaxy as the basic structural unit of the universe.

Leavitt was one of several talented women who made a career in the field of astronomy in the first half of the twentieth century, largely because of the efforts of Pickering to hire women. She was conducting "women's work," at the time limited to the most tedious aspects of science such as recording data and conducting basic calculations. Leavitt was part of a team of female workers at the observatory, and they were occasionally called "Pickering's harem." Some of them, like Leavitt, were physically disabled in some way. A great part of Leavitt's accomplishment is that she did it while confined to what male astronomers would not have considered creative, truly scientific work. The fact that her work gained her a great deal of notoriety is fully justified given the deep and quick impact it had on the world of astronomy. She also became an inspirational icon not only for women scientists but also for the deaf.

See also Astronomical Observatories; Hubble, Edwin; Pickering's Harem; Women

References
Jones, Bessie Z., and Lyle Boyd, *The Harvard College Observatory: The First Four Directorships, 1839–1919.* Cambridge, MA: Harvard University Press, 1971.
North, John. *The Norton History of Astronomy and Cosmology.* New York: W. W. Norton, 1995.
Rossiter, Margaret W. "'Women's Work' in Science." *Isis* 71:3 (1980): 381–398.

Light

Light as we know it—visible light—is only one part of the broader phenomenon of electromagnetic radiation. Although other kinds of electromagnetic radiation would be identified in the twentieth century, this aspect of light had already been identified by James Clerk Maxwell (1831–1879) in the 1870s. The understanding of light changed dramatically during the first part of the twentieth century, while it continued to be used as a tool for understanding the universe. Physicists in the nineteenth century believed that light, like sound, needed a medium through which it could be propagated. A ubiquitous medium known as the "ether," not detectable but theoretically necessary, was invented by scientists to account for the fact that light traveled through seemingly empty (except for the ether) space. But already light appeared to have some special properties. Experiments in the 1880s had demonstrated that light moved at a constant speed in all directions, despite widespread belief that the earth careening through space should have a minute effect on its speed in one direction and not another.

The intensive study of X-rays and radioactivity (the latter produced gamma rays) at the dawn of the twentieth century opened questions about the nature of radiation and the structure of both light and matter at the smallest scales. The work of Albert Einstein (1879–1955) transformed understanding of light in at least two ways. First, he demonstrated that electromagnetic radiation carries momentum. Drawing on recent work by Max Planck (1858–1947) on quantum theory, Einstein believed that light quanta carried momentum and their impacts could be measured. This meant that such quanta (*photons,* light quanta became called) were equivalent to a certain amount of mass. Einstein's famous equation, $E = mc^2$, referred to energy being the product of mass and the square of the speed of light. The finding, that light carried momentum and could exert a small amount of force, appeared to demonstrate the particulate nature of light rather than its wave nature.

Einstein also set the velocity of light as the speed limit of the universe. In his 1905 theory of special relativity, Einstein disposed of the notion of the ether and asserted that no medium was necessary to propagate light. In getting rid of the ubiquitous medium, he also abandoned the whole notion of fixed points in space. He observed that all physical measurements must be made between moving objects, and there is no objective "fixed" point from which to observe motion. At high speeds, objects moving past each other will observe time appearing to slow, mass increasing, and length contracting, in the object flying past. If the speed of light could be reached by anything other than light, time would appear to stop, mass would appear to be infinite, and length would seem to shrink to nothing.

In addition to the concepts of photons and light's constant speed, Einstein proposed with his theory of general relativity that light itself is malleable like the rest of the universe and that it is bent by gravity. British astrophysicist Arthur Eddington (1882–1944) set out to prove this by measuring the effects of starlight in close proximity to the sun. During his 1919 expedition to view a solar eclipse on an island off the coast of Africa, Eddington found that Einstein had been right—the shortest distance between two points, the path always followed by light, could sometimes be a curve.

Relativity was only one of the many manipulations of light in the twentieth century. Light was the principal tool for understanding the cosmos. Before the widespread adoption of radio astronomy after World War II, light had no competitor in the field of astronomy. Henrietta Swan Leavitt (1868–1921), examining photographic plates of the stars at Harvard College Observatory, noted a relationship between the brightness of some flashing stars and the period of the flashing. These "Cepheid variables" became a tool for judging distances, because one needed only to find stars of the same apparent luminosity (brightness) that had differing periods in order to judge rough differences in distance to the earth. Aside from this important finding, by far the most useful application of light phenomena in science was its spectrum, which is produced when light is passed through a prism. Spectral analysis formed the basis of the blackbody problem that sparked the beginning of quantum physics. In astronomy, spectral analysis was very practical. Solids and gases yielded different kinds of spectra: Whereas luminescent solids produce perfect spectra, luminescent gases produce spectral lines rather than all the colors of the rainbow. Each element, in gas form, produces its own "signature" pattern of spectral lines. When gases surround a luminescent body, those lines are absorbed in the light being emitted. Scientists on earth, viewing the spectrum of a star, see black absorption lines, which tell the scientists precisely what kinds of gases surround the star. Using this knowledge, U.S. astronomer Edwin Hubble (1889–1953) in the 1920s found that in the most distant sources of light, the spectral lines were strangely shifted toward to the red end of the spectrum. This "red shift" was a result of elongated light waves, much like the Doppler effect upon sound—that is, the sounds of a siren approaching and receding yield different tones, because of contraction and elongation of sound waves. From this, Hubble determined that the universe was expanding.

The discovery of X-rays and radioactivity at the end of the nineteenth century, along with the twentieth-century development of new theoretical approaches—quantum physics, relativity, quantum mechanics—transformed understanding of light. By midcentury, light was understood as only one kind of electromagnetic radiation, of which X-rays, gamma rays, radio waves, and other kinds of waves, previously thought to be separate entities, were also varieties. The differences among these kinds of waves were wavelength. Although light appeared to act like a wave, the "Copenhagen interpretation" of quantum mechanics, developed and promoted by Werner Heisenberg (1901–1976) and Niels Bohr (1885–1962), proposed something about light that was both startling and paradoxical. They were aware that Einstein also demonstrated the particle-like nature of light, as in the case of the photon. But the principal of complementarity, developed largely by Bohr in 1927, asserted that two seemingly contradictory explanations of phenomena—in this case the wave or particle interpretation—could be perceived as complementary. One did not need to choose between them, because neither one represented a more fundamental "reality" than the other. At the subatomic scale, electromagnetic radiation or even the components of the atom, such as electrons, cannot be conceived of as one or the other, but rather one must accept the wave-particle duality. Light behaved as both a particle and a wave.

Although Percival Lowell believed in its existence, the planet Pluto was discovered after his death at the Lowell Observatory. (Library of Congress)

See also Astronomical Observatories; Cherenkov, Pavel; Einstein, Albert; Philosophy of Science; Physics; Quantum Mechanics; Quantum Theory; Raman, Chandrasekhara; Relativity; X-rays

References

Kragh, Helge. *Quantum Generations: A History of Physics in the Twentieth Century*. Princeton, NJ: Princeton University Press, 1999.

North, John. *The Norton History of Astronomy and Cosmology*. New York: W. W. Norton, 1995.

Pais, Abraham. *Niels Bohr's Times: In Physics, Philosophy, and Polity*. New York: Oxford University Press, 1991.

———. '*Subtle is the Lord . . .*': *The Science and the Life of Albert Einstein*. New York: Oxford University Press, 1982.

Lowell, Percival

(b. Boston, Massachusetts, 1855; d. Flagstaff, Arizona, 1916)

Percival Lowell was an amateur astronomer who built a major private observatory and used it to study what he believed to be the remnants of extraterrestrial beings on Mars; he also predicted the existence of the

planet Pluto. Lowell took a degree in mathematics from Harvard University in 1876. His family was wealthy, and he had the time and resources to pursue his interests in astronomy without taking an advanced degree in the subject. In 1894, he founded the Lowell Observatory in Flagstaff, Arizona, on Mars Hill, at an elevation of some 7,000 feet.

Lowell was fascinated by the "canals" of Mars described in 1878 by the Italian astronomer Giovanni Schiaparelli (1835–1910). He used the telescopes of the observatory to make intricate drawings of the planet and its apparent network of geometric lines. He was convinced that the canals channeled water from the polar regions of Mars to large areas of vegetation, or oases. He concluded that they were constructed by intelligent beings living (at some time) on Mars. He announced these ideas openly in various publications, including his book *Mars and Its Canals* (1906). Although belief in extraterrestrial life was far from widespread, the apparently geometric lines on the red planet were difficult to explain by natural means. But in 1909, Eugene Antoniadi (1870–1944) at Paris's Meudon Observatory debunked the "canal" theory by showing that the canals were simply dark spots on the surface of Mars that only appeared to the human eye to be connected. Using more powerful telescopes allowed greater resolution of the image of Mars, laying the idea of Martian-built canals to rest.

Despite this setback, Lowell continued to observe the heavens and soon found a new obsession. Lowell believed that the planet Uranus behaved inexplicably as it orbited around the sun. He knew that planetary orbits were influenced by the gravitational attraction of the other planets, but he could not account for Uranus's perturbations by the gravitational interference of Neptune alone. He came to believe that there was another planet, as-yet unseen by anyone on earth, that orbited the sun in the outskirts of the solar system. The planet became known as Planet X. Lowell conducted three photographic searches for Planet X with the Lowell Observatory telescope, one each in 1905, 1909, and 1912. By repeatedly photographing the night sky in the region where Lowell thought the planet should be, he hoped to discover evidence that one of the stars moved in a planet-like fashion. But he found nothing. He was still searching for the planet when he died in 1916.

Astronomers at the Lowell Observatory resumed the search for the planet in 1928. The planet that Lowell had sought was finally found in 1930 by another amateur astronomer, Clyde Tombaugh (1906–1997), who later went to college and then pursued an advanced degree. Like the other planets, Planet X was soon named after an ancient Greco-Roman deity. Pluto ("Hades" in the Greek tradition) was the god of the underworld, which seemed fitting for this distant and desolate planet, far from the sun. The first two letters also were Percival Lowell's initials, a final tribute to the astronomer's long search.

See also Astronomical Observatories; Extraterrestrial Life; Science Fiction

References

Dick, Steven J. *The Biological Universe: The Twentieth-Century Extraterrestrial Life Debate and the Limits of Science.* New York: Cambridge University Press, 1996.

Hoyt, William Graves. *Planets X and Pluto.* Tucson: University of Arizona Press, 1980.

Sheehan, William. *The Planet Mars: A History of Observation and Discovery.* Tucson: University of Arizona Press, 1996.

Loyalty

The vast amount of money funneled into science by the U.S. government during World War II created a host of new problems for scientists, among them the requirement of loyalty. In the frosty years of the Cold War, scientists enjoyed healthy patronage from a government that viewed science as a crucial part of U.S. national security. Thus, issues of classification, security, and loyalty of scientists became increasingly important. The loyalty controversy was born in the 1940s and intensified in subsequent years, sparking debates about the academic freedom and political independence of scientists as they assumed new roles in the United States.

Although the most notorious loyalty cases occurred in the 1950s, particularly when J. Robert Oppenheimer (1904–1967) lost his security clearance in 1954, scientists came under suspicion much earlier. In many fields, secret war work was a portent of the postwar years, in which one could not hope to construct much of a career without having access to classified information. For example, the director of the Scripps Institution of Oceanography, the Norwegian Harald Sverdrup (1888–1957), was suspected of having Nazi sympathies and could not gain the proper clearance; because he was director, this proved highly embarrassing, and it could explain his ready departure for his homeland before the decade ended. The best scientists had access to secret information, and involvement in secret work opened up many career opportunities. For example, having participated in the Manhattan Project became a passport to good pay and good jobs for physicists in the postwar era. But because of the need for access, being a good scientist after the war meant being a loyal American. President Harry Truman (1884–1972) institutionalized this in 1947 by issuing an executive order requiring loyalty statements by all federal employees.

In other Western countries, being Communist did not always result in being blacklisted in the scientific community. In France, for example, Frédéric Joliot (1900– 1958) was a Communist and fought in the resistance movement during German occupation. After the war ended, he became chief of France's atomic energy program—the fact that he and his wife, Irène Joliot-Curie (1897–1956), were both Communists was well known. In Britain, some of the most celebrated scientists were Communists, such as J. D. Bernal (1901–1971), Joseph Needham (1900–1995), and J. B. S. Haldane (1892–1964). Bernal in particular argued that capitalism misused science tremendously, holding back social progress. These men's reputations and careers in Britain were never tarnished, even in the early years of the Cold War, to the degree that those of Communists in the United States were.

The professional constraint—that of maintaining a pristine record of loyalty—made scientists vulnerable when their political views clashed with powerful figures; if they could be demonstrated as "disloyal," their careers could be destroyed. The first high-profile scientist to experience this was Edward Condon (1902–1974), head of the National Bureau of Standards. Because of his outspoken support for international cooperation and his opposition to developing the hydrogen bomb, Condon created enemies in Congress who were bent on bringing him down. Beginning in 1948, Condon defended his name repeatedly to members of the House Un-American Activities Committee (HUAC).

The late 1940s were years of paranoia and anxiety, fueled by revelations of spying by high-level bureaucrats such as Alger Hiss and scientists such as Klaus Fuchs (1911–1988). The latter had sold the Soviet Union secrets from the Manhattan Project. In 1949, the Atomic Energy Commission required all of its fellowship holders to take a loyalty oath, regardless of whether classified research was involved. It had been embarrassed when a reporter announced that one of its fellows was an outspoken Communist, and it authorized the Federal Bureau of Investigation to establish the loyalty of its scientists. Soon the fellowship program was discontinued altogether.

When the University of Washington fired a known Communist from its faculty, he traveled to the University of California–Los Angeles, to participate in a debate. Californians were enraged that he should be invited. This was one of many reasons that moved the regents of the University of California in 1949 to require its faculty to swear an oath not just declaring loyalty, but also disavowing ties to organizations that might plan to overthrow the government. In its 1950 version, the oath stated: "I am not a member of the Communist Party or any other organization which advocates the overthrow of the Government by force or violence." The regents made employment at the university contingent on taking the oath. Some of those who had received favorable academic reviews challenged the oath by refusing to sign it. Threatened with dismissal, six decided to sign it and three resigned. More than thirty professors and many other university employees who refused to sign were subsequently fired. Other campuses, such as the University of Washington, had similar experiences in the late 1940s. The California case went to the state supreme court, which in 1952 ordered the professors reinstated.

Some U.S. intellectuals were aghast at the loyalty mania. The American Association for the Advancement of Science showed its solidarity with Condon by electing him its president. At stake, objectors felt, was political freedom for academia. If education rights or privileges were compromised because of political tests, was that not a violation of U.S. principles? Requiring "loyalty" oaths might be the first step in more stringent political control of academic institutions, thus endangering the freedom of thought.

The issues surrounding loyalty and the oaths were not resolved in the 1950s. HUAC linked arms with Senator Joseph McCarthy (1908–1957) and others in the 1950s to expose Communist infiltration at all levels of society. Most of this was focused on the federal government. Because science in the United States increasingly depended on government support, the question of loyalty continued to go hand in hand with scientific research. The 1953 executions of Ethel and Julius Rosenberg, who allegedly had run an atomic spy ring, along with the discovery of spies who had worked on the Manhattan Project during the war, only intensified the dread of those wanting science to be free of any Communist influences.

See also Cold War; Espionage; National Bureau of Standards; Sverdrup, Harald

References

Caute, David. The Great Fear: The Anti-Communist Purge under Truman and Eisenhower. New York: Simon & Schuster, 1978.

Gardner, David. The California Oath Controversy. Berkeley: University of California Press, 1967.

Oreskes, Naomi, and Ronald Rainger. "Science and Security before the Atomic Bomb: The Loyalty Case of Harald U. Sverdrup." Studies in the History and Philosophy of Modern Physics 31 (2000): 309–369.

Rossiter, Margaret. "Science and Public Policy since World War II." Osiris, Second Series, 1 (1985): 273–294.

Stewart, George R. The Year of the Oath: The Fight for Academic Freedom at the University of California. New York: Doubleday, 1950.

Wang, Jessica. American Science in an Age of Anxiety: Scientists, Anticommunism, and the Cold War. Chapel Hill: University of North Carolina Press, 1999.

Wersky, Gary. The Visible College: The Collective Biography of British Scientific Socialists in the 1930s. London: Allen Lane, 1978.

Lysenko, Trofim

(b. Karlivka, Russia, 1898; d. Moscow, USSR, 1976)

Trofim Lysenko was responsible for the predominance of Lamarckian concepts of heredity among Soviet scientists and for the repression of Mendelian genetics. By tying science to political ideas, he made allies among the Communist Party leadership and dominated Soviet biology for decades. Born into a peasant family, Lysenko studied agronomy at the Kiev Agricultural Institute in the 1920s. He came to believe that by manipulating growing conditions, scientists could produce more effective seeds and better productivity in agriculture.

When Soviet leader Joseph Stalin (1879–1953) forced the collectivization of agriculture, beginning in the late 1920s, scientists hoped to find new ways of managing agriculture to increase productivity. Lysenko's contribution to this effort was the concept of vernalization. Lysenko appeared to have shown that freezing seeds made them germinate more rapidly. This technique was not unknown, but Lysenko also claimed that the new trait, stimulated by his own intervention, could be inherited. This seemed to support the Lamarckian style of evolution that insisted that acquired characteristics can be inherited. It also meant that seeds would only need to be "treated" once, and then all future wheat would grow more rapidly and be more abundant. Lysenko capitalized on these results by convincing political leaders that his technique held great promise for Soviet agriculture and that research in plant breeding should be funded for the good of the Soviet Union. He convinced Stalin himself of this, and Lysenko soon found himself in a position of power among the scientific elite: He became a member of the Academy of Sciences (1935) and even a member of the Supreme Soviet (1937). He later claimed to produce rye from wheat plants, cuckoo birds from warblers, and other alleged breeding "successes."

Lysenko and others created the Agriculture

Soviet geneticist, agronomist, and president of the Lenin Academy of Agricultural Sciences, Trofim Lysenko measures the growth of wheat in a collective farm field near Odessa in the Ukraine. (Hulton- Deutsch Collection/Corbis)

Academy, which became the center of research on plant breeding and genetics research. This body enjoyed the patronage and enthusiasm of the Communist Party, which looked to scientists to lead the way to an efficient, centrally planned economy. But by the mid-1930s, lack of progress created disillusionment, particularly against the geneticists who had been supported generously. Lysenko increasingly came to believe that Mendelian genetics stood in conflict with his Lamarckian views of heredity. Further, the simplistic mechanisms of heredity posed by genetics struck him as idealistic, even against Marxist ideology. He began to characterize both genetics and Darwinian natural selection as bourgeois sciences, representative of the competitive outlook of industrial capitalism.

Lamarckism, by contrast, suited Marxism's emphasis on directed, planned progress. Geneticists found themselves being attacked not merely for their scientific views, but for advocating ideas that contradicted Marxism as interpreted by Stalin. The most renowned casualty of these attacks was the leading geneticist Nikolai Vavilov (1887–1943); he was arrested in 1940, and he died in prison in 1943.

Lysenko's attack on the geneticists culminated in a special meeting of the Soviet Agriculture Academy in 1948, after which the views of geneticists were officially condemned, their science branded as bourgeois and anti-proletariat. Research in genetics effectively ended; references to genetics were removed from school curricula; and eminent geneticists were dismissed from their posts on a wide scale. Lysenko's alternative, dubbed *Michurinism* after the earlier Russian Lamarckian biologist I. V. Michurin (1855–1935), became the orthodox view of heredity in the Soviet Union. His influence continued after the death of Stalin in 1953, only abating in the mid-1960s. His eventual fall from power was a result of the fact that he rarely fulfilled any of his promises and his experiments were criticized for their lack of rigor. His apparent successes in increasing crop yields were perceived as the result of his access to better equipment, higher quality seeds, and more efficiently organized peasants working on farms. He was removed from his positions of authority in 1965, after Leonid Brezhnev assumed leadership of the Soviet Union.

See also Academy of Sciences of the USSR; Determinism; Evolution; Genetics; Kammerer, Paul; Philosophy of Science; Soviet Science; Vavilov, Sergei

References

Graham, Loren R. *Science and Philosophy in the Soviet Union.* New York: Alfred A. Knopf, 1972.

Joravsky, David. *The Lysenko Affair.* Cambridge, MA: Harvard University Press, 1970.

Medvedev, Zhores A. *The Rise and Fall of T. D. Lysenko.* New York: Columbia University Press, 1969.

Soyfer, Valery N. *Lysenko and the Tragedy of Soviet Science.* New Brunswick, NJ: Rutgers University Press, 1994.

M

Manhattan Project

The Manhattan Project was the name of the U.S. effort to build an atomic bomb during World War II. It was based on the code name, Manhattan Engineer District, given to the project in 1942 when it was taken over by the Army Corps of Engineers. From that point, the project had three major obstacles. First was the effort to achieve a fission chain reaction, which was accomplished in late 1942. Also there was the technical problem of isolating bomb materials, such as fissionable uranium and plutonium. Most complicated of all was the engineering project to design a weapon that could be dropped onto target cities. The Manhattan Project's work came to fruition in August 1945, when two atomic bombs destroyed the Japanese cities of Hiroshima and Nagasaki.

Although the effort to build the bomb was under way by the early 1940s, it was unclear whether it was a project to which the United States should devote vast resources. The ultimate test of feasibility was the fission chain reaction. The idea of an atomic bomb was based on the possibility of nuclei of heavy atoms splitting, or "fissioning." Fission would result in two lighter atoms, but in the process a small amount of energy would be released. Because this process also ejected neutrons, these free neutrons might collide with nearby atomic nuclei and create other fissions. If that process continued, a chain reaction would occur, leading to continuous fission in many atoms and a large amount of energy. That energy would in fact be so large that it could destroy, as Albert Einstein (1879–1955) said in a 1939 letter to U.S. president Franklin D. Roosevelt (1882–1945), an entire port and most of the surrounding territory. It was in achieving a chain reaction in a laboratory that the U.S. project succeeded where all other countries failed. In late 1942, in a squash court underneath the football field at the University of Chicago, a team of scientists led by Italian immigrant Enrico Fermi (1901–1954) created the first controlled fission chain reaction. After this was achieved, companies such as DuPont decided that they would put their industrial might behind the project, and the U.S. government decided to make the project a major wartime priority.

The United States dedicated a vast amount of resources to the Manhattan Project. Its main organizer was General Leslie Groves (1896–1970), who had been in charge of building the largest office building in the world, the Pentagon. He purchased land for the project and built the first uranium (chemical symbol "U") isotope separation plants in

Oak Ridge, Tennessee. In order to produce bomb material of sufficient quality, the rare isotope U-235 had to be separated from the more abundant U-238. Two methods were used at Oak Ridge: electromagnetic separation and gaseous diffusion. The latter required an unprecedented level of technical sophistication and precision, meaning that most of the equipment had to be specially made; the result was the largest chemical engineering plant ever constructed. Although gaseous diffusion was the most promising long-term method, the early bombs used uranium produced from electromagnetic separation.

For plutonium production, Groves chose another site, this time in Hanford, Washington. Plutonium was an artificial element, made from uranium, produced by collisions of accelerated particles. Like U-235, it promised to be very effective in a bomb. The plants built at Hanford were cooled by the Columbia River, which required that the designers pay close attention to avoiding too much diffusion of radioactive materials, because of risks to the local population.

The final phase of the project was building the bomb itself. A chain reaction had occurred in Chicago, and materials were being prepared at Oak Ridge and Hanford. In 1943 Groves assembled a team of leading physicists, many of them recent immigrants from Europe, in a small town in New Mexico called Los Alamos. This desert locale was far enough from metropolitan areas to provide greater security, although there were some participants who were in fact spying for the Soviet Union, such as Klaus Fuchs (1911–1988). The scientists were led by J. Robert Oppenheimer (1904–1967), and they included such luminaries as Niels Bohr (1885–1962) and Enrico Fermi. Los Alamos also became a breeding ground for the leading physicists of the postwar era, and it was a test case for large-scale collaboration between scientists and the military. Many of the scientists detested Groves and other leaders,

mainly for the policy of compartmentalization. This was a security measure that ensured that scientists worked only on certain aspects of the project, not having knowledge of the whole. Scientists objected to this because it hampered the discussion of ideas and, they argued, the efficiency of the project. Yet for security reasons, Groves still insisted on compartmentalization, keeping scientists and technicians on a need-to-know basis, although often scientists did not respect the rule.

The Los Alamos team produced two different kinds of weapons by 1945. One was a "gun-type" weapon, which used uranium from Oak Ridge. The principle of this weapon was that the bomb itself would fire a piece of uranium into another, with enough force to initiate a chain reaction (and thus an explosion). The other design used plutonium from Hanford, and was an "implosion" weapon, in which conventional explosives were placed over the shell of a spherical bomb; when they detonated, the bomb would be crushed with such intensity that the necessary fission chain reaction would occur. The first atomic device, an implosion weapon made from plutonium, was tested in Alamogordo, New Mexico, on 16 July 1945. The test was code-named Trinity, and the explosion was equivalent to 20,000 tons of dynamite (20 kilotons).

Although the war was already winding down (Germany already had been defeated, and an invasion of Japan was being planned), the United States decided to use the atomic bomb against Japan. Some of the scientists began to have second thoughts, and their movement for social responsibility was born. Leading figures in the Manhattan Project such as James Franck (1882–1964), Leo Szilard (1898–1964), and Niels Bohr (all foreign-born) expressed reservations: They argued that the bomb should not be used, or that the Japanese should be warned first, or that other alternatives should be pursued before unleashing a weapon of such power on

Robert Oppenheimer (left of center), General Leslie Groves (center), and others examine the wreckage of the tower that held the first atomic device. (Corbis)

civilians. Bohr was already thinking about the postwar world, urging the president to share information about the bomb with the Soviet Union to avoid an arms race when the war ended. These objections ultimately were cast aside; on 6 August 1945, a uranium bomb called "Little Boy" was dropped from a B-29 bomber, *Enola Gay,* on the city of Hiroshima. On August 9, *Bock's Car* dropped a plutonium bomb called "Fat Man" on the city of Nagasaki. After years of combining the best scientific and engineering minds with U.S. industry and the Army Corps of Engineers, the Manhattan Project destroyed two cities and instantly killed more than 100,000 people.

See also Artificial Elements; Atomic Bomb; Atomic Energy Commission; Cyclotron; Espionage; Fission; Hiroshima and Nagasaki; Social Responsibility; Uranium

References

Badash, Lawrence. *Scientists and the Development of Nuclear Weapons: From Fission to the Limited Test Ban Treaty, 1939–1963.* Atlantic Highlands, NJ: Humanities Press, 1995.

Groves, Leslie R. *Now It Can Be Told: The Story of the Manhattan Project.* New York: Harper, 1962.

Jungk, Robert. *Brighter than a Thousand Suns: A Personal History of the Atomic Scientists.* San Diego, CA: Harvest, 1956.

Rhodes, Richard. *The Making of the Atomic Bomb.* New York: Simon & Schuster, 1995.

Marconi, Guglielmo

(b. Bologna, Italy, 1874; d. Rome, Italy, 1937)

Guglielmo Marconi, a celebrated inventor, helped to revolutionize communications technology by developing wireless telegraphy and, some years later, shortwave radio. He studied at Livorno Technical Institute before trying to apply recently developed scientific ideas to his technical designs. Building on the nineteenth-century work of James Clerk Maxwell (1831–1879) and Heinrich Hertz (1857–1894), who studied the properties and transmission of electromagnetic waves, Marconi developed instruments to use these waves as a means to communicate.

By 1895, Marconi developed a system that would transmit and record waves without the use of a connecting metal wire. After finding little enthusiasm in the Italian government for his invention, which he presumed to be useful for communications, he took his apparatus to England, where it was received more enthusiastically by the British government. Marconi received the first patent for a wireless telegraphy system in 1896. The next year, he started a company that was renamed Marconi's Wireless Telegraph Company in 1900. By that year, he had demonstrated the effectiveness of wireless telegraphy to various governments and installed a communication system that crossed the English Channel to connect England and France. He also became a celebrity in the United States, where in 1899 he used wireless telegraphy to report on the America's Cup yacht race.

Marconi's most celebrated patent was "Number 7777," taken in 1900. The patent was granted for an instrument that allowed simultaneous transmissions on different frequencies; this allowed for greater flexibility in usage and decreased interference when more than one station was using the system. It also allowed for increases in range. The following year, he set out to prove that wireless telegraphy was not impeded by the curvature of the earth, and he did so by making the first transatlantic signal. Covering more than 2,000 miles, he succeeded in 1901 in communicating via wireless telegraphy from Cornwall, England, to Newfoundland, Canada.

Marconi continued to innovate in the growing science (and industry) of wireless telegraphy, taking out numerous patents in the early 1900s. By 1907, his efforts culminated in a permanent communication service between Nova Scotia and Ireland. Other shorter systems were installed between European countries, and wireless instruments were installed aboard ships as a means of sending distress signals. In the 1920s, several countries, including Great Britain and the United States, instigated broadcasts to the general public to disseminate news, entertainment, and other information. Using what he called the beam system, Marconi used wireless communication for strategic reasons, in cooperation with the British government, to link together the major possessions of the British Empire.

Wireless telegraphy produced a revolutionary change in the mode of communication between countries, not only facilitating economic pursuits but also aiding nations in their competitions and conflicts with one another. The transmission of knowledge through news broadcasts and intelligence reports became powerful tools; obtaining and maintaining networks that exploited wireless technology were of fundamental importance for European powers with far-flung imperial possessions and for any nation with strategic interests around the globe. Wireless communication surpassed not only costly submarine cables but also the unreliable transcontinental telegraph lines that had to be laid and then protected with military force. Marconi's invention coincided with some of the fiercest colonial competition just after the turn of the century. Ship-to-ship radio communications were used as early as the Russo-Japanese War (1905), and during World War I, communications cables routinely were cut.

Marconi pursued technological innovations that would allow communication with shorter and shorter waves. Shortwave radio proved to

Guglielmo Marconi (right), a pioneer in wireless telegraphy, with telegraph equipment. (Library of Congress)

be yet another revolution. In the 1930s, he experimented with radiotelephone communications; the first of these was installed in 1932 to allow the Vatican to communicate with the pope while he was living in his summer residence. He also began experimenting with communication techniques in ship-to-shore and ship-to-ship communication to improve navigation technology. Submarine cables connecting continents lost half of their business because of the effectiveness of shortwave over long distances. In these years, Marconi also pointed out the theoretical possibilities of another major technological revolution, namely, radar (developed after his death).

In 1909, Marconi received the Nobel Prize in Physics. During World War I, Marconi's skills proved useful to both the army and navy of Italy; he held rank in both during the war.

He became part of Italy's delegation to the peace conference that led to the Treaty of Versailles in 1919. He was showered with numerous honors and titles during his lifetime by the governments of several countries. When he died, wireless stations throughout the world observed a two-minute silence in his honor.

See also Colonialism; Radar; World War I
References
Douglas, Susan J. *Inventing American Broadcasting, 1899–1922.* Baltimore, MD: Johns Hopkins University Press, 1987.
Dunlap, Orrin E., Jr. *Marconi: The Man and His Wireless.* New York, Macmillan, 1936.
Headrick, Daniel R. *The Invisible Weapon: Telecommunications and International Politics, 1851–1945.* Oxford: Oxford University Press, 1991.
Marconi, Degna. *My Father, Marconi.* New York: Guernica, 2002.

Marine Biology

Marine biology is concerned with the animal and plant life of the sea. Studies of local flora and fauna often were the typical goals of marine biologists. But new methods developed in the late nineteenth century made marine biology a more interdisciplinary pursuit, combining traditional "provincial" studies of particular species and habitats with large-scale analyses of populations and migrations, closely integrated with ocean dynamics and chemistry. Before the twentieth century, marine biology often was localized in scope, with scientists in different parts of the world describing the sea life in regions closest to them; only comparative studies resulted in greater understanding of biogeography. In addition, marine biologists often did little more than take inventory of the life in the sea. For example, the branch concerned exclusively with fish, ichthyology, was oriented toward understanding the lives of particular fish species or identifying new ones. Part of its task was developing systematic catalogues of known marine species, describing and classifying them. This sort of work struck some scientists as unimaginative—simply a matter of census-taking, not scientific study.

In the twentieth century, however, marine biology acquired a broader emphasis, as scientists sought to integrate the study of marine life with more inclusive studies of the marine environment. In Woods Hole, Massachusetts, the Marine Biological Laboratory (MBL) was founded in 1888, inspired largely by the creation of marine stations in Europe. Under the leadership of Frank Lillie (1870–1947) in the 1920s and later, MBL became one of the principal centers of marine biology in the world. During the summer months, scientists from all over the country converged there to conduct research, and it became the center for the rapid growth of biological studies in the United States. It boasted the most advanced facilities and also a large library, catering to the needs of researchers.

Voyages occasionally enjoyed financial support because of their potential role in understanding fish stocks, which could lead to profit and economic prosperity. In the United States, Henry Bryant Bigelow (1879–1967) combined physical and biological oceanography in his studies of the Gulf of Maine, not only taking salinity and temperature measurements, but also hauling in nets of plankton. Between 1912 and 1928, he made thousands of hauls with towed nets. The voyages of the *Discovery* by the British during the 1920s also resulted in major contributions to marine biology. One of its principal goals was to trace the productivity of whales in the major whaling "grounds" in the vicinity of Antarctica. Understanding the whales' food chain necessitated a more complete analysis of the marine life constituting plankton, upon which the whales fed, and their seasonal migrations.

The study of plankton formed the basis of much marine biological research. By the early twentieth century, scientists in Germany such as Victor Hensen (1835–1924) and Karl Brandt (1854–1931) began the "Kiel school," which tied marine biology firmly to physical oceanography by emphasizing the need to study not only organisms, but dynamics of large populations of organisms such as plankton. Hensen in particular was confident that plankton organisms were distributed uniformly, and thus it was possible to understand entire populations by taking samples and studying them. The Norwegian Haakon H. Gran also played a major role in making such connections. In general, their outlook focused on dynamics and chemistry. In the 1920s, British scientists at the Plymouth Laboratory and U.S. scientists at the Woods Hole Oceanographic Institution followed a similar trajectory, and one of the main problems in marine biology was the phenomenon of the "spring bloom" of plankton in the northern Atlantic and surrounding seas. By the era of World War II and after, marine biology had incorporated this large-scale, population approach, connected firmly to physical and chemical oceanography.

See also Just, Ernest Everett; Oceanic
    Expeditions; Oceanography
References
Bigelow, Henry Bryant. *Memories of a Long and
    Active Life.* Cambridge, MA: Cosmos Press,
    1964.
Hardy, Alister. *Great Waters: A Voyage of Natural
    History to Study Whales, Plankton and the Waters
    of the Southern Ocean.* New York: Harper &
    Row, 1967.
Maienshein, Jane. *100 Years Exploring Life,
    1888–1988: The Marine Biological Laboratory at
    Woods Hole.* Boston, MA: Jones & Bartlett,
    1989.
Mills, Eric L. *Biological Oceanography: An Early
    History, 1870–1960.* Ithaca, NY: Cornell
    University Press, 1989.

Mathematics

Mathematics has two conflicting reputations.
On the one hand, it has been called the queen
of the sciences; on the other, it has been
called science's handmaiden. Whether mathematics exists simply as a tool of science, or
whether it is in fact the purest of all sciences,
is a matter of debate. In the twentieth century, the abstractions of mathematics called
into question its value to science: Should it be
pursued for its own sake, and is it simply a
game of syllogisms (deductions from basic
premises), or does it have bearing on the real
world? Bertrand Russell (1872–1970) once
described mathematics as "the subject in
which we do not know what we are talking
about or whether what we say is true" (Rees
1962, 9). But other mathematicians were
sure about the place of mathematics in the
twentieth century. In 1900, German mathematician David Hilbert (1862–1943) addressed the International Congress of Mathematicians in Paris, advising them that any
field with such an abundance of problems as
mathematics is a science full of life.

One of the greatest problems confronting
mathematics at the turn of the twentieth century was that of internal consistency. Contradictions within mathematical systems were
pointed out in the late nineteenth century
from the work of the German Georg Cantor

(1845–1918), in his theory of sets. A set is
series of numbers. If one describes a set comprising all positive whole numbers, one has
an infinite set (1, 2, 3, . . .). Intuitively, a set
comprising all positive *even* numbers (2, 4,
6, . . .) should contain fewer members than
that comprising all positive whole numbers,
since it will not include the odd numbers included in the first set. Cantor argued, against
intuition, that each of these sets contained
the same number of members. He pointed to
the fact that the "1" in the first set can be partnered with the "2" of the second set, the "2"
of the first set can be partnered with the "4"
of the second set, and so on, leaving no possibility of a number in either set without a
partner. Cantor believed that mathematics
had misused the concept of infinity, and his
work called into question the properties of
infinite sets.

Although Cantor's work was highly controversial, some mathematicians believed
that he was addressing a problem that had
confronted the subject since antiquity,
namely, the paradoxes brought about by concepts of infinity. Centuries before the Christian era, Zeno had proposed a series of paradoxes that seemed to indicate that changes
such as local motion were mathematically
impossible because of the concept of infinitely divisible space. In order to travel from
point A to point B, one must first travel half
the distance, a process that must be repeated
interminably, each time halving the distance
but never quite reaching point B. Zeno had
used the paradoxes to argue against the mathematical possibility of change. But Cantor
saw no contradiction at all and used his set
theory to demonstrate what was called the
actual infinite.

Cantor's work, which was attacked and
criticized from all sides, sparked a major reexamination of the nature of mathematical
reasoning. Bertrand Russell based his logical
work of the early twentieth century on the
notion that Cantor had lifted the veil of confusion and mysticism from the concept of infinity and subjected it to logic. But Russell

and others soon recognized (as did Cantor) that set theory was fraught with its own apparent contradictions. In their *Principia Mathematica* (1910–1913), Russell and Alfred North Whitehead (1861–1947) confronted Cantor's paradoxical set theory and concluded that pure mathematics was no more than an extension of deductive logic, and they insisted that knowledge must rely more on logical analysis. In his 1945 *History of Western Philosophy,* which begins with the ancient Greeks, Russell presented the philosophy of logical analysis in the final chapter as a means to demystify mathematics and remove it from the pedestal upon which it had been placed since the time of the ancient Greeks. Although Russell could be called a mathematician, he may have preferred to be called a logician. He pointed out a paradox from set theory, asking: Is a set, composed of all sets that are not members of themselves, a member of itself? The logical inconsistency of the question, known as Russell's paradox, added further weight to the notion that mathematics' biggest problem was its own number of internal contradictions.

Despite Russell's influence, some mathematicians found logical analysis less than satisfying because it too created numerous contradictions. Frustrations about inconsistencies in mathematical and logical systems led some scholars to attempt to prove that consistencies were possible. The accomplished mathematician David Hilbert was one of these. He attempted to show, through mathematical proofs, that no two theorems could contradict each other if derived from the postulates of arithmetic. He made this one of the great projects of his career, and he revisited the fundamentals of geometry set forth by the ancient Greek Euclid, who first had discussed the now commonplace concepts of axioms, postulates, theorems, and proofs. Hilbert attempted to establish a more rigorous description of these in his 1899 *Foundations of Geometry* and devoted subsequent decades to rigorous analysis of his postulates. Unlike the Greeks, Hilbert insisted that the postulates themselves be proved to be free of any possible contradictions. Hilbert believed that contradictions between mathematical statements should be impossible; if they occurred, one must fault the postulates themselves.

One of Hilbert's challenges was addressed in 1930 by the Austrian Kurt Gödel (1906–1978). To the dismay of his colleagues, he demonstrated that Hilbert's hope of proving the internal consistency of a mathematical system (a system that is governed by predefined axioms) was impossible. There always would be some proposition that could not be proved or disproved on the basis of postulates, regardless of the number of logical deductions made. Gödel's statement became known as the incompleteness theorem, or simply Gödel's theorem. His was a bold assertion that one cannot expect to create a foolproof mathematical system free of paradoxes.

See also Computers; Cybernetics; Game Theory; Gödel, Kurt; Russell, Bertrand
References
Bell, E. T. *Men of Mathematics.* New York: Simon & Schuster, 1965.
Rees, Mina. "The Nature of Mathematics." *Science,* New Series 138:3536 (5 October 1962): 9–12.
Rowe, David E., and John McCleary, eds. *The History of Modern Mathematics.* San Diego, CA: Academic Press, 1989.
Russell, Bertrand. *A History of Western Philosophy.* New York: Simon & Schuster, 1972.

McClintock, Barbara
(b. Hartford, Connecticut, 1902; d. Huntington, New York, 1992)

Barbara McClintock lived a scientific life out of the mainstream. Her contributions to genetics seemed so radical that, although her expertise commanded respect, her colleagues did not appreciate the far-reaching implications of her work for decades. She discovered in maize plants—or one might say that the maize plants revealed to her—the existence of mobile genetic elements.

McClintock took an early interest in the study of cells and their genetic makeup. As a graduate student at Cornell University, she identified the ten chromosomes of maize. Between 1929 and 1935, she and colleagues George W. Beadle (1903–1989), Marcus M. Rhoades (1903–1991), and others at Cornell conducted experiments that explored the relationship between chromosomes and genetics, usually using maize. This period established her as a leading geneticist, and later (1944) she was elected to the National Academy of Sciences, the third woman to have been elected. But she had trouble finding a job in the 1930s. Cornell was kind to graduate students, she found, but had no female professors. McClintock subsisted on fellowships until taking a position as an assistant professor at the University of Missouri in 1936. To her disappointment, colleagues there excluded her from routine departmental activities and let her know that she was unlikely to receive a promotion. She left in 1941 and moved to New York, where she held an appointment at Cold Spring Harbor Laboratory until her death in 1992.

In the 1940s, McClintock became fascinated with chromosome breakage, a topic that led to her most significant contribution. She made an intensive study of the loci of chromosome breakage in maize and found that some kinds of genetic material could move from one site in a cell to another site. She called these transposable elements, a kind of genetic element capable of moving to a different part of the chromosome. They were also called jumping genes, emphasizing the mobility that made them so incredible to other members of the scientific community.

The maize plant had revealed to McClintock a genetic phenomenon that stood against the prevailing assumptions of the time. Evelyn Fox Keller's biography of McClintock suggests that McClintock, through many years of observation, became so well acquainted with maize that she developed a peculiar relationship with it, allowing her to sympathize, listen, and have a "feeling" for the organism. Scholars of gender and science have made much of this, claiming that McClintock was never fully socialized into the masculine world of science, and her success points the way toward a science less dominated by masculine notions of nature and scientific investigation.

McClintock began this work in 1944 and first published her results in 1950 and 1951. But her jumping genes were not readily accepted by everyone. Some objected to her approach, calling it mystical or mad. McClintock recalled that reactions of her paper delivered at the Cold Spring Harbor Symposium of 1951 ranged from perplexed to hostile. Geneticists claimed later that they did not doubt McClintock's findings, but they saw transposable elements as a characteristic of maize, not necessarily as a fundamental principle to be generalized. Indeed, widespread recognition of the importance of her work on gene transpositions was not achieved until the 1970s. Her Nobel Prize in Physiology or Medicine was awarded in 1983. Like Gregor Mendel (1822–1884), the unsung (in his lifetime) founder of genetics, appreciation for McClintock's work was long delayed.

See also Chromosomes; Genetics; Women
References
Fedoroff, Nina V. "Barbara McClintock." In *Biographical Memoirs, National Academy of Sciences,* vol. 68. Washington, DC: National Academy Press, 1996, 211–235.
Keller, Evelyn Fox. *A Feeling for the Organism: The Life and Work of Barbara McClintock.* San Francisco, CA: W. H. Freeman, 1983.

Mead, Margaret
(b. Philadelphia, Pennsylvania, 1901; d. New York, New York, 1978)

Margaret Mead was one of the most well-known anthropologists of the first half of the twentieth century. Her work helped to transform thinking about gender roles, sexuality, and the importance of culture in defining social behavior. She did more than any other anthropologist to extend Franz Boas's

One of the most well-known anthropologists of the first half of the twentieth century, Margaret Mead's work helped to transform thinking about gender roles, sexuality, and the importance of culture in defining social behavior. (Library of Congress)

(1858–1942) cultural deterministic views. She also worked to popularize her findings, making herself a widely recognized scientific celebrity. She attended graduate school at Columbia University, where she studied under Boas, who himself had been a founder of anthropological studies in the United States. Mead traveled extensively during her career, including expeditions to Samoa and New Guinea, where she conducted her most significant fieldwork. She held a position in the Department of Anthropology at the American Museum of Natural History from 1926 until the end of her life.

Mead's mentor, Boas, was an ardent supporter of seeing differences in human populations in terms of environmental and social influences, rather than taking a purely biological (or, more precisely, genetic) approach. Mead's work also reflected this willingness to explore culture as a major anthropological force. Her doctoral research under Boas required her to travel to the Manua islands, in American Samoa, for nine months in 1925–1926, where her observations provided her with evidence that certain phases of human development were culturally, not biologically, determined. In particular, Mead noted that the tumultuous years of adolescence, often thought to be universal, were a cultural by-product resulting from trying to suppress or hide sexuality. She generalized that Samoan culture was casual and easygoing, lacking many of the aggressive tendencies found in Western cultures. In general, she criticized notions that characterized emotions and other behaviors as universal across cultures. Her work became a powerful argument for an emerging group of anthropologists who embraced the concept of cultural determinism.

Mead dedicated most of her professional life to the study of the peoples in the Pacific Ocean region. In 1928, Mead published her doctoral research as a popular book, entitled *Coming of Age in Samoa* (she also penned *Social Organization of Manu'a,* presenting the same evidence for her anthropological peers). It soon became a best seller and established for her a worldwide reputation as a scholar and writer. She used this reputation well and helped to popularize her science through guest appearances on radio programs and through articles in magazines such as *Redbook.* Her work had a great appeal, particularly to those who hoped to improve society by reevaluating which aspects of society were biologically natural and which were culturally determined.

Mead continued to travel and conduct fieldwork among island peoples in the late 1920s and 1930s. She published *Growing Up in New Guinea* in 1930, and followed it with numerous books, some of which emphasized the transmission of culture to children and the roles of women in different cultures. For example, she found that, in the Tchambuli culture of New Guinea, the men (not women) were in charge of the household. Mead's books raised provocative questions about the organization of society, particularly in gender roles. Throughout her career, she authored more than forty books.

Other anthropologists occasionally disagreed with Mead's observations, particularly in regard to the Samoans. Some have attempted to debunk Mead's entire body of work on the grounds that she was prejudiced. Anthropologist Derek Freeman dedicated part of his career to demonstrating that Samoans were more violent and troubled than Mead acknowledged. He observed that Mead simply found in Samoa the results that her mentor wanted to find; according to Free-

man, Boas's deep-seated notions of cultural determinism spoiled Mead's objectivity.

Mead became active in national and international bodies in later years. During the war years, she served on the Committee of Food Habits in the National Research Council. She later advised on mental health questions for the United States Public Health Service and international organizations. She received many honors for her work, including nearly thirty honorary degrees. In the 1970s, she served as president of the American Association for the Advancement of Science. Mead died of cancer in 1978.

See also Anthropology; Boas, Franz
References
Bateson, Mary Catherine. *With a Daughter's Eye: A Memoir of Margaret Mead and Gregory Bateson.* New York: Morrow, 1984.
Foerstel, Lenora, and Angela Gilliam, eds. *Confronting the Margaret Mead Legacy: Scholarship, Empire, and the South Pacific.* Philadelphia, PA: Temple University Press, 1992.
Freeman, Derek. *Margaret Mead and Samoa: The Making and Unmaking of an Anthropological Myth.* Cambridge, MA: Harvard University Press, 1983.
Howard, Jane. *Margaret Mead: A Life.* New York: Simon & Schuster, 1984.

Medicine

The practice of medicine was transformed by changing conceptions of disease—notably the germ theory of disease, developed in the nineteenth century. The work of Louis Pasteur (1822–1895) and Robert Koch (1843–1910) had indicated that tiny microbes were responsible for many diseases and that controlling the reproduction and transmission of such microbes could be the key to modern medicine. Medical practitioners typically emphasized prevention of disease through improved sanitation, or the cure of disease through new drugs.

Much of the medical profession in the early twentieth century was focused on battling infectious diseases. As if carryovers from the nineteenth century such as cholera

Frederick G. Banting, discoverer of insulin. (National Library of Medicine)

role of microbes in spreading disease, the popular conception of dirt as the source of "germs"—and thus disease and death—continued beyond mid-century and is still prevalent today.

Physicians played important roles in shaping social policy, by defining diseases, formulating appropriate therapy, and making recommendations to public health officials. One example was the effort to eliminate tuberculosis, commonly known as consumption, which combined physicians' expertise with state-sponsored social programs. When Britain recruited soldiers for the Boer War at the turn of the twentieth century, for instance, officials were shocked at the numbers of young men who were unfit for service. Armed with the germ theory of disease, physicians proposed a series of measures—rules against spitting in public, strict controls on milk, meat, and other potential breeding grounds for bacteria, and the separation of infected patients in hospital wards. The effort to eliminate tuberculosis continued during the first half of the century as part of larger campaign to improve the overall health of British subjects. Although it was not very effective in combating the disease, the campaign served as a vehicle for enforcing social laws desired by the middle classes. In another example of the connections between physicians and social policy, medical practitioners in Nazi Germany helped to initiate the racial hygiene laws by advising on the language to be used. Thus sterilization, euthanasia (medical killing of "undesirables"), and restrictive marriage laws find their origins not merely in extremist Nazi ideology but also in mainstream medicine of the era. As historian Robert Proctor argues, this active participation of medical professionals served to validate racial policies by providing them with a blessing from the scientific community.

Another important development in medicine was the rise of the pharmaceutical industry. Drug companies were an important source of funding for research in the biomedical sciences, and they benefited from the re-

and tuberculosis were not damaging enough, an influenza epidemic during World War I spread throughout the world, killing more than 21 million people, the work not of a bacterium but of a much smaller organism, a virus. Although studies indicated that better sanitation might decrease the spread of deadly microorganisms, scientists realized that the human body could never be made perfectly sterile. In addition, physicians emphasized the "microbe" nature of disease rather than the "filth" nature of disease. Although cleanliness certainly would help prevent disease, a person infected with tuberculosis could spread the disease to nearby people in even the cleanest of homes. Despite the medical community's recognition of the sults. The German bacteriologist Paul Ehrlich (1854–1915) first developed Salvarsan, a chemical therapy to treat syphilis; it was marketed widely, but the side effects were so painful that many refused to subject themselves to it. Aside from chemically produced therapies, drug companies reproduced naturally occurring fighters of disease. The 1902 discovery of the body's regulating fluids (later called hormones) by British scientists William Bayliss (1860–1924) and Ernest Starling (1866–1927) produced a flurry of interest among pharmaceutical manufacturers. Hormone therapy became a crucial area of research, and products that could use hormones to alter the body's reactions were produced and sold by corporations. The most celebrated of these products was Frederick Banting's (1891–1941) 1921 discovery of insulin, which was the key to treating diabetes. Other products directly fought bacteria and were called antibiotics. During World War II, U.S. pharmaceutical companies such as Pfizer and Merck mass produced penicillin and saved thousands of soldiers' lives and limbs from infection.

Although industries had the potential to develop "miracle drugs," they also presented dangers when the effects of medicines were poorly understood. Some governments tried to respond to such uncertainties with regulations. The United States, for example, passed the Pure Food and Drug Act in 1906 in order to require accurate labeling of drugs; this measure, officials hoped, would enable consumers to make wiser choices. It also established the Food and Drug Administration to ensure proper controls on potentially dangerous substances. In 1938, after hundreds were killed from a toxic form of the drug sulphanilimide, it passed a law (the Food, Drug, and Cosmetics Act) requiring prescriptions for certain drugs. Before, many narcotics could be purchased without permission from a doctor; now, there were distinctions between over-the-counter drugs and those requiring consent of a physician. This put responsibility for medical ethics and risks

directly on the shoulders of government, and it empowered doctors to be the brokers between patients and their cures, not merely recommending treatment but controlling access to potential remedies.

See also Cancer; Eugenics; Hormones; Industry; Microbiology; Penicillin; Public Health; Venereal Disease

References

Bryder, Linda. *Below the Magic Mountain: A Social History of Tuberculosis in Twentieth-Century Britain.* Oxford: Oxford University Press, 1998.

McGowen, Randall. "Identifying Themes in the Social History of Medicine." *Journal of Modern History* 63 (1991): 81–90.

Proctor, Robert N. *Racial Hygiene: Medicine under the Nazis.* Cambridge, MA: Harvard University Press, 1988.

Rosen, George. *Preventive Medicine in the United States, 1900–1975: Trends and Interpretations.* New York: Science History, 1975.

Temin, Peter. *Taking Your Medicine: Drug Regulation in the United States.* Cambridge, MA: Harvard University Press, 1980.

Tomes, Nancy. *The Gospel of Germs: Men, Women, and the Microbe in American Life.* Cambridge, MA: Harvard University Press, 1998.

Meitner, Lise

(b. Vienna, Austria, 1878; d. Cambridge, England, 1968)

Lise Meitner played a critical role in the discovery of nuclear fission prior to World War II. She has been called the "mother of the atomic bomb," but this is not accurate. She was not involved in weapons research; in fact, she was isolated from nuclear research before the war began. Her role in the discovery of fission was covered up and repressed by her long-time colleague, Otto Hahn (1879–1968). He alone received the Nobel Prize for this work, though Meitner probably should have shared it with him. Their decades-long collaboration ended abruptly just prior to Hahn's critical experiments, because Meitner, a Jew, had to flee from Nazi Germany.

Meitner, an Austrian, went to Berlin in 1907 and soon began a fruitful collaboration with Otto Hahn. She was a physicist, and Hahn was an expert in radiochemistry. It was not an equal partnership; for example, Meitner's status as a woman meant that she could not enter the institute through the front door. But in 1913, she was appointed officially to the Kaiser Wilhelm Institute, and a few years later she headed a research section for physics. By 1922 she was teaching at the University of Berlin, a right that was not granted to women until 1918.

The joint work of Meitner and Hahn led to their discovery of the element protactinium in 1918. As biographer Ruth Sime notes, this work was done almost entirely by Meitner, while Hahn served at the front in a chemical warfare unit. The credit, nonetheless, went to both. In subsequent years, Meitner became increasingly interested in nuclear physics, especially after the discovery of the neutron by James Chadwick (1891–1974). Neutron bombardment became a powerful tool in facilitating reactions between atoms, because neutrons had mass but no repellent charge. In 1934, she convinced Hahn to renew their collaboration and investigate the radioactive transmutations of heavy elements.

When Adolf Hitler (1889–1945) came to power in 1933, many Jews resigned their positions or were forced out of work. Meitner was Austrian, which exempted her from German racial laws. But after the *Anschluss*—the annexation of Austria by Germany in 1938—and after strong denunciations by a Nazi colleague at the Kaiser Wilhelm Institute, she fled to Stockholm, Sweden. Later, she said that she regretted staying even as long as she did, surrounded by hostile scientists and forced to watch the ruin of less fortunate colleagues. Yet she was also saddened when she left, uprooted from her home, friends, and colleagues.

When Meitner left Germany in 1938, she and Hahn had been studying chemical reactions in uranium from neutron bombardment. They had also enlisted the help of another colleague, Fritz Strassman (1902–

Lise Meitner and Otto Hahn, ca. 1913. (Otto Hahn, a Scientific Autobiography, *Charles Scribner's Sons, New York, 1966, courtesy AIP Emilio Segrè Visual Archives*)

1980); with Meitner gone, the two found some puzzling results in the laboratory. They thought that their uranium had transformed into an isotope of radium, which was expected, yet it seemed to have the chemical properties of barium, much further down the periodic table. Hahn, wanting to keep up his collaboration with Meitner, asked her in a letter to provide some theoretical explanation. This she did in early 1939, along with her nephew Otto Frisch (1904–1979). The uranium had not decayed into radium, but had split into two atoms of barium and released some energy. Frisch borrowed a term from biology and called it *fission*.

Despite the longtime collaboration of Hahn and Meitner, not to mention the important work of Strassman and Frisch, Hahn alone was recognized as the discoverer of fission. He won the Nobel Prize in Chemistry in 1944. Hahn himself, living in Nazi Germany, feared the consequences of admitting his heavy reliance on a Jewish woman. This is somewhat understandable, although Meitner was surprised that he continued to omit reference to her even after the war. Given the importance of her work to physics, and indeed to the course of the twentieth century, Meitner's life is a strong testament to the power of prejudice in science and society in the first half of the twentieth century.

See also Atomic Structure; Fission; Hahn, Otto; Radioactivity; Physics; Women

References

Rife, Patricia. *Lise Meitner and the Dawn of the Nuclear Age*. Boston, MA: Birkhäuser, 1999.

Sime, Ruth Lewin. *Lise Meitner: A Life in Physics*. Berkeley: University of California Press, 1996.

Mental Health

The study of mental health and the treatment of mental illness have proven controversial for historians ever since the publication of Michel Foucault's 1965 book, *Madness and Civilization*. In that book and others, Foucault argued that institutions or other kinds of incarceration have been forms of social control by elites since the eighteenth century. Certainly many of the controversial figures of history have been labeled mad by political opponents, as in the case of John Brown (1800–1859), the famous slavery abolitionist who sparked a national crisis prior to the American Civil War. But the history of mental health has been more complex than mere social control; in the first half of the twentieth century, it was marked by renewed efforts to treat rather than isolate patients. The means for doing so included psychiatric evaluations, radical surgical procedures, and drug use.

The asylum was the focal point of mental illness treatment in the nineteenth century. But *treatment* is a term loosely used. Seldom did patients leave the asylum cured, and the institutions played a largely custodial role. Around 1875, about 90 percent of mental health patients in the United States lived in public institutions. By the 1920s, institutions provided long-term custodial care for their patients, about half of whom stayed for at least five years. The emphasis on outpatient treatment was a twentieth-century development, resulting largely from pressures from the psychiatric profession. Psychiatrists were part of a "mental hygiene" movement that emphasized prevention, whether it was a patient in a psychiatrist's office or an entire community (or society as a whole). The popularity of psychoanalysis underlined the importance of evaluating each person's own state of mental health, rather than categorizing people as either normal or mad. Such efforts helped to diminish the stigma (but certainly not to eliminate it) attached to seeking help in preventing mental illness.

Desperation for cures of mental illnesses coincided with faith that surgeons could locate the "problem area" of the brain and simply eliminate it. Portuguese neurologist Egas Moniz (1874–1955) designed a special cutting instrument, the leukotome, to be used for cutting the connections between the frontal lobe and the rest of the brain. The first "prefrontal leukotomies" were performed using Moniz's techniques in the mid-1930s, and the results were published in 1936; they appeared to be so effective that he would later (in 1949) win the Nobel Prize in Physiology or Medicine. The most common use of the procedure was in treating schizophrenia. The term *lobotomy* was brought into use by the American Walter Freeman (1895–1972), who promoted the procedure through the media, bringing him instant acclaim. The procedure had many flaws and was dangerous, as in cases when the instrument broke and remnants of it remained in the brain. In 1946, Freeman developed the transorbital lobotomy procedure, involving entry into the brain through the eye socket. But his surgical techniques angered some of his colleagues, and such radical procedures declined dramatically with the development of more effective psychotropic drugs such as chlorpromazine, used to treat schizophrenia.

The treatments for mental illness often

depicted in popular culture—straitjackets, covered bathtubs, chains—were complemented by new techniques developed in the 1930s. Manfred Sakel (1900–1957) of Vienna devised a hypoglycemic coma-inducing injection for schizophrenic patients. Ladislav Von Meduna (1896–1964) of Hungary developed an intravenous therapy for victims of seizures. In 1938, the Italians Ugo Cerletti (1877–1963) and Lucio Bini (1908–1964) developed electric convulsive therapy, which produced seizures of their own but appeared to be an effective treatment for severe depression. These treatments were used to curb symptoms in severely affected patients.

World War II shaped perceptions of treatment because of the many cases of psychological trauma on the battlefield. When some of these cases were treated successfully by psychiatrists, mental health experts increasingly reconsidered the wisdom that keeping patients isolated in institutions was the best kind of treatment. Instead, they began to emphasize the importance of community rather than isolation. This new emphasis had the double effect of taking mental health patients out of the asylum and attracting new people to the field of psychiatry because of the opportunities to study social interactions of patients. In addition, the development of effective psychotropic drugs in the 1950s blurred the boundaries between psychological treatments and somatic (bodily) ones. In the early 1950s, these and other factors prompted widespread demands for reform of existing strategies of care. The 1949 establishment of the National Institute of Mental Health in the United States coincided with these reevaluations and stepped them up further by funding research and development in formulating new treatments for mental health patients.

See also Mental Retardation; Psychology; Public Health

References
Foucault, Michel. *Madness and Civilization: A History of Insanity in the Age of Reason.* New York: Vintage, 1988.

Gamwell, Lynn, and Nancy Tomes. *Madness in America: Cultural and Medical Perceptions of Mental Illness before 1914.* Ithaca, NY: Cornell University Press, 1995.

Grob, Gerald N. *From Asylum to Community: Mental Health Policy in Modern America.* Princeton, NJ: Princeton University Press, 1991.

———. *Mental Illness and American Society, 1875–1940.* Princeton, NJ: Princeton University Press, 1983.

Valenstein, Elliot S. *Great and Desperate Cures: The Rise and Decline of Psychosurgery and Other Radical Treatments for Mental Illness.* New York: Basic Books, 1986.

Mental Retardation

In the twentieth century, mental retardation was perceived as a serious social problem. People with severe cognitive disabilities, or mental retardation, were institutionalized during the nineteenth century as a means to save them from society. Major institutions sprang up in several countries. In isolation, groups of people with mental retardation became objects of scientific studies, leading to new classification schemes and the concept that mental retardation is genetic and connected to some of society's troublesome problems. Sterilizations were widespread during the eugenics movement of the 1920s and 1930s, and efforts to deinstitutionalize people with mental retardation and to include them in mainstream society did not see much success until after World War II.

Historian James W. Trent has argued that the segregation and seclusion of "feeble-minded" children into institutions was, at least in the United States, largely because of the efforts of institution superintendents. These professionals carved a niche for themselves by emphasizing the need for special care and the inappropriateness of caring for such children at home. The result was a large-scale tendency to isolate people diagnosed with mental retardation, supposedly keeping them safe from society and keeping society safe from them.

One carryover from the nineteenth century was the perception that feeblemindedness was linked to other social ills. Richard Dugdale (1841–1883) had written about the Jukes, a family of degenerates whose feeblemindedness turned them into paupers and criminals. The twentieth-century figure who lent credibility to these ideas was the American Henry Herbert Goddard (1866–1957). One of Goddard's goals was to incorporate the intelligence tests invented in France by Alfred Binet (1857–1911), later dubbed the IQ test. Unlike Binet, Goddard wanted to use the system to develop a classificatory scheme that separated individuals with mental retardation into categories of severity: idiot, imbecile, and feebleminded (or, using the word he coined, *moronic*). Goddard's initial studies of immigrant populations labeled vast numbers of them morons. A firm believer in Mendelian genetics, Goddard noted that these categories could be transmitted from one generation to the next. In a 1913 study of a family in rural New Jersey, the Kallikaks (a pseudonym), Goddard traced the origins of the family's apparent moronic tendencies to the union of a man with a moronic wife. He believed that the family's low social position was less a product of social conditions than of their overall genetic feeblemindedness.

People with mental retardation suffered during the eugenics movement. Eugenics (literally: "good breeding") was based on purifying racial groups by eliminating pollutants. Goddard's work contributed to the attitude that the evils of society—namely, poverty, crime, and immorality—went hand in hand with feeblemindedness. Those seeking a long-term cure for mental retardation in society occasionally advocated sterilization to improve "racial stamina" or prevent "race suicide." In efforts to purify society's gene pool, several countries enacted sterilization laws. In the United States alone, more than twenty states passed laws by the mid-1920s requiring sterilization of people with mental retardation, and between 1907 and 1963, more than 60,000 people were sterilized. In the 1927 case of *Buck v. Bell,* the U.S. Supreme Court upheld the practice of involuntary sterilization on the basis that the health of the general public superseded individual rights. Most of the sterilizations were carried out in the 1930s.

The shift in thinking from isolation in institutions toward integration was a slow process, mainly accomplished in the 1950s and after. But as early as the 1920s, institution superintendent Charles Bernstein began to advocate moving patients into smaller homes within the mainstream community, to provide a more normal environment. Moving patients from institutions into group homes constituted an early step in integrating people with mental retardation into society. While allowing residents to live in the mainstream, group homes served as a place to accommodate special needs. Increasingly, experts in "special education" would place more emphasis on including people with cognitive disabilities in schools and community activities rather than secluding them in institutions. People with disabilities began a serious movement for equal access to education, inspired by the 1954 landmark civil rights decision by the Supreme Court, *Brown v. Board of Education.*

See also Eugenics; Genetics; Intelligence Testing; Mental Health; Psychology
References
Gould, Stephen Jay. *The Mismeasure of Man.* New York: W. W. Norton, 1996.
Trent, James W., Jr. *Inventing the Feeble Mind: A History of Mental Retardation in the United States.* Berkeley: University of California Press, 1994.
Tyor, Peter L., and Leland V. Bell. *Caring for the Retarded in America: A History.* Westport, CT: Greenwood Press, 1984.
Winzer, Margaret A. *The History of Special Education: From Isolation to Integration.* Washington, DC: Gallaudet University Press, 1993.

Meteorology

Meteorology is the scientific study of the earth's atmosphere. It focuses on the understanding and prediction of the weather. Twentieth-century theoretical meteorology grew out of the work of the Norwegian scientist Vilhelm Bjerknes (1862–1951). His studies of large-scale circulation processes, which extended mathematical hydrodynamics to include another variable—temperature—made Bjerknes a central figure in the history of both meteorology and oceanography. His interests stemmed from the competing influences of his mathematician father, Carl Bjerknes (1825–1903), and physicist Heinrich Hertz (1857–1894), who had conducted experiments on electromagnetic waves. His theory of circulation, enunciated in 1897, synthesized hydrodynamics and thermodynamics, and his analyses of the interactions between heat and moisture appeared to have important ramifications in meteorology. Bjerknes soon came to see weather forecasting as one of the principal uses of his ideas about circulation.

Studies of meteorology expanded in the first two decades of the century. Bjerknes was personally responsible for starting, with money from the Carnegie Institution, geophysical institutes in Kristiania (later named Oslo), Norway, and Leipzig, Germany. In 1917, he left Germany to return to Norway, where in Bergen he began a meteorological service to attend the country's economic and strategic needs. Around Bjerknes, a new group of leading meteorologists arose in the 1920s, known collectively as the Bergen school. They originated the idea of the weather front, based on their discovery of a major polar front affecting local weather conditions. Tracking its movements became a critical tool of the Norwegians in pioneering the use of weather forecasting for aircraft and for other purposes; this usefulness also gave Norwegian meteorologists considerable influence in their country.

World War I provided a major forum to test the ability of meteorologists to predict weather conditions and thus become useful for the war effort. Armies established flying corps, whose participants had to be trained to handle adverse weather conditions and to know when favorable flying conditions might exist. For example, war meteorologists found that the winds on the Western front tended to come from the west, which was favorable to the Germans if their planes had engine trouble, as they could glide to safety; British, French, or U.S. planes, by contrast, might be obliged to land behind enemy lines or in the no-man's-land between the two sides. Flyers needed to be educated about such variables in order to prevent the Germans from drawing them too far behind enemy lines, especially given the lack of reliable engines.

Meteorology became one of the principal ways in which scientists advised governments. The United States Weather Bureau, for example, relied on the government's Science Advisory Board, which recommended funding air mass analyses, a concept based on the notion of fronts. Air masses were those masses of air of generally the same temperature, moisture, and wind, differing from other masses with different properties. The lines of discontinuity separating them were the "fronts." Polar masses, for example, were cold and dry, whereas equatorial masses were warm and wet. Instead of mixing upon contact, usually the masses remained distinct, often causing storms and other anomalous conditions at the front. These air masses were more intensively studied in the 1930s, and communication systems were developed in order to coordinate studies and facilitate weather forecasting.

Meteorology matured further during World War II. Advanced communications systems ensured faster reporting of local conditions, enabling broader portraits of regional and even world conditions. The U.S. Navy made extensive use of meteorologists when planning combat operations. For example, after the U.S. attacks on the Marshall Islands in the Pacific Ocean in January 1942, naval task forces realized that they would be vulnerable to counterattack by Japanese bombers. But scientists knew that weather fronts provided natural smokescreens, with miserable weather that not only hid ships but also discouraged airplane attacks. By identifying the location of a front heading eastward toward a safe harbor, the Navy simply kept close to the front and enjoyed relative safety from the harassment of Japanese airplanes. Meteorologists also played important roles in selecting the optimal dates of amphibious landings in North Africa and Normandy. One of the requirements for such planning was improved long-term forecasting (as opposed to 24-hour or 36-hour forecasting). Close relationships between the Weather Bureau and the branches of the armed services, the establishment of a network of weather stations, the use of radar to track weather balloons, and frequent storm-scouting by airplanes enabled intensive joint analyses of weather patterns, and the lead-time of wartime forecasts often could be extended to about five days.

See also Bjerknes, Vilhelm; Oceanography; World War I; World War II

References

Fleming, James Rodger, ed. *Historical Essays on Meteorology, 1919–1995.* Boston, MA: American Meteorological Society, 1996.

Friedman, Robert Marc. *Appropriating the Weather: Vilhelm Bjerknes and the Construction of a Modern Meteorology.* Ithaca, NY: Cornell University Press, 1989.

Gregg, Willis Ray. "Progress in Development of the U.S. Weather Service in Line with the Recommendations of the Science Advisory Board." *Science,* New Series 80:2077 (19 October 1934): 349–351.

Van Straten, F. W. "Meteorology Grows Up." *The Scientific Monthly* 63:6 (1946): 413–422.

Ward, Robert de C. "Meteorology and War-Flying, Some Practical Suggestions." *Annals of the Association of American Geographers* 8 (1918): 3–33.

Microbiology

Microbiology concerns itself with the nature and effects of microorganisms. Around the turn of the twentieth century, scientists commonly referred to these as microbes, typically in relation to bacteria causing disease. Many of the celebrated figures in microbiology were called microbe hunters, and came to prominence toward the end of the nineteenth century because their discoveries had far-reaching medical and public health applications. These include Louis Pasteur (1822–1895), Robert Koch (1843–1910), and Paul Ehrlich (1854–1915), each of whom developed treatments for diseases caused by microbes—for example, rabies, anthrax, tuberculosis, and syphilis. Both Koch and Ehrlich were deeply involved in developing public health measures to prevent outbreaks of such diseases in the first decade of the century.

The importance of microorganisms in identifying and combating disease sparked a major reorientation of public health toward the science of microbiology. Whereas nineteenth-century public health measures emphasized better sanitation and social programs for preventive action, the early twentieth century increasingly—although not wholly—relied on microbe hunting as the solution to the problem of major diseases. Ehrlich, arguably the most famous of the microbe hunters, predicted in 1906 that scientists would be able to develop "magic bullets" to combat specific microbes without damaging surrounding tissue. Although the failure to discover a reliable "magic bullet" for major diseases such as syphilis in subsequent decades encouraged a return to the preventive approach, the successes of vaccines and antibiotics during the 1940s in drastically reducing the effects of infectious diseases helped to tip the balance from prevention to cure.

Most of the accomplishments in microbiology in the first half of the century were in identifying the microbes causing specific diseases. For example, Charles Henri Nicolle (1866–1936) discovered in 1909 that typhus fever is transmitted by the body louse. In 1911, American Francis Rous (1879–1970) identified a virus that could cause cancer. In

British microbiologist Alexander Fleming, one of the developers of penicillin, holding a petrie dish. Penicillin's life-saving properties stood next to the atomic bomb and radar as one of the war's greatest scientific accomplishments. (Bettmann / Corbis)

1918, Alice Evans (1881–1975) identified the bacillus responsible for Malta fever, cattle abortion, and swine abortion. The Spanish influenza virus swept the world in 1918–1919 and killed more people (likely between 20 and 40 million) than did the fighting of World War I. The virus finally was identified in 1930. In 1935, Wendell Stanley isolated the tobacco mosaic virus in crystalline form. Research on identification, isolation, and vaccine development for such diseases was funded by various organizations, notably the Rockefeller Institute, which devoted considerable resources to tracking down the microbes responsible for specific ailments. Therapy for such diseases occasionally, as in the case of vaccines, used dead or weakened microbes, although others such as Salvarsan—Ehrlich's treatment of syphilis—were chemical in nature.

Aside from the devastation of infectious diseases, the extraordinary number of complications from simple wound infections during World War I sparked renewed effort to develop ways to combat the effects of microbes in the body. In 1928, British microbiologist Alexander Fleming (1881–1955) observed, quite accidentally, that a certain mold was an effective killer of bacteria. Penicillin, as it was called, was developed for widespread use during World War II to prevent the infections that had been so widespread during the last war. Companies in the United States manufactured penicillin in mass quantities. Its life-saving properties stood next to the atomic bomb and radar as one of the war's greatest scientific accomplishments. Fleming shared the Nobel Prize with penicillin's developers Howard Florey (1898–1968) and Ernst Chain (1906–1979) in 1945.

See also Ehrlich, Paul; Koch, Robert; Medicine; Penicillin; Public Health; Venereal Disease

References

Bigger, Joseph W. *Man against Microbe.* New York: Macmillan, 1939.

Brandt, Allan M. *No Magic Bullet: A Social History of Venereal Disease in the United States since 1880.* New York: Oxford University Press, 1985.

Bulloch, William. *The History of Bacteriology.* New York: Oxford University Press, 1938.

Collard, Patrick. *The Development of Microbiology.* New York: Cambridge University Press, 1976.

Millikan, Robert A.

(b. Morrison, Illinois, 1868; d. San Marino, California, 1953)

Robert A. Millikan was a U.S. physicist whose many contributions served the interests of science, military, and his own institution. He graduated from Oberlin College in 1891 and received a doctorate in physics from Columbia University in 1895. Like many U.S. scientists of his generation, he traveled abroad to leading European institutions after taking his degree. He went to Berlin and Göttingen, Germany, before taking up a position at the University of Chicago. His initial research interest was on light polarization; he later became known for his work on electrons, the photoelectric effect, and cosmic rays. One of Millikan's biographers called his career a microcosm of U.S. science during the first half of the twentieth century. As a scientist, Millikan was not only a researcher, but also a teacher, celebrity, administrator, and a consultant for government and industry.

Millikan came to prominence fairly late in life. He had devoted his early scientific years to teaching, before making the measurement that made him famous. Millikan's many contributions were in observation, measurement, and experiment, helping to lend credibility to many of the theoretical constructs coming from European scientists. For example, in 1910, he measured the charge carried by an electron and demonstrated that the figure was constant for all electrons. He did it by developing the oil-drop method, measuring the pull of gravity on a drop of oil against the force of an electric field. In many respects, he was a relic of the physics community of the late nineteenth century, with its allegedly limited aspirations of simply improving the accuracy of measurements to the next decimal place rather than breaking new ground. His next effort, to test the photoelectric effect, began as an exercise in reining in outrageous new concepts; instead, he ended up strengthening Albert Einstein's (1879–1955) theory about photons. For this and his work on measuring electrons, Millikan was awarded the Nobel Prize in Physics in 1923. He was the first U.S.-born scientist to receive this honor.

During World War I, Millikan played a leading role in putting science in the service of the U.S. military. He replaced Thomas Edison (1847–1931) as the premier "useful" man of science in the country, helping to organize research for submarine detection, antisubmarine warfare, aviation, and other areas of research. He became the executive head of the National Research Council, created during the war to keep the United States at the cutting edge of research for the national interest, in military and other domains.

Despite the honor of winning the Nobel Prize for the oil-drop method and for testing the photoelectric effect, Millikan is equally well known for his work on cosmic rays. In fact, he began a tradition of world leadership in cosmic ray research at Caltech, where in 1921 he became the director of the Norman Bridge Laboratory of Physics. Building on the prewar work of Austrian physicist Victor Hess (1883–1964), Millikan coined the term *cosmic ray* to describe the elementary particles entering the earth's atmosphere from space. These rays often were called Millikan rays by those who were unaware of (or unwilling to acknowledge) the work of Hess, who first identified the phenomenon. Millikan believed that cosmic rays were high-energy photons, but that idea was discarded in the mid-1930s.

Millikan occupied a central place in making Caltech a world-class research institution. As Caltech's chief executive officer, Millikan sought to achieve George Ellery Hale's (1868–1938) goal of having a coterie of top-notch chemists and physicists in the vicinity of the Mount Wilson Observatory. Not all of his efforts were admirable: Like many U.S. academics in the 1920s, he was reluctant to hire Jews. Still, under Millikan's influence, Caltech became a major center for research on cosmic rays, leading to a great deal of important work, including Carl Anderson's (1905–1991) discoveries of the positron and meson (later called muon). In addition, Millikan attracted many leading scientists to Caltech, including geneticist Thomas Hunt Morgan (1866–1945). Caltech flourished in the 1930s despite the hard times of the Great Depression. In fact, Millikan was a staunch opponent of state-supported science, preferring to depend on philanthropy and the cooperation of industry. This sentiment agreed with his anti–New Deal political leanings, but it made him a decidedly pre–Cold War scientific administrator. By the end of World War II, state-supported science was the order of the day. By then, however, Millikan had resigned as head of Caltech. He died in 1953.

See also Cosmic Rays; Physics; World War I
References
Goodstein, Judith R. *Millikan's School: A History of the California Institute of Technology.* New York: W. W. Norton, 1991.
Kargon, Robert H. *The Rise of Robert Millikan: Portrait of a Life in American Science.* Ithaca, NY: Cornell University Press, 1982.
Kevles, Daniel J. *The Physicists: The History of a Scientific Community in Modern America.* Cambridge, MA: Harvard University Press, 1995.
Millikan, Robert A. *The Autobiography of Robert A. Millikan.* New York: Prentice-Hall, 1950.

Missing Link

Evolutionists of the nineteenth century, including Charles Darwin (1809–1882), regretted that the fossil evidence for the theory of evolution was not particularly convincing. Instead, they turned to biological arguments from studies of embryology, breeding, and comparative anatomy. Remains of Neanderthal Man were first discovered in 1856, some three years prior to Darwin's publication of *On the Origin of Species by Means of Natural Selection,* but few looked to these bones as direct evidence of human evolution. Popular views of evolution treated the modern animals in a hierarchy, as if modern man evolved from modern apes (with apes remaining unchanged). Each animal could be classed as "higher" or "lower" on a scale. Darwinists did not hold this view, but they did seek to find in fossils the evidence of common ancestry among apes and men. Such efforts led to popular demands for evidence of the transition between apes and man. Where, the question was posed, were the "missing links" in the chain of evolution?

One of the first naturalists to find evidence of such a missing link was Eugène Dubois (1858–1940), a Dutch surgeon. In his travels in the Dutch colonies in Southeast Asia, particularly the island of Java, he discovered ancient fossils that he believed filled the link. In 1891 and 1892, he found bones—a molar, skullcap, and femur—that appeared to be a cross between and modern ape and a modern man. In Dubois's view, these bones illustrated a transitional phase between apes and man, with a man's femur and a skull too large for an ape but smaller than a man's. Judging from the femur, the man must have walked erect. Dubois called it *Pithecanthropus erectus,* and it became known as Java Man.

Dubois defended his findings against a host of critics, scientists and nonscientists alike. He attempted to withdraw from the debate around 1900, refusing to let anyone see the bones. This seclusion lasted for about twenty years, until he eventually announced that he now believed that the skull was that of a giant ape. Historians recently have reevaluated his pronouncement not as a signal of defeat, but rather as an effort to distinguish Java Man as older than the discoveries being made in the 1920s in China.

Efforts to find the missing link evolved into quests to find the oldest remains of hominids from which modern humans might descend. The most significant findings were in China, at the Zhoukoudian site near Beijing. Called *Sinanthropus Pekinensis,* or Peking Man, the collection of bones came from a deposit revealed by limestone quarrymen in the early 1920s. Excavations continued for many years thereafter. Both Java Man and Peking Man appeared to demonstrate that modern man originated somewhere in Asia, a view widely held until the 1950s, when the work of Louis Leakey (1903–1972) and Mary Leakey (1913–1996) revealed much older hominid remains from the Olduvai Gorge in Africa.

Efforts to find the missing link have been, at times, a bit too vigorous. For example, Piltdown Man was "discovered" in a gravel bed in Sussex, England, in 1912. It had a human skull and an ape's mandible, and after its discoverer Charles Dawson (1864–1916) it was named *Eoanthropus dawsoni.* Its precise age in relation to Java Man and (later) Peking Man was always a subject of controversy, and for forty years, Piltdown Man was England's very own missing link. But in the early 1950s, it was revealed as a forgery, and scientists and historians have been trying to identify the hoaxer ever since.

The idea of the missing link in a hierarchical chain oversimplified an important feature of evolutionary theory, which asserted that apes and men share a common evolutionary ancestry. This is not the same thing as one evolving from the other. Java Man and Peking Man, for example, belonged to species closely related to modern man, but whether or not men descended directly from them was a constant source of controversy.

See also Anthropology; Evolution; Peking Man; Piltdown Hoax; Simpson, George Gaylord

References

Bowler, Peter J. *Evolution: The History of an Idea.* Berkeley: University of California Press, 1989.

Spencer, Frank, and Ian Langham *Piltdown: A Scientific Forgery.* London: Oxford University Press, 1990.

Stocking, George W., Jr., ed. *Bones, Bodies, Behavior: Essays on Biological Anthropology.* Madison: University of Wisconsin Press, 1988.

Theunissen, Bert. *Eugène Dubois and the Ape-Man from Java: The History of the First "Missing Link" and Its Discoverer.* Dordrecht: Kluwer, 1989.

Mohorovičić, Andrija

(b. Volosko, Croatia, 1857; d. Zagreb, Yugoslavia, 1936)

Andrija Mohorovičić, one of the most celebrated scientific figures from the former Yugoslavia, discovered the boundary between the earth's crust and mantle. He studied physics at the University of Prague, and then did his graduate work at the University of Zagreb. Initially he pursued meteorology, founding a meteorological station at Bakar in 1887. He became head of Zagreb's meteorological observatory in 1892, and in 1901, he led all of Croatia's meteorological services. Despite all this work, Mohorovičić is much more widely known for an important discovery in seismology.

A major earthquake in the Kupa Valley in 1909 gave Mohorovičić the opportunity to put some of the latest techniques of seismology to use. The epicenter of the earthquake was some forty kilometers from Zagreb, thus Mohorovičić was in a good position to study it. In doing so, he discerned a major discontinuity in the velocities of seismic waves deep beneath the surface of the earth. Seismic wave velocities changed abruptly, indicating a serious difference in density between the parts of the earth through which the waves passed. Mohorovičić determined that this discontinuity marked the boundary between the earth's mantle (below) and crust (above). That boundary came to be called the Mohorovičić discontinuity, or Moho, for short. The failed project that took form in the late

1950s in the United States, Project Mohole, takes its name from Mohorovičić. The project's goal was to drill a hole to sample the mantle, whose precise constituents were unknown. Mohorovičić was one of several in the early twentieth century to contribute to the model of the earth's interior with evidence based on the speed of seismic waves traveling through different kinds of rock.

See also Geophysics; Seismology
References
Bascom, Willard. *A Hole in the Bottom of the Sea.* New York: Doubleday, 1961.
Oldroyd, David. *Thinking about the Earth: A History of Ideas in Geology.* Cambridge, MA: Harvard University Press, 1996.

Morgan, Thomas Hunt

(b. Lexington, Virginia, 1866; d. Pasadena, California, 1945)

Thomas Hunt Morgan's fruit fly experiments provided the experimental evidence for the relationships among genes, characters (or traits), and chromosomes. His work is the foundation of classical genetics, because it established the material basis of heredity in genes carried by the chromosomes. During his long career he held positions at Bryn Mawr College, Columbia University, and the California Institute of Technology. His early interests were in embryology, but the influence of Hugo De Vries (1848–1935) led him toward genetic variation, particularly mutation. Like other geneticists such as De Vries and William Bateson (1861–1926), Morgan opposed the continuous variation proposed by Darwinian evolution and preferred sudden, discontinuous jumps (mutation). Attempting to observe such mutations, he conducted breeding experiments on animals such as mice, rats, and pigeons, finally settling on the fruit fly.

Because *Drosophila melanogaster,* the fruit fly, has a life cycle of some two weeks, it made an ideal candidate for observing the transmission of characters from one generation to the next. It also has lots of offspring

and is a hardy creature in the same vein as the common cockroach. It also has few chromosomes, making it a relatively simple organism through which to study heredity. Morgan hoped that the fruit fly would yield some readily identifiable mutants, and that he could observe their hereditary transmission. The first mutation he observed was the "white eye" character.

Morgan's studies of the fruit fly persuaded him to reverse his view that chromosomes could not carry the characters of heredity according to Mendelian laws. He previously had held that deviations from these laws proved them inaccurate, especially the deviations suggesting that some characters were inherited together rather than independently. But the fruit fly experiments showed such broad agreement with Mendelian genetics that Morgan upheld the concept of *linkage* to explain the apparent mutual dependence of some characters. If chromosomes carried more than one character (and certainly there were far fewer chromosomes than characters), then the chromosomes would behave according to Mendelian laws. Meanwhile, the characters themselves would show some aberration from the laws because they were linked together (and thus not independent) on the same chromosome. For example, Morgan found that the traits "white eye" and "rudimentary wings" were always inherited together, and thus were an example of linkage.

Morgan also found that linkage was not always complete; in other words, the traits did not always segregate together to be passed on to the next generation. He borrowed a theory from cytology, namely, Frans A. Janssens's (1863–1924) "chiasmatype," which showed the breaking and reformation of chromosomes and the consequent interchange of segments, all in the early stage of cell division. Morgan believed that this process, the crossing-over of characters through the recombination of chromosomes, could explain the observed incompleteness of linkage. With the short-lived and populous fruit flies, Morgan had quantitative and increasingly precise data to demonstrate such crossing-over of characters, along with genetic linkage generally. Between 1910 and 1915, he and colleagues Alfred H. Sturtevant (1891–1970), Calvin B. Bridges (1889–1938), and Hermann J. Muller (1890–1967) observed the linkages and recombination of chromosomes in their fruit flies in the "fly room" at Columbia University. Their studies culminated in the 1915 book, *The Mechanism of Mendelian Inheritance,* now a classic of modern genetics.

Although he had once been an adamant opponent of Mendelian genetics, Morgan now became the center of genetics research as it moved from England to the United States. The focus of this work was the mechanics of chromosomes. In the early 1920s, William Bateson abandoned his own longheld opposition to viewing chromosomes as the carrier of heredity and effectively ceded leadership in the field to Morgan's group. Morgan's status in the United States rose dramatically in subsequent years, and he became president of the National Academy of Sciences in 1927 and president of the American Association for the Advancement of Science in 1930. In 1928, he left Columbia University to take up a position at the California Institute of Technology, where he worked until his death in 1945. In 1933, Morgan won the Nobel Prize in Physiology or Medicine for his work showing the material basis of heredity in chromosomes.

See also Bateson, William; Chromosomes; Genetics; Mutation; Rediscovery of Mendel
References
Allen, Garland E. *Thomas Hunt Morgan: The Man and His Science.* Princeton, NJ: Princeton University Press, 1978.
Bowler, Peter J. *The Mendelian Revolution: The Emergence of Hereditarian Concepts in Modern Science and Society.* Baltimore, MD: Johns Hopkins University Press, 1989.
Kohler, Robert E. *Lords of the Fly:* Drosophila *Genetics and the Experimental Life.* Chicago, IL: University of Chicago Press, 1994.
Magner, Lois N. *A History of the Life Sciences.* New York: Marcel Dekker, 1979.

Mutation

The word *mutation* was adopted by biologists in the early twentieth century to describe significant and sudden changes within species. Its meaning itself mutated over time, from simultaneous "saltation" and subsequent creation of new populations, to the mechanism for random variation in genes. More generally, the word came to mean a drastic metamorphosis of some kind, and its usage is not limited to biology: Institutions and ideas mutate, transforming into entities with only some resemblance to the original. In the era after World War II, the genetic and somatic effects of testing nuclear weapons gave rise to fears of mutation; popular films portrayed giant insects and other mutants brought about by the atomic age.

Dutch botanist Hugo De Vries (1848–1935), one of the scientists who had "rediscovered" Gregor Mendel's (1822–1884) work around the turn of the century, popularized mutation as a means to explain heredity. He favored the idea that changes in species occurred through "saltations," significant and abrupt changes. In *The Mutation Theory* (1901), he forged a new path amid the debate between Darwinian natural selection and Lamarckian willful evolution. Mutation accounted not only for the inheritance of existing characteristics, but also for the creation of new ones. In his view, mutation was sudden and involved many individuals mutating at once, giving rise to a distinct breeding population. This interpretation, later rejected, appeared to solve an outstanding problem of Darwinism, namely, how new characteristics are inherited without being "swamped" by the rest of the population. The mutation theory provided the answer by proposing mutation on a large scale. Mutation was also attractive to those who doubted that Charles Darwin's (1809–1882) version of evolution could fit with the timescale of the earth calculated by physicists. With mutation, perhaps the process was much faster.

De Vries had based his theory on his studies of the evening primrose, *Oenothera lamarckiana;* later, in the 1920s, scientists found that this flower's complex genetic structure would have agreed with his results, but were not truly mutations in the sense that he had thought. Meanwhile, geneticists took up mutation as a tool for attacking Darwinism. One of these was the American Thomas Hunt Morgan (1866–1945), though he would soon abandon his opposition to Darwinian natural selection. Morgan's experiments with fruit flies, or *Drosophila melanogaster,* identified numerous mutations and showed how they contributed to the fruit fly's range of variability. The first observed was the white-eyed fruit fly, followed by others, including ones with bent wings and other characteristics, many of which clearly were not viable in the wild. Mutated genes were transmitted from one generation to the next, obeying Mendelian laws. Morgan's *Mechanism of Mendelian Inheritance* (1915) demonstrated the important role played by mutation in genetics, through its steady introduction of new genes into the population (although not creating a distinct population by simultaneous mutation, as De Vries had thought). This work aided, in the 1920s, in the reconciliation between Darwinism and Mendelism, and mutation became the mechanism for introducing random changes in the process of evolution.

Fruit flies proved useful for tracing mutants, particularly because of their short lifespans. But in the early 1940s, scientists such as George Beadle (1903–1989) and Edward Tatum (1909–1975) turned to bread mold, *Neurospora,* to study biochemical mutations. No longer were mutants confined to understanding heredity. Beadle and Tatum used mutations not to study developmental genetics, looking for the key to evolution, but rather to understand biochemical interactions. For their work on the role of genes in chemical reactions, they won the Nobel Prize in Physiology or Medicine in 1958.

See also Evolution; Genetics; Morgan, Thomas Hunt; Rediscovery of Mendel

References
Allen, Garland E. *Thomas Hunt Morgan: The Man and His Science.* Princeton, NJ: Princeton University Press, 1978.
Bowler, Peter J. *The Mendelian Revolution: The Emergence of Hereditarian Concepts in Modern Science and Society.* Baltimore, MD: Johns Hopkins University Press, 1989.
Kohler, Robert E. *Lords of the Fly:* Drosophila *Genetics and the Experimental Life.* Chicago, IL: University of Chicago Press, 1994.
Stamhuis, Ida H., Onno G. Meijer, and Erik J. A. Zevenhuizen. "Hugo De Vries on Heredity, 1889–1903: Statistics, Mendelian Laws, Pangenes, Mutations." *Isis* 90 (1999): 238–267.

N

National Academy of Sciences

U.S. President Abraham Lincoln (1809–1865) created the National Academy of Sciences in 1863, during the Civil War, to advise the government on scientific matters when needed. Its membership included the nation's leading scientists. The early years of the academy were troubled by personal conflicts among its leading figures and the lack of teeth in policy recommendations. The nineteenth-century opinion of the academy to adopt the metric system, for example, resulted in no official action by the government. The academy found itself, at the turn of the century, to be little more than a mutual admiration society with decisions seldom greater than that of choosing new members. With the creation of new federal bodies toward the turn of the century, such as the National Bureau of Standards, the academy was asked less and less for its advice, and academy leadership did not consider it proper either to offer their unsolicited services or to ask for money.

The academy changed in the second decade of the twentieth century, largely owing to the efforts of one of its members, astronomer George Ellery Hale (1868–1938). When World War I began in Europe, the renowned inventor Thomas Alva Edison (1847–1931) began to work with the secretary of the Navy to help prepare the United States for possible involvement in the war. He headed the Naval Consulting Board, which relied on the National Bureau of Standards and industrial laboratories for scientific advice. Feeling marginalized, Hale wanted to raise the reputation of the academy by reversing its attitudes toward money and practical research. Taking this step would not be easy, because elitist academy members did not see themselves as practical men. The academy had refused membership to figures who had distinguished themselves solely as inventors and engineers, including Edison. But Hale wanted to acquire more money and put the academy back into a place of prominence as the most important scientific body in the nation. He hoped to support individual scientists and research projects, to finance a proceedings journal, and to pay for a new building to establish a permanent home for the academy near the corridors of power in Washington, D.C.

The war itself gave the academy the opportunity to offer its services to the government without the appearance of begging for support. Hale urged President Woodrow Wilson (1856–1924) to support research for defense and to make the academy the leading body to achieve defense-oriented scientific goals. Wilson approved the idea, and thus Hale became the first chairman of the newly created National Research Council (NRC) in 1916 under the auspices of the academy. The NRC's goal was to support both pure and applied research, directed toward the security and welfare of the nation. Not only would it include leading scientists, but also it would embrace engineers, even those from the military. Decried by some as a militaristic step, the creation of the NRC ensured the academy's relevance as the primary scientific consultative body for the U.S. government. It also began a precedent of large-scale involvement of scientists in war work, such as submarine detection. When the war ended, Hale and like-minded colleagues continued these ideas by creating the International Research Council, which excluded scientists from the defeated powers of the war.

The 1920s did not herald a new age of increasing demands for the academy's advice. Its permanent building was finished in 1924. But the academy continued its strict elitism of membership, even rejecting a prominent mathematician who allegedly had broken the prohibition laws, which was too undignified for an academy member. The average age of members was near sixty, meaning that most of them had their productive years behind them. One critic called the new building a "marble mausoleum," and as historian Daniel J. Kevles has written, many of Hale's most ambitious scientific dreams were interred there.

Nevertheless, the academy continued to play a major role in U.S. science in subsequent years. The NRC's committees published bulletins that periodically presented the "state of the field" in various disciplines, and it established a fund for prestigious postdoctoral fellowships. In the 1930s, the academy had to fight for influence again, against the upstart Science Advisory Board. Academy president William Wallace Campbell (1862–1938) distrusted the new body, which was not completely composed of academy members and seemed likely to sway under political influence, which helped to kill it after only two years. Later, during World War II, the academy failed to take the reins of research in the national interest. Vannevar Bush (1890–1974), who created the National Defense Research Committee and enjoyed money from the president's emergency funds, avoided academy influence (and interference) like the plague. Science soon sprouted up throughout the government, out of the academy's hands, especially after the war when scientific research proved useful to numerous agencies, particularly the Department of Defense and the Atomic Energy Commission. The degree of the academy's relevance to the postwar world is debatable; it certainly sponsored many study committees, but the center of scientific activity, funding, and even advising might have rested elsewhere.

See also Elitism; Hale, George Ellery; International Research Council; World War I; World War II

References

Cochrane, Rexmond C. *The National Academy of Sciences: The First Hundred Years, 1863–1963.* Washington, DC: National Academy of Sciences, 1978.

Kevles, Daniel J. *The Physicists: The History of a Scientific Community in Modern America.* Cambridge, MA: Harvard University Press, 1995.

National Bureau of Standards

The National Bureau of Standards was created when, at the turn of the twentieth century, the U.S. economy seemed to be entering a new phase. The Spanish-American War recently had provided Americans with possessions beyond its shores, and exports finally were beginning to exceed imports for the first time. To compete with European nations, some businessmen urged the federal government to follow their lead by funding scientific laboratories devoted to determining standard physical and chemical units. As it stood, commercial firms had to depend on measuring devices imported from European scientific agencies. Led by physicist Samuel Wesley Stratton (1861–1931), scientists eagerly promoted a U.S. equivalent, and the bill to create the National Bureau of Standards passed in 1901.

The bureau was part watchdog and part scientific establishment. Its watchdog role was most evident early on. In 1909, bureau scientists visited shops and markets in all of the states, and they reported a large amount of marketplace cheating through false weights and measures, which prompted many states to increase their regulations. Such activities were clearly needed, but they required people and facilities. Stratton convinced the government to provide money for them. Thus the bureau became a home for men (women were explicitly barred from the bureau for some years) to conduct studies of anything that might have a bearing on standards, weights, and measures. To some extent, these golden years for science at the bureau ended when Stratton resigned the directorship in 1923. During most of the 1920s, the bureau focused on more practical studies that were in line with Republicans' restrained attitudes toward government spending. After the start of World War II, the bureau's director Lyman Briggs (1874–1963) became the chairman of the Uranium Committee, charged with deciding what work (if any) should be done on building an atomic bomb. The director of the bureau was considered to be the nation's top government physicist. This unique prestige would fade during the war years, as physicists began to fill government ranks and take on more responsibilities in national policy decisions.

In 1948, the bureau came into wide public consciousness when its director, Edward Condon (1902–1974), was accused by a subcommittee of the House Un-American Activities Committee (HUAC) as being "one of the weakest links in our atomic security." The bureau became the site of the first major scientific casualty of anticommunist paranoia in the late 1940s and early 1950s. Condon was active in support of civilian control of atomic energy and international cooperation in science, and he was hostile to excessive policies of secrecy. He was an easy target for Republican congressman J. Parnell Thomas (1895–1970), who held opposite views; his distrust of scientists led him to launch an effort to strip Condon of his security clearance and his job. His effort failed, but similar periodic attacks convinced Condon to resign in favor of a lucrative job in the private sector. The bureau thus entered the 1950s as the centerpiece of the clash between scientists and Congress during the anticommunist "Red Scare."

See also Loyalty; Patronage; Physics; Manhattan Project

References
Hewlett, Richard G., and Oscar E. Anderson, Jr. *The New World: A History of the United States Atomic Energy Commission, vol. I, 1939–1946.* Berkeley: University of California Press, 1990.
Kevles, Daniel J. *The Physicists: The History of a Scientific Community in Modern America.* Cambridge, MA: Harvard University Press, 1995.
Wang, Jessica. "Science, Security, and the Cold War: The Case of E. U. Condon." *Isis* 83 (1992): 238–269.

National Science Foundation

During World War II, the importance of science to national well-being was demonstrated by such marvels as penicillin, radar, and the atomic bomb. The head of the wartime Office of Scientific Research and Development (OSRD), Vannevar Bush (1890–1974), published a vision for postwar research in his *Science—The Endless Frontier* (1945). Bush argued for a national foundation that would support "basic" research, with no preconceived applications, as the capital for new technology. The foundation would act as a clearinghouse for all governmental scientific patronage. As the war ended, science found an enthusiastic patron in the U.S. military establishment, especially new bodies such as the Office of Naval Research (ONR) (founded in 1946). But no agency emerged to funnel all research money to promote the health of the nation. Instead, scientists applied for funds for various sources, primarily military ones. In 1949, for example, 96 percent of federal funding for university research in the physical sciences came from either the Atomic Energy Commission (AEC) or the Department of Defense.

Concerns about military control of science convinced many scientists that a single federal agency for science was necessary. However, they disagreed about how it should be organized and who should control it. Senator Harley Kilgore (1893–1956) authored a bill making the president responsible for appointing (or firing) the foundation's director and making all discoveries and inventions the property of the government. Senator Warren Magnuson (1905–1989) introduced a competing bill that left patent policies to a board of governors and gave less control of the foundation to the president. He also wanted to put control in civilian hands, whereas Kilgore favored military dominance. Some debate centered around whether or not to include social sciences. Eventually a compromise bill was passed, and President Harry Truman (1884–1972) promptly vetoed it in 1947, primarily because it did not give the president enough power over the foundation's director.

When the National Science Foundation (NSF) finally was created in 1950, it was much weaker than scientists had hoped. For one, legislation limited its annual budget to $15 million, but appropriations were in fact much less. Other agencies, such as the Public Health Service, the AEC, and ONR, refused to concede control of scientific research to the foundation. Thus the original purpose of the foundation was defeated, because it did not become the single agency charged with equalizing the civilian and military aspects of science. Former ONR chief scientist Alan T. Waterman became the NSF's first director, but he was unwilling to challenge the status quo by insisting that the Department of Defense transfer its programs to the fledgling

foundation. The delay in creating the National Science Foundation helped other agencies such as ONR and AEC establish their own credibility for supporting science and made a balance between civilian and military interests far less likely.

See also Elitism; Office of Naval Research; Patronage

References

England, J. Merton. *A Patron for Pure Science: The National Science Foundation's Formative Years, 1945–57.* Washington, DC: National Science Foundation, 1982.

Kevles, Daniel J. *The Physicists: The History of a Scientific Community in Modern America.* Cambridge, MA: Harvard University Press, 1995.

Lomask, Milton. *A Minor Miracle: An Informal History of the National Science Foundation.* Washington, DC: National Science Foundation, 1976.

Nationalism

Although science has a reputation for being above politics, science often was pursued in the interests of individual nations. Some efforts were aimed primarily at prestige, to connect scientific achievement to national greatness. This especially was true in oceanic and Antarctic expeditions. Americans were ecstatic, for example, when Robert Peary (1856–1920) was the first to reach the North Pole in 1909; Norwegians and British cheered Roald Amundsen (1872–1928) and Robert Scott (1868–1912), hoping that their national hero would reach the South Pole first (Amundsen did, in 1911). All such expeditions claimed to be scientific, but really they were aimed at national prestige. Science was a symbol of cultural sophistication and perhaps of technological achievement, and it became useful as something to showcase to other nations. For example, the German government between the two world wars was forbidden to send its warships abroad. Instead, it sent the scientific vessel *Meteor* to foreign ports, to show a German presence and advertise Germany's cultural importance to other nations through science.

One way that leading scientists showed their national pride was in nominating candidates for the Nobel Prize, the most prestigious honor for a scientist, in Physics, Chemistry, and Physiology or Medicine. Historical studies of the nominating process have revealed that scientists tended to recommend their own country's most respected scientist rather than choose the most eminent scientist from any country. Those awarding the prizes, bodies composed of leading scientists in Sweden, were skeptical of nationalistic impulses and tended to choose candidates who received significant numbers of nominations from scientists outside their own countries.

Scientists, like other citizens of any country, could be patriotic. Their actions stood out, however, because of the supposed universality of science; scientists' activities were not supposed to be political. The most egregious breakdown of this attitude occurred during World War I, when scientists on both sides became very nationalistic. For example, many German intellectuals signed a manifesto asserting their common cause in the war, claiming that accusations of Germany's "militarism" were unjust. The 1914 Appeal to the Civilized World, or the Appeal of the Ninety-Three Intellectuals, as it was called, provoked unprecedented bitterness among the Allies not only against Germany, but also against its intellectual establishment. Included among the signatories were Max Planck (1858–1947), Fritz Haber (1868–1934), Walther Nernst (1864–1941), and other leading scientists. Some, like Planck, later recanted, but many others refused to do so, believing that any harm they might have done had been repaid many times by the harsh treaty conditions against Germany at war's end. Allied scientists condemned German scientists personally, not only during the war, but also afterward. The International Research Council, created at the end of the war, barred Germans from participation and was designed to isolate German scientists from the rest of the scientific community. That "international" body replaced the International Association of Academies, which had been dominated by Germans.

By the mid-1920s, the rivalries from the war began to subside, and the institutional barriers between scientists gradually decreased. But the idea of "national" science did not. It resurfaced not merely as patriotism, but as a particular kind of science. As early as 1920, Germans such as Paul Weyland (1888–1972) singled out Albert Einstein (1879–1955) for corrupting pure science with what amounted to scientific Dadaism (a postwar artistic movement that emphasized meaninglessness). Relativity, and later quantum mechanics, became symbols of a kind of theoretical science to which many Germans were hostile. In the 1930s, many experimental physicists in Germany began to speak of "German physics" as distinct from degenerate, "Jewish" physics. Philipp Lenard (1862–1947), who won the Nobel Prize in Physics in 1905 for his work on cathode rays, believed that only experimental physics (as opposed to theoretical) could lead to knowledge about the real world. Abstract theory, or reliance on mathematics, struck Lenard and his younger compatriot and fellow Nobelist, Johannes Stark (1874–1957), as anti-German. This brand of nationalism was wrapped firmly in anti-Semitism.

After World War II, nationalism in science took on new meaning because science no longer was considered primarily a cultural product. Although science and technology had been used for military purposes in the past, the role of the atomic bomb promised to bring about an elevated status for science. Americans used atomic bombs in war against Japan in 1945, and the Soviet Union unexpectedly tested their own atomic device in 1949, after an intensive scientific and industrial effort of which no other country thought it was capable. "American" science and "Soviet" science came to denote the separate research communities of the Cold War that often existed under a veil of secrecy, one strengthening the capitalist, democratic world, and the other trying to strengthen the

Communist world. Funds for science came increasingly from the governments of these two nations, primarily through their military establishments, as political leaders looked to science to build new weapons and define global strategy. Scientists were not just nationalistic but played a leading role in national strength and security, finding a lasting place within what became known as the military-industrial complex.

See also Cold War; International Research Council; Nazi Science; Nobel Prize; Oceanic Expeditions; Polar Expeditions; World War I; World War II

References

Badash, Lawrence. *Scientists and the Development of Nuclear Weapons: From Fission to the Limited Test Ban Treaty, 1939–1963*. Atlantic Highlands, NJ: Humanities Press, 1995.

Cock, A. G. "Chauvinism and Internationalism in Science: The International Research Council, 1919–1926." *Notes and Records of the Royal Society of London* 37:2 (1983): 249–288.

Crawford, Elisabeth. *Nationalism and Internationalism in Science, 1880–1939: Four Studies of the Nobel Population*. New York: Cambridge University Press, 1992.

Heilbron, J. L. *The Dilemmas of an Upright Man: Max Planck and the Fortunes of German Science*. Cambridge, MA: Harvard University Press, 2000.

Kirwan, L. P. *A History of Polar Exploration*. New York: W. W. Norton, 1960.

Mills, Eric L. "Socializing Solenoids: The Acceptance of Dynamic Oceanography in Germany around the Time of the 'Meteor' Expedition." *Historisch-Meereskundliches Jahrbuch* 5 (1998): 11–26.

Nazi Science

The German National Socialist Party, or the Nazis for short, incorporated a number of scientific ideas to bolster their racist agenda. The field of eugenics, born in Britain in the nineteenth century, seemed to suggest that populations could be improved or strengthened through selective breeding. During the first decades of the twentieth century, eugenicists lobbied their governments to take measures to discourage procreation among social *undesirables,* a term with flexible mean-

ing. When the Nazis came to power in Germany in 1933, they sought to put such ideas into practice, purifying the Aryan race through legislative actions, many of them long sought by eugenicists in the United States and Britain. Among these laws were forced sterilization of those with mental retardation and ultimately the Nuremberg Laws (1935), which discriminated against Jews and forbade intermarriage between Aryans and Jews.

Nazi racial theories had odd consequences in the realm of health. Many of the Nazis were hostile to smoking, drinking, or even eating meat. Purity of health went hand in hand with racial purity. Because of the aggressive anti-Semitic policies of the regime, leading to mass killings of Jews during World War II, it is easy to overlook the public health measures taken by the Nazis. They conducted surveys and compiled data intending to assess the causes of cancer and to eliminate it from their society, leading to policies restricting carcinogens and overexposure to X-rays, as well as a major public campaign against smoking. Because educated Jews were fired in great numbers during the 1930s, the scientific establishment lost its ability to focus on curing diseases such as cancer. This forced the government to concentrate on public health measures for prevention. None of these policies aimed to improve the health of society as a whole; rather, they were part of a broad campaign to improve the vigor of the Aryan race.

Despite its respect for science, the Nazi Party crippled its own scientific community through its anti-Semitism. Although some fields prospered under the Nazis, particularly biology, physics literally fell apart. Jews in academic fields lost their positions after 1933, initiating a massive intellectual migration from Germany, Austria, and Italy, to Britain and the United States. Such leading minds as Albert Einstein (1879–1955), Enrico Fermi (1901–1954), Niels Bohr (1885–1962), Erwin Schrödinger (1887– 1961), and many others found homes elsewhere. One of those who remained, Max Planck (1858–1947), tried to use his leadership position in the Kaiser Wilhelm Society to protect his Jewish colleagues when possible. But even among leading scientists there was no unity of opinion. Physicists Johannes Stark (1874–1957) and Philipp Lenard (1862–1947), for example, criticized theoretical physics as "Jewish physics." Its focus on abstract, intangible ideas struck them as a corruption of the true nature of positivist, experiment-based science. They equated experimental physics with "German physics" and demanded that Jewish influences be uprooted. Thus even non-Jews such as Planck and Werner Heisenberg (1901–1976) were targeted as practitioners of Jewish science.

One of the most controversial aspects of Nazi science was its wartime work. Rocket scientists such as Wernher Von Braun (1912–1977) developed long-range rockets to deliver bombs to their targets without the need for bomber aircraft. The first such rocket successfully used was the V-2, which was launched against London in 1944. U.S. and Soviet armies were pleased to capitalize on Nazi accomplishments in this field, and they captured rocket scientists for their countries' own purposes. Von Braun, for example, went on to play a leading role in the U.S. space program. In addition, German physicists were charged with developing an atomic bomb. One of the principal founders of quantum mechanics, Werner Heisenberg, led this project. It was controversial because Heisenberg later claimed that he had purposefully slowed the project, never truly wanting Adolf Hitler (1889–1945) to have the bomb. But based on transcripts from secret recordings of Heisenberg after he and his colleagues were captured toward the end of the war, some historians have argued that Heisenberg's team simply never found a workable method to develop the weapon, despite sincere efforts to do so.

In the medical profession, Jews also were persecuted and pushed out. Those who re-

One of the most controversial aspects of Nazi science was its wartime work. Rocket scientists such as Wernher Von Braun (pictured here) developed long-range rockets to deliver bombs to their targets without the need for bomber aircraft. The first such rocket successfully used was the V-2, which was launched against London in 1944. (Library of Congress)

mained presided over involuntary sterilization of a host of undesired individuals and euthanasia (medical killing) of asylum inmates. Psychiatrists played an active role in justifying such practices on scientific grounds. Jews in institutions were killed regardless of their specific condition, requiring no special justification. After the war began, concentration camps such as Auschwitz (in Poland) became the sites of a more generalized policy of genocide. Doctors continued to play a role, selecting healthy new arrivals at the camps and separating them for work duty, and taking part in human experimentation. Physicians forged bonds with the Nazi leadership in great numbers, nearly half of them becoming party members at some time during the Nazi regime.

See also Cancer; Eugenics; Human
  Experimentation; Patronage; Race; World
  War II

References

Geuter, Ulfried. *The Professionalization of
  Psychology in Nazi Germany*. New York:
  Cambridge University Press, 1992.

Heilbron, J. L. *The Dilemmas of an Upright Man:
  Max Planck and the Fortunes of German Science*.
  Cambridge, MA: Harvard University Press,
  2000.

Kater, Michael H. *Doctors under Hitler*. Chapel
  Hill: University of North Carolina Press,
  1989.

Kühl, Stefan. *The Nazi Connection: Eugenics,
  American Racism, and German National Socialism*.
  New York: Oxford University Press, 1993.

Lifton, Robert Jay. *The Nazi Doctors: Medical
  Killing and the Psychology of Genocide*. New
  York: Basic Books, 1986.

Macrakis, Kristie. *Surviving the Swastika: Scientific
  Research in Nazi Germany*. New York: Oxford
  University Press, 1993.

Proctor, Robert N. *The Nazi War on Cancer*.
  Princeton, NJ: Princeton University Press,
  1999.

Rose, Paul Lawrence. *Heisenberg and the Nazi
  Atomic Bomb Project: A Study in German Culture*.
  Berkeley: University of California Press, 1998.

Nobel Prize

In the twentieth century, scientists could receive no higher recognition for their accomplishments than winning the Nobel Prize. Actually there were five categories for the prize, of which three were scientific: physics, chemistry, and physiology or medicine. The other two were literature and peace. The prize was named after Alfred Nobel (1833–1896), the Swede who invented dynamite and patented it in 1867. His companies manufactured and sold explosives for decades, and he died a very wealthy man in 1896. In his will, he stipulated that large sums of his money should be set aside, and that the interest should be awarded each year to individuals who had made significant contributions in the above categories. After some legal wrangling, the Nobel Foundation was created in 1900 to manage Nobel's assets, and various Swedish institutions had the task of deciding the prize recipients. In physics and chemistry, this responsibility fell to the Royal Swedish Academy of Sciences, and the Royal Caroline Medico-Surgical Institute chose the winners in physiology or medicine. The first prizes were awarded in 1901.

Nobel stipulated that country of origin should not be a factor in deciding who received the prizes. Initially, some Swedish scientists decried this as unpatriotic (why should the Swedish Academy go out of its way to select and honor foreign scientists?). But eventually the Nobel Prizes became symbols of science's internationalism. However, nominations for the prize were certainly subject to nationalist impulses. In her study of the Nobel candidates and nominators in chemistry and physics, Elisabeth Crawford notes how much World War I dampened sentiments of scientific internationalism. The postwar years saw an increase in nominators choosing candidates from their own countries, especially in Britain, Germany, and the United States. French scientists also tended to nominate their own, but that was so even prior to the war. Between 1916 and 1920, less than 2 percent of the nominators selected scientists from enemy (or former enemy) countries. The national bias continued, although on a less drastic scale, after the war, especially in chemistry, a field in which the German chemical warfare specialist Fritz Haber (1868–1934) won in 1918, much to the dismay of scientists in Allied countries. In physics, nationalism was less a factor, because Germany had a candidate who appealed to scientists in all countries, the pacifist Albert Einstein (1879–1955), who received the prize in 1921.

While nominating practices reveal quite a bit about nationalism and internationalism in science, a cursory glance at the actual prizewinners also reveals a severe imbalance of gender. One of the famous cases of a woman who should have won, but did not, was that of Lise Meitner (1878–1968). Meitner was conspicuously missing from Otto Hahn's (1879–1968) 1944 Nobel Prize in Chemistry for the discovery of nuclear fission, a product of Hahn and Meitner's long-time collaboration. But even leaving aside that specific case, the Nobel Prizes are symbols not only of excellence, but also of elitism and inequity. The small number of women recipients highlights the exclusionary character of scientific activity in the first half of the twentieth century. Most of the Nobel Prize–winning women have received their awards in either literature or peace. Science is another story. Between 1901 and 1950, only Marie Curie (1867– 1934) won a prize in physics (1903), and she shared it with two men: Henri Becquerel (1852–1908) received one-half, and Marie Curie and her husband, Pierre Curie (1859–1906), each received one-quarter. In chemistry, Marie Curie won another prize (1911), this time her own. She was the only woman in the first half of the twentieth century to win a whole prize, not splitting it with someone else. Marie's daughter, Irène Joliot-Curie (1897–1956), also won in chemistry (1935), splitting it equally with her husband, Frédéric Joliot (1900– 1958). No non-Curie woman won a prize in physics or chemistry during the first half of the century. In physiology or medicine, Gerty Cori (1896–1957) won in 1947, but she shared the prize with her husband, Carl Cori, who received one-quarter, and Bernardo Houssay (1887–1971), who received one-half. She received one-quarter of the prize for herself. Tallying up whole, one-half, and one-quarter prizes, the total number of Nobel Prizes in the sciences for women between 1900 and 1950 was two. Subtracting those awarded to Marie Curie, the total was three-quarters.

See also Arrhenius, Svante; Elitism; Women
References
Crawford, Elisabeth. *The Beginnings of the Nobel Institution: The Science Prizes, 1901–1915.* New York: Cambridge University Press, 1984.
———. *Nationalism and Internationalism in Science, 1880–1939: Four Studies of the Nobel Population.* New York: Cambridge University Press, 1992.
Nobel e-Museum. http://www.nobel.se.
Wilhelm, Peter. *The Nobel Prize.* London: Springwood Books, 1983.

Nutrition

Nutrition is the process by which living organisms use food for growth and energy. Knowledge of human nutrition and its effects was rather limited prior to the twentieth century. The industrial revolution of the nineteenth century demonstrated that physical attributes such as height could not be attributed to heredity alone. Working-class people's dietary makeup differed markedly from that of the wealthier classes, and their body sizes reflected it. When young men signed up for military service for the Great War in 1914, governments were embarrassed to find many of the recruits failing some of the basic medical requirements. They had to conclude that malnutrition was to blame.

Around 1900, the study of nutrition (or more commonly during that era, "metabolism") emphasized how carbohydrates, fats,

Scurvy victim. In the early twentieth century, some of the most common diseases such as rickets and scurvy were attributed to deficiencies in diet, and malnutrition appeared linked to a host of other ailments, diseases, infections, and general poor health. (National Library of Medicine)

and proteins contributed to energy. Some of the most common diseases such as rickets and scurvy were attributed to deficiencies in diet, and malnutrition appeared linked to a host of other ailments, diseases, infections, and general poor health. In Britain, these connections were taken very seriously, and before World War I the government allocated funds for medical research, leading researcher Edward Mellanby (1884–1955) to conclude that rickets was the result of a vitamin deficiency. Rickets, which often caused pelvic deformities in women, increased the likelihood of death during childbirth. The British studies before and after the war, which demonstrated the ability to cure rickets by making up for the deficiency with butter or cod-liver oil, seemed to point to nutrition studies as the road to curing many diseases. It also revealed that most people, particularly the poor ones, satiated themselves primarily with foods rich in carbohydrates and often neglected the vitamin-rich foods such as milk, fruits, vegetables, and meat.

Studies in the 1920s affirmed the relationship between vitamin-rich foods and strong, healthy bodies. Research showed that boys drinking extra milk each day grew heavier and taller. Britain's Medical Research Council attempted to educate the public about the minimum food requirements and about methods for ensuring that foods were cooked properly to avoid losing the benefits. But as scholar Madeleine Mayhew has argued, these efforts proved controversial in government by the 1930s, as they made explicit reference to the minimum cost of staying healthy. That cost, some government officials feared, would be higher than many working-class people could afford, and the official "minimum" could be used as a tool by some to insist on better wages and handouts from the government. Nutrition, it seemed, had potentially disastrous political consequences.

Some of what was known about nutrition deficiencies came from observations in colonial regions. Studies of African nutrition began as investigations of livestock, but soon British scientists took an interest in the comparative nutrition of human tribes. A study begun in 1926 by John Boyd Orr (1880–1971) and John Langton Gilks compared, for example, an agricultural tribe to a pastoral tribe in Kenya. They concluded that the pastoral tribe, subsisting on meat and milk, was far better off than the tribe depending on only vegetables for sustenance. This emphasis on protein by investigators was not uncommon. In 1931, an English physician in West Africa identified the disease kwashiorkor as the result of a lack of sufficient protein. In this case, physicians and local inhabitants searched for other sources to make up for the deficiency, concluding that foods such as fish and soybeans could add the needed protein to the human diet. By mid-century, the lack of protein sources was credited for most of the world's nutrition shortfalls by scientists and international agencies. Other causes, such as insufficient caloric consumption, were recognized in the 1960s as a major contributor to malnutrition in many areas of the world.

See also Colonialism; Medicine; Public Health

References

Brantley, Cynthia. "Kikuyu-Maasai Nutrition and Colonial Science: The Orr and Gilks Study in Late 1920s Kenya Revisited." *International Journal of African Historical Studies* 30 (1997): 49–86.

Carpenter, Kenneth J. *Protein and Energy: A Study of Changing Ideas in Nutrition.* New York: Cambridge University Press, 1994.

Mayhew, Madeleine. "The 1930s Nutrition Controversy." *Journal of Contemporary History* 23 (1988): 445–464.

Worboys, Michael. "The Discovery of Colonial Malnutrition between the Wars." In David Arnold, ed., *Imperial Medicine and Indigenous Societies.* Manchester: Manchester University Press, 1988, 208–225.

O

Oceanic Expeditions

The nineteenth century saw some of the most ambitious oceanic voyages, such as Britain's worldwide *Challenger* expedition. In the early twentieth century, funds for such expeditions were scarce. Harvard University's Alexander Agassiz financed many of his own oceanic expeditions, taking the ship *Albatross* to the Atlantic and Pacific Oceans in the late nineteenth and early twentieth centuries. But deep-sea expeditions were rare, and their existence usually was tied to some economic need, especially fishing and whaling.

The need to develop fisheries spurred some expeditions, particularly ones that targeted specific areas. At the turn of the century, countries of northern Europe such as Britain, Germany, and Norway pooled resources to found the International Council (later the International Council for the Exploration of the Sea), a body of ocean researchers who sought to help illuminate the commercial problem of fisheries, especially in the North Sea. Individual countries were responsible for funding their own ships, but their cruises at sea were closely coordinated, to provide a coherent and useful picture of entire regions. Similarly, in 1912, Harvard zoologist Henry Bryant Bigelow (1879–

1967) initiated a twelve-year intensive investigation of the Gulf of Maine. The work was supported by the United States Bureau of Fisheries, which believed that Bigelow's expeditions might, by shedding light on the physical and biological characteristics of the sea in that region, help to enhance the ability to exploit the sea for fish. His work began to appear in print in the 1920s, establishing fundamental ideas not only about fish but, more generally, about the circulation of currents. He later became the first director of the Woods Hole Oceanographic Institution in Massachusetts. Although long-range oceanic voyages did not come to an end during these years, scientists were learning the value of intensive area studies, particularly because of the opportunity to acquire funding from commercial or government sources.

Deep-sea oceanic expeditions were financed for several reasons, some scientific and some economic. In 1925, Britain commissioned the vessel *Discovery* to begin deep-sea studies of whales and other commercially important resources. But the *Discovery* also conducted more general scientific work, in the tradition of the *Challenger* expedition of the previous century. Its 1925–1927 expedition targeted areas near Antarctica, specifically in the Falkland Islands area off South America; it surveyed these fertile whaling areas to understand local plankton populations and the properties of water masses and currents. The ship was replaced in 1929 by a converted steamship, the *Discovery II,* whose longer range helped British scientists to broaden the scope of the work, to better comprehend wide-ranging migration patterns of whales, as well as large-scale oceanic processes. Throughout these expeditions, the British worked through a long-term Discovery Committee that planned and coordinated oceanic expeditions for the sake of science and national economic interests in the deep ocean.

Conducting oceanic expeditions also was a way to bring prestige to one's country. In 1933–1934, an Egyptian fishing vessel sponsored by the wealthy Englishman Sir John Murray, the *Mabahiss,* steamed throughout the Indian Ocean, making observations at more than 200 stations. Aboard the ship were British and Egyptian scientists and crew members, and the expedition stimulated respect for Egyptian science and fostered the growth of oceanography in Egypt (these efforts, unfortunately, were curtailed by economic troubles during the Depression and wartime conditions in nearby Ethiopia). Germany also used expeditions to promote itself during these years; forbidden by the Treaty of Versailles to send naval vessels abroad, Germany sent the *Meteor* on a scientific voyage instead, to demonstrate the country's scientific prowess. Oceanic expeditions were international as well. The Second International Polar Year (IPY) was organized by scientists of several nations, to take place in 1932–1933. These joint investigations were facilitated by advances in radio communication, and they produced considerable data about the continent itself and the surrounding ocean. Many of the results, however, remained underutilized for many years, even beyond World War II, calling attention to the fact that expensive expeditions might have limited scientific value.

See also International Cooperation; Nationalism; Oceanography; Polar Expeditions

References

Barry, R. G. "Arctic Ocean Ice and Climate: Perspectives on a Century of Polar Research." *Annals of the Association of American Geographers* 73:4 (1983): 485–501.

Brosco, Jeffrey P. "Henry Bryant Bigelow, the U.S. Bureau of Fisheries, and Intensive Area Study." *Social Studies of Science* 19:2 (1989): 239–264.

Mackintosh, N. A. "The Work of the Discovery Committee." *Proceedings of the Royal Society of London, Series A, Mathematical and Physical Sciences* 202:1068 (22 June 1950): 1–16.

Rice, A. L., ed. *Deep-Sea Challenge: The John Murray*/Mabahiss *Expedition to the Indian Ocean, 1933–34.* Paris: UNESCO, 1986.

Rozwadowski, Helen. *The Sea Knows No Boundaries: A Century of Marine Science under ICES.* Seattle: University of Washington Press, 2002.

Oceanography

Oceanography has been, if nothing else, a very expensive pursuit throughout its history, requiring not merely a working space in an institution but usually a ship, well-equipped and sufficiently provisioned to make oceanic voyages. It should come as no surprise that oceanography flourished most when its practitioners managed to convince rich patrons that it was worth doing. Although some oceanographers enjoyed the support of private philanthropic foundations, most of the large-scale efforts in oceanography were sponsored by organizations tied either to commercial fisheries or to national navies. Oceanography in the twentieth century was connected to both, and support for it exploded after World War II, when its military implications became abundantly clear to the United States Navy.

At the turn of the twentieth century, knowledge of oceanography was rudimentary and organizations devoted solely to it were scarce. The British *Challenger* expedition, 1873–1876, was the first circumnavigation of the world by a ship with a primarily scientific purpose, making hundreds of "stations" throughout the world. A station was the name given to efforts to conduct a series

Aerial View of the Woods Hole Oceanographic Institute. (Bettmann/Corbis)

of observations in any given spot; such observations included depth, bottom water sampling, fauna collection, and other activities. But such expeditions were rare, as they required a great deal of money and political support. Most efforts were confined to local waters. After World War I, several countries mounted expeditions for scientific and political reasons (such voyages were a measure of prestige); for example, Britain sent the *Discovery II,* Germany sent the *Meteor,* and the United States sent the *Atlantis* to various waters and ports. The *Atlantis* was managed by Henry Bryant Bigelow (1879–1967), the first director of the Woods Hole Oceanographic Institution, in Massachusetts, founded in 1930. The *Atlantis* became that institution's primary instrument for conducting intensive, long-term, yet geographi-

cally confined, area studies, which Bigelow believed would yield more useful information (for science and especially for fisheries) than long-range expeditions.

The limitations imposed on scientists because of high costs, and the fact that its object of study existed outside national borders and often entailed hostile environments, made oceanography well suited to international cooperation. The International Council (later called the International Council for the Exploration of the Sea) was founded in 1902, by countries in northern Europe, such as Great Britain and Norway, that had a great interest in improving the efficiency of commercial fisheries. Providing regional regulatory efforts with sound scientific principles was one of its primary goals. Countries bordering on the Pacific Ocean also tried to combine their efforts, meeting in a series of conferences organized by the Pacific Science Association, beginning in the 1920s.

Oceanography in general was dominated by biological and dynamical oceanography. One traditional biological pursuit was comparative marine zoology, often the most productive pursuit for those lacking access to ships or resources. As for dynamical oceanography, the work of the Norwegian Vilhelm Bjerknes (1862–1951) on atmospheric dynamics inspired oceanographers to apply his principles to the sea, in order to understand the motions of currents and the relationships between water masses. Plankton studies combined biology with dynamical oceanography. They also promised to help with fisheries, thus contributing to the health of a nation's economy. Plankton became a major focus of research at the Plymouth Laboratory in Britain and at Woods Hole in the United States, both of which used chemical methods to comprehend the nutrient cycles of these sea organisms that were the key food source for so many ocean animals. Thus by the 1920s, oceanography was an interdisciplinary pursuit, combining principles drawn from chemistry, biology, and physics, and increasingly its practitioners fell under the influence of ecological approaches that emphasized a systemic approach to scientific problems.

During the 1930s and especially during World War II, oceanography's focus turned increasingly toward military interests. This was due largely to advances in understanding the propagation of sound in the sea, from experiments made by the American W. Maurice Ewing and others with explosives. Such studies had enormous ramifications for undersea warfare, particularly as the navies of the world relied on submarine vessels in times of war. Submarine detection had great military value, and thus the military became an important patron for oceanographers. U.S. and British oceanographers played crucial roles also in general weather forecasting for ships at sea and in planning amphibious troop landings, such as the D-day invasion of Normandy on 6 June 1944. U.S. scientists learned the value of oceanographic studies the hard way, after ignorance of tide conditions contributed to the death and wounding of thousands of U.S. marines attacking Pacific islands.

After World War II, the field of oceanography expanded dramatically. The field was no longer so small that most researchers knew each other fairly well. Government agencies such as the Office of Naval Research (in the United States) provided unprecedented levels of funding for sciences across numerous disciplines—the one most likely to provide information about the Navy's workplace, oceanography, received a great deal of money. U.S. institutions such as the Scripps Institution of Oceanography (founded in 1903) and Woods Hole both benefited handily from this, as did other universities and several corporate laboratories. Oceanographers in Europe and the United States began to conceptualize deep-sea expeditions to use recently developed instruments. For example, the Swedish Deep-Sea Expedition 1947–1948 used the sediment corer developed by the Swede Borje Kullenberg. British oceanographers tried in vain to mount a long-range oceanographic expedition, but the days of northern European dominance in oceanography had ended, and the United States began to fund studies of the deep sea in earnest. The United States launched its Mid-Pacific expedition in 1950, and many more similar voyages would follow. These studies yielded important information about the seafloor that led to the development of plate tectonics in subsequent years. In addition to enjoying greater patronage, oceanography increasingly was defined more broadly, focused less exclusively on dynamical or biological oceanography and taking in studies of geophysics, marine geology, and magnetism.

See also Bjerknes, Vilhelm; Cold War; Oceanic Expeditions; Patronage; Sverdrup, Harald; World War II

References

Brosco, Jeffrey P. "Henry Bryant Bigelow, the U.S. Bureau of Fisheries, and Intensive Area Studies." *Social Studies of Science* 19:2 (1989): 239–264.

Mills, Eric L. *Biological Oceanography: An Early History, 1870–1960.* Ithaca, NY: Cornell University Press, 1989.

Rozwadowski, Helen M. *The Sea Knows No Boundaries: A Century of Marine Science under ICES.* Seattle: University of Washington Press, 2002.

Schlee, Susan. *The Edge of an Unfamiliar World: A History of Oceanography.* New York: Dutton, 1973.

———. *On Almost Any Wind: The Saga of the Oceanographic Research Vessel "Atlantis."* Ithaca, NY: Cornell University Press, 1978.

Office of Naval Research

The Office of Naval Research (ONR), created in the United States in 1946, considered itself to be a very sophisticated part of the armed services. Informed by the vision of science set forth in Vannevar Bush's (1890–1974) wartime report, *Science—The Endless Frontier,* ONR scientists supported the notion that the strength of the nation depended at least as much on generous support of basic research as on applied research or engineering. Its reputation to this effect goes back to its first few years, before the creation of the National Science Foundation in 1950, when it was the leading government agency actively promoting basic research.

Why did ONR, part of the United States Navy, fund basic research after World War II? One might say that it was the first to recognize the relationship between scientific strength and national security, and it embedded this concept into its own military mission. It accepted Bush's notion that basic research is the capital with which others would build new technology. Although there is some truth to this interpretation, the Navy was no great supporter of science before the creation of ONR. In his book about ONR, Harvey Sapolsky calls its generosity a "bureaucratic accident" resulting from the early

retirement of its first director and the subsequent purposelessness of the agency.

ONR's creation was born from wartime experience. Admiral Julius A. Furer, the naval member of the Office of Scientific Research and Development (OSRD), had established connections with scientists through his staff of science Ph.D.s. These "Bird Dogs," as they called themselves, acted as the liaison between the Navy and the scientific community; consequently, they were the people most capable of turning scientific knowledge into military technology. They wanted a postwar agency to find and give money to researchers, civilian and military, who might produce the research to develop new weapons in the coming years. Their advocacy convinced the Navy and Congress to create the office in 1946. But ONR's first director, Admiral Harold G. Bowen, saw ONR primarily as an opportunity to steal nuclear scientists from the Army's Manhattan Project, to facilitate the development of a nuclear navy. Ultimately, ONR scientists did much more, turning the office into a generous funding agency for pure science, an anomaly that went unchecked by the Navy's top brass until war erupted in Korea in 1950 and tight budgets forced ONR to economize.

ONR's activities and unapologetic financial support for civilian scientists in U.S. universities raised many eyebrows. *Newsweek* called the Navy "the Santa Claus of basic physical science." ONR scientists had unprecedented freedom to award money and they did so as if they were philanthropists rather than military officials. Its support was not limited to nuclear physics. Other fields included electronics, chemistry, computing, and all kinds of oceanographic research. The science it funded did not need to be related to weapons research. The Army soon imitated the Navy, leading some to wonder if the military would dominate and control the country's scientific activities. Vannevar Bush did not oppose military funding, but he hoped that his 1945 report would lead to a civilian-controlled science foundation, not a free-for-all of military sponsorship. ONR saw no need to stop supporting research while legislation for such a foundation stalled in Congress for half a decade. Still, some scientists refused to accept ONR's money; these and others rejoiced when the National Science Foundation finally came into being in 1950. However, it was no victory for civilian control, because ONR continued to support scientific research on a massive scale.

See also Cold War; National Science Foundation; Patronage

References

Sapolsky, Harvey M. *Science and the Navy: The History of the Office of Naval Research.* Princeton, NJ: Princeton University Press, 1990.

Zachary, G. Pascal. *Endless Frontier: Vannevar Bush, Engineer of the American Century.* Cambridge, MA: MIT Press, 1999.

Oort, Jan Hendrik

(b. Franeker, Netherlands, 1900; d. Leiden, Netherlands, 1992)

Jan Hendrik Oort claimed to have been captivated at the age of seventeen by the lectures of the eminent Dutch astronomer Jacobus Kapteyn (1851–1922). Deciding to pursue astronomy at a more advanced level, Oort became one of Kapteyn's last students at the University of Groningen. Oort's early career in the 1920s focused on the velocity distribution of high-velocity stars. In 1927 he added the observational evidence for Bertil Lindblad's (1895–1965) hypothesis that the galaxy was rotating. Oort did so by measuring the radial velocities of stars, the results of which suggested that our galactic system was indeed in differential rotation. Oort was active in bringing Dutch astronomy firmly into the era of radio astronomy, and he played a major role in the production of the first radio maps of the galaxy.

One of the results of Oort's work in the 1920s was his recognition that the galaxy must be much more massive than previously believed if it were to have the gravitational power it seemed to have. In other words, there must be a great deal of matter in the galaxy that cannot be seen, amounting perhaps to as much as two hundred times the mass of the visible stars. This premise provoked the idea of "dark matter." But where is the missing mass? Some of it took the form of dust and small particles, but soon astronomers concluded that some stars of low mass and low luminosity cannot be seen, and there are stars of such great mass that even light cannot escape their gravitational effects. In 1932, Oort calculated a value for the density of space, at roughly one solar mass for every ten cubic parsecs (a parsec, the unit of computing distances between stars, is the distance light travels in 3.26 years). This became known as the Oort limit, and Oort believed that much of the density was made up of invisible stars and gas.

Oort became interested in comets from one of his graduate students, and he made one of his best-known contributions on this topic in 1950. He proposed that long-period comets originate in a reservoir of debris at the far reaches of the solar system. Comets of long periods appeared to be subject to perturbation by planets. Oort reasoned that if this were so, the orbits of these long-period comets should have been altered long ago. Thus, he concluded that the long-period comets in fact were entering the solar system for the first time, rather than simply repeating a cycle in a long-standing orbit. Because these long-period comets did not appear to originate from any particular place, he posited the existence of a vast cloud of debris just beyond the solar system. This swarm of comets became known as the Oort Cloud. The cloud moves, according to Oort, in elliptical orbits around the solar system, but at far greater distances from the sun than planets, ranging from 6 trillion to 21 trillion kilometers. The cloud is subject to significant influence not only by the sun, but also by other stars.

In the 1930s, Oort was the Netherlands' rising star in astronomy, and he became the director of the Leiden Observatory, a post he held until his death in 1992. He took a leading role in the International Astronomical

Professor Jan H. Oort reads research papers in his backyard. He sits next to a sundial shaped like the Westerbork radio telescope, a gift from the staff of the University of Leiden on his 70th birthday. (Jonathan Blair / Corbis)

Union, becoming its general secretary in 1935 (in the late 1950s, he served as the president). In the years following World War II, he advocated international collaboration among observatories. Much of his later career was devoted to capitalizing on new advances in radio astronomy, and trying to determine distances to high-velocity clouds.

See also Astronomical Observatories; Astrophysics; Kapteyn, Jacobus

References

North, John. *The Norton History of Astronomy and Cosmology.* New York: W. W. Norton, 1995.

Oort, J. H. "Some Notes on My Life as an Astronomer." *Annual Review of Astronomy and Astrophysics* 19 (1981): 1–5.

Van Woerden, Hugo, Willem N. Broew, and Hendrik C. Van de Hulst, eds. *Oort and the Universe: A Sketch of Oort's Research and Person.* Dordrecht: D. Reidel, 1980.

Origin of Life

In the eighteenth and nineteenth centuries, the debate about life's origins could be viewed as a conflict between vitalists and mechanists. The vitalists insisted on a special life-giving force or substance that provides life to an otherwise mechanical organism, whereas mechanists reduced all processes—including life itself—to mechanical action. Toward the end of the nineteenth century, Charles Darwin's (1809–1882) publications on evolution by means of natural selection showed organisms developing as a process, provoking questions about the fundamental beginning of life in the expanse of time. One of the central questions was of spontaneous generation: Can life arise from inorganic matter without the action or presence of some kind of preexisting living organism? French scientist Louis Pasteur (1822–1895) answered this authoritatively, conducting experiments with liquids in carefully prepared tubes that suggested that spontaneous generation did not occur. German scientist Rudolf Virchow (1821–1902) concurred, claiming that all cells come from preexisting cells.

The spontaneous generation question was raised again in the early twentieth century, but couched in terms of trying to discern the origin of life. Because the prevailing theory of life was based on continuity, from one cell to another (as opposed to spontaneous generation), renowned Swedish chemist Svante Arrhenius (1859–1927) postulated that the earth had been "seeded" somehow by some unknown extraterrestrial source. Others came to believe that, although continuity of cells was crucial, there was some point in the distant past when nonlife became life. British biologist J. B. S. Haldane (1892–1964) observed in 1923 that life could have originated in a virus that developed from organic chemicals. He and Russian biologist Aleksandr Oparin (1894–1980), who initially believed that the first living cell came from an accidental combination of particles in suspension, both still preserved the notion of spontaneous generation, long after it supposedly had been discredited by experimenters such as Pasteur.

The most comprehensive account of the origins of life in the first half of the twentieth century was written by Oparin in 1936. Unlike his earlier work, Oparin now proposed a theory that did not rely on spontaneous generation, but rather emphasized biochemical continuity. Oparin assumed that the earth was once a very hot mass of liquid and that its surface cooled and formed complex molecules under the influence of ultraviolet light. In this view he borrowed considerably from Haldane. These molecules increased in number and, owing to Darwinian natural selection, some survived at the expense of others, encouraging complexity. Oparin's innovation was in seeing natural selection operating on "nonliving" substances; even molecules that could not be considered organisms were subject to Darwinian rules. For Oparin, evolution as a whole could be considered in terms of biochemical complexity. Primitive cells were formed from these molecules, and complexity typically involved a change in biochemical processes, particularly in metabolism. Even in the most complex forms, Oparin reasoned, changes in the metabolism of matter and energy were the guideposts for true change.

Oparin's views were not universally well received. For example, critics noted that Darwinian evolution does not always lead to greater complexity, which Oparin seemed to suggest. In addition, laboratory experiments on droplets of molecules, like those described by Oparin, did not entirely support his theory. Some geneticists criticized him for allowing Marxist ideas, which also emphasized progressive evolution over time, to influence and thus pollute his theory. But unlike his compatriot Trofim Lysenko (1898–1976), who was reviled outside of the Soviet Union for crippling genetics in his country, Oparin was generally praised even by non-Russians for providing a persuasive and unique possible explanation of the origins of life. His concept that life evolved from complex molecules, giving rise to tiny organic beings, had considerable influence in Europe and North America into the 1950s.

See also Arrhenius, Svante; Cosmology; Evolution; Soviet Science

References

Arrhenius, Svante. *Worlds in the Making: The Evolution of the Universe.* New York: Harper, 1908.

Bowler, Peter J. *The Mendelian Revolution: The Emergence of Hereditarian Concepts in Modern Science and Society.* Baltimore, MD: Johns Hopkins University Press, 1989.

Farley, John. *The Spontaneous Generation Controversy from Descartes to Oparin.* Baltimore, MD: Johns Hopkins University Press, 1977.

Oparin, A. I. *Origin of Life.* New York: Dover, 1953.

P

Patronage

Before the twentieth century, science was traditionally a pursuit for the philosophically minded rich. Large-scale projects or laboratories, however, depended on patronage— financial support by wealthy philanthropists, by corporations, or by governments. One of the principal ways in which governments funded research during the nineteenth century was through geological surveys and coast surveys, which promised not only to help resolve territorial disputes, but also aided in the economic exploitation of the countryside. In general, however, the same laissez-faire attitude adopted by governments toward industry was transferred to governmental relations with scientists and their laboratories. Gradual increases in patronage, first by private foundations and industry, and eventually by governments, resulted in the most well-funded scientific communities in history by mid-century.

Most patronage for science before World War I came from private individuals or organizations. Wealthy entrepreneurs and their heirs, particularly in the United States, took an active role in funding science. Two of the primary supporters of natural science were the Rockefeller Foundation and the Carnegie Corporation. The Carnegie Institution of Washington, founded in 1902, was an early example of an effort to make such support on a significant scale. Yet already some of the serious difficulties with patronage emerged, such as the freedom of scientists to pursue the kinds of problems that they deemed important (rather than subjecting their research agendas to manipulation by the sponsor).

Patronage on a large scale came from industry as well. In the United States, corporations such as AT&T sponsored physical laboratories in order to establish and develop patents. German industries, particularly strong in the development and production of dyes, also were important patrons of science in the field of chemistry. The pharmaceutical industry was another important patron of scientific research in both Europe and North America in the interwar years and after. Governments occasionally promoted such connections among science and the national economy and public health. For example, the British government took steps to provide the country with a technically savvy workforce to help them compete with continental Europeans, by founding the Imperial College of Science and Technology in 1907. It already had founded the National Physical Laboratory (1899), and it formed the Medical Research Committee (later Medical Research Council) in 1912.

World War I ushered in a new phase in government relations with scientists, but this did not always translate into patronage. Instead, most recognized the need for increased coordination, so that scientists could contribute more efficiently to the war effort. Britain's Department of Scientific and Industrial Research was founded in 1922 to ensure proper coordination and funding of research in strategically important industries. In the United States, astronomer George Ellery Hale (1868–1938) convinced President Woodrow Wilson (1856–1924) to form the National Research Council to coordinate and deploy scientists to serve national needs. Because most of the leading chemists before the war had been Germans, and Germany was the first to develop and use chemical weapons, most observers believed that Allied governments had been remiss before the war and that steps should to be taken to encourage the growth of scientific brainpower in their own countries.

Science philanthropy continued in the interwar years, especially through the efforts of the Rockefeller Foundation. Two of its most active arms, the General Education Board and the International Education Board, were led by Wickliffe Rose (1862–1931) in the 1920s. He ensured that research projects were both "pure" and practical; for example, although the foundation wished to support the areas of science in need of development, it also hoped to use science to eradicate some of the main agents of disease in the world. As such, a great deal of its focus was upon public health and sanitation research. Like most organizations, the Rockefeller Foundation had to tighten its belt considerably during the Depressions years of the 1930s, because of financial constraints. Under the leadership of Rose's successor, Warren Weaver (1894–1978), the Rockefeller Foundation still continued to be a significant patron of research, but it focused its mission on the biological sciences.

The highly technical military projects of World War II convinced governments, particularly in the United States and Soviet Union, that science was going to be an increasingly important aspect of national strength. The atomic bomb alone set the tone for the postwar geopolitical arena, and nuclear physicists had little difficulty in securing patronage for their work. The most influential patron for science was the U.S. military establishment, which supported research through its various arms, such as the Office of Naval Research (established in 1946). This body e supported research in a number of fields, hoping for short- or long-term impact on military technology or operations. Because its long-term vision appeared to give scientists free reign over their own projects rather than force them to develop specific weapons systems, the Navy acquired the reputation of having a sophisticated approach to basic research.

Civilians also saw the need to support research, particularly if military officers were so keen to do it, to maintain national strength without subjecting scientists to military control. Despite claims by military officials that they were pleased to give scientists freedom to pursue their interests, civilians worried about excessive military involvement. The wartime director of the Office of Scientific Research and Development, Vannevar Bush (1890–1974), published a civilian vision of postwar patronage in his report, *Science—The Endless Frontier* (1945). In it, he proposed a national foundation to act as the principal source of federal patronage of science. Political negotiations stalled its creation until 1950. The resulting National Science Foundation was smaller than envisioned, and it complemented rather than replaced military sources of patronage. Overall, federal patronage for science between 1945 and 1950 tended to favor mathematics, physics, and engineering.

In the Soviet Union, scientists initially were viewed with suspicion by the government; in the 1920s, the Bolsheviks believed they might be hostile to socialism. But under Joseph Stalin's (1879–1953) leadership, a new class of scientists and technical specialists emerged who were not only loyal but

A worker from the Rockefeller Foundation hunts for lice in a girl's hair in a Naples slum. The Rockefeller Foundation hoped to use science to eradicate disease, often focusing its efforts on public health. (Bettmann / Corbis)

committed to Stalin's goals of accelerated industrialization and efficient agricultural collectivization. The government subsidized scientists' work in an effort to meet these objectives, and scientists themselves enjoyed a relatively high social status. On the other hand, many scientists were victims of Stalin's purges in 1936–1938. Stalin's desire to have an atomic weapon provided physicists with a major commitment of support beginning in 1943 and, after World War II, enormous levels of financial resources as well as political pressure to keep abreast of the most recent work in nuclear physics.

Increased patronage for science in the years after World War II led to a period that many historians call Big Science. The influx of funding made possible larger laboratories, bigger research teams, and collaboration among universities, agencies, and even countries. But such levels of patronage provoked tough questions for scientists. Did patronage amount to control of science? Should scientists exercise restraint in taking money from sponsors, particularly from those expecting to develop more destructive weapons? By mid-century, such questions often were muted by the political exigencies of the Cold War and the expectation that the outcome of a new, more destructive war would depend greatly on the government's investment in scientific expertise.

See also Cold War; Great Depression; Industry; National Science Foundation; Office of Naval Research; World War I; World War II

References

Alter, Peter. *The Reluctant Patron: Science and the State in Britain, 1850–1920.* New York: St. Martin's Press, 1987.

England, J. Merton. *A Patron for Pure Science: The National Science Foundation's Formative Years, 1945–57.* Washington, DC: National Science Foundation, 1982.

Hamblin, Jacob Darwin. "The Navy's 'Sophisticated' Pursuit of Science: Undersea Warfare, the Limits of Internationalism, and the Utility of Basic Research, 1945–1956." *Isis* 93 (2002): 1–27.

Holloway, David. "The Politics of Soviet Science and Technology." *Social Studies of Science* 11 (1981): 259–274.

Kohler, Robert E. *Partners in Science: Foundations and Natural Scientists, 1900–1945.* Chicago, IL: University of Chicago Press, 1991.

Pavlov, Ivan

(b. Ryazan, Russia, 1849; d. Leningrad, USSR, 1936)

Ivan Pavlov began a career in physiology in the 1870s, but was not successful in finding a university appointment. However, the astounding public health implications of physiological research in Europe, notably Louis Pasteur's (1822–1895) rabies vaccine and Robert Koch's (1843–1910) tuberculin, convinced eminent aristocrats to found an institute in which comparable work might be accomplished in Russia. One of these, Prince Aleksandr P. Ol'denburgskii (1844–1932), founded the Imperial Institute of Experimental Medicine, and Pavlov was appointed chief of its division on physiology in 1891. In their efforts to modernize Russian medicine, officials encouraged doctors to spend parts of their training in a laboratory environment; this gave Pavlov an uncommonly talented coterie of workers from which to draw.

Pavlov's laboratory housed not only a large number of workers but also a full-scale kennel for the experimental animals. Pavlov was interested not simply in dissecting or vivisecting animals, but in observing their long-term physiological processes. This required keeping them alive and healthy in order to conduct chronic experiments, as he called them. These were experiments over time, designed to understand the normal functions of animals. This was a new kind of study, because previously experiments had been "acute," meaning that the dog was vivisected and ultimately killed in the process.

Pavlov's most famous experiments were on dogs, whose digestion he studied by implanting in them fistulas (tube-like devices that connect internal organs to the surface of the body). Historian Daniel Todes has argued that Pavlov's dogs were themselves laboratory technologies rather than just objects of study. The surgical innovations were designed to acquire certain fluids (such as pancreatic, salivary, or gastric secretions) from the dog but not to kill the dog in the process. The goal was to keep the dogs in full health (as nearly as possible), to observe the normal functioning of its digestive system. Only in this way could Pavlov pursue "chronic" experimentation of the digestive system. Through these experiments Pavlov drew strong connections between the digestive system and the nervous system, showing the latter's influence over the former.

Pavlov did not fit the mold of the heroic scientist striving alone in his laboratory. Instead, he credited his entire laboratory for his work. Pavlov was nominated for the Nobel Prize in Physiology or Medicine in four successive years beginning in 1901, but his nominations were not specific to any discovery. They were based on a variety of laboratory findings; this, combined with his explicit acknowledgment of the role of the entire laboratory (rather than his taking all the credit), made him a difficult choice for the Nobel Prize, itself a symbol of the personal nature of scientific discovery. Eventually the prize committee overlooked this, and it awarded the prize to Pavlov in 1904.

Perhaps Pavlov is best known, despite his Nobel Prize for earlier work, for his research on conditioning. The phrase "like Pavlov's

Despite his Nobel Prize for earlier work, Ivan Pavlov is best known for his research on conditioning. The phrase "like Pavlov's dog" has entered common speech to indicate a conditioned response. (National Library of Medicine)

a bell. His work on such conditioning became the basis of animal behaviorism. It provided an avenue for the study of animal response to any environmental stimulus, which influenced the direction of work in other fields, notably human psychology. Pavlov was lucky in that his success under czarist Russia did not falter after the Bolshevik Revolution in 1917. In fact, Soviet leader Vladimir Lenin (1870–1924) publicly affirmed Pavlov's importance to the cause of the working class in a special decree in 1921. The Soviet Union, under Pavlov's influence, became a major center of activity in the study of animal physiology.

See also Medicine; Psychology; Skinner, Burrhus Frederic; Soviet Science
References
Kozulin, Alex. *Psychology in Utopia: Toward a Social History of Soviet Psychology.* Cambridge, MA: MIT Press, 1984.
Todes, Daniel P. "Pavlov's Physiology Factory." *Isis* 88:2 (1997): 205–246.

dog" has entered common speech to indicate a conditioned response. His previous studies of secretions by bodily organs led him to an interest in "psychic" bodily secretions, namely, those that occurred from stimuli not directly connected to the animal. For example, saliva was secreted when dogs saw food. Eventually he rejected the notion of a "psychic" secretion and affirmed that even these secretions were uncontrolled, often temporary reflexes. Pavlov proposed two kinds of reflexes, unconditional and conditional. Conditional reflexes were those connected to specific stimuli and had been developed over time. The connection between sensory input (for example, seeing food) and motor output (for example, salivation) was called signalization. He set forth many of these ideas as early as 1903, noting that reflexes can be both physiological and psychological.

Pavlov's subsequent research located such reflexes in the cerebral cortex, and over the next several years he and others continued research on artificial conditioned reflexes, such as making a dog salivate at the sound of

Peking Man

Peking Man, discovered over a period of years in the 1920s and 1930s, was a set of bones that belonged to human beings living hundreds of thousands of years ago. Because the bones were human (or hominid) but not Homo sapiens, Peking Man became one of the principal pieces of fossilized evidence for those who sought to strengthen the theory of human evolution. The bones themselves were lost or destroyed during World War II.

In 1918, workers uncovered a fossil deposit in a limestone quarry near the village of Zhoukoudian, some fifty kilometers from Beijing, China (Westerners then referred to Beijing as Peking). The Chinese government's mining advisor, Johan G. Andersson (1874–1960), noted the numerous bird and mammal bones found there and determined that it was of fairly recent origin. He returned to the site with scientists Walter Granger (1872–1941) and Otto Zdansky (1894–1988) in 1921, and the quarrymen led them to another, larger fossil site recently uncovered. There they found pieces of white quartz, suggesting that the deposits were very old; they decided to excavate, in the hopes of finding human remains.

The Zhoukoudian site thus began a long period of excavation that continued sporadically for decades. Many scientists, including the priest and scholar Pierre Teilhard de Chardin (1881–1955), conducted research on the bones found there. In 1923, Zdansky discovered a molar tooth of a hominid, and soon laboratory investigations of samples uncovered other teeth. Andersson announced the initial discoveries in 1926. After arousing the interest in the Chinese government, a major excavation was organized by the Geological Survey of China and the Peking Union Medical College, funded largely by a grant from the Rockefeller Foundation. The excavations took considerable time, because the deposit turned out to be larger than expected. But piece by piece, remains were uncovered that appeared to be those of human beings, including a complete skull.

Anatomist Davidson Black (1884–1934), a faculty member at the Peking Union Medical College, concluded that the deposits represented the remains of a large, ancient cave that had been filled over time by sedimentation in the course of being occupied continuously by men and beasts. The community of humans, he and others concluded based on the tools found at the site, had not learned their crafts from other cultures and seemed autonomous. Black conferred the name *Sinanthropus pekinensis* on the hominids found at the site, and he proclaimed that they certainly belonged to the family of humans. Peking Man is the name that took hold. Peking Man was actually a collection of bones, some of which were part of the female anatomy. They were not one person but several, whose age was determined to be somewhere between 200,000 and 700,000 years.

Scientists concluded that Peking Man walked erect and that he or she was closer to modern humans than Neanderthal Man.

Peking Man was used frequently as empirical evidence for the evolution of species. After Piltdown Man was revealed as a forgery in the early 1950s, opponents of evolution began to accuse scientists of fabricating Peking Man as well. Such accusations were buoyed by the fact that the original bones disappeared during World War II; their fate has been enshrouded in mystery ever since. With Japanese occupation of Beijing imminent, Chinese archaeologists had hastened to protect the bones by sending them to the United States. But what actually happened—the fate of Peking Man—remains unknown. Certainly the chaos of war provided a host of opportunities to lose the important artifacts. Perhaps the bones were mistook as trash and thrown away, or perhaps they were hidden by someone who then perished in the war.

See also Anthropology; Missing Link; Piltdown Hoax; Teilhard de Chardin, Pierre; World War II

References

Black, Davidson. "The Croonian Lecture: On the Discovery, Morphology, and Environment of Sinanthropus Pekinensis." *Philosophical Transactions of the Royal Society of London, Series B, Containing Papers of a Biological Character* 223 (1934): 57–120.

Lanpo, Jia, and Huan Weiwen, *The Story of Peking Man: From Archaeology to Mystery*. New York: Oxford University Press, 1990.

Shapiro, Harry L. *Peking Man*. New York: Simon & Schuster, 1975.

Penicillin

The discovery and commercial production of penicillin can be classed as one of the most significant public health achievements of the first half of the twentieth century. Discovered by a British scientist in 1928, penicillin was used as a therapeutic antibiotic by the Allies during World War II, saving countless lives and limbs from the effects of wound infection and infectious diseases. Penicillin was medicine's version of the atomic bomb, in the sense that it required an unprecedented level of cooperation among scientists, industry, and government, and resulted in an equally unprecedented scale of production.

In 1928, British biologist Alexander Fleming (1881–1955) discovered that a culture plate of bacteria had been contaminated by spores of the mold penicillium. Although the contamination was not uncommon, Fleming noted that the mold spores grew into a large colony over time and seemed to have a deleterious effect on the colonies of staphylococcus, which had inhabited the plate prior to contamination. Fleming realized that the contaminating mold of penicillium had an antibacterial effect, and he named the active agent penicillin. He was not the first to recognize that some microorganisms inhibit the growth of others, but the later importance of penicillin lent extraordinary significance to his work.

Work on penicillin progressed little until the late 1930s, when the Rockefeller Foundation financed a team of British scientists led by Ernst Chain (1906–1979) and Howard Florey (1898–1968) to investigate the antibacterial effects of microorganisms. By 1940 the term *antibiotic* came into use to describe those living substances that could be used to kill other living substances such as bacteria. Chain and Florey produced small quantities of penicillin that could be used in humans to combat infections and certain infectious diseases. Eventually, penicillin went into widespread use during the war by the British and Americans, and it would later stand together with radar and the atomic bomb as one of the great scientific and industrial achievements of the war. For their role in discovering penicillin and developing it as a therapy, Fleming, Chain, and Florey shared the Nobel Prize in Physiology or Medicine in 1945.

Although the drug was discovered by British scientists, penicillin's large-scale manufacture resulted from U.S. involvement. Lacking sufficient industrial resources in wartime Britain, Florey and colleague Norman Heatley traveled to the United States in 1941 and contacted a number of pharmaceutical manufacturers. These companies, among them Pfizer and Merck, worked with scientists and the federal government to find ways to mass-produce the drug. This required extensive research on the properties of the mold and more specific data on the range of its therapeutic uses. Most importantly, a great deal of money was needed for developing the means to mass-produce it. The Office of Scientific Research and Development (OSRD) allocated over $2 million to the research, development, and production costs of penicillin. Other breakthroughs contributed: A laboratory of the Department of Agriculture found a way to increase the yield of penicillin by growing it in corn steep liquor, and it also discovered a more productive strand of penicillin than that offered by the British scientists. Wartime necessity dramatically shortened the time for testing and production, and penicillin soon became widely available. By 1945 the price of penicillin dropped from $20 per vial of 100,000 units to less than a dollar. The price dropped even more in ensuing years.

One effective use of penicillin was in the control of venereal disease. The U.S. Health Service received strains of penicillin from British scientists in 1943, and it began to inject the antibiotic into rabbits infected with syphilis. The symptoms and microscopic evidence of the disease disappeared. Experimenting on humans, the rates of cure were between 90 and 97 percent. Penicillin was also used to treat other diseases, such as gonorrhea, and by the mid-1950s these venereal diseases no longer carried the serious threat that they had posed prior to the development of penicillin.

See also Industry; Medicine; Venereal Disease; World War II

References

Brandt, Allan M. *No Magic Bullet: A Social History of Venereal Disease in the United States since 1880, with a New Chapter on AIDS.* New York: Oxford University Press, 1987.

Landsberg, H. "Prelude to the Discovery of Penicillin." *Isis* 40 (1949): 225–227.

Liebenau, Jonathan. "The British Success with Penicillin." *Social Studies of Science* 17 (1987): 69–86.

Macfarlane, Gwyn. *Alexander Fleming: The Man and the Myth.* Cambridge, MA: Harvard University Press, 1984.

Neushul, Peter. "Science, Government, and the Mass Production of Penicillin," *Journal of the History of Medicine and Allied Sciences* 48 (1993): 371–395.

Sheehan, John C. *The Enchanted Ring: The Untold Story of Penicillin.* Cambridge, MA: MIT Press, 1982.

Pesticides

Around the turn of the twentieth century, the growth of commercial farming led large-scale businesses to seek ways to control insects. These pests could destroy large swaths of profitable crops. Entomologists and farm- ers developed a few key chemical compounds to destroy them, such as lead arsenate and calcium arsenate. To protect humans from these and other toxins, the United States in 1906 passed the Pure Food and Drug Act, which began the regulatory efforts to limit the poisons that might turn harmless consumable substances into toxic materials. The use of pesticides was widespread by the 1920s, and in 1938 a federal law instituted even more strict controls on hazardous materials such as lead arsenate. But laws and attitudes about pesticides generally looked at the short-term effects of exposure as a means to determine safety. When pesticides were sprayed, did people get sick? If not, they were safe. Such reasoning ignored any long-term damage that had no visible symptoms. Although scientists at Harvard University warned against the cumulative effects of pes-

Sacks and drums of pesticides are neatly stored in a warehouse. The use of pesticides was widespread by the 1920s, and in 1938 a federal law instituted even more strict controls on hazardous materials such as lead arsenate. (National Library of Medicine)

ticides as early as 1927, most assumed that the residues left by these chemicals would have no harmful effects if people did not exhibit symptoms of poisoning.

New insecticides were introduced in the 1940s, most notoriously DDT, which was very effective against malaria and typhus. Invented by Paul Müller (1899–1965) in Switzerland in 1939, DDT seemed to be a wondrous chemical: It eradicated dangerous insects more effectively than previous pesticides and seemed to pose less risk of sickness in humans. It was an important weapon during World War II, because it helped to reduce soldiers' danger of contracting insect-borne diseases. In the short term, DDT saved lives. Controversy about DDT began when, after the war, it was used in cities to control the gypsy moth and it began to eradicate birds as well. More extensive discoveries of the effects of DDT on wildlife and humans did not occur until the 1950s and 1960s. Only then did it become clear that the long-term DDT residues could contribute to death, reproductive damage, and even species extinction. After Rachel Carson (1907–1964) published her celebrated 1962 book *Silent Spring,* detailing many of the dangers, activists and scientists succeeded in getting DDT's use banned in the United States.

See also Public Health

References
Dunlap, Thomas R. *DDT: Scientists, Citizens, and Public Policy.* Princeton, NJ: Princeton University Press, 1981.
Whorton, James. *Before Silent Spring: Pesticides and Public Health in Pre-DDT America.* Princeton, NJ: Princeton University Press, 1975.

Philosophy of Science

The philosophy of science is concerned with epistemology, which is the study of knowledge. It asks, "how does one know something?" Around the turn of century, many philosophers and scientists were concerned about the consequences of drastic changes in scientific ideas in the nineteenth century, from Darwinian evolution to thermodynamics and electromagnetism. The philosophy of science increasingly was debated, particularly in Europe, where thinkers concerned themselves with the mechanics of knowledge production and the impediments to truly "scientific knowledge."

Although philosophers occasionally took divergent views, most acknowledged that science appeared to be a cumulative process. Pierre Duhem (1861–1916), the French historian and philosopher, stressed the evolutionary (rather than revolutionary) nature of scientific change, going back to the medieval universities. In his 1906 *La Theorie Physique,* Duhem described the history of physics as slow accumulation, with imperceptible improvements amidst a morass of misconceptions. Mathematician and philosopher Bertrand Russell (1872–1970) noted that improvements could be made more easily by restating problems in precise logical form; Russell's career was marked by impatience with irrelevant influences in philosophical and scientific ideas, which he believed tarnished science. Russell's logical analysis attempted to avoid conceptual prejudices, by explicitly stating that some kinds of knowledge will probably never be known, because they exist beyond the human capacity to understand them. Scientists should not attempt proofs of things that are not provable (i.e., the existence of God), and should stick to what Russell called more objective methods.

One of the most influential philosophers of science during the late nineteenth and early twentieth centuries was the German Ernst Mach (1838–1916). Himself a physicist and mathematician, Mach fashioned a philosophy of science that favored empirical study above all other forms of knowledge. Physics, he wrote, should confine itself to the description of facts, avoiding reliance on ideas and concepts not directly measurable. In his *History of Mechanics* (1883), Mach attempted to show that too much reliance on

One of the most influential philosophers of science during the late nineteenth and early twentieth centuries was the German scientist Ernst Mach. (Corbis)

theory and rationality, at the expense of observation, welcomes false and harmful ideas. Mach's philosophy held wide appeal, because it seemed to promise a science free of *metaphysics*, a word often used to describe a host of preconceived notions that frame (or prejudice) one's beliefs about the natural world. In the first decade or so of the twentieth century, Mach was critical of contemporary theories about atomic structure and relativity. Founder of quantum physics Max Planck (1858–1947) was an ardent critic of Mach, largely because the latter's rejection of atoms and his insistence that knowledge must come from direct sensory experience rather than imaginary (or theoretical) constructs.

Despite the widespread critique of him, Mach influenced a group of thinkers who in the 1920s called themselves the Vienna Circle. The circle included Moritz Schlick

(1882–1936), Rudolf Carnap (1891–1970), Kurt Gödel (1906–1978), W. V. Quine (1908–2000), and sometimes Ludwig Wittgenstein (1889–1951). By separating observable facts from theoretical interpretation, they wanted to get rid of metaphysics and preserve the objectivity of science. The pursuit of objectivity in science and the belief that knowledge can be increased through empirical methods culminated in a philosophy called logical positivism. The term *positivism* was used by the nineteenth-century philosopher Auguste Comte, who had urged the abandonment of religion in scientific thinking; knowledge could grow (hence the "positive") with an acceptance only of observable causes and effects. The method that drew the Vienna Circle together was that of "verifiability." They believed that objectivity was best preserved if scientific theories could be presented in such a way that they could be verified by observation or experiment.

The "verifiability" of the Vienna Circle had a number of critics, including many of its own members. One critic was a philosopher who was not a member but who was instrumental in promoting its views, Karl Popper (1902–1994). In treatises such as his 1935 *Logik der Forschung* (literally "Logic of Research," though in English it has been published as *Logic of Scientific Discovery*), Popper revised logical positivism and turned "verifiability" upside down. He replaced it by developing the concept of *falsification*. According to Popper, an idea gains scientific legitimacy only if it conceivably can be falsified, that is, proven incorrect by some observation or experiment. If one cannot subject an idea to such scrutiny, it does not deserve to be called science; it cannot lead to increased knowledge, and must forever be consigned to the realm of speculation. Instead of verification, which attempts to find a means of proving something to be true, Popper emphasized the importance of understanding what conditions or experiments could conceivably discredit a theory. This method could still lead to the "positive" accumulation of knowledge, which is why falsification became the basic method for the new logical positivists.

Other philosophies of science flourished in totalitarian countries, where leaders expected philosophy to complement political doctrines. In the Soviet Union, for example, scientists and philosophers alike attacked the "idealism" of relativity, quantum mechanics, and genetics. Bolsheviks blended the nineteenth-century philosophies of Georg Hegel (1770–1831) with the political philosophies of Karl Marx (1818–1883) and Friedrich Engels (1820–1895) and then added their own changes. This resulted in a branch of the philosophy of science that eventually came to be known as Stalinist science. In other totalitarian countries, philosophies of science were not as well defined. In Nazi Germany, for example, theoretical physics typically was dubbed a Jewish science. But aside from the new element of racism, German physics was not much more than a prejudice in favor of experimental physics, which was a carryover of Machist empiricism.

For the most part, logical positivism as reformed by Popper remained the dominant philosophy of science for decades, among philosophers and scientists; it did not receive a significant challenge until the concept of paradigm shifts was raised by Thomas Kuhn (1922–1996) in the early 1960s.

See also Gödel, Kurt; Popper, Karl; Nazi Science; Soviet Science

References

Bradley, J. *Mach's Philosophy of Science*. New York: Oxford University Press, 1971.

Heilbron, J. L. *The Dilemmas of an Upright Man: Max Planck and the Fortunes of German Science*. Cambridge, MA: Harvard University Press, 2000.

Kraft, Victor. *The Vienna Circle: The Origins of Neo-Positivism*. New York: Philosophical Library, 1953.

Losee, John. *A Historical Introduction to the Philosophy of Science*. New York: Oxford University Press, 1972.

Russell, Bertrand. *A History of Western Philosophy*. New York: Simon & Schuster, 1945.

Physics

Physics in the twentieth century evolved from a finished field, in which all questions seemed answered, to one of the most exciting frontiers of discovery. This was due primarily to a surge of interest in physics after Wilhelm Röntgen's (1845–1923) discovery of X-rays in 1895, which inspired experiments that led to the discovery of radioactivity. With radioactive particles, scientists could experiment with subatomic projectiles and theorize about the nature of matter itself. Most of the important early activity in physics took place in France, Britain, and Germany. The last was the birthplace of quantum theory, relativity, quantum mechanics, and even the discovery of atomic fission. By the 1940s, however, Europeans ceded their dominance in physics to the United States, where facilities were well funded and there was a growing appreciation of physics by government and military sponsors.

The discovery of radioactivity sparked a revolution in experimental physics. Henri Becquerel (1852–1908) had discovered the phenomenon in France in 1896, and his compatriots Pierre and Marie Curie (1867–1934) conducted the early fundamental work on this subject. Marie Curie coined the term *radioactivity* and discovered the radioactive elements polonium and radium. The presence of these and other radioactive elements in the ore pitchblende prompted New Zealander Ernest Rutherford (1871–1937) to identify the series of elements produced in radioactive decay. Experiments with radioactive particles compelled Rutherford to change his conception of the structure of the atom, and in 1911 he developed a new model that included not only electrons but also a nucleus. By the end of World War I, the most promising experimentalists worked with Rutherford in England, using his model of the atom for experimental and theoretical work.

Experiments with subatomic particles were conducted widely in Europe and North

America. British scientists discovered the neutron and achieved the first artificial atomic disintegration with a particle accelerator, both in 1932. That same year, U.S. physicist Ernest Lawrence (1901–1958) developed an accelerator that forced charged particles to travel in a spiral, increasing their energy before being ejected. This "cyclotron" became the most effective kind of accelerator, designed to produce highly energized particles in an experimental setting. Other Americans conducted important work in physics, particularly the cosmic ray researchers at Caltech who worked with the poor man's particles, so-called because radiation from space did not require expensive accelerators to generate them. Another important center of research was developing in Italy, under the leadership of Enrico Fermi (1901–1954), whose team subjected an array of elements to neutron bombardment in order to understand nuclear reactions. Other fields of inquiry were in solid-state physics, in which experimenters such as the Braggs— father and son, William Henry Bragg (1862–1942) and William Lawrence Bragg (1890–1971)—and Max Von Laue (1879–1960) developed the field of X-ray crystallography from their studies of X-ray diffraction in crystals in the 1910s. Efforts to understand the interactions of electrons in crystalline matter became crucial after the development of quantum mechanics in the 1920s, and solid-state physicists were responsible for the electronic technological inventions of the postwar era, such as transistor radios.

Although radioactivity proved crucial for understanding the nature of the atom, some of the most controversial work in physics was theoretical. In Germany in 1900, Max Planck (1858–1947) proposed that energy is distributed in packets, rather than in a continuous stream. These packets, called quanta, became the basis of quantum physics. One of the first major applications of quantum theory was made by Danish physicist Niels Bohr (1885–1962), who believed he could "save"

Rutherford's atomic model from most of its outstanding problems by claiming that electrons exist in quantum orbits, moving only when atoms absorb or release energy in discrete quanta. Other important theoretical contributions were Albert Einstein's (1879–1955) special and general theories of relativity. The first (1905) disposed of the nineteenth-century concept of the ether and redefined the variables space, time, and mass. The second (1915) provided a theoretical explanation of gravitation, equating it with acceleration; it also proposed that light might be bent by gravity. In the 1920s, the most exciting theoretical problem was that of mechanics—how should the quantum theory change classical mechanics? A number of theoreticians, including Werner Heisenberg (1901–1976) and Erwin Schrödinger (1887–1961), proposed versions of quantum mechanics. The most controversial aspect of Heisenberg's work was his denial of determinism; he concluded that at quantum scales, one could not increase knowledge of one variable without decreasing knowledge of others. Also disturbing was the seeming incompatibility of different versions of quantum mechanics, some interpreting the physical world in terms of particles and others in terms of waves. Both were correct, yet incomplete on their own; Bohr referred to this as the wave-particle duality. Such questions were hotly debated, primarily in Europe and, more specifically, in the universities in Berlin and Göttingen.

The center of activity in physics shifted in the 1930s and World War II from Europe to the United States. In Germany, anti-Semitic racial policies made it impossible for Jews to begin careers in science, and many lost existing jobs. A number of leading scientists were exempt from these rules, because of impressive contributions or past service, but many chose to leave their posts and to leave Germany. The most famous of these was Albert Einstein, who was out of Germany when Adolf Hitler (1889–1945) rose to power in 1933; he decided not to return. Many non-

Jewish physicists, such as Werner Heisenberg and Max Planck, deplored such practices, claiming that Nazi policies would kill German physics. Others, such as Philipp Lenard (1862–1947), believed that Jews were corrupting physics anyway, and that theoretical physics had become so detached from reality that it no longer was fitting to be called physics at all. He and others referred to Einstein's relativity as Jewish physics. By the end of the 1930s, most leading Jewish scientists had fled, aided by organizations, such as the Academic Assistance Council (Britain), that helped refugee scientists to relocate in Britain and the United States.

The development of atomic bombs permanently altered the character of physics. Atomic fission was discovered in Germany in late 1938; soon scientists of many countries realized that it might lead to bombs of extraordinary explosive power. Most of the work became secret in 1939, after the commencement of World War II, and bomb projects were set up by governments in several countries. Many of the Jewish scientists who had fled Europe ended up in the United States, where they contributed to the Manhattan Project. The project cost $2 billion and forged close ties among scientists, the military, and government. When the bombs were first used, it seemed clear to all that the United States would continue to rely on physicists to exercise its influence in the world. More than any other discipline, physics seemed directly relevant to national security, and henceforth it began to receive ample funding from governments and enormous prestige. Because of this importance, for example, physics in the Soviet Union managed to avoid the excesses that befell the biological community during the Trofim Lysenko (1898–1976) affair; Joseph Stalin's (1879–1953) desire to build a bomb trumped his efforts to ensure ideological conformity. In the United States, physicists from the "Los Alamos generation" became the leading scientific figures in academic, industrial, and military circles. Both Britain and France also generously supported physics, in efforts to develop nuclear power and nuclear weapons. The use of expensive particle accelerators, reactors, and new experimental designs required more money than ever before, at levels only governments could afford. Thus physics led the way into what often was called the era of Big Science, with large teams of researchers, large sums of money, expensive equipment, and often collaboration across university, military, and government lines.

See also Atomic Bomb; Atomic Structure; Cavendish Laboratory; Cloud Chamber; Cosmic Rays; Cyclotron; Fission; Light; Manhattan Project; Nazi Science; Quantum Mechanics; Quantum Theory; Radioactivity; Relativity; X-rays

References

Heilbron, J. L., and Bruce R. Wheaton, *Literature on the History of Physics in the Twentieth Century.* Berkeley: University of California Press, 1981.

Hoddeson, Lillian, Ernest Braun, Jürgen Teichmann, and Spencer Weart, eds. *Out of the Crystal Maze: Chapters from the History of Solid-State Physics.* Oxford: Oxford University Press, 1992.

Kevles, Daniel J. *The Physicists: The History of a Scientific Community in Modern America.* Cambridge, MA: Harvard University Press, 1995.

Kragh, Helge. *Quantum Generations: A History of Physics in the Twentieth Century.* Princeton, NJ: Princeton University Press, 1999.

Weart, Spencer R., and Melba Phillips, eds. *History of Physics.* New York: American Institute of Physics, 1985.

Piaget, Jean

(b. Neuchâtel, Switzerland, 1896; d. Geneva, Switzerland, 1980)

Jean Piaget became a celebrated psychologist for his work on the cognitive development of children. His early interests were in natural history, and he published scientific works on mollusks before turning to psychology. He received a doctorate from the University of Neuchâtel in 1918. He worked in France in an intellectual milieu that favored the intelligence testing methods recently developed by Alfred Binet (1857–1911). While working on intelligence testing, he started a long-standing interest in the development of cognition in children and its relationship to intellectual development generally. He became the research director of the Institut Jean-Jacques Rousseau in Geneva in 1921. In subsequent years he held several positions in Switzerland, alternating his interests between psychology, sociology, and the history and philosophy of science.

Although Piaget was concerned with the child's early intellectual and moral development, his wider interests were in epistemology, the study of knowledge itself. He was fascinated by the growth of knowledge, and this motivated his interests in the cognitive development of children. He believed that knowledge growth was a process of adaptation from practical intelligence, from the earliest stage when the child directly assimilates the external environment into his or her own activities. While developing intellectual schemata (interrelated concepts and ideas) to make use of the environment's symbols and images, the child also develops concepts that help him or her to explain the external world itself. External stimuli are either assimilated into existing schemata or the child accommodates new stimuli by creating new schemata. According to Piaget, more external influences tend to make links between the external environment and the child's cognitive schemata more numerous, more complex, and thus less subjective (in other words, less in relation solely to his or her practical situation). This process was crucial for the development of higher knowledge, which, in Piaget's estimation, meant increasing objectivity about the surrounding world.

Piaget devised a number of stages of development through which children pass in order to achieve adult intelligence. He emphasized the importance of perception, mental imagery, and memory. The result of the cumulative early stages of the child's development is a higher state of knowledge; as the title of one of his books put it, the process is the "construction" of reality. The first stage, the "sensorimotor," comprises approximately the first two years of life, when the child's knowledge consists of practical intelligence. More advanced thought occurs in the "preoperational" stage of the next five years, when the child can mentally imagine objects that he or she cannot see. "Concrete operations" are possible in the next stage, such as deductive reasoning. Abstract thought occurs in the final stage of "formal operations."

The salient feature of Piaget's work was his belief in what he called genetic epistemology. The growth of knowledge stems not simply from what is taught to the child; instead, it is built on the foundations laid during the early stages of development. This makes knowledge growth truly developmental; like biological evolution, there are distinct relationships between mature intelligence and the more primitive childlike intelligence. The phrase *genetic epistemology* draws attention to the importance of this developmental aspect of intelligence. At the same time, Piaget did not diminish the crucial role of external influences; although intelligence is based on the past, it can only grow from interaction with the environment.

Piaget's influence was generally confined to the French-speaking world until after World War II. His works on the construction of intelligence were first published as *La Naissance de L'Intelligence chez L'Enfant* (1936), *La Construction du Reel chez L'Enfant* (1937), and *La Formation du Symbole chez L'Enfant* (1946). These were all translated into English by the 1950s. His work became widely known in the English-speaking world in the 1960s when John Flavell published, in 1963, *The Developmental Psychology of Jean Piaget*. In 1955, Piaget created in Geneva the International Center for Genetic Epistemology. Merging his scientific and historical interests, he hoped to generate links between his work in child psychology and the history of the development of scientific thought. He was its director until his death in 1980.

See also Philosophy of Science; Psychology
References
Chapman, Michael. *Constructive Evolution: Origins and Development of Piaget's Thought.* New York: Cambridge University Press, 1988.
Kitchener, Richard F. *Piaget's Theory of Knowledge: Genetic Epistemology and Scientific Reason.* New Haven, CT: Yale University Press, 1986.
Piaget, Jean. *The Construction of Reality in the Child.* New York: Ballantine, 1954.
Vidal, Fernando. *Piaget before Piaget.* Cambridge, MA: Harvard University Press, 1994.

Pickering's Harem

The rather dated term *Pickering's Harem,* used among astronomers around the turn of the century, referred to a group of women astronomers who were employed at the Harvard College Observatory under the directorship of Edward Pickering (1846–1919). Pickering began to employ women to do astronomical calculations for several reasons in the 1880s: because he believed that women should go into advanced study, because he was irritated at the inefficiency of his male employees, and because many of them could be hired at a comparatively lower wage than men. He continued the practice until his death in 1919. Most were not college graduates, and many had physical disabilities: both Annie Jump Cannon (1863–1941) and Henrietta Swan Leavitt (1868–1921), the most notable of the group, were partly deaf.

Pickering and his first hire, Wilhelmina Fleming (1857–1911), tried to use the success of the Harvard group to promote such hires at other observatories. The skills involved in this work were stereotyped as "women's work," of a routine nature, requiring the kind of patience and perseverance that (some argued) were particularly suited to women. By the 1920s, almost every large astronomical laboratory in the country hired such women. They were a convenient and reliable workforce, requiring no high salaries and few promotions, and thus no threat to men already competing for prestigious positions.

Although these women were charged with making calculations—they were called computers—several of them made fundamental contributions to astronomy. Part of the reason for this was that much of Pickering's effort was devoted to classifying photographic plates rather than just making laborious calculations. This required cheap and intelligent labor. The women charged with doing this became leading experts on astrophysical phenomena, knowledge that was required for correctly identifying stars. Working for years in the same job with no chance for advancement, aside from the negative impact on women's opportunities in astronomy, had the positive result of creating a few scientists with unsurpassed levels of specialized knowledge in star identification. One of them was Annie Jump Cannon, a graduate of Wellesley College, who led the effort to compile the Henry Draper Star Catalog, published between 1918 and 1924. Another leading women astronomer from Harvard was Margaret Harwood (1885–1979), who went on to direct the Maria Mitchell Observatory on Nantucket Island. The most famous was Henrietta Swan Leavitt, a graduate of Radcliffe College, who used the photographic plates at Harvard to study Cepheid variable stars, and in 1912 discovered a relationship between their period and luminosity. Leavitt's discovery was the key to determining stellar distances, and it became a major tool for astronomers in the next decades, leading astronomers such as Harlow Shapley (1885–1972) and Edwin Hubble (1889–1953) to reevaluate the sizes and dynamics of galaxies.

See also Astronomical Observatories; Leavitt, Henrietta Swan; Shapley, Harlow; Women
References
Haramundanis, Katherine, ed. *Cecilia Payne-Gaposchkin: An Autobiography and Other Recollections.* New York: Cambridge University Press, 1984.
Jones, Bessie Z., and Lyle Boyd. *The Harvard College Observatory: The First Four Directorships, 1839–1919.* Cambridge, MA: Harvard University Press, 1971.

Rossiter, Margaret W. "'Women's Work' in Science, 1880–1910." *Isis* 71:3 (1980): 381–398.

Welther, Barbara L. "Pickering's Harem." *Isis* 73:1 (1982): 94.

Piltdown Hoax

The forgery of scientific evidence and the subsequent shaping of entire fields of inquiry provide a fascinating context to understand the often-tenuous nature of scientific knowledge. The Piltdown Man was so named because it was found in a gravel bed in Piltdown, Sussex, England, in 1912. The fossilized remains consisted of a human skull and a jawbone that resembled that of an ape. The discovery was monumental, because it seemed to be the "missing link" necessitated by the prevailing interpretation of evolution. According to popular understandings of evolution, species progress from lower forms to more advanced forms; thus, one should expect to find transitional species between modern apes and modern men (this is not the Darwinian view). The Piltdown Man appeared to be exactly that: an apelike jaw attached to an otherwise modern-seeming man.

Seeing the Piltdown Man as the "missing link" depended on agreement that the skull and jawbone in fact belonged together, and indeed originated in Piltdown. Doubt about the veracity of the claim appeared as early as 1914 by William K. Gregory, who noted that the fossils could have been planted at Piltdown to fool the scientists. Others chose not to see Piltdown Man as a link in the evolutionary chain. Some, including Henry Fairfield Osborn (1857–1935), viewed him as an anomaly from which no modern species evolved. But the majority of scientists accepted Piltdown Man as evidence for the progressive view of evolution so popular at the time.

With the skull and jawbone in hand, lawyer and amateur archaeologist Charles Dawson (1864–1916) and British Museum geologist Arthur Smith Woodward announced the discovery of a new and early hominid. In subsequent years more fragments—teeth, some stone and bone tools, and bones of other fauna—were discovered in the gravel pits that appeared to substantiate the claim. The Piltdown Man appeared to be so unique that a new binomial appellation was created to classify it. It became the first known (and only) evidence of the prehistoric existence of *Eoanthropus dawsoni,* named in honor of Charles Dawson. Dawson himself died in 1916, before his hope to be elected to the Royal Society could be fulfilled.

In the early 1950s, Piltdown Man was uncovered as a fraud by Joseph Weiner (1915–1982) and Kenneth Oakley (1911–1981). They showed that the skull belonged to a modern man, though it was unusually thick. The jawbone belonged to a modern orangutan. Both had been stained artificially in order to look ancient, and the teeth had been filed down. Some of the local rocks originated from elsewhere, and the local animal bones came from faraway lands in North Africa.

Who planted the evidence? Weiner was the first to write an authoritative account, in 1955, called *Piltdown Forgery*. Dawson himself has been the most cited suspect, and in retrospect is not hard to imagine why. Dawson also claimed to have discovered other curiosities, such as a mummified toad. But because of the scientific expertise required to make the forgery convincing, it is doubtful that he acted alone. Most of those associated with the initial discovery have found themselves (posthumously) under suspicion by various historical detectives, including by a twist of fate the creator of Sherlock Holmes, Sir Arthur Conan Doyle (1859–1930). Other high-profile suspects include Catholic priest, writer, and paleontologist Pierre Teilhard de Chardin (1881–1955). Researcher Ian Langham concluded that Sir Arthur Keith (1866–1955), a prominent British physical anthropologist, was among the guilty. The plethora of other candidates and the endless array of circumstantial evidence, despite the

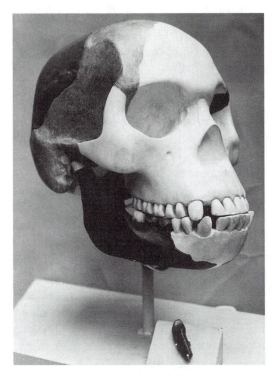

The cast of the first construction of the "Piltdown Man" attracted considerable attention, but was revealed later as a hoax. (Bettmann/Corbis)

scandalous effect on biology, have led a number of historians to give up on what they consider to be pointless detective work, and to let the Piltdown Man lie.

See also Anthropology; Missing Link; Peking Man; Teilhard de Chardin, Pierre

References

Blinderman, Charles. *The Piltdown Inquest*. Buffalo, NY: Prometheus Books, 1986.

Gould, Stephen Jay. *The Panda's Thumb*. New York: W. W. Norton, 1992.

Spencer, Frank, and Ian Langham. *Piltdown: A Scientific Forgery*. London: Oxford University Press, 1990.

Planck, Max

(b. Kiel, Germany, 1858; d. Göttingen, Germany, 1947)

In 1900, Max Planck became the unwitting founder of quantum physics. This field, more than any other, became a symbol of science in the first half of the twentieth century for its novelty and its controversial and revolutionary character. Planck was catapulted into a long-standing period of fame and scientific leadership, spanning the golden age of physics in Germany and extending well into the Nazi period. His pivotal role in facilitating the rise of quantum mechanics was tempered by his inability to control the disintegration of science under Adolf Hitler (1889–1945).

Planck's reputation was not initially in theoretical physics, but rather chemical thermodynamics. He firmly believed in the universal applicability of the law of entropy and was fascinated by the irreversible processes of nature. This led him to study blackbody radiation. A blackbody is a body whose radiation is dependent only on temperature (and independent of the nature of the body). Wilhelm Wien (1864–1928) determined in 1894 that if the spectral distribution of the radiation was known at one temperature, it could be deduced at any other temperature. However, Wien's radiation law proved incorrect at long wavelengths, and at best it provided only for approximations. Planck, who had supported Wien's law, was as surprised as anyone. He tried to improve on Wien's law by introducing, in 1900, the concept that the total energy was composed of finite equal parts. He claimed that the energy of blackbody radiation could be determined by multiplying frequency by a very small constant of nature (h).

Planck's innovation was to calculate radiation as discontinuous. Energy was not emitted in a constant stream; instead, it was composed of discrete packets, individual tiny amounts called quanta. Neither Planck nor others recognized immediately that his new radiation law required a break with classical physics. He was simply smoothing out some problems with Wien's law. Only slowly did the implications for the world of physics become clear. The strongest advocate of quantum physics would be Albert Einstein (1879–1955), who in 1905 published on the photon—the quantized unit of light—in a paper no less influential than that outlining his special theory of relativity (the same year).

Aside from his scientific work, Planck had enormous influence among scientists and other intellectuals. Like many of his countrymen, Planck reveled in the patriotic fervor accompanying the commencement of hostilities in Europe in 1914. He then made a decision he soon regretted, namely, signing the document, "An die Kulturwelt! Ein Aufruf" (this can be translated as "To the Civilized World! An Appeal"). This was a manifesto of ninety-three intellectuals defending Germany against the charges of the Entente powers and declaring solidarity with the German army. Although nationalistic sentiments were not limited to Germany, scientists in Britain, France, and the United States used the manifesto to identify German scientists with the brutality of the Kaiser's regime and to exclude Germany from postwar science.

Planck was a spokesman for German science in its years of isolation following World War I. He traveled to Sweden to accept a Nobel Prize in Physics; in 1918, no prize had been awarded, and after the war it went to Planck. He received it officially in 1919, the same year that two other Germans—experimental physicist Johannes Stark (1874–1957) and chemist Fritz Haber (1868–1934)—received prizes. The choice of these three infuriated many who felt that German science should not be celebrated, especially since Haber had spearheaded chemical warfare projects during the war. Planck moved to ease tensions, first by recanting his adherence to the manifesto of intellectuals, then by attempting to steer German science back into international organizations. It was not easy; the International Research Council formed in the wake of the war explicitly barred Germans from participation in international scientific gatherings. Planck emphasized science's internationalism and deplored its politicization. Ultimately, the place of German science in the international community was helped most, not by Planck or any or-

ganizations, but by the fact that most of the leading physicists of relativity, quantum physics, and quantum mechanics were German.

When Hitler became Reich chancellor in 1933, Planck was president of the Kaiser Wilhelm Society and secretary of the Berlin Academy of Sciences. This began a tragic period in the lives of Planck and others, told by John Heilbron in his *The Dilemmas of an Upright Man*. The upright man was Planck himself, holding on to his sense of duty to the state, his countrymen, and to science in the shadow of Nazism. Planck met with Hitler, to convince him that alienating Jews would hurt German science; Hitler responded that the racial laws were intended to protect eminent Jews, who could take advantage of exceptions for veterans and men of eminence. Planck's hopes that the Nazis would have to moderate their views, in order to remain in power, were in vain. Hitler's policies crippled German science. Ultimately, Planck's efforts to salvage some of it by defending Jews and their ideas (such as Einstein's theories of relativity) brought him under fire from physicists such as Stark, who gave Hitler his fullest support and trumpeted the cause of an "Aryan physics" bereft of Jewish influences.

Planck resigned his posts in 1938 and lectured on religion and science, espousing science's universalism and refusing to abandon "un-German" concepts such as relativity and quantum mechanics. Although he did not speak out explicitly against the regime, he had become something of an outsider. His worst personal tragedy occurred during the war; not only did he lose his house to an air raid, but also his son Erwin was executed for complicity in a plot to assassinate Hitler. Health complications increasingly pained him, and his hometown became such a battlefield that he and his wife slept in haystacks. He continued to lecture after the war; he died of a stroke in 1947.

See also Kaiser Wilhelm Society; Nazi Science; Physics; Quantum Theory; Race; World War I; World War II

References

Cline, Barbara Lovett. *The Questioners: Physicists and the Quantum Theory*. New York: Thomas Y. Crowell, 1965.

Heilbron, J. L. *The Dilemmas of an Upright Man: Max Planck and the Fortunes of German Science*. Cambridge, MA: Harvard University Press, 2000.

Kragh, Helge. *Quantum Generations: A History of Physics in the Twentieth Century*. Princeton, NJ: Princeton University Press, 1999.

Polar Expeditions

Polar expeditions in the twentieth century combined scientific discovery with adventures and exploration. Before the nineteenth century, economic considerations drove most of the initial explorations of the northern polar region by Europeans, as intrepid ship captains hoped to find a way to penetrate into Asian markets by a shorter route. In the nineteenth century, polar expeditions gained a quasi-scientific status because of the thrill of discovering new resources and the true geographic and magnetic poles. By the twentieth century, the marriage of science and adventure was encapsulated by such expeditions to study the relatively unknown geologic, oceanographic, climatic, and meteorological conditions at the poles. Although most scientific expeditions were undertaken under national auspices, with strategic or propaganda objectives in mind, they often served the interests of science as well.

The thrill of adventure brought polar expeditions into public consciousness, and expedition leaders became heroes. Many scientists were drawn to the expeditions, including the founder of the theory of continental drift, Alfred Wegener (1880–1930), who died on the Greenland Ice Sheet in 1930. The American Robert Peary (1856–1920) made celestial and magnetic observations on his journey to the North Pole; when he reached it in 1909, he proclaimed that he had finally gotten the North Pole out of his system. He became an instant celebrity and disappointed the hopes of other

After reaching the North Pole, American Robert Peary became an instant celebrity. However, historians have pointed out defects in his scientific observation techniques and some have suggested that these defects help to show that Peary fell short of reaching the pole by a hundred miles. (Library of Congress)

nations to achieve this "first." However, historians have pointed out defects in his scientific observation techniques; indeed, some have suggested that these defects help to show that Peary never really reached the pole at all, and fell short by a hundred miles.

With the North Pole evidently reached, the South Pole on the continent of Antarctica became the next objective. In fact, Peary's announcement spoiled the plans of the Norwegian Roald Amundsen (1872–1928) to go reach the pole. Amundsen quickly revised his plans and headed south, where a British expedition already was planned, led by Robert Scott (1868–1912). Amundsen argued that because the British claimed to be conducting scientific research, he did not feel that there was a competition. But the scientific elements of the expedition, in Britain as well as in Norway, were secondary to achieving the goal of being "first." In this case, the famed competition between the Norwegian and British teams ended in tragedy. Amundsen's team was first to reach the pole, in 1911, and returned alive. Scott's team arrived at the pole thirty-four days later, finding the remnants of Amundsen's camp. The entire party died on the return trip, in 1912.

In popular culture, polar explorers were the great national heroes of their time, combining virile endurance with scientific discovery. In 1914, Irishman Ernest Shackleton (1874–1922) embarked on an expedition to cross the Antarctic, but in 1915 his ship was caught in the ice and crushed. For five months, he and his men survived and they became British celebrities for their endurance. In 1926, Richard Byrd (1888–1957) became a U.S. celebrity for flying a plane to the North Pole. In 1937, Soviet pilot Valerii Chkalov (1904–1938) became a celebrity (in his own country) for being the first to fly over the North Pole from Moscow and land safely in the United States (in the state of Washington), breaking the nonstop long-distance flying record. That same year, Mikhail Gromov (1899–1985) followed up by making the same journey and landing in California. These scientists attained celebrity status upon their return, and the Soviet government hailed them as conquering heroes.

Scientific study received less attention from the press than exploration. Anthropologist Vilhjalmar Stefansson (1879–1962), for example, began a series of ice-floe experiments and other explorations in the Canadian Arctic around the time of these more widely known efforts in Antarctica. His studies yielded a great deal of knowledge about the environment of the arctic and the nearby inhabitants, but his fame paled next to the explorers. Scientific research often entailed international cooperation rather than competition. In 1932–1933, several countries organized the Second International Polar Year (the first was held fifty years earlier), designed to have numerous scientists take similar measurements in various places at roughly the same time. Magnetic, aurora, and meteorological observations dominated the scientific work during the Polar Year, conducted by a network of stations in the Arctic and Antarctic. The International Meteorological Committee, which planned much of the year, hoped that the scientific work would have practical applications related to terrestrial magnetism, marine and air navigation, wireless telegraphy, and the forecasting of weather.

See also Nationalism; Oceanic Expeditions
References
Kirwan, L. P. *A History of Polar Exploration*. New York: W. W. Norton, 1960.
McCannon, John. "Positive Heroes at the Pole: Celebrity Status, Socialist-Realist Ideals and the Soviet Myth of the Arctic, 1932–1939." *The Russian Review* 56 (1997): 346–365.
Rawlins, Dennis. *Perry at the North Pole: Fact or Fiction?* New York: Robert B. Luce, 1973.

Popper, Karl
(b. Vienna, Austria, 1902; d. London, England, 1994)

Karl Popper was one of the most influential philosophers of science in the first half of the twentieth century, rivaled only by Ernst Mach (1838–1916). The latter was noted for his strict empiricism, favoring experimentation over abstract ideas. Mach's point of view was critiqued by the Vienna Circle, a group of philosophers that included mathematicians and logicians such as Rudolf Carnap (1891–1970). The Vienna Circle philosophers tried to reevaluate Mach's philosophy in ways that provided room for deductive logic and rational choice between competing theories. The philosophy of science dubbed logical positivism was the result. Popper was critical of both Mach and the Vienna Circle and of their beliefs that the positivist approach could lead to greater certainty in science. For Popper, theories could never be absolutely certain, and any method that claimed to be able to verify theories beyond doubt was fundamentally flawed.

Popper's critique of Mach and the Vienna Circle compelled him to develop new criteria for legitimate, "scientific" knowledge, which he published in his *Logik der Forschung* (its English edition was *Logic of Scientific Discovery*) in 1935. Some members of the Vienna Circle had argued that ideas needed to be proven, or "verified." Popper suggested a replacement for verification, namely, falsification. Instead of leaving it to scientists (or any producers of knowledge) to prove an idea to be true, he proposed that an idea can only be legitimate if there is a conceivable way to test it, to prove it false. Falsifiability thus became the critical criterion for the legitimacy of new ideas. Popper's method began with a problem, requiring the investigator to propose a hypothesis, which would be subjected to refutation by theoretical criticism and a critical experimental or observational test. This method differed from logical positivism in that it was not inductive—it did not begin with the facts, but rather with a problem. In addition, Popper's method made no claim of establishing truth through logical means; instead, it established a system for eliminating false theories or hypotheses, leaving the ones with the best explanatory power (though still subject to critique). Although *falsification* is the term most often associated with Popper, his overall philosophy also was called critical rationalism. Aside from its value in providing a noninductive method of scientific discovery, its strength as a philosophy was its ability to draw a boundary between science and pseudoscience—one falsifiable and the other not.

Popper began his career in Austria, where he was in close proximity to the leading philosophers of the Vienna Circle. He took his doctorate in philosophy from the University of Vienna in 1928. Because he was a Jew, he left Austria during the rise of Nazism in the 1930s, taking a university position in New Zealand. After World War II, he was invited to become a professor of philosophy at the London School of Economics, and he accepted. He stayed there until he retired. Popper also critiqued political ideologies and was particularly hostile to Marxism (although he was himself a socialist as a teenager). He published *The Open Society and Its Enemies* in 1945. It closely resembled his philosophic thought, warning that one should be wary of ideas that claimed absolute certainty; he noted that the best political (and scientific) systems were ones that permitted critique and discussion. Popper's contributions to the philosophy of science were recognized with a knighthood in 1965. In broad terms, Popper's outlook toward scientific activity was problem-oriented rather than topic-oriented, focusing on criticism of ideas rather than the random accumulation of data. This view proved to be the most widely accepted theory of the positive growth of knowledge until the 1960s, when it met a major challenge in the philosophy of Thomas Kuhn (1922–1996).

See also Philosophy of Science

References

Magee, Bryan. *Karl Popper*. New York: Viking Press, 1973.

O'Hear, Anthony, ed., *Karl Popper: Philosophy and Problems*. Cambridge: Cambridge University Press, 1995.

Raphael, Frederic. *Popper*. New York: Routledge, 1999.

Watkins, John. "Karl Raimund Popper, 1902–1994." *Proceedings of the British Academy* 94 (1997): 645–684.

Psychoanalysis

Although physicians studied mental disorders such as hysteria in the late nineteenth century, modern psychoanalysis began with Sigmund Freud (1856–1939), who published *The Interpretation of Dreams* in 1900. Freud determined in the late nineteenth century that many mental problems, which he routinely saw through his own private practice, were connected to conflicts residing in the unconscious mind. He and his colleague Josef Breuer (1842–1925) had conducted hypnosis experiments that seemed capable of bringing to light forgotten memories in patients, which had the "cathartic" effect of occasionally ending the neurotic behaviors simply by making the patient conscious of their probable origins in a memory from the distant past. Although he had some success with hypnosis, Freud believed that analysis of dreams and fantasies could help bring these problems to light by moving them from the unconscious to the conscious mind. Hallmarks of psychoanalysis were concepts such as the *repression* of memories and the *sublimation* of erotic or aggressive instincts—that is, channeling them in other ways, such as through art or music. The purpose of psychoanalysis was to understand how the mind interacts with itself and with the environment, often creating mental problems (neuroses).

By the early 1920s, Freud had refined his version of psychoanalysis. In his effort to create a new science of the mind, he developed a conception of mental dynamics based on fundamental human instincts—Eros and Thanatos, or life and death. The life instinct tended toward self-preservation and sexuality, driven by the libido (sexual energy). The death instinct was the cause of self-destructive or aggressive behavior. These instincts made up the id, which was balanced by the superego, which reined in one's instincts because of norms of behavior acquired from external influences (such as moral values learned from parents or teachers). The conscious mind, the ego, was a product of these two forces in tension.

Although Freud's emphasis on sexuality, the overarching motivation of the life instinct, scandalized some scholars, psychoanalysis provided a tool to engage neuroses in a scientific way. Critics of Freud were taken aback by his willingness to ascribe human character to the development of the libido, which gave a paramount importance to erotic impulses. Yet the systematic analysis of dreams and memories had great appeal to physicians and psychiatrists—and patients—for whom strictly physiological studies had provided little aid in a clinical setting.

Freud's ideas attracted numerous scholars in Europe and North America; the first International Psychoanalytical Congress took place in 1908 in Salzburg, Austria, which facilitated its spread.

Psychoanalysis became popular in the United States after 1909, when Freud and his protégés Alfred Adler (1870–1937) and Carl Jung (1875–1961) were invited to lecture at U.S. universities. One scholar has argued that somatic (body-oriented) approaches to mental disorders had reached a crisis point among many U.S. psychologists, and Freudian psychoanalysis established a new paradigm of thinking about the nature of neuroses and how to treat them. In Russia, psychoanalysis took root among intellectuals, and thus was widely appreciated by early revolutionaries such as Leon Trotsky. Under Joseph Stalin (1879–1953), however, psychoanalysis was condemned because of its focus on the past. Stalin and other Soviet leaders believed that the implicit belief that the past inescapably shapes one's life forever challenged Communist ideology in the sense that it seemed to deny the possibility of radical improvement or change in behavior. Accepting the possibility of such changes was crucial to ensuring the life of the Soviet state. Psychoanalysis itself acquired a reputation as a degenerate, immoral philosophy inextricably tied to Western capitalism.

Despite Freud's foundational work, psychoanalysis evolved in others' hands. Both Adler and Jung, for example, broke with Freud. Jung became renowned for his theories of collective unconscious memories, or archetypes. Like others, he was disturbed by Freud's emphasis on sexuality, which he viewed as a potentially destructive instinct; Jung founded "analytical psychology" in the second decade of the twentieth century, breaking with Freudian psychoanalysis.

The creation of psychoanalysis continues to be controversial. Freud initially believed that most neuroses, especially in women, were a result of repressed memories of trauma, specifically sexual assault during childhood. However, he later revised his view, claiming that unconscious memories were only part of many elements of the unconscious, of which fantasies also played a crucial role. One author, J. M. Masson, has made a controversial claim that Freud in fact made an important discovery—that child abuse was far more prevalent than anyone realized—but the negative reactions of his peers convinced him to abandon his premise. However, psychoanalysis ultimately embraced far more than repressed memories or even fantasies; it studied desires, guilt, shame, paranoia, aggression, self-loathing, and a host of other instincts and learned values that—according to psychoanalysis—underpin the conscious self.

See also Freud, Sigmund; Psychology
References

Burnham, John C. "The Reception of Psychoanalysis in Western Cultures: An Afterword on Its Comparative History." *Comparative Studies in Society and History* 24:4 (1982): 603–610.

Fine, Reuben. *A History of Psychoanalysis.* New York: Columbia University Press, 1981.

Hale, Nathan G. *The Rise and Crisis of Psychoanalysis in America: Freud and the Americans, 1917–1985.* New York: Oxford University Press, 1995.

Masson, Jeffrey Moussaieff. *The Assault on Truth: Freud's Suppression of the Seduction Theory.* New York: Farrar, Strauss, and Giroux, 1984.

Miller, Martin A. *Freud and the Bolsheviks: Psychoanalysis in Imperial Russia and the Soviet Union.* New Haven: Yale University Press, 1998.

Schwartz, Joseph. *Cassandra's Daughter: A History of Psychoanalysis.* New York: Penguin, 2001.

Psychology

Psychology is the science of the mind and its processes. Although theories about how the human mind functions have been prevalent since ancient times, modern psychology was born in the late nineteenth century with efforts to turn it from a philosophical pursuit into a rigorous scientific discipline. Even by the turn of the century, "psychology" was not a separate field at all and was pursued by re-searchers trained (and often working in) philosophy departments. Several schools of thought emerged among academic researchers by the early twentieth century, notably structuralism and functionalism, only to be dominated in the mid-twentieth century by behavioral psychology. Meanwhile, other outlooks captured the public imagination, particularly the work in psychoanalysis.

The most well-known names associated with psychology in the twentieth century were Sigmund Freud (1856–1939) and Carl Jung (1875–1961). However, their "applied" psychology (working with clients to help with mental problems) was not entirely well received among researching psychologists. Freud published his *Interpretation of Dreams* in 1900, and in subsequent years he developed his most influential ideas about the unconscious mind, identifying categories such as the id, ego, and superego. He founded psychoanalysis, which focused on the individual and sought to address the unconscious mind as the interplay of sexual drive and past experiences. His goal in therapy was to bring these unconscious aspects into the conscious realm. Freud's most famous student, Carl Jung, eventually broke away from psychoanalysis and its emphasis on sexuality, and developed his own method of personality development called analytical psychology. One of his innovations was the controversial theory of the "collective unconscious" shared by society as a whole. He believed in the existence of a series of universal archetypes, images holding symbolic meaning, in the dreams and unconscious minds of all people.

Mainstream research in psychology was less connected to personality development and much less focused on variables—such as dreams—open to a wide range of interpretation. One of modern psychology's principal founders, Wilhelm Wundt (1832–1920), began in Leipzig in the 1870s a school of physiological psychology in which he attempted to study human behavior in the context of very detailed studies of anatomy. He emphasized the role of organs such as those

One of modern psychology's principal founders, Wilhelm Wundt was an experimentalist who encouraged psychologists to frame their questions as a scientist would and to impose rigorous tests to support their findings. (National Library of Medicine)

found in the central nervous system and attempted to locate the areas of important activity. Wundt was an experimentalist who encouraged psychologists to frame their questions as a scientist would and to impose rigorous tests to support their findings. Psychology's tenuous status as a science was strengthened by his efforts, and the impulse to establish experimental guidelines to ensure reliability proved extremely influential on psychologists in Europe. His approach, often called introspectionism, called on researchers to break down the consciousness into its most crucial elements and to analyze perceptions as a cluster of these elements.

One of Wundt's doctoral students, Edward B. Titchener (1876–1927), took a position at Cornell University and was a major figure in spreading Wundt's influence to the United States. Titchener was the key figure in developing one of the important trends in

psychology, namely, structuralism. Set forth in Titchener's *An Outline of Psychology* (1896), structuralism borrowed heavily from Wundt's viewpoint and accepted the positivist philosophy of German epistemologist Ernst Mach (1838–1916). His goal was to analyze mental experiences by breaking them down into component parts and to ascertain the laws governing their interaction. Like positivists in other scientific fields, he emphasized the importance of accumulated factual evidence rather than abstract theorizing. He constructed a system that categorized all processes as either sensation (received from sensory organs) or affection, or some combination of the two. Titchener had enormous influence, especially in the United States, until his death in 1927. He opposed many currents in psychology, including mental testing, applied psychology (which, because it often focused on children or people with mental problems, detracted from his goal of establishing the laws governing the average adult mind), and especially behaviorism.

Another school of thought in psychology was functionalism. Influenced by the U.S. psychologist William James (1842–1910), functionalism grew from a critique of Wundt's and Titchener's work, particularly their constrained approach. James noted that their efforts to identify elements of human consciousness implied universality. Claiming that this was not necessarily so, James argued in the late nineteenth century that humans adapt to their environments in different ways. He developed the concept of the stream of consciousness, observing that the mind is constantly adapting. It would be useful, then, to study different populations, such as children or people with disabilities. James had a strong influence upon John Dewey (1859–1952) and James Rowland Angell (1869–1949), who formed what became known as the Chicago school, emphasizing adaptation, the cornerstone of functionalist psychology. Titchener was explicit in separating structuralism from functionalism, and the two schools developed separately and in opposition in the first years of the century. The key difference was in the goal of their studies; while structuralism attempted to analyze and describe the laws governing psychological processes, functionalism was more teleological, attempting to ascertain "why" as much as "how." What exactly was the function of human consciousness (what was its purpose)?

Although functionalism had a strong following, behaviorism became the dominant trend in psychology. It drew upon the two existing schools of thought and also the experiments of Russian scientist Ivan Pavlov (1849–1936), who had demonstrated the power of behavioral conditioning in dogs. Behaviorism's first outspoken advocate, the American John B. Watson (1878–1958), began as a functionalist under Angell's mentorship. In 1913, he published an article in *Psychological Review* entitled "Psychology as the Behaviorist Views It," which was widely influential in setting forth the basic tenets of the new school. Behaviorism treated the subject as a machine responding to stimuli; thus, behavioral psychology became not only a method for understanding mental processes, but also a tool for shaping them. Radical behaviorists such as Watson believed that any person, given a well-designed environment, could be shaped through behavioral stimuli into becoming a certain kind of person with particular proficiencies and desires. U.S. psychologist B. F. Skinner (1904–1990) demonstrated the primacy of behaviorism in the 1940s and later with a series of experiments showing his concept of operant conditioning, in which certain operations (such as pulling a handle) were reinforced by a positive outcome (such as the release of food), thus leading to a learned behavior. Radical behaviorists believed that this is the basic dynamic of all psychology (and thus should inform education policy). Behaviorism was a very contentious theory because it provoked fears of mind control. In fact, Skinner became a controversial figure in the 1940s when he took his theories to their logical extension and

published a utopian novel, *Walden Two* (1948), arguing for a carefully planned and controlled society.

Both functionalism and behaviorism were U.S. reactions to the introspection approach of Wundt and Titchener. In Europe, another major critique arose, called *Gestalt* psychology. This German word means form, and in psychology it refers to the efforts to reaffirm the importance of a holistic perception (of the whole form rather than its parts), as opposed to breaking down perceptions into elements, as Wundt had proposed. This was in part a reaction against Wundt's atomistic outlook about isolated elements of perception, but also it was an effort to ensure that psychology did not alienate the philosophers, whose approach was more holistic. Because philosophers and psychologists usually belonged to the same academic departments, and philosophers had attempted in 1912 to ban experimental psychologists from academic chairs, some accommodation was wise. Holism was also trendy in Weimar Germany, because it eschewed barren, mechanical outlooks in favor of vitalized, quasi-Romantic views. The key figure in the *Gestalt* movement was Max Wertheimer (1880–1943), the cofounder of the journal *Psychological Research*. In 1924, Wertheimer gave an address explaining *Gestalt* psychology, decrying efforts to conceive the practice of science solely in terms of breaking down complexes into component parts; instead, he argued that some wholes cannot be explained as the sum of their parts, and that *Gestalt* psychology addressed this by probing the mind as a whole, analyzing issues that were inherent only to the whole and not to the parts. Wertheimer and other prominent *Gestalt* psychologists such as Wolfgang Köhler (1887–1967) and Kurt Koffka (1886–1941) continued to be influential, but the *Gestalt* movement never proved as powerful as behaviorism, particularly in the United States. Its lasting contribution was the philosophical one that challenged the conventional atomistic view of science.

See also Freud, Sigmund; Jung, Carl; Mental Health; Mental Retardation; Piaget, Jean; Psychoanalysis; Skinner, Burrhus Frederic; Vygotsky, Lev

References

Ash, Mitchell G. *Gestalt Psychology in German Culture, 1890–1967: Holism and the Quest for Objectivity.* New York: Cambridge University Press, 1985.

Benjafield, John G. *A History of Psychology.* Boston: Allyn & Bacon, 1996.

Ferguson, Kyle E., and William O'Donahue. *The Psychology of B. F. Skinner.* Thousand Oaks, CA: Sage Publications, 2001.

Leahy, Thomas Hardy. *A History of Psychology: Main Currents in Psychological Thought.* Englewood Cliffs, NJ: Prentice-Hall, 1987.

Public Health

Public health describes efforts to keep entire populations in good health; it requires the cooperation of scientists, physicians, and public officials, who together make policies to ensure the health of the general populace. At the turn of the twentieth century, public health was in a period of change owing to advances in bacteriology. In the nineteenth century, public health typically concentrated on controlling infectious disease by the improvement of sanitation in urban areas where diseases such as cholera often had flourished. Such efforts entailed bringing clean water into cities and building sufficient sewage facilities, as well as providing for garbage collection and disposal. The basic assumption of these measures was that dirt caused disease. But in the final decades of the nineteenth century, urban laboratories identified specific bacteria and developed the means to combat individual diseases. Instead of sanitation, public health measures emphasized therapy, using antitoxins developed by scientists in laboratories. For example, boards of health in U.S. cities such as Newark, New Jersey, began to use antitoxins against diphtheria in the late 1890s. Although this benefited many of those suffering from disease, it also de-emphasized the importance of public sanitation and diminished the sense of public responsibility for the state of the urban environment. Boards of health in Europe and North America used the discoveries of German scientists Paul Ehrlich (1854–1915) and Robert Koch (1843–1910), notably Koch's isolation of the bacterium causing tuberculosis, to focus instead on curing specific diseases with specific remedies. These therapeutic remedies would act, Ehrlich famously hoped, as magic bullets, identifying and killing the deadly organism without harming any other part of the victim's body.

Despite the new methods, a more holistic approach to public health did not completely disappear in the twentieth century. This was partly because of the practical difficulty of isolating carriers of disease. Doing so would have required the state to impose its will on seemingly healthy individuals who, according to urine tests, happened to be carrying deadly diseases such as typhoid fever. Naturally those people who were suffering from the illness had to be treated differently than those who were living healthy lives while carrying the microbe. U.S. public health official Charles Chapin (1856–1941), in his book *Sources and Modes of Infection* (1910), argued that other approaches were required, such as training carriers to work in jobs that would be least likely to spread the disease (for example, carriers would avoid working in the food industry).

Public health was often connected to other impulses in society, such as social purification. The eugenics movement, popular in the United States, Britain, and Germany, sought to improve the health of the race as a whole by promoting social reforms to encourage or discourage breeding within the disparate groups of the population. In Germany, such efforts at *Rassenhygiene* (racial hygiene) were not part of an extreme political movement, but were integrated into mainstream public health measures. These were not, as later associations with Nazi Germany might suggest, necessarily designed to promote the Nordic or Aryan ethnic groups; in fact, many eugenicists opposed designing laws that would

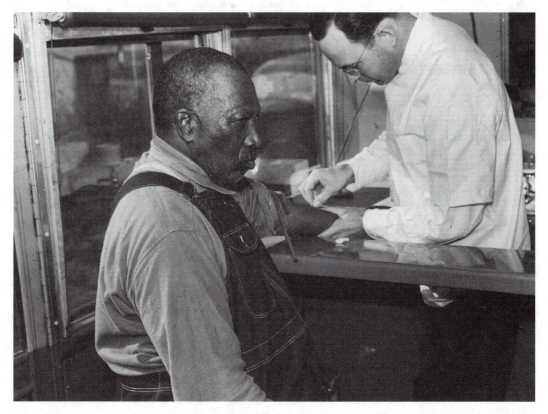

Patient receiving treatment for syphilis in public health service clinic-on-wheels, Wadesboro, North Carolina. (Library of Congress)

shape the nation exclusively to favor these groups. Instead, they hoped to promote measures to encourage breeding among the most socially or culturally productive people, which made eugenics in Germany (as in the case of the United States and Britain) primarily a class-based social reform effort. However, by the 1930s, eugenics in Germany acquired a strong racial basis. Aside from efforts to limit the influence of Jews and isolate them (and later, to murder them on a massive scale), Germany under Adolf Hitler (1889–1945) concerned itself with broad public health issues. Hitler, himself a vegetarian, tried to curb smoking and to eliminate environmental causes of cancer, with mixed successes. His government's efforts to quarantine and eliminate Jews—a racial hygiene measure—was closely tied to his outlook about public health in general.

Some of the deadliest diseases that menaced public health until the development of penicillin were venereal diseases. No "magic bullet" was discovered for diseases such as syphilis, and the problem of controlling its spread was compounded by the fact that it had many negative social connotations. In the United States, Progressive-era public health officials attempted to combat the disease by closing down brothels, arresting prostitutes, and instigating propaganda campaigns against vice (the last was deemed necessary to persuade U.S. soldiers not to venture into French brothels during World War I). Venereal diseases were connected to sin, as products of sexual promiscuity, and often the victims were too ashamed to admit they had the disease and were not treated. The most effective treatment for syphilis was a chemical therapy called Salvarsan. But ostensibly for public health reasons, this treatment was not always used. A notorious example occurred between 1932 and 1972, when a group of African American men in Tuskegee, Alabama, became unwitting participants in an experiment sponsored in part by the U.S. Public Health Service to observe the effects of untreated syphilis. None of them were informed that they had the disease, but were instead told that they had bad blood; they were not treated, either, even after the development of penicillin.

The greatest threats to public health, namely, infectious diseases, were curbed dramatically during World War II. British and U.S. soldiers benefited from the development of penicillin, which was an uncommonly effective killer of bacteria. Although discovered before the war, penicillin began to be mass-produced in the United States in the 1940s and helped to eliminate a host of diseases. In the postwar period, penicillin and other antibiotics became the most effective and commonly implemented weapons against diseases, to prevent widespread breakdowns in public health.

See also Ehrlich, Paul; Medicine; Nazi Science; Patronage; Penicillin; Venereal Disease

References
Brandt, Allan M. *No Magic Bullet: A Social History of Venereal Disease in the United States since 1880.* New York: Oxford University Press, 1985.
Galishoff, Stuart. *Safeguarding the Public Health: Newark, 1895–1918.* Westport, CT: Greenwood Press, 1975.
Jones, James H. *Bad Blood: The Tuskegee Syphilis Experiment.* New York: The Free Press, 1981.
Leavitt, Judith Walzer. "'Typhoid Mary' Strikes Back: Bacteriological Theory and Practice in Early Twentieth-Century Public Health." *Isis* 83:4 (1992): 608–629.
Proctor, Robert N. *The Nazi War on Cancer.* Princeton, NJ: Princeton University Press, 1999.
Rosen, George. *A History of Public Health.* New York: MD Publications, 1958.
Weiss, Sheila Faith. "The Racial Hygiene Movement in Germany." *Osiris* 3 (1987): 193–236.

Q

Quantum Mechanics

Because Max Planck's (1858–1947) quantum theory revised the notion of energy, by describing it in chunks or packets rather than a continuous stream, theoretical physicists tried to develop an entirely new system of mechanics. The development of quantum mechanics in the 1920s, despite being less well known than the theories of relativity, was the most fundamental innovation in physical science in the first half of the twentieth century, because of the establishment of a new system of physics and the construction of a philosophical worldview that appeared to deny the possibility of a complete understanding of reality.

One of the principal figures in the development of the new mechanics was German physicist Werner Heisenberg (1901–1976). His version of quantum mechanics, often called matrix mechanics, was expressed in equations and emphasized mathematical relationships. Heisenberg's matrix mechanics described the natural world in terms of statistics and probabilities. It had some curious characteristics, such as the fact that the traditional notion of commutative multiplication did not appear to apply (in matrix mechanics, the results of multiplication depends on the order in which two variables are multiplied). The

details of matrix mechanics were worked out, beginning in 1925, not only by Heisenberg, but also by German colleagues such as Max Born (1882–1970) and by the Englishman Paul Dirac (1902–1984), who expressed quantum relationships in a more coherent mathematical formulation than did the others.

Newtonian mechanics appeared to be a special case of the more inclusive quantum mechanics. Isaac Newton's (1642–1727) laws did not apply to the subatomic scale. Knowledge of the velocity of an electron, for example, would give an imperfect picture of where its location might be over time, because motion occurs in "jumps." Thus precise, deterministic notions of cause and effect were cast aside in favor of probabilistic equations. In 1927, Heisenberg added to his quantum mechanics the principal of uncertainty, which extended the lack of determinism to the present. Not only could physicists not accurately predict future events, they also could not precisely understand present conditions. Because light itself carries inertia, it is impossible to construct an experiment in which human beings can observe nature at the subatomic scale. Light photons, needed for human observation, would disrupt the experiment by colliding with electrons and kicking them. The uncertainty principle noted that the crucial variables in quantum mechanics, such as the position and momentum of electrons, could never be known simultaneously. The more one knew about position, for example, the less one knew about momentum, and vice versa. No precise picture of present reality can be determined (it is also called the indeterminacy principle).

Matrix mechanics dissatisfied many physicists because of its abstract, mathematical character. It was a description of mechanics that appeared to represent nothing real. Austrian physicist Erwin Schrödinger (1887–1961) proposed instead a new mechanics, called wave mechanics, which immediately enticed other physicists. Here was a description of energy and matter that was proposed in comprehensible terms, using a construct (waves) to which physicists were already accustomed. It built on the notion, proposed by French physicist Louis De Broglie (1892–1987), that electrons might behave as waves, just as light waves appeared—according to Albert Einstein's (1879–1955) notion of photons—to act like particles. Schrödinger was part of the older generation and did this remarkable work in his late thirties. Although wave mechanics certainly broke with traditional physics, it was conservative in the sense that it was expressed in terms that could be visualized.

The two pictures of mechanics, matrix mechanics and wave mechanics, appeared to be opposed fundamentally. Matrix mechanics built upon Einstein's notion that light acted like a particle, and all the mathematical variables were stated as if they were particles (using variables such as momentum and position). Wave mechanics, by contrast, was based on the idea that all matter and energy behave as waves, not particles. More startling than the contradiction, however, was the fact that both descriptions of mechanics equally appeared to be capable of describing the quantum world. Neither theory was clearly superior to the other. To Heisen-

berg's disdain, many physicists preferred Schrödinger's formulation because wave mechanics at least had the merit that it could be visualized.

The "Copenhagen interpretation" of quantum mechanics, named because it was dominated by the views of Danish physicist Niels Bohr (1885–1962), tried to reconcile the two systems of mechanics. Bohr proposed, in what was called the principle of complementarity, that each was equally valid. Neither was completely correct, but together they could explain various phenomena. Electrons appeared to have both properties, but never at the same time. The two outlooks complemented each other. Like Heisenberg's uncertainty principle, Bohr's principle appeared to challenge the most fundamental notions of science. It required physicists to accept the counterintuitive idea that matter has a dual nature.

Bohr and Heisenberg came to the conclusion that the principles of uncertainty and complementarity were closely related. Uncertainty was based on the idea that more precise knowledge of an electron's position gave less precise knowledge of its momentum, and vice versa. Similarly, waves do not themselves have an exact position, but they do have momentum. In this sense, the more one knew about the wave properties of matter, the less clear the particle notion became, and vice versa. In fact, the wave and particle natures would never—and could never—be observed at the same time. What we observe in nature, Bohr asserted, is simply an answer to a question posed by ourselves, based on our conceptions. If the experiment is designed based on notions of waves, we will see waves. If we are looking for particles, we will see particles. Bohr went on to conclude that it is meaningless to describe reality as it exists separately from human experiment. This interpretation provoked sharp disagreement among physicists and philosophers, because it appeared to abandon the idea that theories could be developed to describe physical reality.

See also Bohr, Niels; Determinism; Heisenberg, Werner; Physics; Quantum Theory; Schrödinger, Erwin; Uncertainty Principle

References

Gribbin, John. *In Search of Schrödinger's Cat: Quantum Physics and Reality.* New York: Bantam, 1984.

Hendry, John. *The Creation of Quantum Mechanics and the Bohr-Pauli Dialogue.* Dordrecht: D. Reidel, 1984.

Jammer, Max. *The Conceptual Development of Quantum Mechanics.* New York: McGraw-Hill, 1966.

Kragh, Helge. *Quantum Generations: A History of Physics in the Twentieth Century.* Princeton, NJ: Princeton University Press, 1999.

Quantum Theory

Quantum theory was first formulated in 1900 by German physicist Max Planck (1858–1947). Although Planck was seeking a way to consolidate classical physics with the laws of thermodynamics developed in the nineteenth century, he in fact discovered a phenomenon about the nature of energy that challenged the foundations of classical physics. The ideas of quantum physics transformed physicists' understanding of nature, leading (among many things) to a comprehensible model of atomic structure, to a new system of mechanics, and to atomic fission—and thus to nuclear weapons and energy.

The quantum initially was proposed as a solution to the blackbody problem in physics. A blackbody is a solid that, when heated, emits light that will produce a perfect spectrum. According to classical physics, in which the amount of energy released is proportional to frequency of radiation, higher frequencies would lead to extraordinarily high energy releases in the ultraviolet portion of the spectrum (where frequencies were high). But here theory did agree with observation, because such levels of energy did not really occur. This was often referred to as the ultraviolet catastrophe, a theoretical conundrum in which vast amounts of energy were ex-

pected yet none were actually present. To avoid the catastrophe—that is, to develop a theoretical explanation that agreed with observation—Planck employed a mathematical device, a constant. In Planck's formulation, energy (E) was proportional to frequency (v), multiplied by a very small constant (h). His equation $E = hv$ represented that relationship. The constant, h, would ensure that energy itself would always be a multiple of a small, indivisible amount. In other words, Planck wished to avoid allowing nature to release energy continuously.

Planck's mathematical solution, if taken as a measure of reality and not simply as a device to fix a mathematical equation, pointed to a significant ramification: Energy is not continuous, but is instead separated into discrete, indivisible chunks, or packets. Planck's constant, h, required a reformulation of some of the fundamental precepts of classical physics, taking into account the discontinuous nature of the new "quantum" physics. One of the first to realize the consequences was Albert Einstein (1879–1955), who assessed Planck's innovation as one of the most important in physics. In 1905, Einstein used quantum theory to formulate his own conception of light quanta, known as photons. Conceptualizing light as broken up into photons, Einstein determined that light carries inertia and thus must be equivalent to a certain amount of mass. His recognition of this relationship, impossible without the advent of quantum theory, formed the basis of his famous equation, $E = mc^2$, which later became the theoretical underpinning of the development of nuclear weapons and energy.

An application of quantum theory to an outstanding problem left unsolved by classical physics was made by Danish physicist Niels Bohr (1885–1962) in 1913. Since the discovery of radioactivity in the last years of the nineteenth century, scientists had debated the structure of the atom itself. A new model of the atom was proposed by Ernest

Quantum theory was first formulated in 1900 by German physicist Max Planck, pictured here. (Library of Congress)

Rutherford (1871–1937), who suggested the existence of an atomic nucleus in 1911. He formulated a "planetary" model with a tiny nucleus surrounded by orbiting electrons. The outstanding critique of this model was that these electrons should release energy continuously as they moved in orbit, and the electrons themselves would simply collapse into the nucleus. Bohr "saved" the Rutherford atom by noting that with quantum physics, the atom only radiates energy when an electron shifts its orbit. The orbits, arranged like layers of a shell surrounding the nucleus, were separated by quantum distances. When an electron moves closer to the nucleus, it shifts quantum orbits, and the atom releases energy.

As the atom gains energy (as by heating), the electron is obliged to shift to an outer orbit to accommodate the energy. The electron will come no nearer to the nucleus than the innermost quantum orbit, thus disallowing collapse even in atoms that have released the vast majority of their energy.

Bohr's work firmly established both quantum theory and the Rutherford atom, and he was recognized for his efforts by being awarded the Nobel Prize in Physics in 1922. But quantum theory was far from complete, and far from widely accepted. One of the principal ways in which the idea spread was through international conferences, particularly those sponsored by Belgian industrialist Ernest Solvay (1838–1922). Planck and

Einstein became the principal figures in a new brand of theoretical physics, although the connections between quantum theory and Einstein's theories of relativity were rarely clear, and the two occasionally sparred intellectually. Physicists recognized that a new system of mechanics would be required to replace classical physics, because classical physics assumed that energy was continuous. Accommodating discontinuity at the subatomic scale became the principal concern of those physicists in the 1920s who attempted to develop quantum mechanics.

See also Atomic Structure; Bohr, Niels; Physics; Planck, Max; Quantum Mechanics; Solvay Conferences

References

Bohr, Niels. "The Structure of the Atom" (Nobel Lecture, 11 December 1922). In *Nobel Lectures, Physics, 1922–1941*. Amsterdam: Elsevier, 1964–1970, 7–43.

Heilbron, J. L. *The Dilemmas of an Upright Man: Max Planck and the Fortunes of German Sciences.* Cambridge, MA: Harvard University Press, 2000.

Kragh, Helge. *Quantum Generations: A History of Physics in the Twentieth Century.* Princeton, NJ: Princeton University Press, 1999.

R

Race

Although racism has existed for centuries, the nineteenth century saw new disciplines such as anthropology and ethnology providing systematic "scientific" studies that seemed to reinforce racism. Such studies included measuring skulls, weighing brains, and describing facial shapes. By categorizing the world's races, either hierarchically or according to some other index, one could differentiate between them and assign superiority and inferiority to them. Despite the importance of Darwinian evolution—which denied hierarchy—these conceptions proved powerful until the middle of the twentieth century and beyond. The widespread belief among whites in their own racial superiority was, in the first half of the twentieth century, supported by pronouncements of white scientists.

There were some exceptions to this rule. Celebrated U.S. anthropologist Franz Boas (1858–1942), for example, spoke out strongly against stereotypes of Africans and against widespread fears of racial corruption from immigration. In 1906, he gave an address on achievements by Africans at Atlanta University, a speech that was an inspiration to African American intellectual W. E. B. Du Bois (1868–1963). Yet although Boas was an egalitarian politically, he felt constrained by his belief, based on nineteenth-century findings, that there were indeed inequalities in mental capacity based on undeniable physical differences. Still, he was opposed to the idea prevalent among Americans that the influx of immigrants was a challenge to pure racial stocks, and he pointed out that there had been plenty of mixing among races in Europe and elsewhere before they ever got to the United States, and that the existence of "pure" types in Europe was a myth. The threat to racial purity in the United States, he said, was imaginary.

Despite the efforts of Boas and others to dispel racist attitudes, scientific racism was reinforced in a number of ways. One of them was intelligence testing. The intelligence quotient (IQ) tests, developed by French psychologist Alfred Binet (1857–1911) to identify children in need of special attention, were used by U.S. scientists to gauge the comparative intelligence of entire peoples. In the 1910s, Henry Herbert Goddard (1866–1957) drew some far-reaching implications from his IQ studies of recent immigrants to Ellis Island. He concluded that the vast majority of immigrants from Eastern Europe, many of them Jews, were "feeble-minded." Were these people racially inferior to native-born Americans whose ancestry came from western and northern parts of Europe? Although today scientists would emphasize the importance of education, language ability, and general familiarity with the nature of the test (for example, some of the subjects had never before held a pencil), at the time the results appeared to present a scientific finding of the utmost importance. Even given the tenuous nature of the conclusions, IQ tests lent credence to attitudes that these new immigrants were of an inferior racial type.

Eugenics appeared to provide powerful evidence that some peoples were inferior, and it provided a method for stopping the negative effects of inferior races on large populations. Karl Pearson (1857–1936), who was a leading proponent of population studies and was a Darwinist, believed that there was plenty of room in evolution for the existence of lower races, even if one abandoned the concept of hierarchy. Eugenics was a movement that emphasized good breeding, urging not only scientific study, but also an entire social program designed to eliminate elements that corrupted society as a whole. Pearson and others believed that such policies would increase the average character of the race as a whole, making it more likely that new babies would have desirable traits. The population could be improved, Pearson argued, by preventing undesirable elements of society from procreating. Eugenics laws were enacted in several countries, including the United States and Britain, to compel sterilization of those believed to pose a risk to the population. These typically did not target races, however, but rather people with mental retardation.

Eugenics and social Darwinism posed a paradoxical vision of the world that was both stark and hopeful. Social Darwinists took Charles Darwin's (1809–1882) nineteenth-century concept of natural selection and applied it to human societies, insisting on racial competition for the world's resources. In addition to this brutal competition, eugenics offered the promise of racial improvement, to increase the likelihood that one's race could compete effectively. Both ideas informed the thinking of leading scientists and social reformers in Europe and North America, but they saw their clearest and most destructive manifestation in the policies of Nazi Germany. Nazis such as Adolf Hitler (1889–1945) believed that Jews were corrupting the health of the German people as a whole, presenting a challenge to their ability to compete with other nations of the world. In an effort to bar Jews from full participation in German society, the Nuremberg Laws, passed in 1935, created new requirements for German citizenship along racial lines. These laws denied the rights of Jews to vote or hold public office and forbade German citizens from marrying Jews or having any sexual relations with them; the laws also forbade Jews from employing female household staff, all in the name of protecting German blood and honor.

Toward the late 1930s, two developments challenged prevailing views about racial categories. One was the acceptance in the 1920s of a new synthesis of Mendelian genetics and Darwinian natural selection—the modern synthesis, it has been called. Darwin's strongest critics in the scientific community now found a way to accommodate his ideas, and the idea of racial hierarchy (along with the idea of a chain of being that puts all living organisms on a ladder of superiority and inferiority) seemed more spurious than ever. The other cause of the change was the appalling racial laws and harsh treatment of Jews in Nazi Germany, which some believed to be an extreme result of an excess zeal for "scientific" racial theories. After World War II, the American Jewish Congress used new findings from the social sciences to attempt to end discrimination in the United States. But the uncertainties of research on race made this difficult; still, drawing on such efforts and the more-successful efforts of the National Association for the Advancement of Colored People, the Supreme Court ruled against racial segregation in the landmark 1954 case *Brown v. Board of Education*. Yet despite these events, the idea that one's attitudes about racial superiority were supported by science and that belief in absolute equality was a naive wish continued well into the second half of the twentieth century.

See also Anthropology; Boas, Franz; Eugenics; Intelligence Testing; Just, Ernest Everett; Nazi Science; Public Health

References

Boas, Franz. "Race Problems in America." *Science,* New Series, 29:752 (28 May 1909): 839–849.

Gould, Stephen Jay. *The Mismeasure of Man.* New York: W. W. Norton, 1996.

Jackson, John P., Jr. "Blind Law and Powerless Science: The American Jewish Congress, the NAACP, and the Scientific Case against Discrimination, 1945–1950." *Isis* 91:1 (2000): 89–116.

Noakes, Jeremy, and Geoffrey Pridham. *Documents on Nazism, 1919–1945.* New York: Viking Press, 1974.

Stepan, Nancy. *The Idea of Race in Science: Great Britain, 1800–1960.* Hamden, CT: Archon, 1982.

Williams, Vernon J., Jr. *Rethinking Race: Franz Boas and His Contemporaries.* Lexington: University Press of Kentucky, 1996.

Radar

Although the atomic bomb was the best-known scientific accomplishment of World War II, radar was far more decisive in winning the war itself. Radar was used to determine the distance and direction of unseen objects by the reflection of radio waves. The name means "radio detection and ranging." Unlike the atomic bomb, radar technology existed prior to the war and already had been developed for military use, using radio waves that were more than a meter long. The possibilities of using radio waves for detection had been predicted by Guglielmo Marconi (1874–1937), the pioneer of communication technology.

Workable radar systems were pursued prior to and during World War II by at least eight countries: the United States, Britain, France, Germany, the Soviet Union, the Netherlands, Italy, and Japan. The Naval Research Laboratory was one of a few U.S. institutions that developed long-wave radar in the 1930s. British physicist Robert Watson-Watt (1892–1973) typically receives credit for the "invention" of radar, because in 1935 he proposed detecting aircraft with radio methods, and his work led to the British "Chain Home" system of stations that were built to detect enemy aircraft, installed just prior to the outbreak of war. In Britain, interest in radar largely came from the Royal Air Force as it expanded, prompted by the efforts of physicist Sir Henry Tizard (1885–1959). He and other physicists, notably Edward G. Bowen (1911–1991), began an intensive effort in 1935 to use radar technology for detecting airplanes. In 1936 at Bawsey Manor, a British team developed the first successful airborne detection system. Improvements over the next several years were interrupted by the war in 1939. The team repeatedly was obliged to change headquarters, and the future of this militarily important technology became uncertain. The radar system put into place prior to the war helped ensure British success in the summer of 1940 during the Battle of Britain, the struggle for air superiority between the German Luftwaffe and the Royal Air Force.

Also in the summer of 1940, British prime minister Winston Churchill (1874–1965) sent a group of defense scientists to Canada and the United States to promote the idea of scientific and technical cooperation. This team was led by Tizard, who visited numerous high-level American scientists and administrators. One of them was the wealthy financier Alfred L. Loomis (1887–1975), who was well aware that radar technology could prove very useful should the United States enter the war. At the time, most radar sets operated with long wavelengths of a meter or more. British scientists had been developing radar on the scale of centimeters. Also called "microwave" systems because the wavelengths were much shorter, centimetric radar promised to be more accurate and required a smaller apparatus. It would be ideal for locating aircraft and differentiating between targets. When the Tizard Mission arrived, Loomis's group realized that the British secretly had developed precisely what the Americans lacked, namely, a device capable of generating microwaves of sufficient energy. In Loomis's estimation, this device, called the cavity magnetron, saved the U.S. group two years of work. In exchange, the Americans helped the British to develop the magnetron into a full-scale airborne intercept system to defend against German planes in Britain.

The Tizard Mission began a fruitful cooperation between the British and the Americans, and a U.S. radar project was established under the National Defense Research Committee. Centered at the falsely named Radiation Laboratory (Rad Lab) of the Massachusetts Institute of Technology, the radar project broke away from the U.S. military's long-wave system and concentrated on microwaves. The U.S. team developed a working air-to-surface system in August 1941, to help planes detect surface ships (and submarines when coming up for air). The Navy adapted this system for use aboard their destroyers. Once the United States entered the war in late 1941, both the Navy and the Air Corps were using the Rad Lab's system. This technology proved decisive in turning the tide of war at sea, where German submarines initially had reigned supreme.

Physicists during the war gradually were redirected to other projects, notably the effort to build an atomic bomb, but radar had proven its worth and was developed vigorously throughout the war. It was used far more effectively in the Atlantic theater of operations than in the Pacific, where the U.S. fleet initially was equipped with only the longer-wave systems. When the fighting ended, research on radar continued and revolutionized other scientific fields. For example, it stimulated the growth of radio astronomy, the study of nuclear magnetic resonance, and the development of transistors.

See also Marconi, Guglielmo; Patronage; World War II

References

Bowen, E. G. *Radar Days*. Bristol: Adam Hilger, 1987.

Brown, Louis. *A Radar History of World War II: Technical and Military Imperatives*. Philadelphia, PA: Institute for Physics Publishing, 1999.

Buderi, Robert. *The Invention that Changed the World: How a Small Group of Radar Pioneers Won the Second World War and Launched a Technological Revolution*. New York: Simon & Schuster, 1996.

Guerlac, Henry E. *Radar in World War II*. New York: American Institute of Physics, 1987.

Kevles, Daniel J. *The Physicists: The History of a Scientific Community in Modern America*. Cambridge, MA: Harvard University Press, 1995.

Radiation Protection

The discovery of X-rays and radioactivity at the close of the nineteenth century seemed not only to revolutionize physics but also to aid in medical technology. But soon it became clear that too much radiation could be harmful, as doctors who routinely were exposed to X-rays developed debilitating illnesses. Marie Curie's (1867–1934) isolation of radium produced a fascination with the mysterious glowing substance; radium was painted onto watch dials to glow in the dark, and charlatans mixed radium into tonics and marketed them as miracle drugs for wealthy but gullible buyers. In the 1920s, a group of watch dial painters, soon dubbed the radium girls, developed cancers in their mouths, having moistened the tips of their paintbrushes with their own saliva. A radium elixir, Radithor, killed a wealthy socialite named Eben M. Byers (1880–1932) in 1932.

Radiation was harmful, scientists soon concluded, whenever it had enough energy to ionize atoms in the material it contacts. There are many kinds of radiation, including ordinary light; scientists were concerned only with ionizing radiation, such as X-rays or the products of radioactive decay, which clearly had serious but poorly understood biological effects. The widespread use of X-rays during World War I led British scientists to form a radiation protection committee in 1921; its goal was to educate the public and all users of radiation-producing instruments about the potential hazards. In 1928, these and other scientists established the first international body to advise on—not to regulate—the levels of radiation that might be deemed safe, the International X-Ray and Radiation Protection Committee. The following year, U.S. scientists formed a national counterpart, the Advisory Committee on X-Ray and Radium Protection. Headed by physicist Lauristen S. Taylor (1902–), who was the U.S. representative to the international committee, the U.S. body closely reflected the international consensus.

The problem of radiation protection rose to a high level in the Manhattan Project during World War II, when it became clear that the world increasingly was going to rely on nuclear weapons and nuclear power, meaning a host of problems related to radiation protection. In the United States, a new breed of expert arose called the health physicist, whose knowledge had to include genetic and somatic effects of radiation, necessitating a broad understanding of physics, biology, and even politics. These experts worked in the national laboratories, first built to serve nuclear production facilities in the United States; soon such experts appeared in other countries with nuclear ambitions, such as the Soviet Union, Britain, and France.

Although U.S. standards for radiation protection had followed international recommendations prior to the war, the reverse was true in the postwar period. The Americans broadened and renamed their committee the National Committee on Radiation Protection in 1946; the international body, renamed the International Commission on Radiological Protection, followed the Americans' lead. The NCRP worked closely with the Atomic Energy Commission (AEC), and, despite declarations that the two bodies worked independently, individuals in one often served in the other. The two bodies came into conflict often in the postwar era, as in the case of the AEC's efforts to review a 1947 report about the harmful effects of medical X-rays, prior to publication. Although the AEC ostensibly was concerned that the NCRP might inadvertently disclose classified information, it was also hoping to avoid publication of statements that might alarm the public about its practices.

The NCRP revised the terminology of radiation protection to reflect changes in attitudes. For example, geneticists asserted that any amount of radiation would cause genetic mutations to occur, and these mutant genes would be inherited from one generation to the next. Although somatic (bodily) damage might not be discernable, the genetic effects of radiation exposure could threaten human descendants. Thus the NCRP concluded that no amount of radiation was absolutely harmless, and the recommended thresholds of exposure were no longer called *tolerance doses.* That term had implied that there was a threshold of exposure, below which people could consider themselves safe. Instead, the NCRP adopted the term *permissible dose,* a decision that admitted harmful effects and treated exposure as a matter of balancing risks. By the late 1940s, the NCRP established a new weekly permissible dose based on exposure to vulnerable body tissue such as those of the gonads and eye lenses.

Defining permissible doses became controversial because various groups—the AEC, the NCRP, the ICRP, and others—rarely agreed on the technical parameters required to ensure safety and never agreed on whether the statistical costs of biological damage were worth the benefits of maintaining U.S. leadership in an arms race with the Soviet Union. Mistrust of the government struck many leaders as misplaced, because the whole purpose of having nuclear weapons and developing nuclear power was for the greater good of the nation. Yet the fears proved well founded. In the 1990s the U.S. government acknowledged that, in its efforts to identify safe levels of radiation exposure, the AEC secretly had injected unknowing patients with plutonium in the late 1940s. Patients in hospitals in New York, Illinois, California, and Tennessee became unwitting experimental subjects. The AEC scientists believed none of the patients were at risk of getting cancer, because they chose patients who were near death anyway. In most cases they were correct, but to the AEC's surprise, many of the patients lived for many years. The AEC also conducted experiments in which it fed radioactive iron or calcium to children at a school for the mentally retarded. The consent forms misled parents to believe that the experiments might improve their children's condition, but in reality they were part of a larger effort to measure how the human body absorbs dangerous radiation.

See also Atomic Energy Commission; Cold War; Public Health; Radioactivity; X-rays

References

Badash, Lawrence. *Radioactivity in America: Growth and Decay of a Science.* Baltimore, MD: The Johns Hopkins University Press, 1979.

Clark, Claudia. *Radium Girls: Women and Industrial Health Reform, 1910–1935.* Chapel Hill: University of North Carolina Press, 1997.

Walker, J. Samuel. *Permissible Dose: A History of Radiation Protection in the Twentieth Century.* Berkeley: University of California Press, 2000.

Welsome, Eileen. *The Plutonium Files: America's Secret Medical Experiments in the Cold War.* New York: Dell, 1999.

Radio Astronomy

Radio astronomy changed the way that scientists could peer into the heavens. In past centuries, they relied on light, which could be enhanced through the use of telescopes and observed with the naked eye. But other nonvisible sources of radiation were emitted by objects in space, and in the 1930s, a few individual scientists discovered this radiation and realized that receivers could be built to record it. For the first time in human history, astronomers could expand knowledge of the universe using a phenomenon other than light and using tools other than the human eye.

Karl G. Jansky (1905–1950) discovered the fundamental radio astronomical technique in 1932. Since 1928, he had been helping a major company study a practical problem: the sources of radio interference in the atmosphere and the effects on long-distance communication. While conducting research at one of the stations of Bell Telephone Laboratories in New Jersey, Jansky discerned that some of the signals received could only originate from beyond earth's atmosphere. Not only that, but these signals appeared to come from fixed points somewhere in space. Although he could not pinpoint the location with much precision, Jansky believed that the signals came from somewhere near the center of the Milky Way, the same galaxy to which the sun (and the earth) belong.

In subsequent years, radio astronomy did not become a major field of inquiry. However, some researchers continued to explore its unknown characteristics. Amateur astronomer Grote Reber (1911–2002), working from his own equipment at his home in Wheaton, Illinois, provided further evidence that the center of the Milky Way was a major source of radio waves, causing what then was simply called static interference. Although Reber was not part of the academic community (during the Great Depression such jobs were few), his findings had a widespread influence. In the Netherlands, the eminent astronomer Jan Oort (1900–1992) took notice of his work and determined that radio sources could help scientists "see" farther into the galaxy, beyond the points at which further resolution of light sources had proved impossible. Increasingly astronomers realized that some phenomena in space left radio residues that had no visual representation and could only be studied with radio astronomy.

Radio astronomy's development after World War II was due largely to technological innovations during the war years. New electronic components, invented for other purposes such as radar equipment, allowed scientists to improve on the receivers used by Jansky and Reber. Radio astronomy soon

Astronomer Professor Bernard Lovell at work at his desk at Jodrell Bank Experimental Station, with a view of part of a radio telescope through his window. (Hulton-Deutsch Collection / Corbis)

became a major new field of science; instead of relying on telescopes, postwar researchers could also use radio waves as a technique to learn the properties of distant star systems. Soon theoretical predictions helped to solidify the undertaking, as when Dutch astronomer Hendrik C. Van de Hulst (1918–2000), one of Oort's students, predicted in 1945 that scientists should be able to discern hydrogen radiation of a certain wavelength (roughly 21 centimeters). When

that radiation was detected as predicted, in 1951 by Harvard University scientists Harold I. Ewan and Edward M. Purcell (1912–1997), radio astronomy took a serious step toward turning galactic "noise" into specific, identifiable signals.

Scientists in several countries began to build the first radio telescopes. British astronomer Alfred Charles Bernard Lovell (1913–) played a major role in developing a community of radio astronomers at Jodrell

Bank Observatory (in England), which soon became a major center for radio astronomy. In 1947, Jodrell Bank built the largest radio telescope in the world, a 218-foot parabolic reflector. New radio telescopes were built in numerous locations, the most powerful in Britain, the Netherlands, Australia, and the United States; these were the countries in which the strongest communities of radio astronomers emerged. The Soviet Union also pursued radio astronomy, but its activities were largely unknown to the rest of the world. The 1950s would see a surge in activity in radio astronomy, culminating in major institutions devoted to its study, such as the National Radio Astronomy Observatory in the United States.

See also Astronomical Observatories; Oort, Jan Hendrik; Patronage; Radar

References
Edge, David O., and Michael J. Mulkay, *Astronomy Transformed: The Emergence of Radio Astronomy in Britain.* New York: John Wiley & Sons, 1976.
Emberson, Richard M. "National Radio Astronomy Observatory." *Science* 130:3385 (13 November 1959): 1307–1318.
Malphrus, Benjamin K. *The History of Radio Astronomy and the National Radio Astronomy Observatory: Evolution toward Big Science.* Malabar, FL: Krieger Publishing, 1996.
Sullivan, W. T., III, ed. *The Early Years of Radio Astronomy: Reflection Fifty Years after Jansky's Discovery.* Cambridge: Cambridge University Press, 1984.

Radioactivity

Radioactivity was discovered in 1896 by French physicist Henri Becquerel (1852–1908), who had been inspired by German physicist Wilhelm Röntgen's (1845–1923) discovery of X-rays. X-rays were created with cathode-ray tubes, using electrical discharge. Becquerel's studies revealed that certain materials, such as uranium, were capable of blackening photographic plates like X-rays, but without the presence of an electrical charge or even the influence of the sun. Thus there was something inherent in the

material itself that was emitting some kind of radiation. At first these were called uranium rays, because they were thought to be analogous to the recently discovered X-rays. The term *radioactivity* was coined shortly afterward by Marie Curie (1867–1934), whose research with her husband, Pierre, yielded the discovery of new elements that also were radioactive, and it was soon clear that X-rays and uranium rays were quite distinct phenomena. Polonium was the first of these new elements, which Marie Curie named for her homeland, Poland, followed by radium. Curie's methods were chemical; she laboriously separated these elements from the radioactive ore, pitchblende, and measured atomic weight in order to confirm that indeed she had found previously unknown elements.

The nature of these strange "radiations" was explored further in the first years of the twentieth century. Intensive study of radioactivity became possible in 1898 when New Zealander Ernest Rutherford (1871–1937) isolated two different kinds of "radiations" that appeared to be ejected from radioactive substances: alpha and beta radiations. Gamma radiations, a form of electromagnetic radiation, were discovered by Paul Villard (1860–1934) in 1900. With more specific knowledge of what seemed to be ejected from these substances, Rutherford tried to incorporate beta and alpha radiations—which he believed to be particles—into the prevailing model of the atom. Eventually he developed a new model based on the existence of electrons (beta particles) orbiting a nucleus.

Rutherford and British chemist Frederick Soddy (1877–1956) believed that the emissions of alpha and beta particles accompanied something more significant. The loss of these particles, they argued, was causing the transformation of elements. These two, along with U.S. scientist Bertram Boltwood (1870–1927) and others, attempted to identify the relationships among all of the known radioelements, substances that exhibited radioactive properties. This required new concepts. For example, radioelements lost alpha or beta particles, and thus "decayed," forming a different element. When this should occur seemed uncertain, but each of the radioelements appeared to have a half-life, the amount of time during which about half of the atoms in a given sample would undergo decay. Thus radioactivity was based on statistical and probabilistic conceptions, challenging the deterministic principle that one could identify precisely what a given atom will do at any given time. In 1911, Soddy published his identification of all the radioelements, in their proper order of decay (there were three major decays series, beginning with the radioelements uranium I, thorium, and actinouranium).

What made these researches remarkable was the implication that radioactivity was something akin to alchemy, the age-old effort to find a means to turn base metals such as lead and copper into precious ones such as silver and gold. Radioactivity was not merely the ejection of particles, but the transmutation of elements. When an isotope of uranium ejected an alpha particle, it ceased to be uranium; now, it was an isotope of thorium. Different isotopes of thorium became protactinium when they ejected beta particles, or radium when they ejected alpha particles. Radioelements in nature were clumped together, in ores called pitchblende. These pitchblende ores also contained lead, a nonradioactive element that scientists suspected was the final, stable end-stage of radioactive decay. Ironically, radioactivity reversed the efforts of the ancients: Nature was trying to turn elements into lead. The task of identifying the precise stages of this process and the reasons behind them, that is, discovering the mechanism of the "decay series," became a principal activity of radioactivity studies.

The problem of identifying the stages was compounded by the fact that few of the radioelements observed had even been identified chemically. The work that connected these radioelements to chemical elements in the periodic table was accomplished by Polish chemist Kasimir Fajans (1887–1975) and Frederick Soddy in February 1913. Although Soddy built upon Fajans's work, which was published first, it was Soddy alone who later (1921) won the Nobel Prize in Chemistry for his work on the radioelements. Margaret Todd, a colleague of Soddy's, coined the term *isotope* to describe such substances that are chemically identifiable as certain elements (same atomic number) but that have slightly different atomic weights. Soddy's principles about the isotopes, known as the group displacement laws, set forth rules for the transformation of elements—their displacement on the periodic table of elements. By 1913 it was clear to chemists and physicists alike that the crucial distinction was the charge of the atom, rather than just its atomic weight, which the concept of isotopes had demonstrated was somewhat more flexible than previously believed. The loss of an alpha particle, for example, would result in a shift two places lower on the periodic table; the loss of a beta particle would result in a shift one place higher. Such changes affect the charge of the atom, not merely the atom's weight.

After the discovery of the group displacement laws, radioactivity was used primarily as a tool for investigating the interior of the atom. Radioactive substances acted as ideal firing mechanisms with which scientists could bombard other atoms with alpha and beta particles. The important work in atomic physics in the 1920s and after the discovery of the neutron in the early 1930s was made possible through the use of radioactive substances. Yet the chemical study of radioactivity itself languished; as historian Lawrence Badash has noted, the early successes of Becquerel, the Curies, Rutherford, and Soddy were astounding, but the field had a "suicidal success." All of its main problems appeared solved after the development of the group displacement laws, and radioactive substances simply were appropriated as tools for other fields of inquiry.

See also Becquerel, Henri; Boltwood, Bertram; Curie, Marie; Radiation Protection; Rutherford, Ernest; Uranium; X-rays

References

Badash, Lawrence. *Radioactivity in America: Growth and Decay of a Science.* Baltimore, MD: Johns Hopkins University Press, 1979.

Keller, Alex. *The Infancy of Atomic Physics.* New York: Clarendon, 1983.

Romer, Alfred, ed. *The Discovery of Radioactivity and Transmutation.* New York: Dover, 1964.

Trenn, Thaddeus J., ed. *Radioactivity and Atomic Theory.* New York: Halsted Press, 1975.

Raman, Chandrasekhara Venkata

(b. Trichinopoly, India, 1888; d. Bangalore, India, 1970)

C. V. Raman is best known for the "Raman effect" and for creating a robust physics community in India. He attended Presidency College in Madras, India, where he won the gold medal in physics. He did not pursue a scientific career; because he had not taken his undergraduate degree in a British university, the academic path was not open to him. He entered the Indian Financial Civil Service instead and moved to Calcutta. There he became acquainted with a non-British body of scientists, the Indian Association for the Cultivation of Science, which gave Raman the contacts and laboratory space needed to pursue science. He pursued scientific research on his own time, specifically the study of acoustics and optics, subjects that would occupy him throughout his career. In 1917, he was sufficiently accomplished to be offered an endowed chair in physics at Calcutta University.

In the early 1920s, Raman began investigations of the molecular diffraction of light, to find out how light behaves when it passes through different media. These studies culminated in the discovery of a major effect of light, and radiation generally. In 1928, Raman announced that scattered light from a transparent medium does not have the same wavelength as the incident radiation. Using a spectroscope, he measured distinct changes in wavelength after diffraction. The shift, Raman explained, was owing to the exchange of energy between the incident radiation and the molecules in the medium itself. This peculiar finding immediately came to be known as the Raman effect. The discovery prompted a wave of experiments using spectroscopes to measure light wavelengths and polyatomic molecules.

The discovery also elicited a surge of respect for Indian science. Raman's efforts brought science in India to a new position in the international community of scientists. Raman already had begun to construct a strong physics community in India. For example, he established the *Indian Journal of Physics* in 1926, and he later (1934) founded the Indian Academy of Sciences. But it was winning the Nobel Prize in Physics, in 1930, that brought India to an entirely new level of prestige. For the first time, an Indian was recognized as having made a world-class and far-reaching contribution to science.

At home, Raman used his influence and wealth to steer a fledgling physics community. In 1933 he became the director of the Indian Institute of Science at Bangalore, but soon came into conflict with the faculty because of his autocratic style. He championed physics, but some felt that he did so at the expense of other fields. Opposition to his leadership even led to an embarrassing incident when Raman invited German physicist Max Born (1882–1970) to take a position at the institute, only to have his colleagues unite to prevent it. In 1937 he was compelled to step down as director. He continued to be a professor there through the war years, but in 1948 he left to found and endow the Raman Institute of Research. Over the course of his long career, he trained more than a hundred physicists.

Raman's influence in India had both positive and negative effects. On the one hand, he was responsible for strengthening physics and he personally had a hand in creating many of India's scientific institutions. But he also used his influence to prevent some scientific ideas from coming to light in India. Scholar Abha Sur has noted that Raman often was reluctant to confront evidence against his own theories. Ironically, the main controversy of his scientific life was with Max Born, on the subject of lattice dynamics. Raman, according to Sur, used his influence and position to prevent free discussion of the scientific clash in his institute and even in scientific journals, which might have severely limited the scope of the development of optical spectroscopy in the 1950s and beyond.

See also Colonialism; Light; Physics
References
Brand, J. C. D. "The Discovery of the Raman Effect." *Notes and Records of the Royal Society of London* 43:1 (1989): 1–23.
Sur, Abha. "Aesthetics, Authority, and Control in an Indian Laboratory: The Raman-Born Controversy on Lattice Dynamics," *Isis* 90:1 (1999): 25–49.

Rediscovery of Mendel

The rediscovery of Gregor Mendel's (1822–1884) experiments on garden peas sparked the twentieth-century science of genetics, a branch of biology that deals with the principles of heredity and variation in organisms. Born into a family of peasants in 1822, Mendel joined an Augustinian monastery in Moravia, in a town called Brünn (later called Brno). In the early 1850s, he traveled to Austria to study at the University of Vienna. There he learned about cytology, a kind of biology that deals with the formation, structure, and function of cells. His teacher, the eminent biologist Franz Unger, advocated evolution, then a controversial idea. At the time, scientists had begun to criticize evolution because it did not fit with recent hybridization experiments. Hybrids, or crossbreeds, are the offspring of two different kinds of organisms (for example, a mule is a hybrid of a female horse and a male donkey). These experiments had suggested that species were stable and that change was a product of crossbreeding rather than evolution. Mendel's interests were piqued by the apparent

contradictions between evolutionary ideas and hybridization experiments. But this was as close as Mendel ever got to a mainstream scientific community. He never completed his examinations at the University of Vienna, because of a nervous illness. Instead, he returned to Brno to teach in a local school.

Using peas grown in the monastery's garden, Mendel experimented with hybrids using techniques in artificial fertilization. He grouped the peas according to seven different characteristics, such as shape, color of various parts of the seed or pod, texture, kinds of flowers, and height. In the first hybrid generation, he found that one characteristic showed up in all the plants, while its opposite appeared to disappear. For example, when short and tall pea plants were crossed, all of the resulting hybrids were tall. One characteristic seemed to be dominant compared with the other. But then, when Mendel crossed the seeds from the hybrid group with other seeds from the hybrid group, he found something quite different: Most plants from this second generation were tall, but some were short. The "short" characteristic had come back, after having disappeared in the first generation. In fact, Mendel observed a 3:1 ratio of tall to short.

Mendel's work suggested that scientists had to consider at least two characteristics present in each plant, inherited from the parents; if the inherited characteristics differed, the dominant one would be visible, without any blending (i.e., plants should appear tall or short, not medium height). By assuming that each seed's characteristics existed in pairs, Mendel was able to simplify his results and express them mathematically, as a process of elementary multiplication. This became a powerful tool for geneticists in the twentieth century.

Mendel did not achieve any great status as a scientist in his lifetime, nor did he make any contribution to the dispute about evolution while he lived. In 1866, he published his results in the journal of the Brno Natural History Society, a bit outside the mainstream of scientific literature. In 1868, he became abbot of his monastery and a few years later, he stopped his experiments. The greatest arguments of his later years were with the government, over the monastery's taxes. After his death in 1884, some sixteen years passed before Mendel's ideas were resurrected as a means to lay the foundations of the laws of heredity. Early geneticists, such as Hugo De Vries (1848–1935), Carl Correns (1864–1933), Erich Von Tschermak (1871–1962), and William Bateson (1861–1926), hailed Mendel as a man ahead of his time. De Vries, Correns, and Von Tschermak typically receive credit for having "resurrected" Mendel from obscurity in 1900. Later scientists blamed the obscurity of the journal in which Mendel published or his lack of standing in the scientific community for the long neglect of his ideas.

But was Mendel truly ahead of his time? He did not go to his grave believing that he had created laws of heredity and that no one had listened. His own view of his contribution was that he had found a way to understand the development of new species, through hybridization rather than through evolution. He was not, as his "rediscoverers" were, primarily interested in the inheritance of characteristics. Instead, the founding work on genetics was for many years what Mendel intended, an effort to replace evolution with hybridization. Only when a new problem needed solving, namely, the principles of heredity, did Mendel emerge as the prophet of genetics.

See also Bateson, William; Evolution; Genetics

References
Bowler, Peter J. *The Mendelian Revolution: The Emergence of Hereditarian Concepts in Modern Science and Society.* Baltimore, MD: Johns Hopkins University Press, 1989.
Keller, Evelyn Fox. *The Century of the Gene.* Cambridge, MA: Harvard University Press, 2000.
Stern, Curt, and Eva R. Sherwood, eds. *The Origins of Genetics: A Mendel Source Book.* London: W. H. Freeman, 1966.
Sturtevant, A. H. *A History of Genetics.* New York: Harper & Row, 1965.

Relativity

The theory of relativity was developed by German theoretical physicist Albert Einstein (1879–1955) in the first two decades of the twentieth century. Not only did it present a new, highly abstract, and difficult concept, but also it required the revision of all known ideas in physics along relativistic lines. The theory actually had two main parts, one special and the other general. Einstein published his theory of special relativity in 1905, the same year he published other seminal works, such as those on Brownian motion and the photoelectric effect (the latter won him the Nobel Prize in Physics in 1921). His theory of general relativity came later, in 1915.

Einstein's concept of relativity was based, in part, on the notion that there are no fixed points. Physicists in the nineteenth century believed in the presence of an ether, an omnipresent substance that could not be recognized with human senses, connecting all things. That ether was fixed, much as an x-y axis is "fixed" in any elementary geometry student's mind. In such a view, all movement is considered in relation to something that is at rest. In his special theory of relativity, Einstein challenged physicists to abandon the ether altogether and to accept that absolute rest (i.e., fixed points) cannot exist. Instead, all physical relationships must be considered in relative terms—as one moving object relates to another moving object.

The theory of special relativity was controversial because the credit for it was typically given solely to Einstein. It is true that by 1905 two others, Dutch physicist Henrik A. Lorentz (1853–1928) and French mathematician Henri Poincaré (1854–1912), had developed relativistic solutions to a prevailing problem of the time: the speed of light. In the 1880s, experiments had shown that light appeared to move at a constant speed, no matter what direction it traveled. Because of belief in the ether, which was supposed to propagate light, the speed of light should have varied in different directions as the earth sped along its path through the ether. Yet despite the construction of sensitive measuring devices, this effect was not observed. Lorentz proposed that molecules themselves, in this case those of the measuring apparatus, might be contracted by the ether in the direction of motion, thus rendering accurate measurement impossible. Lorentz's solution was to invent equations to transform variables such as space and time to fit the point of view of a moving object. Working from Lorentz's claim, Poincaré tried to expand it to include optical phenomena in general. Both men were concerned with the movement of objects relative to the ether.

Special relativity certainly built on concepts that were not entirely new. In the seventeenth century, for example, Galileo had demonstrated that if an object is dropped from the top of a mast on a sailing ship, the path of the object will differ, depending on the point of view of the observer. Anyone on the ship will see the object fall in a straight line, to the bottom of the mast, whereas an observer from the shore would see the object (which is not only falling, but moving as the ship moves) follow a curved path. Thus, the shortest distance between two points is occasionally a curve! But special relativity took this a step further, because Galileo's example still assumed the possibility of a fixed position (in this case, the shore) from which to view the moving object. Einstein proposed that more variables than space were at risk when all motion was relative to other moving objects: Instead of simply a curved path, as in Galileo's example, both time and mass would be affected, too.

Einstein's theory of special relativity thus proposed the following. The speed of light is constant. There is no ether, and thus no possibility of absolute rest. The laws of physics appear "classical" to each observer in his own situation (i.e., the ship's crew will see the object drop from the top of the mast and follow a straight line). The measurable variables of physics—space, time, and mass—will differ according to the observer's vantage point, as in the case of the object appearing to follow a curve from the point of view of the shore. The outside observer will measure time moving more slowly, mass increasing, and space contracting. At great speeds, these differences become more pronounced. Were it possible for an object with more than zero mass to achieve the speed of light, its mass would appear infinite to outside observers, its length would contract to nothing, and its time would stop.

This aspect of Einstein's theory was just a special case. It did not apply to objects that were accelerating or decelerating, only those moving at uniform velocities. In a moving vehicle moving at a constant speed, life inside the vehicle seems to approximate absolute rest. But as anyone who has carried a hot beverage in a moving vehicle knows, sudden acceleration changes physical relationships.

Over the next decade Einstein attempted to generalize his theory, taking into account nonuniform velocity and other phenomena, especially gravitation. Isaac Newton (1642–1727) had established laws of a gravitational force during the seventeenth century, but he never had explained the nature of that force. With general relativity, proposed in 1915, Einstein argued that gravity was equivalent to acceleration. The two forces appeared to exert the same kind of influence. As long as a body continues to accelerate, it will be subject to a force in the opposite direction, just as gravity attracts objects toward the earth. Both acceleration and gravity seemed to elicit curved motion. An object moving in an accelerating vehicle, for example, will not follow a straight line, but instead will travel on a curved path. Einstein believed that the presence of massive bodies necessitates the curvature of space, which was in fact the nature of a gravitational field, a curvature that all objects followed near massive bodies. A 1919 expedition to an island off the coast of Africa to observe a solar eclipse, conducted by Arthur Eddington (1882–1944), confirmed that even light from stars was bent by massive objects (in this case, the sun). Relativity thus became the new prevailing concept upon which subsequent scientists had to base all notions of the structure of the universe.

See also Eddington, Arthur Stanley; Einstein, Albert; Physics; Solvay Conferences

References

Cassidy, David. *Einstein and Our World.* Atlantic Highlands, NJ: Humanities Press, 1995.

Einstein, Albert. *Relativity: The Special and General Theory.* New York: Bonanza Books, 1952.

Goldberg, Stanley. *Understanding Relativity: Origin and Impact of a Scientific Revolution.* Cambridge: Birkhäuser, 1984.

Pyenson, Lewis. *The Young Einstein: The Advent of Relativity.* Boston: Adam Hilger, 1985.

Religion

The relationship between science and religion often has been treated as a clash of armies in a cosmic battle over whose truth should win the day. Such conflicts are rooted deeply in time, hearkening back at least as far as the struggle to integrate the works of Aristotle into university curricula during the Middle Ages. The bearer of the modern version of this conflict was Charles Darwin (1809–1882), whose 1859 *On the Origin of Species by Means of Natural Selection* fueled preexisting disagreements about the evolution of humans from lower species. The "conflict" is not so clear cut, however; there have been pious scientists or churchgoers with a respect for science, too. In the twentieth century, the myth of a clash of armies has been enhanced by high-profile contests and virulent critiques by partisans from both sides of the debate.

Darwinian evolution, with its random, purposeless variation, raised philosophical issues as well as religious ones. If there is no purpose in nature, is evolution not progressive? Even those who did not cast their opposition in religious terms were unsettled by the lack of design in nature. Some were more specific: The Bible says that man was created by God, not descended from an ape. Although geologists in the nineteenth century already had abandoned the literal interpretation of the six days of creation, Darwin became an icon of conflict, as if he were the first to challenge the literal meaning of scripture.

Some saw in evolution a threat to the very fabric of spiritual, God-fearing society. The most famous example of this was the Scopes trial of 1925. When substitute teacher John T. Scopes (1900–1970) was arrested for teaching evolution in a school, against the laws of Tennessee, the American Civil Liberties Union was there to defend him. The ensuing trial pitted the well-known politician William Jennings Bryan (1860–1925), on the side of the fundamentalists, against the criminal trial lawyer Clarence Darrow (1857–1938). Bryan was afraid that if the evolutionist view took hold, it would rob society of its moral compass. Bryan won the case, but at a high cost, because it brought nationwide attention to the narrow, anti-intellectual views of the people in the U.S. South. The case served to polarize some views, and it ignited a wave of commentaries on the relationship between science and religion.

Although evolution was the centerpiece of conversations about science and religion, it was not the only scientific concept to provoke thought about the relationship. In physics, Albert Einstein's (1879–1955) general theory of relativity challenged the mechanistic, Newtonian universe. One of the great popularizers of relativity, Arthur Eddington (1882–1944), wrote that physics described relationships among things, not their nature. The idea that the real world might be beyond the reach of science found many supporters, Eddington included, who saw a place for both God and science in their lives. This view was only amplified by the development of quantum mechanics, particularly the uncertainty principle, which insisted that some facts simply were unknowable, and the complementarity principle, which admitted that sometimes contradictory viewpoints can describe the same phenomena with equal accuracy. Whereas evolutionary biology seemed to be confident and arrogant in its claims, physics was becoming more humble.

In the first half of the century, religion largely reconciled itself to science. Historian Richard Hofstadter argued that in the United States, scientists were the aggressive ones, spreading evolutionist thought into the best universities in the country. It is possible that, as Neal Gillespie argued, the real conflict was not between science and religion, but rather between theological reconciliation and positivism. Positivism, a philosophy of science that emphasizes the positive accumulation of knowledge strictly through empirical study, rejected supernatural explanations. Positivism became popular among scientists in the early twentieth century owing to the writings of European philosophers, notably Ernst Mach (1838–1916). While some scientists had sought to reconcile scientific findings with religious teachings, positivists called such efforts anti-scientific. Positivism among scientists, as in the case of Christian fundamentalism, ruled out any reconciliation between science and religion.

Churchmen did not rally together to form a united front against science. On the whole, the fundamentalist Protestant sects, particularly those in the U.S. South, insisted on interpreting the Bible literally. They were horrified at the prospect of being descended from an ape, and they treated the uncertainties announced by physicists as a triumph of God-fearing people over the corrupting influence of intellectuals. Owing to the ferocity of their opposition, fundamentalist Christians came to symbolize religion's antiscience character. But antiscience views were not universal. An Anglican, William R. Inge (1860–1954), dean of St. Paul's Cathedral in London, wrote in the journal *Science* that the fundamentalists were wrong to dream of routing the enemy. Religious people should come to terms with scientific views. "Christ never wished us to outrage our scientific conscience as a condition of being His disciples"

(Stevens 1926, 282), he said. Catholics were against evolution, as decreed by Pope Pius X (1835–1914) in 1907. But they were certainly not antiscience, if the activities of the Jesuits are any indication. In the early twentieth century, the Jesuits revived their focus on scientific activities by operating seismograph stations throughout the world and contributing to the new field of seismology. Their efforts testify to the importance placed on scientific research in U.S. culture generally, despite arguments between scientists and churchmen.

If science seemed poised to fulfill the material aspirations of society in the 1920s, it seemed far less so in the Great Depression of the 1930s. Religious figures renewed their critiques, citing the lack of morals or ethics in science. Catholic writer Gilbert K. Chesterton (1874–1936) noted that electricity is a fine product of science, but that man should not be considered as a god with a thunderbolt but rather as a savage who has been hit by a lightning bolt. One must not forget, he and others argued, the real source of science's wonders. In addition, science without values would never put society back on its feet; critics argued that putting one's faith in science led only to despair, as revealed by the calamitous state of the world in the 1930s.

The struggle for primacy between science and religion did not end during the first half of the twentieth century. The destructiveness of the fruits of science, such as the atomic bomb during World War II, reinforced the belief that moral and ethical values were much needed in the face of scientific progress. If people were turning more to science for knowledge about the universe, they looked increasingly to religion to anchor them in other realms in which science had proved lacking, such as ethics, morality, and social meaning.

See also Eddington, Arthur Stanley; Evolution; Great Depression; Origin of Life; Russell, Bertrand; Scientism; Scopes Trial; Teilhard de Chardin, Pierre

References

Geschwind, Carl-Henry. "Embracing Science and
    Research: Early Twentieth-Century Jesuits and
    Seismology in the United States." *Isis* 89
    (1998): 27–49.
Gillespie, Neal C. *Charles Darwin and the Problem of
    Creation.* Chicago: University of Chicago Press,
    1979.
Hofstadter, Richard. *Social Darwinism in American
    Thought.* Boston: Beacon, 1955.
Kevles, Daniel J. *The Physicists: The History of a
    Scientific Community in Modern America.*
    Cambridge, MA: Harvard University Press,
    1995.
Larson, Edward J. *Summer for the Gods: The Scopes
    Trial and America's Continuing Debate over Science
    and Religion.* New York: Basic Books, 1997.
Numbers, Ronald L. "Science and Religion."
    Osiris, Second Series, 1 (1985): 59–80.
Stevens, Neil E. "Dean Inge on the Relation
    between Science and Religion Today." *Science,*
    New Series, 63:1628 (12 March 1926):
    281–282.

Charles F. Richter used seismographs such as this one to detect and measure earthquakes. (Bettmann / Corbis)

Richter Scale

The Richter scale was developed in 1935 by seismologist Charles F. Richter (1900–1985), then working with Beno Gutenberg (1889–1960) at the California Institute of Technology. Richter's scale used a logarithm of wave amplitudes measured from seismographs to measure the intensity of the earthquake at the epicenter. The epicenter was the place on the surface of the earth directly above the origin of the earthquake. Richter's scale was numeric, based on whole numbers and decimal fractions. For example, a strong earthquake might measure 5.7 on the Richter scale. But the scale did not increase arithmetically; instead, each whole number marked a tenfold increase. Thus, a magnitude 7 earthquake would be ten times stronger than a magnitude 6 earthquake.

The most commonly used scale before this was the modified Mercalli scale, named for Guiseppe Mercalli (1850–1914), the Italian geologist who developed the original scale in 1902. Adapted in 1931 for use in North America, it classified earthquakes by Roman numerals numbered I through XII, each increasing in magnitude. The modified Mercalli scale was essentially qualitative, basing its categories on the effects on people and structures such as buildings and bridges. At level VII, for example, everyone runs outdoors, and the earthquake is noticed by people driving cars. At level XI, few structures remain standing, and there are broad fissures on the ground. The most glaring shortcoming of the modified Mercalli scale was that it described the earthquake by its effects at the point of measurement, and thus estimations by different observers often were inconsistent. The Richter scale, rather than arbitrarily categorizing earthquakes according to damage caused, provided a quantitative means of assessing how much the earth had moved at the epicenter.

See also Geophysics; Gutenberg, Beno;
    Seismology

References

Oldroyd, David R. *Thinking about the Earth: A History of Ideas in Geology.* Cambridge, MA: Harvard University Press, 1996.

Richter, C. F. "New Dimensions in Seismology." *Science,* New Series, 128:3317 (25 July 1958): 175–182.

Rockets

Although the history of rocketry arguably can be traced back for centuries, the late nineteenth century marks the beginning of intensive research into the possibilities of flying tube-shaped objects with combustible fuel. But using gunpowder and other solid fuels to launch objects into the sky met with limited success. Rockets were used during World War I but were not nearly as effective as conventional artillery, and rockets launched from biplanes were unreliable. The war did see the birth of the first guided missiles, although they crashed often and made no serious contribution to the war effort.

The most influential figure in the history of early rocketry was Robert H. Goddard (1882–1945), who envisioned liquid-fuel rockets and multistage rockets and showed that rockets can function in a vacuum. He received funding from the Smithsonian Institution for experiments, and the Smithsonian also published "A Method for Reaching Extreme Altitudes," 1919 paper that enumerated the results of Goddard's work at Clark University on powder rockets. His early work considered the fundamentals of design, such as the optimal nozzle shapes and velocity. Subsequent work improved on this by developing various kinds of propulsion. He launched the first successful liquid-fuel rocket in 1926 (it flew for only a few seconds); a few years later, Goddard launched the first "scientific" rocket (it carried a barometer and camera). Goddard was prolific in ideas and was credited with more than 200 patents.

In the United States, scientists and military officers established a fruitful partnership in the field of rocketry. The partnership went back at least to World War II, when the Navy sought the aid of scientists and inventors to help the United States catch up to European advances in military ordnance during the war. During World War II, leading U.S. scientists such as Lawrence Hafstad (1904–1993) and Charles Lauritsen (1892–1968) visited England during the Battle of Britain, providing stark evidence of the potency of Germany's available ordnance. When they returned home, the Navy put extraordinary pressure on scientists at the California Institute of Technology to design rocket-oriented weapons such as the barrage rockets later used by ships to support amphibious assaults. In 1943, the Navy established the Naval Ordnance Test Station in Inyokern, California. Its creation was intended to prevent the scientific-military collaboration from falling into relative disuse in the postwar period, as had been the case after World War I.

Scientists in Nazi Germany experienced phenomenal success in rocket technology, and they were the first to create a viable long-range ballistic missile. Their secret weapons center was located in Peenemünde. Adolf Hitler (1889–1945) was skeptical that weapons designed by "rocket club" enthusiasts such as Wernher Von Braun (1912–1977) could make much difference in the war, and initially he did not give it much priority. The V-2, the first large ballistic missile, was tested successfully in October 1942. By 1943, Adolf Hitler personally congratulated Von Braun and gave the project the highest priority, believing that it could decide the war. Although the rockets did not prove to be war-winning weapons, they were indeed used; the V-2 was launched against London and other cities held by the Allies, beginning in 1944.

After the war, both the United States and the Soviet Union enhanced their own rocket programs by not only stealing design technology from the Germans but also by acquiring the scientists themselves. The U.S. version,

Professor Robert H. Goddard, instructor of physics at Clark College in Worcester, Massachussetts, is shown with the first rocket he experimented with, charged with gun cotton, in 1924. (Bettmann/Corbis)

conventional bombers, and the postwar years saw the development of a fleet of strategic bombers under the Air Force's Strategic Air Command (SAC). But by the 1950s, intensive research projects on rocketry were under way, with the goals of orbiting artificial communications satellites and developing long-range ballistic missiles to target enemy cities for nuclear attack without the need for strategic bombers. Such missiles, named intercontinental ballistic missiles (ICBMs), would not only be nearly impossible to shoot down (unlike bombers), but they would shorten warning time dramatically. The successful orbiting of an artificial satellite by a rocket, achieved by the Soviet Union in 1957, heralded not only the space age but also a new phase of the nuclear arms race.

See also Cold War; Nazi Science; World War II
References

Christman, Albert B. *Sailors, Scientists, and Rockets: Origins of the Navy Rocket Program and of the Naval Ordnance Test Station, Inyokern.* Washington, DC: Naval History Division, 1971.

Goddard, Robert H. *Rockets.* New York: American Rocket Society, 1946.

Hall, R. Cargill, ed. *History of Rocketry and Astronautics: Proceedings of the Third through the Sixth History Symposia of the International Academy of Astronautics.* San Diego, CA: American Astronautical Society, 1986.

Naimark, Norman M. *The Russians in Germany: A History of the Soviet Zone of Occupation, 1945–1949.* Cambridge, MA: Harvard University Press, 1995.

Neufeld, Michael J. "Hitler, the V-2, and the Battle for Priority, 1939–1943." *The Journal of Military History* 57 (1993): 511–538.

Von Braun, Wernher, and Frederick I. Ordway II. *History of Rocketry and Space Travel.* New York: Thomas Y. Crowell, 1969.

begun in 1945, was called Project Paperclip, and hundreds of Nazi scientists and technicians were transferred to the United States to be used in the defense industries of the Cold War era. Among these were Von Braun and other members of the Peenemünde team that had developed the V-2 rocket. The Soviets also attempted to gather as many rocket (and atomic) scientists as they could, including one of Von Braun's collaborators, Helmut Gröttrup. The Soviets lured many rocket scientists into the Soviet zone of occupation with generous compensation packages, and the Americans did the same.

Rockets posed a special threat because of their potential as delivery vehicles for atomic bombs, developed by the United States during the war. The first bombs used against Hiroshima and Nagasaki were dropped from

Royal Society of London

The Royal Society, based in London, England, was among the most well-known national scientific bodies, with an elite selected membership. Established during Restoration England in the seventeenth century, it was one of the oldest government-sponsored scientific bodies. Although some of its scientific role was supplanted by the British Association for the Advancement of Science, founded in the nineteenth century, the Royal Society retained an important role in supporting and directing scientific activity in Britain during the early twentieth century. It strove, with mixed success, to weather the storms of two world wars and to adapt to changing conditions for science in Britain.

Although taking part in early efforts at international intellectual cooperation, the Royal Society became a victim of divisive geopolitics. Under the guidance of Arthur Schuster (1851–1934), it promoted the International Association of Academies around the turn of the twentieth century, trying to create inclusive international arrangements rather than bilateral agreements between countries. Schuster, like many of his colleagues in Britain and elsewhere, believed that men of science could help society rise above political antagonism and take over where diplomacy failed. However, during World War I, the Royal Society became a hotbed of anti-German sentiment, and it reversed course; it took the lead in denouncing German militarism and German scientists alike. Schuster himself, of German descent, was attacked by both the press and his colleagues. When the war ended, Royal Society members helped to ensure that German scientists were blocked from postwar international scientific bodies.

The Royal Society began to work closely with the government during World War II, a practice that would continue for a long time and bring scientists into close proximity to politics. Based largely on recommendations by physicist Henry Tizard (1885–1959), the Royal Society took an active role in making sure that Britain's scientific manpower was put to use in the war against Germany. The Royal Society's president, Sir William Henry Bragg (1862–1942), nominated Nobel laureate biologist A. V. Hill to serve on a government advisory council to organize "emergency" work. This became the Scientific Advisory Committee to the War Cabinet. The Royal Society's offer of assistance ultimately put leading scientists in touch with every kind of defense work, including radar, continuing a long-standing relationship between the scientific elite and national strength and security in Britain.

After the war, the Royal Society continued to seek ways to marshal the empire's scientific strength in the service of Britain. In 1946, the king formally opened the Royal Society Empire Scientific Conference. Now that World War II had ended, part of the society's task was to find ways to put science (and scientists) in the service of maintaining the strength of the British Empire despite postwar economic strains. Scientists from imperial possessions (such as Australia, New Zealand, India, etc.) met in Cambridge and Oxford to assess the state of research in their countries, to outline potential applications, such as relationships between chemistry and industry, and between biology and the control of disease. By mid-century, the Royal Society embraced its role of putting science to work for the good of the state and the empire.

See also Colonialism; Elitism; Nationalism; Patronage; World War I; World War II

References
Alter, Peter. "The Royal Society and the International Association of Academies, 1897–1919." *Notes and Records of the Royal Society of London* 34:2 (1980): 241–264.
Badash, Lawrence. "British and American Views of the German Menace in World War I." *Notes and Records of the Royal Society of London* 34:1 (1979): 91–121.
McGucken, William. "The Royal Society and the Genesis of the Scientific Advisory Committee to Britain's War Cabinet, 1939–1940." *Notes and Records of the Royal Society of London* 33:1 (1978): 87–115.
"The Royal Society Empire Scientific Conference." *Notes and Records of the Royal Society of London* 4:2 (1946): 162–167.

Russell, Bertrand

(b. Trelleck, Wales, 1872; d. Merioneth, Wales, 1970)

Bertrand Russell was a British mathematician, logician, and philosopher. His views on logical analysis shaped not only the world of mathematics but also society at large. A controversial figure, he was an outspoken opponent of dogma and fanaticism from World War I to the Cold War era. He started his career when he went to Cambridge University in 1890, where he met his future friend and collaborator Alfred North Whitehead (1861–1947). His studies of mathematics and philosophy led to a teaching position at Trinity College in 1908. An aristocrat, Russell became the third Earl Russell in 1931.

Russell's contributions to logic included the discovery in 1901 of a paradox in set theory (now called Russell's paradox), which showed that all statements in classical logic are in some sense contradictory. He spent the next several years developing improvements, resulting in his theory of types. During this period Russell published a book, *The Principles of Mathematics* (1903), which he conceived as a multiple-volume treatment. Because Whitehead was working on a similar project, the two decided to collaborate, resulting in the first volume of their *Principia Mathematica* in 1910. The next volumes of the *Principia Mathematica* were published in 1912 and 1913. They were remarkably popular, and they were evocative of connections not only between mathematics and logic but also between logical analysis and both philosophy and the production of knowledge in general. This work reduced many mathematical problems to logical ones, a theme that recurred in much of Russell's writing. The primacy of logical analysis certainly informed his philosophical views, which emphasized logic at the expense of traditional concepts of ethics, morality, or religious dogma.

World War I also helped to shape Russell's philosophical outlook, turning him away from optimistic progressivism toward more emphasis on skepticism and logic. He became more outspoken in his views; disgusted and disillusioned by the war, he refused to fight and spent several months in jail for his pacifism. His efforts did little to abate the militarism he felt pervaded his society, and he considered his wartime activism to be a complete failure. Because of his wartime jail sentence, he lost his position at Cambridge.

The emphasis on logic appeared to free Russell from other influences, such as traditional concepts of morality and religion. He was widely regarded as a freethinker and he believed also in free love, which attracted a great deal of criticism. He also promoted women's suffrage. Russell's opposition to dogma made him many enemies when he published *Why I Am Not a Christian* in 1927, in which he outlined the logical arguments for and against the existence of God, including an analysis of the supposed moral imperative of accepting it on faith. He also pointed out what he considered a number of defects in the teachings of Christ, as well as the retardation of human progress by religion. He challenged his readers to look at the world around them and to put human intelligence to work in understanding it.

Russell's views made him both celebrated and widely unpopular. He traveled to China and Russia and returned to England to form an experimental school. He ran for public office several times, without success. In the 1930s, he lived in the United States, teaching at a number of universities, including the University of California–Los Angeles. Russell stirred up controversy after he was appointed to be professor of philosophy at the College of the City of New York in 1940. He was denounced by his many opponents, particularly the Episcopal Church and the Catholic Church, as an enemy of religion and morality. A scandal ensued, which resulted in Russell being without a job, his philosophy decreed to be against the penal codes of the state. The U.S. scholarly community was horrified at this major assault on academic freedom, but Russell did not get his position back. He returned to England in 1944.

During the war years, Russell composed his work, *The History of Western Philosophy* (1945), which traced scientific and philosophical ideas from the pre-Socratic Greeks to his own influences in the philosophy of logical analysis around the turn of the twentieth century. He argued that philosophers' injection of moral considerations has impeded the progress of knowledge for centuries. Efforts to use science and philosophy to prove or disprove religious dogmas, such as the existence of God, all "had to falsify logic, to make mathematics mystical, and to pretend that deep-seated prejudices were heaven-sent intuitions" (Russell 1945, 835). Logical analysis, the school of philosophy to which Russell himself adhered, rejected the notion that humans could find conclusive answers to such questions; science should concern itself with improving the current state of knowledge by building on the present rather than assuming there is a "higher" way of knowing.

The years after World War II were paradoxical ones for Russell. Against communism, he was convinced that the Soviet Union posed the greatest risk of war; later, after the death of Joseph Stalin (1879–1953) and the U.S. development of hydrogen bombs, he would reverse this opinion. He would become one of the principal figures in the disarmament movement during the late 1950s, but in the late 1940s he was associated with a more dire prescription: preventive warfare. Familiar with the principles of game theory, which some had applied to the confrontation between the United States and the Soviet Union, Russell argued that the best course of action was simply to declare war. If such a war was inevitable, he believed, it was more humane to do it while nuclear stockpiles were relatively low.

Russell received the Nobel Prize in Literature in 1950, for his various writings that promoted the freedom of thought and humanitarian values. He died of influenza, after nearly ten decades of life.

See also Cold War; Game Theory; Mathematics; Philosophy of Science; Religion

References

Clark, Ronald William. *The Life of Bertrand Russell.* London: J. Cape, 1975.

Dewey, John, Horace M. Kallen. *The Bertrand Russell Case.* New York: Viking Press, 1941.

Grattan-Guinness, Ivor. "Bertrand Russell (1872–1970) after Twenty Years." *Notes and Records of the Royal Society of London* 44 (1990): 280–306.

Poundstone, William. *Prisoner's Dilemma: John Von Neumann, Game Theory, and the Puzzle of the Bomb.* New York: Anchor Books, 1992.

Russell, Bertrand. *A History of Western Philosophy.* New York: Simon & Schuster, 1945.

———. *The Autobiography of Bertrand Russell.* New York: Routledge, 2000.

Wood, Alan. *Bertrand Russell: The Passionate Skeptic.* New York: Simon & Schuster, 1958.

Rutherford, Ernest

(b. Nelson, New Zealand, 1871; d. Cambridge, England, 1937)

Ernest Rutherford was the leading figure in the science of radioactivity. He conducted the seminal experiments on radioactive emission and helped to develop a theory of radioactive decay and atomic transformation. In addition to this work, which already established his reputation, he revised the model of the atom through his discovery of the atomic nucleus. Born and raised in New Zealand, Rutherford attended Canterbury College in Christchurch. But in 1895 he moved to England to pursue graduate work at Cambridge University. There he worked under J. J. Thomson (1856–1940), who soon made an exceptional discovery that would shape both men's lives: the electron, the first subatomic particle to be discovered.

After the 1896 discovery of radioactivity by Henri Becquerel (1852–1908), Rutherford made radioactive decay and atomic physics his principal object of study. In 1898 he identified two different kinds of emission from radioactive substances, which he named alpha and beta rays. Alpha rays appeared to be positively charged, whereas beta rays were negatively charged. The beta rays, he surmised, were composed of Thomson's

The leading figure in the science of radioactivity, Ernest Rutherford conducted seminal experiments on radioactive emission and helped to develop a theory of radioactive decay and atomic transformation. (National Library of Medicine)

electrons. Rutherford was convinced that alpha rays held the key to understanding the nature of radioactivity, and he conducted numerous experiments on their emission.

In 1898, Rutherford left Cambridge to take a position at McGill University in Montreal, Canada. While at McGill, Rutherford began collaborate with chemist Frederick Soddy (1877–1956) to understand the meaning of alpha and beta emissions. They determined that such emissions actually were processes of "disintegration" and that the atoms transformed into lighter elements upon ejecting them. Thus radioactivity acquired a reputation as a naturally occurring kind of alchemy, because elements transmuted into others. Radioactive elements were those that were unstable enough to

decay—in other words, to release an alpha particle; the process would continue until a stable state was reached. This "transformation theory" was a fundamental innovation, particularly because Rutherford insisted that the changes were of an atomic rather than molecular character; the resulting element is completely independent in character from the previous one, having ejected an alpha particle and thus having become a different element. He and Soddy identified the elements of the decay series—a difficult process given the number of transmutations and the varying periods required for different elements to decay. Rutherford gained worldwide acclaim for his work on radioactivity, and he was awarded the Nobel Prize in Chemistry in 1908, which Rutherford found ironic because he considered himself a physicist.

Rutherford moved again in 1907, back to England to become Langworthy Professor of Physics at Manchester University. There he determined the nature of the mysterious alpha particles, noting that they were simply helium atoms that no longer had electrons. One of his assistants, Hans Geiger (1882–1945), ultimately gave his name to an instrument the two developed, enabling the counting of individual atomic disintegrations (the Geiger counter). While at Manchester, Rutherford conducted a series of experiments that transformed physicists' understanding of the nature of the atom. He placed an alpha emitter (a radioactive element emitting alpha particles) near a sheet of metal and noticed that the beam of particles seemed to scatter upon passing through the metal. Curious about the effect, he set Geiger and an undergraduate assistant, Ernest Marsden (1888–1970), to the task of measuring scattering angles. The prevailing model of the atom at the time, Thomson's "plum pudding" atom, assumed that electrons swam freely in a dispersed, positively charged medium. The positive charge of the medium would have some effect on the alpha particles as they passed, and the scattering angles could be observed in a cloud chamber. This instrument had been designed by C. T. R. Wilson (1869–1959) to allow humans to view vapor trails of tiny particles. But the experiments showed unexpected scattering angles. Not only were most of the alpha particles deflected as expected, a small few of them appeared to be bouncing back, at angles of more than ninety degrees. From this, Rutherford concluded that there must be another particle in the atom, where all the mass and all of the positive charge resided. This would be strong enough to deflect most alpha particles as they passed nearby, and even to reverse the course of some that managed to collide with the particle. This particle Rutherford called the atom's nucleus.

In 1911, Rutherford used his findings to propose a different atom from the model that J. J. Thomson had proposed. Instead of electrons swimming in a positive medium, there were electrons orbiting around a nucleus. Not without its own problems, this model stood the test of time largely because of the efforts of Danish physicist Niels Bohr (1885–1962), who used recent concepts in quantum physics to provide the atom with stability. It is upon Rutherford's and Bohr's atom that all modern nuclear physics is based.

Rutherford had become the world's leading experimental physicist. He was knighted in 1914 and, during World War I, he worked for the British Admiralty on submarine detection research using acoustic methods. When the war ended, he succeeded J. J. Thomson as the director of Cambridge University's Cavendish Laboratory in 1919. Aspiring physicists from abroad flocked to the Cavendish Laboratory to work under his guidance. Numerous important discoveries were made there under his leadership, such as James Chadwick's (1891–1974) discovery of the neutron, and John Cockcroft (1897–1967) and Ernest Walton's (1903–1995) development of the first particle accelerator, both in 1932. In 1931, Rutherford was elevated to the British peerage, thus becoming Lord Rutherford of Nelson. He died in 1937 and was buried in Westminster Abbey.

See also Atomic Structure; Bohr, Niels; Cavendish Laboratory; Chadwick, James; Cockcroft, John; Radioactivity; Thomson, Joseph John

References
Badash, Lawrence. *Radioactivity in America: Growth and Decay of a Science.* Baltimore, MD: Johns Hopkins University Press, 1979.
Campbell, John. *Rutherford: Scientist Supreme.* Christchurch, England: AAS, 1999.
Moon, P. B. *Ernest Rutherford and the Atom.* London: Priory Press, 1974.
Wilson, David. *Rutherford: Simple Genius.* Cambridge, MA: MIT Press, 1983.

S

Schrödinger, Erwin

(b. Vienna, Austria, 1887; d. Vienna, Austria, 1961)

Erwin Schrödinger received his doctorate in 1910 from the University of Vienna, specializing in theoretical physics. He served in the army during World War I, and in 1921 he took a professorship of physics at the University of Zurich. He was raised speaking English and German, and as an adult he could also speak French. In many ways he did not fit the mold of the theoretical physicist in its "golden years" of the 1920s. He was nearly forty years old when he made his singular contribution, and he did not accept the widely influential Copenhagen interpretation of quantum mechanics.

Schrödinger was best known for his formulation of wave mechanics in 1925–1926. He was aware of Werner Heisenberg's (1901–1976) work on matrix mechanics, but he found it difficult because it could not be visualized. He favored instead the work of French physicist Louis De Broglie (1892–1987), whose 1924 thesis had emphasized the duality of waves and particles. Inspired, Schrödinger devised a wave equation for the hydrogen atom and soon established an entire dynamical system in several papers in *Annalen der Physik* in 1926. His wave mechanics had some advantages, not the least of which was that it was easy to make calculations with it. In addition, its emphasis on waves was less abstract than Heisenberg's matrix mechanics. What puzzled physicists was the fact that wave mechanics achieved the same mathematical results as matrix mechanics, despite the fact that one was premised on matter being made up of waves, and the other conceived of matter as particles. The paradox of the wave and particle duality became the cornerstone of quantum mechanics, both in its physical and in its philosophical ramifications. For his work, Schrödinger in 1933 shared the Nobel Prize in Physics with Paul Dirac (1902–1984), whose work integrated wave mechanics with Albert Einstein's (1879–1955) theory of special relativity.

Schrödinger's physics had far-reaching philosophical implications. Although he was responsible, through wave mechanics, for affirming the wave and particle duality of nature, he disliked how some physicists interpreted it. For example, he stood firmly against the Copenhagen interpretation of quantum mechanics, based on the indeterminacy and complementarity principles of Werner Heisenberg and Niels Bohr (1885–1962). In particular, Bohr's complementarity used the wave and particle duality as

Erwin Schrödinger, pictured here, developed wave mechanics as a rival to Werner Heisenberg's matrix mechanics. (Bettmann/Corbis)

evidence of opposing systems providing the same results, indicating that science can neither describe reality nor even precisely define physical systems.

Schrödinger was disturbed by the abandonment of notions such as cause and effect, and his famous thought problem, "Schrödinger's cat," highlights the problems of abandoning determinism and describing reality with probabilities. In 1935, he presented the following scenario: An unfortunate cat is enclosed in a box, along with some radioactive material and a device that will release fatal poison if a single radioactive atom decays. After an hour, how does one know if the cat has been poisoned? Quantum mechanics would describe the system "by having in it the living and the dead cat (pardon the expression) mixed or smeared out in equal parts" (Kragh 1999, 217). But clearly, logic compels us to accept that the cat is not an equation; it is either dead or alive. This thought problem, although not taken seriously at the time as a legitimate critique, has become a popular example of the difficulties inherent in reducing reality to probabilistic equations.

Although he was not a Jew, Schrödinger was dissatisfied with the rise of the Nazis, and he decided to leave Germany when they came to power in 1933. He first went to Oxford, England, then home to Austria to take a university chair in Graz. After the annexation of Austria by Germany, Schrödinger left again, ultimately finding his way to Dublin, Ireland. There he worked on a number of theoretical problems, including a unified field theory of gravitation and electromagnetism. He failed to achieve this monumental task, as did Einstein and all others who attempted it.

Schrödinger's influence extended beyond physics and into biology. In 1944 he published a slim volume entitled *What Is Life?*, which discussed chromosomes as the biological code of an organism. The book made no major contribution to biology, but it did inspire many young researchers with an interest in physics to turn their attention to the important problems of biology. Scientists who later credited this book with changing their careers include Francis Crick (1916–) and James Watson (1928–), the duo that would work out the structure of DNA in 1953.

What stimulated Schrödinger's thought? His biographer Walter Moore has connected Schrödinger's most ingenious ideas with the passions of his love affairs. He and his wife lived a remarkably free sexual existence, which scandalized their friends and colleagues. He complemented these sexual escapades with an abiding interest in Indian mysticism, based on the ancient Hindu Veda. After he retired from his position in Dublin, he returned to his native Austria, where he died in 1961.

See also Determinism; Heisenberg, Werner; Quantum Mechanics

References

Kragh, Helge. *Quantum Generations: A History of Physics in the Twentieth Century.* Princeton, NJ: Princeton University Press, 1999.

Moore, Walter. *Schrödinger: Life and Thought.* New York: Cambridge University Press, 1989.

Science Fiction

Science fiction has not always occupied a prominent place on library bookshelves amidst other genre categories such as mysteries and Westerns. Like fantasy, it is a genre in which the reader is expected to suspend one's disbelief; it is different from fantasy in that the ideas behind the story are inspired by science and technology. As literature, science fiction was used long before the twentieth century. For example, Mary Shelley's (1797–1851) *Frankenstein, or, The Modern Prometheus* (1818) marked the nineteenth century as one attuned to the contradictions between the eighteenth-century Age of Reason and the nature-oriented emotionalism of the Romantic era. Science fiction has also been a means to present utopian societies, using recent changes in science and technology as tools to speculate about the future. Because of this, science fiction provides insights about the hopes, expectations, and fears brought about by science and its fruits.

The most celebrated science fiction author of the early twentieth century actually penned many of his stories in the nineteenth. English author H. G. Wells (1866–1946) set standards for science fiction with his frightening depictions of future biological evolution (*The Time Machine*, 1895), mad or overly ambitious scientists (*The Island of Dr. Moreau*, 1896; *The Invisible Man*, 1897), and highly advanced alien races who choose to invade the earth (*The War of the Worlds*, 1898). His *The World Set Free* (1914) speculated that by the 1950s wars would be fought with terrible new weapons harnessing the power of the atom. *The War of the Worlds* later became a topic of serious study by sociologists because of a bizarre event decades after its first publication. In 1938, actor Orson Welles (1915–1985) and his Mercury Theater group performed a radio adaptation of *The War of the Worlds*, setting it in a real U.S. town, Grover's Mill, New Jersey. Thousands of the listeners, hearing symphony music interrupted by fake new bulletins, believed that an invasion from Mars actually was taking place. The event immediately became a source of fascination for researchers interested in the manifestations of mass hysteria.

Like Wells, many of the authors who chose scientific or technical topics used them to satirize their own societies. Among these were Sinclair Lewis's (1885–1951) *Arrowsmith*, Aldous Huxley's (1894–1963) *Brave New World*, and George Orwell's (1903–1950) *1984*. *Arrowsmith* (1925) tells the story of a physician during times of plague whose efforts to help society make him an outcast; *Brave New World* (1932) is a futuristic glance at a society that depends on drugs and social engineering to make its residents happy; *1984* (1949) describes a society whose people are constantly being watched and policed. The latter added a few phrases, such as *Big Brother* and *Orwellian,* to the English lexicon. These classics of the science fiction genre exemplify the use of science-inspired fantastical elements to satirize or otherwise comment about contemporary times.

Science fiction as a distinct and popular genre took off in the 1920s, through short stories in magazines. One of the most popular of these was *Amazing Stories,* founded in 1926 by Hugo Gernsback (1884–1967). Science-based fiction soon became a specialized kind of genre fiction, generating a "pulp" market with predictable stories and themes. Some magazines, such as *Astounding Science Fiction,* founded in 1937 by John Campbell (1910–1971), attempted to publish less-formulaic stories. Science fiction gained enormous popularity in the postwar period, as the atomic bomb and the unknown effects of radioactive fallout

Actor Orson Welles explains the radio broadcast of H.G. Wells's The War of the Worlds *to reporters after it caused widespread panic. (Bettmann/Corbis)*

provided a limitless resource for both creative and repetitious fiction. But by mid-century the genre had many critics. Aside from defects in originality, many scientists complained at the lack of plausibility and sophistication in the science itself. In 1951, one critic in the research journal *Science* complained that science fiction too often was built "around a bag of standard magic tricks," such as time travel and robots. Critics complained periodically that even the more original fiction hardly seemed like "science" fiction, because it often contained no science at all, but rather confined itself to social commentary.

See also Atomic Bomb; Extraterrestrial Life; Lowell, Percival

References

Cantril, Hadley. *The Invasion from Mars: A Study in the Psychology of Panic, with the Complete Script of the Famous Orson Welles Broadcast.* Princeton, NJ: Princeton University Press, 1940.

Carter, Paul A. *The Creation of Tomorrow: Fifty Years of Magazine Science Fiction.* New York: Columbia University Press, 1977.

Philmus, Robert M. *Into the Unknown: The Evolution of Science Fiction from Francis Godwin to H. G. Wells.* Berkeley: University of California Press, 1970.

Pierce, J. R. "Science and Literature." *Science,* New Series, 113:2938 (20 April 1951): 431–434.

Smith, David C. *H. G. Wells: Desperately Mortal, A Biography.* New Haven, CT: Yale University Press, 1986.

Scientism

Efforts to import the tools of natural science into other realms to make them more rational, analytical, and "scientific" belonged to a trend that now is often called scientism. Its outlook privileged theory and facts, and for every idea it demanded empirical data as evidence. Adherents of scientism shared a fundamental goal: objectivity. In fact, scientism is often called objectivism. Objectivity implies complete detachment between the observer and the observed, dispassionate analysis, and a lack of bias. To intellectuals of the late nineteenth and early twentieth centuries, the tools of the natural sciences appeared to promise much-needed objectivity in an era fraught with uncertainty.

Social scientists embraced scientism as they transformed their subjects in the early twentieth century into disciplines demanding fieldwork and quantitative analysis of data. Sociologists hoped to make their discipline more rigorous in this way, with textbooks such as Franklin Giddings's (1855–1931) *Inductive Sociology* (1901). Inspired by the philosophy of science promoted by Ernst Mach (1838–1916), such sociologists believed that the only way to ensure the veracity of their claims was to make their work more empirical. Still, critics suggested that sociologists were simply trying to add credibility to their ideas by clothing them in mathematics. It is true that one important result of scientism was the justification it gave to calls for action by social reformers. Other fields such as anthropology were moved by scientism as well. The rise of functionalist anthropology in the 1920s is directly related to the perceived need, by scholars such as Bronislaw Malinoswki, to make the subject less conjectural and more rigorous.

Scientism had a wide-ranging influence. In Europe and North America, it helped to transform the social reform impulse in the first decades of the century, in form if not in substance. Whereas reformers previously justified their efforts for humanitarian or for moral reasons, these motivations were replaced by the language of scientism: efficiency and social control. Rather than pointing to traditional authorities ("Jesus taught that . . ."), reformers reinforced their proposals with scientific findings ("Studies have shown that . . ."). Although faith in science (or particularly its product, technology) was shaken by the mechanized killing of World War I, scientism continued to be influential among reformers for whom no other methodology seemed able to provide an objective, nonreligious path to progressive change. Scientism provided a secular outlook to replace traditional values, without removing the promise of social progress. This was not just true in the West. The revolutionaries who led Sun Yat-sen's (1867–1925) postimperial China, for example, looked to objective science as a guidepost to replace the dominance of Confucianism as a philosophy of living. The enormous growth in the authority of science, or statements couched in scientific terms, was one of the principal historical developments of the early twentieth century.

See also Elitism; Patronage; Social Progress; Technocracy

References

Bannister, Robert C. *Sociology and Scientism: The American Quest for Objectivity, 1880–1940.* Chapel Hill: University of North Carolina Press, 1987.

Kwok, D. W. Y. *Scientism in Chinese Thought, 1900–1950.* New Haven, CT: Yale University Press, 1965.

Sorell, Tom. *Scientism: Philosophy and the Infatuation with Science.* New York: Routledge, 1991.

Scopes Trial

William Jennings Bryan (1860–1925) believed in 1925 that evolution and Christianity were dueling to the death. In the most famous contest between science and religion in the United States, the former presidential candidate represented fundamentalist Christians

Clarence Darrow at the Scopes evolution trial, Dayton, Tennessee, July 1925. (Library of Congress)

who refused to believe in Darwinian evolution. The idea that man evolved from apes was so repugnant to many religious-minded Americans that believing Charles Darwin's (1809–1882) ideas appeared to preclude belief in God. In 1925, the state of Tennessee passed the Butler Act, a law prohibiting the teaching of evolution in schools. Tennessee's governor Austin Peay said that its intent was to protest an alarming tendency to exalt science at the expense of the Bible. It was not intended to be enforced rigorously, but its existence tempted some to test its legality by arresting someone for the crime. That person was John T. Scopes (1900–1970).

Scopes himself taught chemistry and algebra, hardly typical venues to introduce evolution. But he also had been a substitute

teacher for two weeks in a biology class, during which time he assigned readings from a textbook that accepted Darwinian evolution. The American Civil Liberties Union (ACLU) had offered to finance a case to test the constitutionality of Tennessee's law, and several prominent citizens of Dayton thought it was a perfect opportunity to provide some publicity for their town. They asked Scopes to be the guinea pig for a trial, and he agreed.

The trial turned into a contest between the ACLU and the World's Christian Fundamentals Association. In the eyes of the world, on one side was agnostic criminal lawyer Clarence Darrow (1857–1938), retained by the ACLU; on the other was Bryan himself, an outspoken fundamentalist Christian. The verdict was expected: Scopes was found guilty. But many felt that the trial actually hurt the fundamentalists' cause. Journalists reported Tennessee "yokels" rejoicing in their lack of education and their belief that all scripture should be taken literally. Called to the witness stand, Bryan conceded that he did not take every part of the Bible literally, an admission that hurt his cause. The judge, however, was an unapologetic fundamentalist, and so perhaps were the jury members: They took less than ten minutes deliberating their verdict. Throughout the country, people wondered about this curious event taking place in the middle of nowhere, a debate that was part farce, part comedy, and part circus.

Scopes himself had to pay a fine of $100. Darrow appealed the case, but the ruling was upheld by the Tennessee Supreme Court. Although the ACLU no doubt wanted to take it to the U.S. Supreme Court, the state's top court prevented this by reversing the original decision to impose a fine on Scopes. The case ended, but the matter did not go away. Scopes retained a lifelong reputation as the "Monkey Trial" defendant. Tennessee, and the South in general, kept the reputation implied by the name "Bible Belt." The trial raised important questions: Did Americans want to be part of the spectacle? Should governments pass laws to eradicate controversial ideas because they offend religious doctrine? Perhaps the most lasting result of the trial was psychological, because it dramatized an alliance between fundamentalism and all varieties of ignorance and intolerance. Although Scopes lost the trial, fundamentalists no doubt lost many sympathizers. Still, the debate between science and religion continued long after memories of the Scopes trial faded.

See also Evolution; Religion

References

Gatewood, Willard B., Jr., ed. *Controversy in the Twenties: Fundamentalism, Modernism, and Evolution.* Nashville, TN: Vanderbilt University Press, 1969.

Larson, Edward J. *Summer for the Gods: The Scopes Trial and America's Continuing Debate over Science and Religion.* New York: Basic Books, 1997.

Seismology

Seismology is the study of earthquakes, typically through analysis of the waves produced by motions of the earth's crust. This scientific specialization started in the late nineteenth century in earthquake-plagued Japan and concentrated on measuring the waves using sensitive instruments. By the turn of the century, seismographs produced in Germany could record quakes originating continents away. Using the seismographs, scientists began to speculate on what the waves could reveal about the earth's interior. Because waves traveled at different velocities through different media, time-travel plots revealed the variety in density in the rocks beneath the earth's surface. For example, it became clear early on that the earth's core was denser, but less rigid, than the surrounding material known as the mantle.

Once the techniques for identifying and measuring differences came into use, natural disasters hastened their exploitation for seismological study. In 1906, a massive earthquake rocked San Francisco; although it was a tragic disaster for the residents, it led to the acquisition of seismographs by many leading U.S. universities. Other earthquakes led to new discoveries, such as one in 1909 in the Kupa Valley, in Croatia. The wave measurements from that quake led Andrija Mohorovičić (1857–1936) to propose a distinct boundary between the crust and the mantle, the two layers composed of very different material. The boundary became known as the Mohorovičić discontinuity.

Seismological work could not be completed in a single laboratory. Instead, it required recording stations in various distant parts of the earth. In the United States, seismology attracted the Society of Jesus, a Catholic association typically known as the Jesuits, who wanted to be involved in scientific studies but wanted to avoid controversial topics such as evolution. Because the Jesuits were spread out in institutions throughout the country, they were ideally suited to create an early network of seismological stations. As scholar Carl-Henry Geschwind has argued, the need to coordinate information within a highly structured organization of colleges was reminiscent of the role the Jesuits had played in science in early modern Europe. At the same time, the Jesuits' efforts to engage in science, so valued by contemporary U.S. society, helped in their process of adaptation to twentieth-century cultural values.

Despite the recording activities by the Jesuits in the United States, the center of seismological research remained in Germany, and most recording stations sent their measurements to leading German specialists for analysis. Beno Gutenberg (1889–1960) had become one of the most renowned seismologists by World War I, when he established the depth of the earth's core to be at about 2900 kilometers. He conducted work at the International Association of Seismology, in Strassburg, Germany. Most seismograph stations sent their results to Strassburg, which Gutenberg and others used as a central clearinghouse for records of the earth's movements. The war interrupted this work, and Strassburg itself became a French city (Strasbourg); Gutenberg was not rehired.

In the 1930s, the United States began to take over the leadership role in seismology. One reason was that Gutenberg accepted a post at the California Institute of Technology, where he was surrounded by other talents such as Hugo Benioff (1899–1968) and Charles F. Richter (1900–1985). Caltech, situated in Pasadena, California, also sat atop one of the most seismically active regions in the country. The Caltech group improved methods for identifying epicenters, the points on the earth's surface directly above the origin of an earthquake. Richter became well known for his development, in 1935, of the Richter scale, which measured the intensities of individual earthquakes on a numeric scale, using the logarithm of wave amplitudes measured from seismographs. The Richter scale, which measured the magnitude at the epicenter, was an improvement on the Mer-

calli scale, which measured magnitude at the point where the waves were measured.

After the development of the atomic bomb, seismologists took on a role of increasing importance for national security. Although the United States enjoyed a monopoly on the bomb for a few years after World War II, its leaders fully expected other nations to follow suit in developing the powerful weapon. In 1947, the Atomic Energy Commission realized that detecting a foreign test of such a weapon would be very difficult. It determined that, along with radiological monitoring of the environment, sensitive seismic techniques could be used for this purpose. Although the successful detection of atomic bombs detonated by the Soviet Union depended on finding traces of radioactivity in the atmosphere, the ensuing years saw increasing focus on seismology as the key to detecting nuclear blasts to ensure adherence to arms control treaties.

See also Atomic Energy Commission; Geophysics; Gutenberg, Beno; Mohorovičić, Andrija; Richter Scale

References
Geschwind, Carl-Henry. "Embracing Science and Research: Early Twentieth-Century Jesuits and Seismology in the United States." *Isis* 89 (1998): 27–49.
Oldroyd, David R. *Thinking about the Earth: A History of Ideas in Geology.* Cambridge, MA: Harvard University Press, 1996.
Shor, George G., Jr., and Elizabeth Noble Shor. "Gutenberg, Beno." In Charles Coulston Gillispie, ed., *Dictionary of Scientific Biography,* vol. V. New York: Charles Scribner's Sons, 1972, 596–597.
Ziegler, Charles. "Waiting for Joe–1: Decisions Leading to the Detection of Russia's First Atomic Bomb Test." *Social Studies of Science* 18 (1988): 197–229.

Shapley, Harlow

(b. Nashville, Missouri, 1885; d. Boulder, Colorado, 1972)

Harlow Shapley was a leading U.S. astronomer and science writer, director of the

Harvard College Observatory, and for some years was the representative of the view that the universe is not populated by distinct galaxies. He initially planned a career in journalism, but then turned to astronomy. He took an undergraduate degree at the University of Missouri before going to Princeton University, where he pursued astronomy when the department was headed by Henry Norris Russell (1877–1957). After taking his doctorate in 1914, he took a position at the Mount Wilson Observatory, in southern California, where he distinguished himself by calculating the size of the Milky Way and determining its center.

Shapley was deeply involved in one of the crucial controversies in astronomy in the first decades of the century. The controversy surrounded the nebulae, vast clouds in space that some believed were in fact distant "island universes." This had been debated in the nineteenth century, but in 1912 astronomers determined that spiral nebulae were moving at extremely high velocities, and in 1917 faint stars were detected in them. Both developments seemed to suggest that the nebulae might be distinct entities from our own "universe." The term *galaxy* is most commonly used to differentiate such systems from each other. In the same year, however, Shapley put forth his own theory of the universe's structure; our own galaxy, he noted, was about one hundred times larger than anyone previously believed. That would mean that the "island universes" were simply distant parts of our own system, rather than separate galaxies.

While working at the Mount Wilson Observatory, Shapley became part of a celebrated debate with Lick Observatory astronomer Heber D. Curtis (1872–1942) in 1920 at the National Academy of Sciences. Curtis defended the view that the nebulae were indeed distinct from our own galaxy. Taking the opposite view, Shapley argued that the spiral nebulae were moving very fast, and their speed precluded them from being as large as our own galaxy (otherwise one would have to conclude that parts of the galaxy exceeded the speed of light, which was deemed impossible). It followed, then, that they were smaller, component entities of the vast Milky Way. The rotation velocities of nebulae, upon which Shapley's objection was based, had been measured in 1916 by Shapley's friend and colleague Dutch astronomer Adriaan Van Maanen (1884–1946). Shapley later argued that their close relationship—and their mutual distaste for Edwin Hubble (1889–1953), one of Van Maanen's critics—probably clouded his judgment.

A lot was at stake for both Shapley and Curtis in the 1920 debate, aside from the acceptance of their theories. At one level, it was a duel between the two major observatories in California—Mount Wilson and Lick. But in addition, careers were made from the debate. Curtis made a very technical presentation and swayed most of the audience toward his view; he came out very well and soon took up a position as director of Allegheny Observatory. Shapley, the younger of the two, was hoping to impress his colleagues enough to attract an offer for the directorship of the Harvard College Observatory, a position that was offered to him after the debate (he became director in 1921). He was a personable speaker, with a certain charisma. His popular presentation was persuasive enough to convince many astronomers—fuller texts presenting both sides were later published by both men—that the island universe theory was incorrect.

Shapley's views about the status of the nebulae were decisively reversed in 1924, when Edwin Hubble at the Mount Wilson Observatory determined the distances to the spiral nebulae using Cepheid variable stars. The distances were so vast that even Shapley's grandiose estimates of the Milky Way's size could not possibly include them. Shapley had been wrong. Soon, Hubble would determine that not only were these nebulae distant galaxies, they were moving away from the earth at high velocities, and the universe was expanding.

Despite the fact that Shapley had been wrong about the nebulae, his status as a leading U.S. astronomer was not diminished. His work at Harvard continued, and it included important studies of the Magellanic Clouds. His open advocacy of international scientific cooperation with the Soviet Union provoked accusations that he was a Communist, and he and his colleagues were investigated by the Federal Bureau of Investigation and criticized by Senator Joseph McCarthy (1908–1957). Shapley retired from the directorship of the Harvard College Observatory in 1952, but continued in a research position for a few more years. His writings were both scientific and popular, including articles and books such as *Galaxies* (1943), *Of Stars and Men* (1958), and *The View from a Distant Star* (1963), which explore not only scientific concepts but also the place of mankind in a vast universe that scientists were only beginning to understand.

See also Astronomical Observatories; Astrophysics; Cold War; Cosmology; Hubble, Edwin; Loyalty

References

Hoskin, Michael A. "The 'Great Debate': What Really Happened." *Journal for the History of Astronomy* 7 (1976): 169–182.

Shapley, Harlow S. *Through Rugged Ways to the Stars.* New York: Scribner's, 1969.

Smith, Robert W. *The Expanding Universe: Astronomy's 'Great Debate' 1900–1931.* Cambridge: Cambridge University Press, 1982.

Wang, Jessica. *American Science in an Age of Anxiety: Scientists, Anticommunism, and the Cold War.* Chapel Hill: University of North Carolina Press, 1999.

Simpson, George Gaylord

(b. Chicago, Illinois, 1902; d. Tucson, Arizona, 1984)

George Gaylord Simpson, a paleontologist and biologist, was one of the principal figures in reinforcing the evolutionary synthesis by mid-century. Although others (especially population geneticists) were responsible for joining Mendelian genetics with Darwinian natural selection, Simpson's analysis of fossil records provided long-term empirical evidence for evolutionary theories. Simpson received his doctorate from Yale University in 1926 and became a postdoctoral fellow at the British Museum. When he returned, he took an appointment at the American Museum of Natural History, an institution with which he remained affiliated for more than thirty years.

Simpson's field research on the fossils of mammals took him to Patagonia in the early 1930s and other locales in South America in subsequent years. Combining samples from his extensive travels with the collections at the American Museum of Natural History, Simpson had access to an impressive body of fossils. He was the author of several works, the most significant of which was published during World War II, in 1944. This text, *Tempo and Mode in Evolution,* incorporated the evolutionary synthesis with evidence from paleontology. Simpson noted a broad agreement between the inheritance described by geneticists, from laboratory evidence, and the fossil evidence over huge spans of time. Simpson's work placed paleontology within the emerging synthesis of evolutionary biology, not only strengthening the validity of the latter but also opening new avenues for using paleontological evidence.

One important contribution of Simpson's book was its acknowledgment of different rates of evolution. He credited variability in the rapidity of change with many of the anomalous patterns in the known fossil record. He also tied the fossil record directly to Darwinian adaptive evolution rather than Lamarckian directed evolution; evolution in the former sense likened species to branches of a tree, whereas evolution in the latter sense ranked species as "higher" and "lower" on a scale, one evolving from the other. Simpson's version was adaptive, placing less influence on the need to search the fossil record for the "missing link" between apes and men. He popularized his ideas in his 1949 book, *The Meaning of Evolution*.

During World War II, Simpson served in the U.S. Army. When the war ended, Simpson's work helped to hasten the acceptance of what became known as the "modern synthesis" between genetics and evolution. In subsequent years he held positions at Columbia University, Harvard University, and the University of Arizona. He was a very private person, and he preferred the written word to conversation. He proved this by publishing numerous books in the postwar era, including his 1978 autobiography, *Concession to the Improbable*.

See also Evolution; Genetics; Missing Link
References
Bowler, Peter J. *Theories of Human Evolution: A Century of Debate, 1844–1944*. Baltimore: Johns Hopkins University Press, 1986.

Laporte, Léo F., ed. *Simple Curiosity: Letters from George Gaylord Simpson to His Family, 1921–1970*. Berkeley: University of California Press, 1987.

Simpson, George Gaylord. *Concession to the Improbable: An Unconventional Autobiography*. New Haven, CT: Yale University Press, 1978.

Smocovitis, V. Betty. *Unifying Biology: The Evolutionary Synthesis and Evolutionary Biology*. Princeton, NJ: Princeton University Press, 1996.

Skinner, Burrhus Frederic

(b. Susquehanna, Pennsylvania, 1904; d. Cambridge, Massachusetts, 1990)

B. F. Skinner was one of the principal founders of behavioral psychology. He attempted to provide a scientific basis upon which to develop principles of education. As a psychologist, he was second only to Sigmund Freud (1856–1939) in influence and notoriety in the twentieth century. Skinner was raised in Susquehanna, Pennsylvania, and he grew up with ambitions to become a writer. After graduating from Hamilton College, he attempted to devote serious attention to fiction writing, but failed to make any progress. He turned instead to psychology,

doing his doctoral work in that subject at Harvard University, finishing in 1931. His life's work in behavioral psychology was deterministic, emphasizing the power of behavioral science to plan and predict human behaviors.

Skinner's most influential conceptual contribution to behavioral psychology was "operant conditioning." He believed it was possible to design an environment so closely, with such attention to detail, that the subject living within the environment naturally would condition itself according to the designer's plan. Certain behaviors (operants) could be reinforced by a positive reward from the environment, such as food. For example, Skinner devised a contraption (called the Skinner Box) in which a small animal such as a rat could be placed, with a pedal inside that dispensed food. Once the rat accidentally pressed the pedal, the food was released and, now that the operant behavior had been reinforced, the animal repeated it. It was "conditioned" to do so by the planned environment, through reinforcement of behavior. He discussed the concept of operant conditioning in his 1938 book, *The Behavior of Organisms*.

Similar experiments with pigeons resulted in an interesting finding about superstition. Skinner found that operant behaviors could also be illusory, and a pigeon would replicate activities that it happened to be doing as the food was delivered. In this case, the food arrived at predetermined intervals, with no regard to the pigeon's action. Yet the bird believed that its own actions—a particular motion or activity such as turning around or standing in one part of the cage—had an effect. Skinner drew a parallel to human ritualistic activity, in which human beings believe erroneously that their activities, which may have coincided with the desired goal in the past, would bring about the desired outcome.

One of Skinner's most well-known and perhaps notorious experiments involved the creation of a "baby tender," a climate-controlled glass container that was designed to

Two pigeons in a box developed by psychologist B. F. Skinner are studied as part of research into operant conditioning. (Bettmann/Corbis)

make infants more comfortable. Skinner and his wife, Yvonne, believed it was possible to improve on the traditional crib, with its potentially entangling bars. Skinner abandoned the crib and used his new device as the home of one of his own daughters, Deborah, for more than two years. He wrote about the device in an article in *Ladies' Home Journal;* the editor chose an unfortunate title, "Baby in a Box." Skinner had called it the Aircrib, and it earned him scorn, because some believed that his faith in radical behaviorism had blinded him to the dubious ethicalities of using one's own child as an experimental subject.

Skinner's views about the value of behavioral psychology alienated some, particularly because his general outlook scratched against the grain of U.S. cultural values. His empha-

sis on conditioning and planning smacked of mind control; they appeared to fit more closely a totalitarian society rather than a free one. Skinner's 1948 *Walden Two* set forth the idea that behavioral sciences were society's best hope for improvement. The novel described a utopian society in which behavioral scientists engaged in social engineering, trying to shape communities according to their principals. It was possible, he argued, to influence people to pursue good. In later years he gained further notoriety and even hostility with his provocative *Beyond Freedom and Dignity,* which challenged U.S. culture to recognize the limits of such ideals as individual liberty and autonomy.

Skinner had a long and successful, if controversial, career. In the 1930s, he taught at the University of Minnesota, and in 1945 he became the chair of Indiana University's psychology department. During World War II, he helped the U.S. military establishment by designing a secret project to use behaviorally conditioned pigeons to guide bombs toward their targets (the project allegedly was discontinued before being finished). In 1947, he gave a series of lectures at Harvard University, and in 1948, that university invited him to return in a full-time position. His most controversial years were ahead of him, as his ideas about individual liberties and social engineering would be widely debated in the late 1960s and early 1970s. He died of leukemia in 1990.

See also Determinism; Psychology

References

Bjork, Daniel W. *B. F. Skinner: A Life*. New York: Basic Books, 1993.

Nye, Robert D. *The Legacy of B. F. Skinner: Concepts and Perspectives, Controversies and Misunderstandings*. New York: Wadsworth, 1992.

Skinner, B. F. *The Shaping of a Behaviorist: Part Two of an Autobiography*. New York: Knopf, 1979.

————. *Walden Two*. New York: Macmillan, 1948.

Social Progress

Science often has carried the connotation of social progress. As early as the Enlightenment, philosophers, economists, and political thinkers sought to "naturalize" their theories and make them more rational, like the sciences. Social reformers sometimes sought to remove traditional forms of power in society, such as political or religious elites, and instead build a society on scientific principles. The concept was challenged during World War I, because of scientists' role in creating new weapons for the world's most destructive war. But belief that scientific research was the cornerstone of human progress endured, especially in the United States.

In the United States at the turn of the twentieth century, the progressive movement hoped to base its reform agenda on rational policy choices, drawn from disinterested technical expertise. Thus science played an important role in their idea of social progress. Scientists who were part of this movement emphasized their role of training new technically minded people to fill important jobs in society. University presidents, such as future U.S. president Woodrow Wilson (1856–1924), then heading Princeton University, encouraged science education as a way to convert knowledge into prosperity for the country as a whole. The progressive James McKeen Cattell (1860–1944) bought the journal *Science* in 1895, and in 1900 he made it the official arm of the American Association for the Advancement of Science; that year he also became the editor of *Popular Science Monthly*. Leading U.S. science publications like his increasingly drew attention to the connections among science, economic growth, and other social benefits.

No event did more to shatter the progressive spirit of science than World War I. The image of scientists working for the progress of civilization was difficult to reconcile with the destructive fruits of science and technology: machine guns, mass-produced explosives, and the new innovation of poison gas. Much of this work began in Germany under the leadership of chemist Fritz Haber (1868–1934), whose research on nitrogen also allowed the German army to use synthetic explosives rather than rely on imported saltpeter. But scientists became deeply involved in war work on both sides of the conflict, helping their armies develop weapons, improve communications, and detect enemy ships with scientific methods. The end of the war brought some disillusionment not only with the rationale of the war, but also with science and its alleged role in promoting social progress.

Although some questioned anew the connections between science and progress, others capitalized on the relationship. In the struggle between civilizations, one could argue, scientists had shown themselves to play a crucial role. After World War I, science advocates increasingly demanded more financial support for science and recognition of scientists. In Britain, university researchers, industrial scientists, and government technicians banded together to form the National Union of Scientific Workers in 1918. This union insisted that social progress owed a great debt to scientists, and their prestige and compensation ought to reflect their important role. The solidarity among scientists consolidated science not merely as a community of intellectuals but also as a profession.

Evolutionary ideas shaped thinking about the possibility of social progress. One of the reasons for the pervasive belief in Lamarckian evolution, at the expense of Darwinian natural selection, was Jean-Baptiste Lamarck's (1744–1829) emphasis on willful evolution. Because Lamarck believed that evolutionary change occurred through the action or inaction of the organism, his ideas resonated with social progressives wishing to change society for the better. If species evolved in a progressive way, perhaps society as a whole could do so as well. Partly this view was due to beliefs that civilization was a biological product that could be inherited, rather than a product of being educated with specific social and intellectual norms. Also, some reformers saw natural selection's emphasis on purposeless variation and competition for resources as a reinforcement of the evils of industrial capitalism. If Lamarck's views were invalid, efforts to ensure civilization's progress through educational, political, and social reforms would fail. Playwright George Bernard Shaw often used his creative works, such as *Back to Methuselah,* as vehicles to repudiate natural selection, because it seemed to disallow social progress. The need to ensure willful, permanent progress in both the human organism and his society also stimulated Marxist scientists in the Soviet Union, such as Trofim Lysenko (1898–1976), to reject the random, competitive character of Darwinian evolution and to embrace Lamarckism instead.

In the United States, scientists actively promoted their role in social progress. Massachusetts Institute of Technology president Karl T. Compton (1887–1954) wrote in 1937 that human comfort and social progress could no longer be maintained by national acquisitiveness, such as expansion or conquering territory. Instead, intensive scientific research was required to make the wisest use of existing resources. World War II saw renewed appeals to connect science to social progress. Alan Gregg of the Rockefeller Foundation warned in 1944 that unless Americans were willing to support research in all fields, the United States would lose its strength. The American way of living—its processed foods, its transportation systems, its medical capabilities, and its national defense—depended largely on technology, and thus social progress increasingly was being defined by changes in technology, many of which could not be predicted prior to development. Gregg argued that the most progressive societies had to support fundamental research in such fields as medicine, chemistry, physics, and a host of others if the United States expected to ensure its progress.

The argument for fundamental research—scientific work with no known connection to practical technology—became commonplace in the postwar period. Scientists observed that their research could be used in unforeseen ways for the well-being and progress of society. The director of wartime research in the United States, Vannevar Bush (1890–1974), wrote *Science—The Endless Frontier* (1945) with the same plea for government support of fundamental research. This report was influential in persuading military and civilian arms of government to pursue progress through scientific research, leading eventually to the formation of the National Science Foundation in 1950.

See also Chemical Warfare; Elitism; Evolution; Great Depression; Patronage; Scientism; Soviet Science; Technocracy; World War I; World War II

References

Bowler, Peter J. *Evolution: The History of an Idea.* Berkeley: University of California Press, 1989.

Compton, Karl T. "Science in an American Program for Social Progress." *The Scientific Monthly* 44:1 (1937): 5–12.

Gregg, Alan. "The Essential Need of Fundamental Research for Social Progress." *Science,* New Series 101:2620 (16 March 1945): 257–259.

Kevles, Daniel J. *The Physicists: The History of a Scientific Community in Modern America.* Cambridge, MA: Harvard University Press, 1995.

Macleod, Roy, and Kay Macleod. "The Contradictions of Professionalism: Scientists, Trade Unionism and the First World War." *Social Studies of Science* 9 (1979): 1–32.

Social Responsibility

Social responsibility in science was a widely debated idea that scientists bore considerable ethical responsibility for the applications made from their ideas. In the realm of medicine, researchers became heroes to society because their contributions led to better public health measures, to vaccines, and to useful drugs of various kinds. But when scientific ideas were put to destructive use, such as the development and use of chemical weapons during World War I, praise for scientists turned into critique. For example, the German chemists Fritz Haber (1868–1934) and Walther Nernst (1864–1941) were criticized during and after the war for their roles in developing weapons to blind, asphyxiate, or kill enemy combatants. But it was not until World War II and the development of the atomic bomb, which shaped the postwar geopolitical situation, that social responsibility became a major issue among scientists.

During the Manhattan Project, the U.S. effort to build the bomb, numerous scientists developed a keen sense of personal responsibility for large-scale destruction. Scientists who had joined the project in the hope of helping the United States to deter Adolf Hitler's (1889–1945) use of atomic bombs now doubted whether the bombs should be used against Japan. Scientists such as James Franck (1882–1964) and Leo Szilard (1898–1964) attempted to convince their government superiors not to use them without warning. They stepped out of their roles as scientists and strode into the realm of politics and war, believing that their role in creating the weapon gave them some understanding of the ramifications of its use, and an obligation to advocate one course of action over another. Forming the Federation of Atomic Scientists (later Federation of American Scientists), these scientists published the *Bulletin of Atomic Scientists* to alert the world to the true ramifications of a nuclear era, specifically pointing to the dangers of an arms race with the Soviet Union and the need for international control of atomic energy. Sometimes called the scientists' movement, such actions irritated many political leaders, who believed that scientists should be "on tap" rather than "on top." In other words, scientists were a resource to be used by the country's leaders, not individuals who should try to shape policies themselves.

Not all scientists agreed with those wanting to assert—in the name of social responsibility—a larger role in public affairs. In 1946, Harvard physicist and recent Nobel laureate Percy Bridgman (1882–1961) discussed the topic in an address at the annual meeting of the American Association for the Advancement of Science. He noted that the development of the atomic bomb had made the public acutely aware of the relationship between science and society, and scientists were in the process of developing their own philosophy of their new role in civilization. But Bridgman cautioned against hasty conclusions. He believed that, although scientists might be deemed scientifically responsible for the uses of science, they should never be seen as morally responsible for every single use of their work. Just as the miner of iron ore should not be held responsible for the use of iron in weapons, the scientist should not be held responsible when scientific ideas are used for whatever purpose, even building weapons of mass destruction.

Part of the problem in the debate was the fact that the military became the primary supporter of scientific activity after the war. The Office of Naval Research began to fund scientific projects generously after it was founded in 1946, and a major civilian funding agency would not be created until 1950 (the National Science Foundation). Even if one took Bridgman's view, the mere fact that one's work was funded by the military implied a potential destructive application of one's work. A researcher taking money from the Navy could not claim ignorance of its possible future use. Still, restricting research struck many, including Bridgman, as a blow against the freedom of scientific inquiry.

One of the most eminent scientists to lend his name to the movement was Albert Einstein (1879–1955), who joined the Society for Social Responsibility in Science in 1950. A longtime pacifist, Einstein believed that scientists and engineers carried particular moral obligations because weapons of mass destruction came largely from scientific efforts. At the very least, an organization was needed for the discussion of moral problems brought about by science, to help scientists—individually and collectively—follow not only their academic interests but also their consciences.

See also Atomic Bomb; Chemical Warfare; Einstein, Albert; Federation of Atomic Scientists; Manhattan Project; Office of Naval Research; Patronage; Szilard, Leo

References
Bridgman, P. W. "Scientists and Social Responsibility." *The Scientific Monthly* 65:2 (1947): 148–154.
Einstein, Albert. "Social Responsibility in Science." *Science,* New Series, 112:2921 (22 December 1950): 760–761.
Smith, Alice Kimball. *A Peril and a Hope: The Scientists' Movement in America, 1945–1947.* Chicago, IL: University of Chicago Press, 1965.

Solvay Conferences

The Solvay Conferences take their name from Ernest Solvay (1838–1922), a Belgian chemist whose development of a process for producing sodium carbonate allowed him to amass a great deal of money in soda production in the nineteenth century. Solvay's wealth and influence prompted him to help found a number of research institutes, including the Solvay Institutes for Physics and Chemistry. The Solvay Institutes sponsored a series of research conferences; the Solvay Conference on Physics, inaugurated in Brussels in 1911, became a major forum for discussion, debate, and defining scientists' theoretical positions during crucial years in the history of physics. Solvay himself died in 1922, but the conferences continued (periodically—they were not annual) to shape debate among physicists long after his death.

The Solvay Conferences, and the fruitfulness of the discussions that occurred during them, demonstrated to leading physicists the importance of establishing person-to-person contacts across national borders. Journals no longer seemed to suffice as vehicles for exchanging ideas, especially at a time of such rapid change and intense debate among the top physicists. The Solvay Conferences played an important role particularly in the years 1911–1913 and 1924–1930, periods during which crucial concepts such as quantum physics, relativity, and quantum mechanics were openly debated. On these occasions, the conferences did not serve merely to present new evidence, they also served to discuss and argue broad, far-reaching interpretations and deep theoretical foundations. The importance of these discussions was reinforced when the proceedings volumes of the conferences included not only texts of papers presented, but also synopses of the discussions.

One of the best-known uses of the forum was the debate, in 1927 (and again in 1930), between Albert Einstein (1879–1955) and

Some of the leading figures in physics, including Albert Einstein (standing second from right), attending the Solvay Conference. (Hulton-Deutsch Collection/Corbis)

Niels Bohr (1885–1962). With the development of quantum mechanics, and in particular the announcement by Werner Heisenberg (1901–1976) of the uncertainty principle, scientists began seriously to consider the physical and philosophical ramifications of a science that abandoned old notions of cause and effect. Einstein abhorred indeterminacy, famously noting that God does not play dice; Bohr, conversely, embraced the disturbing limitations on knowledge implied by Heisenberg's principle and his own complementarity principle. Uncertainty and complementarity both helped to shape what became known as the Copenhagen interpretation of quantum mechanics (Bohr was a Dane). The Solvay Conferences in these years played the pivotal role of showcasing the discussions between these two eminent theoreticians. Einstein appeared to have lost on both occasions, perhaps hastening the acceptance of the Copenhagen interpretation over Einstein's own determinism.

See also Bohr, Niels; Determinism; Einstein, Albert; Physics; Quantum Theory; Relativity; Uncertainty Principle

References

Kragh, Helge. *Quantum Generations: A History of Physics in the Twentieth Century.* Princeton, NJ: Princeton University Press, 1999.

Marage, P., and G. Wallenborn, eds. *The Solvay Council and the Birth of Modern Physics.* Basel, Switzerland: Birkhäuser, 1998.

Mehra, Jagdish. *The Solvay Conferences on Physics: Aspects of the Development of Physics since 1911.* Dordrecht: Reidel, 1975.

Soviet Science

In the Soviet Union, science played an important role in shaping society. Since the time of Peter the Great in the eighteenth century, Russians used and adapted science and technology to the local situation but typically imported it from European countries. Russian leaders often were suspicious of the foreign influences inherent in science. The Russian word for science, *nauka,* actually meant natural science and secular learning, making it the work of all nonreligious intellectuals. Thus it was viewed as a potentially subversive influence by the czars. Even after the Russian Revolution, the identification of science with the West proved problematic, because communism in Russia was explicitly against the degrading influence of industrial technology in the West. But science was transformed into a powerful tool under the leadership of Joseph Stalin (1879–1953) in the 1930s and 1940s. The Soviet regime was totalitarian; this meant not only active loyalty by all the people to the regime's goals, but also active participation in all aspects of society by the state. Science and scientists were not immune to this, and they were tied very closely to the activities and outlooks of the Soviet state.

One of the threats allegedly posed by scientists was the philosophy at the root of scientific practice, which appeared to be at variance with Bolshevik ideology. Science required freedom of inquiry and was a kind of individualism with which many Soviet leaders were uncomfortable. Between 1928 and 1931, Stalin addressed such tensions by promoting the careers of younger scientists and engineers so that, by the mid-1930s, a much stronger sense of cooperation existed between them and the government. Because science and technology played such a crucial part of Stalin's plan for a new Communist Russia, scientists and engineers acquired a special cherished status in society. The Soviet government spent far more money on research and development in the 1930s, as percentage of gross national product, than

did the United States. The Communist Party took an active role in trying to implement scientists' recommendations. At the same time, Stalin intervened personally to ensure that science and technology served his political goals. For example, in the 1930s, he directed research on aircraft toward maximizing speed, in order to have Russian engineers receive credit for record-breaking flights. But Stalin's primary goal in the 1930s was to use science to help the country industrialize quickly and collectivize its agriculture efficiently. Such efforts were noticed outside the Soviet Union and attracted the admiration of some Western intellectuals, such as British scientist John D. Bernal (1901–1971), who advocated for more active governmental involvement in science in his own country.

The most renowned individual to capitalize on the political aspects of the Soviet Union was the agronomist Trofim Lysenko (1898–1976), who achieved official support for his own non-Mendelian approach to plant breeding. Lysenko and others believed that genetics was too formalistic and that Darwinian evolution implied randomness in nature that did not leave room for engineering. His own outlook, that nature could be transformed by design and changed permanently for the better, fit well with the progressive goals of the Soviet state. He managed to convince Stalin and others of this and imposed conformity of opinion among biologists to suit his own theories and to support a Stalinist vision of nature. Although Lysenko's dominance might be attributable in part to wishful thinking and the influence of ideology over science, it also was because of the fact that no one else proposed to increase crop yields at the enormous levels promised by Lysenko. Defending genetics and opposing Lysenko seemed to be the same as attacking Stalin's policy of collectivization. After 1948, after a major meeting of agronomists at the Leningrad Academy of Sciences, genetics virtually disappeared from the Soviet scientific landscape.

When Soviet ideology was integrated into science, it was called *dialectical materialism.* This was a term that borrowed from the ideas of German philosopher Georg Hegel (1770–1831), who had proposed that progress is generated by "dialectic," a dynamical system of struggling opposites. Friedrich Engels (1820–1895) and Karl Marx (1818–1883) added *materialism,* believing that all struggle concerned material things. The meaning of dialectical materialism was developed in the late nineteenth and early twentieth centuries, by socialist thinkers and by the leaders of the Russian Revolution. Both Vladimir Lenin (1870–1924) and Joseph Stalin described dialectical materialism as the only theory of knowledge compatible with communism, and scientists were obliged to ensure that their theories abided by it. As a rule of thumb, scientists avoided theories that ignored materialistic forces or that seemed too idealistic, lest they be subjected (as in the case of Lysenko's enemies) to political critique and possibly punishment. Scientific debate in the Soviet Union was never far removed from political conflict.

Stalin's scientific priority after World War II was developing an atomic bomb to rival that of the United States. In fact, his desire to maximize this effort shielded physicists from the ideological perils faced by biologists. The Soviet atomic bomb project began after the battle of Stalingrad, in early 1943. Once the German advance had been halted, Stalin was able to plan some long-term projects. The team to build the bomb was led by physicist Igor Kurchatov (1903–1960). Thanks in part to spies in the U.S. atomic bomb project, the Soviet Union tested its first atomic device in 1949, years before most Westerners had predicted they would. By mid-century, then, the Soviet Union had made a serious commitment to keeping apace with the United States in science and technology and continued to envision scientists and engineers as crucial parts of a prosperous Communist society that could defend itself against Western powers.

See also Academy of Sciences of the USSR; Cold War; Lysenko, Trofim; Nationalism; Patronage; Philosophy of Science; Social Progress; Vavilov, Sergei

References
Bailes, Kendall E. *Technology and Society under Lenin and Stalin: Origins of the Soviet Technical Intelligentsia, 1917–1941.* Princeton, NJ: Princeton University Press, 1978.
Holloway, David. "Science in Russian and Soviet Society." *Science Studies* 3 (1973): 61–87.
———. *Stalin and the Bomb: The Soviet Union and Atomic Energy, 1939–1956.* New Haven, CT: Yale University Press, 1994.
Lewis, Robert. *Science and Industrialization in the USSR: Industrial Research and Development, 1917–1940.* London: Macmillan, 1979.

Sverdrup, Harald
(b. Sogndal, Norway, 1888; d. Oslo, Norway, 1957)

The Norwegian Harald Sverdrup was a meteorologist and oceanographer who helped to transform oceanography in the United States and build a strong institutional support for physical oceanography. Sverdrup was heavily influenced by another Norwegian, the meteorologist Vilhelm Bjerknes (1862–1951), and he took his doctorate under the direction of Bjerknes in 1917. Like his mentor, his early interests were in the atmospheric sciences. He broadened this to include dynamic interactions of many kinds, from winds to ocean currents.

In 1918, Sverdrup became the scientific leader of Roald Amundsen's (1872–1928) Arctic expedition, which lasted for over seven years. During the expedition he spent several months living in Siberia and also visited the United States, meeting scientists at the Scripps Institution of Oceanography, in La Jolla, California. He was developing an abiding interest in the oceans, particularly in tidal currents, and his studies made substantial contributions to dynamical oceanography. Upon his return from the expedition, Sverdrup took a chair in meteorology at the University of Bergen and was fast becoming the leading figure in studies of the sea and atmosphere.

Sverdrup helped to transform U.S. oceanography by orienting it toward interdisciplinary studies with a strong emphasis on physical oceanography and meteorology. Marine science in the United States had been primarily devoted to biological studies such as the role of plankton; the trend in Norway, which many Americans wished to emulate, was moving toward studies of air and sea dynamics. Sverdrup's experience with physical oceanography and his close affiliation with Bjerknes convinced scientists at Scripps to invite him to become their director. He moved to the United States in 1936 and brought his dynamic approach to bear on the institution's research. During the war years, Sverdrup and colleagues Richard Fleming and Martin Johnson published *The Oceans: Their Physics, Chemistry, and General Biology* (1942), which became a widely used textbook for oceanographers.

Sverdrup's position as director was supposed to terminate in 1940 but was extended because of the war. During these years, Scripps developed a long-lasting relationship with the military, owing largely to a good working relationship between Sverdrup and Navy scientist Roger Revelle, who later became director of Scripps. Under Sverdrup's leadership, oceanographers contributed to the war effort in numerous ways, developing new methods in antisubmarine warfare and amphibious troop landings. Recently historians have uncovered evidence that Sverdrup was suspected during the war of being a Nazi sympathizer. He was repeatedly denied security clearance, which was embarrassing given the close coordination between the Navy and the oceanographers at Scripps. He was barred from being involved in many of his own institution's projects. Ultimately he decided to leave the United States and resettle in his native Norway, where he took up the directorship of the Norwegian Polar Institute. In the 1950s, he continued his involvement in oceanographic affairs, spearheading major oceanographic efforts for the International Geophysical Year (1957–1958). He died before these preparations were implemented.

See also Bjerknes, Vilhelm; Loyalty; Oceanic Expeditions; Oceanography

References

Mills, Eric L. "The Oceanography of the Pacific: George F. McEwen, H. U. Sverdrup and the Origin of Physical Oceanography on the West Coast of North America." *Annals of Science* 48:3 (1991): 241–266.

Nierenberg, William A. "Harald Ulrik Sverdrup." *Biographical Memoirs of the National Academy of Sciences* 69 (1996): 339–374.

Oreskes, Naomi, and Ronald Rainger. "Science and Security before the Atomic Bomb: The Loyalty Case of Harald U. Sverdrup." *Studies in the History and Philosophy of Modern Physics* 31 (2000): 309–369.

Spjeldnaes, Nils. "Harald Ulrik Sverdrup." In Charles Coulston Gillispie, ed., *Dictionary of Scientific Biography,* vol. 13. New York: Charles Scribner's Sons, 1976, 166–167.

Szilard, Leo

(b. Budapest, Hungary, 1898; d. La Jolla, California, 1964)

Leo Szilard had more ideas than he could pursue. Colleagues often liked having him around simply for his insights and fresh ideas. But he was also a difficult man, very outspoken and indifferent to authority. He alienated people easily. Consequently, his scientific contributions occasionally have been brushed aside, particularly by his enemies. But Leo Szilard, a Hungarian-born physicist, biologist, and political activist, played a central role in the development of atomic energy in the United States. He was one of a handful of Hungarian-born physicists who immigrated to the United States and became scientific celebrities, a group that included Edward Teller (1908–2003) and John Von Neumann (1903–1957).

Szilard's flighty scientific habits were reflected in the way he lived. He spent most of his nights in hotels, with his belongings barely unpacked and ready for travel on short notice. He was not a disciplined worker: His ideas came to him as he soaked for hours in the bathtub. He carried a sense of spontaneity throughout his life, undoubtedly strength-

Hungarian physicist Leo Szilard posing with a newspaper after the United States detected the first Soviet atomic test. (Bettmann / Corbis)

ened by the fact that in 1933 he left Nazi Germany by train one night before an anti-Jewish boycott would have made such a journey very difficult or impossible (Szilard was a Jew).

In the 1930s, Szilard was obsessed with the possibility of achieving a nuclear chain reaction. Soon after the discovery of the neutron in 1932, he believed that a free neutron could trigger a fission reaction and release energy, while ejecting more than one neutron; these new neutrons could initiate the same process in other atoms, thus creating a chain reaction. He took out a patent on the idea, although he identified beryllium, not uranium, as the element that could be split. Initially, he was unable to convince anyone that his ideas were possible. He asked the esteemed Ernest Rutherford (1871–1937) if he could work on his ideas at his laboratory (the

Cavendish Laboratory), and Rutherford practically threw him out of his office.

During these years, Szilard published research on isotope separation and neutron absorption, establishing his name in physics and making it difficult for more eminent scientists to ignore him. He also moved to the United States. Italian physicist Enrico Fermi (1901–1954) had been working on neutron reactions as well and received the Nobel Prize for his efforts in 1938; Fermi traveled from the prize ceremonies to the United States, where he and his wife just happened to check in to the New York hotel where Szilard was staying. Fermi took an appointment at Columbia University, and Szilard, with no appointment, often came by to enlighten—or harass—the physicists there with ideas of his own.

In 1936, Szilard urged his fellow physicists outside Germany not to publish their neutron research. He believed that a war was coming, and he feared that atomic bombs would determine its outcome. It made no sense to aid Adolf Hitler's (1889–1945) scientists in their effort. He asked the British Admiralty to put his patent under secrecy, which they did. Other scientists, including Fermi, disliked Szilard's meddling with the established process of scientific research and publishing. Nonetheless, Szilard's continued urging for self-censorship, especially after the discovery of fission in 1939 by scientists in Germany, convinced some U.S. scientists not to publish.

Szilard wanted the United States to help its scientists explore fission and its military uses before the Nazis did. His work on neutrons indicated the increasing likelihood of sustaining a chain reaction. One of the outstanding technical problems was the need for a moderator amidst the uranium to *slow down* neutrons to make fission more probable *but not to capture them,* thus rendering them useless in a chain reaction. But an even bigger problem was not technical at all. Szilard needed to convince the U.S. government to start research on fission reactions and to begin stockpiling uranium.

Leo Szilard visited Albert Einstein (1879–1955), a friend and colleague from their days in Berlin in the 1920s, by now the most famous scientist in the world. He convinced Einstein to write a letter to President Franklin D. Roosevelt (1882–1945), in August 1939, urging him to take action in fission research and uranium acquisition. Einstein later said he was simply acting as Szilard's mailbox; Einstein's letter is now famous for having commenced the U.S. effort to build the atomic bomb. Szilard soon was a key part of Enrico Fermi's team, which, by 1942, created the first atom "pile," a reactor that sustained the kind of nuclear chain reaction that Szilard had envisioned nearly a decade earlier.

Szilard made numerous enemies during these years, particularly General Leslie Groves (1896–1970), the military leader of the Manhattan Project. Throughout the war, Groves tried to discredit Szilard by making him appear to be a security threat. Szilard was equally hostile to Groves: He often complained about Army mismanagement and excessive secrecy and was routinely disrespectful of Groves's authority. Szilard became concerned about the postwar world and the possibility of an arms race between the United States and the Soviet Union. He tried to stop the United States from using the bomb, which Szilard saw as the first step in a new postwar conflict rather than the final act of the present war. He joined a committee headed by physicist James Franck (1882–1964), which issued a report outlining alternatives to using the bomb. The Franck report suggested demonstrating the bomb to the Japanese first, to give them a chance to surrender. The Franck report stands as a testament to Szilard's and Franck's (as well as others') efforts to assert a sense of social responsibility in scientific activity. The committee's recommendations, however, were rejected, and President Harry Truman (1884–1972) decided to drop two bombs, one on Hiroshima and one on Nagasaki.

Szilard's agitation, against the wishes of Groves and even scientific leaders such as J. Robert Oppenheimer (1904–1967), became more pronounced when the war ended. He opposed the postwar legislation that would have given the military greater control of atomic energy; Szilard advocated openly for the McMahon Bill, which provided for greater civilian control. He was a founding member of the Federation of Atomic Scientists, a body that sponsored the *Bulletin of the Atomic Scientists* to air scientists' often controversial views on science policy.

Groves retaliated to Szilard's actions during these years by pushing him out of atomic energy altogether. Groves tried to prevent public shows of appreciation for Szilard, even writing to officials to prevent something as basic as a Certificate of Appreciation from the

Army. He criticized books about the atomic bomb that emphasized Szilard's importance, evening belittling encyclopedia entries. In his words, Szilard was a "parasite living on the brains of others" (Lanouette 1992, 313).

Szilard made the best of a bad situation, scientifically by pursuing biology and politically by advocating arms control. When Jacques Monod (1910–1976) shared a Nobel Prize in 1965 for his work on enzymes, he credited Szilard for his productive ideas in this area. He regularly put his scientific celebrity to use, in the Pugwash conferences between scientists of the East and West, the Council for a Livable World (which he founded in 1962), and even by meeting personally with Soviet premier Nikita Khrushchev (1894–1971) in 1960 to discuss ways of easing global tensions. Throughout his life, Szilard and his fellow Hungarian physicists were dubbed Martians for their bizarre genius, unorthodox views, and outsider status. More than the others, Szilard was both at the center of activity and on the periphery, never quite belonging. As his biographer called him, he truly was a genius in the shadows.

See also Atomic Bomb; Brain Drain; Federation of Atomic Scientists; Manhattan Project; Social Responsibility

References

Lanouette, William, with Bela Szilard. *Genius in the Shadows: A Biography of Leo Szilard, the Man behind the Bomb.* Chicago, IL: University of Chicago Press, 1992.

Marx, George. *The Voice of the Martians.* Budapest: Akademiai Kiado, 1997.

Szilard, Leo, Spencer R. Weart, and Trude Weiss Szilard, eds. *Leo Szilard: His Version of the Facts.* Cambridge, MA: MIT Press, 1980.

T

Technocracy

Rule by scientific elite was a fantasy of Francis Bacon at the turn of the seventeenth century. After the industrial revolution of the nineteenth century, some thought that the progress of civilization would be ensured if governments were entrusted to the brightest minds trained in technical fields. But the twentieth century saw both endorsement and fear of technocracy. Should society put its faith in science and technology? Many people in the United States, Europe, and Russia thought so. Some of that faith faltered during World War I, which saw the first use of chemical weapons and other destructive products of superior technology. It stumbled further during the Great Depression, when advances in technology favored automation in a time of rising unemployment. Only World War II provided a boost to the idea that governments should be influenced by— if not run by—a set of specialists in science and technology.

After the Bolshevik Revolution of 1917, when Russia's aristocratic elites disappeared from government, mid-level technical specialists rose to the higher ranks. The process of replacing old regime functionaries with engineers, scientists, and physicians led to an emphasis on technical expertise in state planning, and it also provided a new means of social mobility in the Soviet Union. This process was encouraged by the new political leaders who hoped to fully transform the government from a czarist bureaucracy to a Communist one. The technocrats, previously a sub-elite class, became a major source of support for Joseph Stalin (1879–1953) in the 1930s during his reforms of agriculture, industry, and education. Both Vladimir Lenin (1870–1924) and Stalin tied Communist progress closely to technological progress, whether it was Lenin's effort to provide widespread access to electricity or Stalin's dogged pursuit of the atomic bomb.

Other countries saw science and technology as a mark of civilization as well, but technological innovation also sparked hostile reactions. U.S. president Herbert Hoover's (1874–1964) Committee on Recent Social Trends concluded in 1933 that social innovation was far behind technical innovation. Some argued that permanent unemployment would be the natural result, unless technocrats took over to create social reforms to keep pace with technological change. The technocratic movement in the United States was brief, but its basis was readily comprehensible: Men were being replaced by machines in the workforce, and without specialists in charge, this would continue unabated. But entrusting scientists with society's broad problems did not appeal to everyone; if science was destroying jobs, scientists could not be trusted to run the country. The Great Depression saw, instead, a rise in religious groups' hostility to science and the increasing influence of the humanities on university campuses.

World War II, with its scientific and technological miracles of radar, penicillin, and the atomic bomb, revived the concept of technocracy in the United States. Vannevar Bush (1890–1974), the head of the wartime Office of Scientific Research and Development, tried to secure permanent recognition of science as the cornerstone of the nation's welfare by publishing a report on postwar research called *Science—The Endless Frontier*. His efforts, along with Manhattan Project scientists' attempts to shape postwar atomic energy policy by lobbying for civilian control, provoked hostility by political and military leaders who felt that scientists should be "on tap" rather than "on top." The early Cold War period would mark a high point in the power of scientists in government, and the dream—or nightmare—of technocracy seemed to be at hand. Although scientists' recommendations were often ignored, as in the case of a top advisory panel's 1949 recommendation not to develop the hydrogen bomb, scientists entered the corridors of power increasingly in the 1950s. This was owing largely to U.S. and Soviet efforts to compete for scientific and technological superiority while pursuing research on new weapons. By the end of that decade, President Eisenhower would warn the nation about the growing power of the military-industrial complex in which scientists often played the leading role.

See also Elitism; Great Depression; Patronage; Scientism; Social Progress; World War II

References

Akin, William E. *Technocracy and the American Dream: The Technocrat Movement, 1900–1941.* Berkeley: University of California Press, 1977.

Rowney, Don K. *Transition to Technocracy: The Structural Origins of the Soviet Administrative State.* Ithaca, NY: Cornell University Press, 1989.

Zachary, G. Pascal. *Endless Frontier: Vannevar Bush, Engineer of the American Century.* Cambridge, MA: MIT Press, 1999.

Teilhard de Chardin, Pierre
(b. Orcines, France, 1881; d. New York, New York, 1955)

Pierre Teilhard de Chardin was a Jesuit scholar and philosopher who was associated with many of the famous paleontological findings of his time and who developed a controversial spiritual philosophy that attempted to bring together science and religion. Taking an interest in paleontology, geology, and anthropology, Teilhard's scientific efforts gained some notoriety because he was among the first researchers to analyze the bone fragments in Zhoukoudian, near Beijing, China, known as Peking Man. His philosophical writings attempted to incorporate the notion of evolution into his faith. A Frenchman, Teilhard attended the Jesuit College of Mongré at Villefranche-sur-Saône, joined the Society of Jesus (the Jesuit Order), and lived in England and Egypt from 1904 to 1912 while completing religious studies. Upon his return to France, he became interested in evolution when studying under French paleontologist Marcellin Boule.

Teilhard's initial notoriety came when he traveled to England in 1913 and distinguished himself by discovering a human tooth at the famous Piltdown site. However, in the 1950s, the whole site was decreed fraudulent by scientists, posing a giant question mark on Teilhard's integrity. He published on other subjects, such as Eocene mammals. After serving his country in North Africa during World War I, Teilhard returned to France and received his doctorate from the Sorbonne in 1922. He then became a professor of geology at the Institut

Pierre Teilhard de Chardin (right) gained a reputation as a paleontologist and later tried to incorporate evolutionary theory into Catholic theology. (Bettmann / Corbis)

Catholique de Paris. But his nascent ideas about the relationship between science and religion proved intolerable to the Jesuit Order; he lost his position and was obliged not to publish his ideas. He did continue his scientific writing, however. After hearing about the discovery of a major deposit of ancient remains in China, he traveled to Beijing and became one of the early writers on Peking Man. He also traveled to eastern Africa and helped to excavate areas along the rail lines. He returned to China to become the scientific adviser to the Geological Survey, where he worked until the outbreak of war. He also conducted fieldwork in India, Burma, Java, and South Africa.

His embrace of scientific ideas compelled Teilhard to reshape his own faith. In the 1920s and 1930s he developed a spiritual philosophy that would, after his death, inspire many followers. His philosophy incorporated evolution not merely into biology but also into human consciousness. Teilhard's tendency toward Lamarckism is evident in the fact that his conception of universal evolution, or "cosmogenesis," is directed toward a goal. In Darwinian evolution, evolution is random, branching out in many directions without a goal. Lamarckian evolution is goal-oriented, directed toward increasingly more useful forms. For Teilhard, the human consciousness evolves in the Larmarckian sense toward increasing complexity, sophistication, and—perhaps most important—spiritualization. In his view, the randomness of Darwinian evolution is tempered by the will of human consciousness to evolve toward greater complexity. Teilhard wanted to emphasize that the evolutionary process created (and continues to create) the mind, not just the body. Eventually, at some point in our evolutionary future, all human minds will unite into a single entity or personality, at what Teilhard called the Omega point. This concept underlies Teilhard's spiritual monism, or his conviction that all human beings are linked as parts of a single spirit. These ideas were set forth in his book, *The Phenomenon of Man,* written in the 1920s and 1930s but not published until after his death.

Although Teilhard gained many followers through his philosophical efforts, he created numerous enemies. Some scientists thought he used the doctrines of evolution selectively; clergymen made the same charge about the doctrines of the Catholic faith. In general, the Catholic Church took a dim view of Teilhard's efforts to refashion the nature of faith, especially his notion that the goal of the future was the unity of the mind. His reputation as a philosopher, for the most part, was achieved posthumously. Many scientists criticized his work and sought to discredit him. Writer and evolutionary biologist Stephen Jay Gould went so far as to suggest that Teilhard was responsible for the infamous Piltdown forgery.

See also Evolution; Peking Man; Piltdown Hoax; Religion

References

Birx, H. James. *Pierre Teilhard de Chardin's Philosophy of Evolution.* Springfield, IL: Charles C. Thomas, 1972.

Cuénot, Claude. *Pierre Teilhard de Chardin.* Paris: Seuil, 1958.

Dodson, Edward O. *The Phenomenon of Man Revisited: A Biological Viewpoint on Teilhard de Chardin.* New York: Columbia University Press, 1984.

Movius, Hallam L., Jr. "Pierre Teilhard de Chardin, S. J., 1881–1955." *American Anthropologist,* New Series 58:1 (1956): 147–150.

Thomson, Joseph John

(b. Manchester, England, 1856; d. Cambridge, England, 1940)

Joseph John Thomson, or J. J. Thomson as he was more commonly known, discovered the electron at the end of the nineteenth century. His discovery began a period of intensive scientific activity in subatomic physics, which became the most significant field of physics in the first half of the twentieth century. He became Cavendish Professor at Cambridge University's Cavendish Laboratory, a position previously held by James Clerk Maxwell (1831–1879) and John William Strutt, known better as Lord Rayleigh (1842–1919). He proposed the first widely accepted model of atomic structure. Under his guidance, the Cavendish Laboratory became a center of research on rays, atomic structure, radioactivity, and subatomic particles.

In the late nineteenth century, many physicists began to investigate the behavior of electricity in evacuated tubes. The conduction of electricity through a near vacuum appeared to produce a kind of ray, lighting up the inside of the tube. According to then-accepted physical concepts, these "cathode rays" could be propagated through a vacuum by the "ether," an ill-defined substance that allegedly was omnipresent but not detectable. The cathode rays appeared to behave like light, and thus some concluded that they were waves. Other evidence indicated that the rays were in fact material in nature. In the 1890s, experiments showed that the rays could be manipulated by magnets and that they carried a negative charge. In 1897, German physicist Emil Wiechert (1861–1928) determined that the ratio of mass to charge was much smaller (by more than a thousand times) than the accepted ratio for the smallest charged atom. The implication was that if the rays were made up of particles, these particles would need to be far tinier than anything yet known.

Thomson was also deeply involved in exploring cathode rays. His lectures to physicists in the United States culminated in his 1897 book, *Discharge of Electricity through Gases.* But it was upon his return to England that he made a more important discovery, that the "rays" were indeed built up of particles and that they were the constituents of all atoms. He believed that his experimental evidence, by electromagnetic deflection and by measuring the kinetic energy of the rays, had proven it. When he discovered it in 1897, he did not initially call it the electron, but chose instead the word *corpuscle* to emphasize the material nature of the particle. The *electron* had already been proposed theoretically by others, notably Henrik Lorentz (1853–1928) and Joseph Larmor (1857–1946), but these physicists had conceived of electrons as charges without matter. For Thomson, the new particle was very much matter, and indeed he believed it was the fundamental form of matter in atoms. Despite Thomson's choice, the name *electron* proved more lasting.

After making his discovery, Thomson went on to devise a model of atomic structure. He envisioned small electrons surrounded by positively charged, fluidlike matter. The visual image of electrons swimming in a charged soup earned Thomson's concept the name "plum pudding" atom. Between 1904 and 1909, Thomson wrote frequently of his model, with electrons rotating about the atom, bound to it by a homogenous mass of positively charged electricity. This model proved empirically problematic for several reasons. The most decisive one was that the "plum pudding" atom failed to account for the experiments on alpha scattering taking place at the University of Manchester during these years. Ernest Rutherford (1871–1937), who presided over these experiments, introduced a replacement—the nuclear atom—for Thomson's model in 1911, and after a few years the "plum pudding" atom lost its adherents.

The inadequacy of Thomson's atom did not detract from his monumental accomplishment of identifying the electron as a subatomic particle. In 1906, Thomson received

the Nobel Prize in Physics. His role at the Cavendish was taken over by Rutherford himself, who assumed the mantle of leadership in subatomic physics in the second decade of the century. Thomson resigned from Cavendish Laboratory in 1919 to become master of Trinity College (Rutherford became the director of the Cavendish). His son, George Paget Thomson, also became a world-renowned physicist for his work on electrons (and in 1937, he also won a Nobel Prize for his efforts).

See also Atomic Structure; Cavendish Laboratory; Physics

References

Dahl, Per F. *Flash of the Cathode Rays: A History of J. J. Thomson's Electron.* London: Institute of Physics, 1997.

Davis, E. A., and Isabel Falconer. *J. J. Thomson and the Discovery of the Electron.* Bristol, PA: Taylor & Francis, 1997.

Kragh, Helge. *Quantum Generations: A History of Physics in the Twentieth Century.* Princeton, NJ: Princeton University Press, 1999.

Thompson, George P. *J. J. Thomson and the Cavendish Laboratory in His Day.* London: Thomas Nelson and Sons, 1964.

Turing, Alan

(b. London, England, 1912; d. Manchester, England, 1954)

Alan Turing was a central figure in the development of computation theory, digital computers, and artificial intelligence. He attended Cambridge University and was influenced by the dilemmas in mathematical logic posed by theorists such as Bertrand Russell (1872–1970) and Kurt Gödel (1906–1978). In attempting to address the question of the possibility of definitive mathematical proofs, which Gödel had proclaimed impossible in his "incompleteness" theorem, Turing developed a theoretical computation machine. He worked for British intelligence during World War II, helping it to break the German military's radio codes, and later designed one of the first computers. He also became a pioneer in the field of artificial intelligence. He died of cyanide poisoning in 1954, likely a suicide.

Turing came to prominence in 1936 after publishing a mathematical article that outlined the concept of a computation machine. This Turing machine, as it was later called, became the theoretical basis of digital computing. The machine made calculations based on sequences of binary digits set down on a piece of tape. Theoretically, the tape could be made as long as needed. This device used simple mathematics, which meant that complex problems would require extensive amounts of tape to be worked out completely. Turing's device was quite different from another kind of computer being conceptualized in the 1930s in the United States by Vannevar Bush (1890–1974), the differential analyzer. Turing's theoretical device was digital, whereas Bush's was analog. Digital computers had the disadvantage of requiring long, basic calculations, whereas analog computers made calculations that modeled the particular problem.

During World War II, Turing became a central figure in Britain's cryptanalysis efforts. He helped to develop devices that could make extensive calculations automatically, which were well suited to deciphering codes that had been mechanically produced by the Germans' encryption devices, including the Enigma cipher machine. At Bletchley Park, north of London, Turing and others worked to decode these messages; their secret successes allowed the British to have detailed knowledge of German movements, allowing for a decisive advantage in the war. The work at Bletchley Park established the foundation of the study of cryptanalysis. When the war ended, Turing continued his involvement in computer design, working for the National Physical Laboratory. There he designed an early electronic digital computer, the Automatic Computing Engine, which could be used to make calculations in ballistics and navigation. Such a machine was soon developed, the Pilot ACE, to put the theoretical possibilities of high-speed computation into practice.

Turing's interests turned increasingly toward artificial intelligence, which posed the question of whether a computer could learn. He left the National Physical Laboratory and focused on artificial intelligence when he moved to Manchester University in 1948. In 1950, he wrote an article in the journal *Mind* entitled "Computing Machinery and Intelligence," in which he proposed that human beings themselves were information processors, and thus were computers, and the world around us is information to be processed. Computers themselves would eventually "learn" in the same ways that humans do. He developed what became known as the Turing Test. This was simply a way to judge whether a computer was intelligent or not; it required a person to interrogate a human being and a computer, without knowing which one was which. Based on the responses, the person would guess who was who; a wrong guess meant that the computer was intelligent.

Although Turing was a pioneer in mathematics, computers, and artificial intelligence, he also was alienated because of his homosexuality after he moved from Cambridge to Manchester. He was arrested in 1952 because of his sexual activities, and was forced to submit to estrogen therapy to reduce his desire to have sex with men. His biographer, Andrew Hodges, suggests that this might have contributed to his suicide.

See also Computers; Gödel, Kurt; Mathematics; Russell, Bertrand; World War II
References
Hinsley, F. H., and Alan Stripp, eds. *Codebreakers: The Inside Story at Bletchley Park.* New York: Oxford University Press, 1993.
Hodges, Andrew. *Alan Turing: The Enigma.* New York: Simon & Schuster, 1983.
Yates, David. *Turing's Legacy: A History of Computing at the National Physical Laboratory, 1945–1995.* London: National Museum of Science and Industry, 1997.

U

Uncertainty Principle

The uncertainty principle was one of the most celebrated concepts developed in the twentieth century. Enunciated by Werner Heisenberg (1901–1976) as a result of his work in creating quantum mechanics in the 1920s, the uncertainty principle challenged the fundamental notions of physical reality and the project of science itself, by denying determinism and casting doubt on the commensurability of classical variables in physics, such as momentum and position. Uncertainty became part of the Copenhagen interpretation of quantum mechanics, a view that was both physical and philosophical.

After developing his matrix mechanics, which formed the basis of quantum mechanics, Heisenberg was struck by the implications of the quantum in such simple relationships as the momentum and position of an electron. Heisenberg knew that, in order to describe such variables, one had to devise a way to observe them; unfortunately, at such tiny scales, the interplay of light quanta (which carry inertia) alters the system as soon as they are introduced and before they are perceived by the experimenter. Therefore some indeterminacy must be attached to these variables, and indeed quantum mechanics is based on probabilistic relationships that

require a unit for indeterminacy. A statistician would see the quantitative unit for indeterminacy as equivalent to standard deviation. Indeterminacy might be expressed as ?, the position could be q, momentum could be p, and Planck's constant—the basis of quantum physics—is h. Heisenberg developed an equation to represent the relationship between the amount of indeterminacy for position and momentum at the quantum scale, $?q?p = h/4p$. All one needs to notice here is that the product of the two uncertainties cannot be less than $h/4p$, which itself is constant (and extremely small). Should the value attempt to fall below this, decreasing either $?q$ or $?p$ will simply increase the other. In other words, more precision about an electron's position would simply make its momentum less clear, and vice versa. These became known as the uncertainty relations, which Heisenberg proposed in 1927. Heisenberg announced that one could no longer count on science to give predictions based on causality, because the present dynamics of any system could never be determined with precision.

Also called the indeterminacy principle, the uncertainty principle posed a fundamental question about the nature of knowledge. Scientists long had been accustomed to the possibility that future conditions could only be predicted probabilistically and that it might not be possible to determine cause and effect. But this kind of indeterminacy was, physicists such as Albert Einstein (1879–1955) reasoned, owing to limitations on scientists' knowledge of present conditions. It did not negate the fact that scientists should be determinists—that is, concerned with understanding the dynamics of nature to such an extent that ultimately causes and effects could be identified and calculated, and thus future conditions accurately predicted. True, the future could only be determined probabilistically, but only because of present imperfections in knowledge, not the impossibility of such knowledge. Heisenberg's uncertainty principle asserted that the probabilistic nature of knowledge could never be overcome, because present conditions could never be identified precisely (or at least, all of the variables could not be identified with precision simultaneously). It struck a blow against determinism in science. The uncertainty principle also emphasized the connection between reality and human measurement, which Heisenberg did not see as separate. The notion that there is no independent reality from a scientist's observation stood against the assumptions of classical physics.

The uncertainty principle became part of a larger body of ideas called the Copenhagen interpretation of quantum mechanics, named after the home city of theoretical physicist Niels Bohr (1885–1962). More than just a series of equations or a framework for understanding physical relationships, this interpretation drew out some of the epistemological ramifications of the uncertainty principle. Heisenberg's premise was that knowledge of one variable (such as momentum) cannot simultaneously be known with another (such as position). The more one knows of one, the less one knows of the other; precise knowledge of them both, at the same time, is impossible. Bohr extended this interpretation to include understandings of entire systems of nature, particularly the competing visions

Werner Heisenberg developed the uncertainty principle, which became a cornerstone of the Copenhagen interpretation of quantum mechanics. (Bettmann/Corbis)

of what matter was composed of—waves or particles. In 1929, Bohr claimed that both are necessary, because a complete picture is impossible without both. Yet one cannot observe a system without disturbing it, and one's results will depend greatly on the way the scientific questions are posed. Although a more precise understanding of physics can be achieved if the investigator perceives the system as one composed of waves, such investigations yield less precise understanding of questions framed from the particle interpretation. The two frameworks complement each other—and Bohr chose the name *complementarity principle* to reflect that. The uncertainty principle, Bohr determined, was a specific case of the broader complementarity principle, both of which denied the separation of the natural world from the action and expectations of the observer.

See also Bohr, Niels; Determinism; Einstein, Albert; Heisenberg, Werner; Philosophy of Science

References

Cassidy, David C. *Uncertainty: The Life and Science of Werner Heisenberg.* New York: W. H. Freeman, 1992.

Gribbin, John. *In Search of Schrödinger's Cat: Quantum Physics and Reality.* New York: Bantam, 1984.

Kragh, Helge. *Quantum Generations: A History of Physics in the Twentieth Century.* Princeton, NJ: Princeton University Press, 1999.

Price, William C., and Seymour S. Chissick, eds. *The Uncertainty Principle and Foundations of Quantum Mechanics: A Fifty Years' Survey.* New York: Wiley-Interscience, 1977.

Uranium

Uranium is a natural element that is present in ores called pitchblende. It is an unstable, "radioactive" element that emits alpha particles as part of its normal radioactive decay. Its two main isotopes, U-235 and U-238, are the parents in decay series that ultimately result, after many transmutations, in the production of stable isotopes of lead. Pitchblende was used in the early twentieth century by scientists such as Marie Curie (1867–1934) to isolate the elements radium and polonium, both of which are parts of the decay series. A great number of radioactive elements exist in pitchblende, many of them not for very long, as they continue to emit radiation and thus transmute to other elements. Until the late 1930s, uranium's importance to science was confined to its experimental value as a source of alpha particles and to its role in the decay series. Researchers such as Ernest Rutherford (1871–1937), Frederick Soddy (1877–1956), and Bertram Boltwood (1870–1927), all of whom sought to identify the elements of the decay series, saw the process as modern alchemy (although in this case, lead was the end product, not the beginning one). But by 1939, uranium assumed a new importance as an element that could be split—atomic fission made uranium the most sought-after element on the planet because it was needed to make atomic bombs.

Albert Einstein's (1879–1955) famous 1939 letter to President Franklin Roosevelt (1882–1945), widely credited for catalyzing the U.S. bomb project, did not suggest that the United States build a bomb, but rather that the United States take care to ensure an ample supply of uranium. This element, he suspected, might be turned into an important source of energy and a new kind of bomb. Although the United States had some uranium mines, the most important sources of uranium were in Canada and in Africa, in Belgian-controlled Congo. He alerted the president that sales of uranium from Czechoslovakia (then controlled by Germany) had stopped—a possible sign that a German bomb project was well under way.

In the 1930s, it was not clear that uranium would be the best material to "fission" and cause a chain reaction, both necessary to create a bomb. Hungarian-born Leo Szilard (1898–1964), who was the most constant believer in the possibility of fission, had believed that the element beryllium might work best. But after the laboratory discovery of fission in 1939 by German scientists Otto Hahn (1879–1968) and Fritz Strassman (1902–1980), and the subsequent theoretical explanation by Lise Meitner (1878–1968) and Otto Frisch (1904–1979), uranium was clearly capable of fission. Fission split the atom in two, creating lighter elements and converting a small portion of the mass from the uranium into energy. Others, such as Szilard and Enrico Fermi (1901–1954) in the United States, believed that if fission could be sustained among many atoms—a chain reaction—then an extraordinary amount of energy could be released. In 1942, they presided over the first major experimental success of the Manhattan Project (the name of the atomic bomb project in the United States), achieving a controlled chain reaction.

The task of turning uranium into bomb material was complex. Only one of uranium's isotopes was suitable for chain reactions, namely, uranium with an atomic weight of 235, or U-235. This was a rare isotope, swamped by the ubiquitous presence of its more common relative, U-238. Processes of separating, or "enriching," uranium, to produce larger samples of U-235, became the major industrial focus of the bomb project. Several methods of isotope separation were attempted, including electromagnetic separation and gaseous diffusion. Another element, plutonium, was especially suitable for fission as well, and it was created artificially from uranium. Ultimately these manipulations produced enough uranium and plutonium to build atomic weapons. The bombs used by the United States against Japan in August 1945 were very inefficient, and caused only a fraction of the material to fission. Throughout the postwar era, scientists in the United States worked to maximize the explosive power of uranium and to test the biological effects of its harmful radiation.

In order to match the power of the United States, the Soviet Union also needed the precious commodity. Soviets began to seek out uranium from any source it could find. The military leader of the Manhattan Project, General Leslie Groves (1896–1970), made it a priority to establish an Anglo-American monopoly on uranium, to prevent the Soviet Union from having enough material to build a bomb for at least a decade. But as the war ended, the Red Army seized about one hundred tons of uranium from Auer, a German company. The Soviet Union also began an intensive effort to mine uranium, and it did so with little concern about the health effects. Although it had small sources of uranium within its own borders, a much larger supply was found in the western Erzgebirge (Ore Mountains), in Germany. The Red Army occupied this area in 1945, and the next year it transformed its towns and villages into a complex with the single purpose of mining uranium; it soon became one of the world's leading producers. The Soviets drafted workers for the mines and even used work in the mines as punishment for criminals. The region's population increased dramatically, owing to the influx of forced labor, creating poor living conditions to complement the bad working conditions. Miners routinely breathed radioactive dust and waded in radioactive sludge. Efforts to recruit miners frightened thousands of workers into leaving the Soviet zone of occupation. Although these conditions were not paralleled in the United States, both countries often prioritized the demands of national security over individual human health. By 1949, the three most productive uranium mines in the world were the Eldorado mine in Canada, the Shinkolobwe mine in the Belgian Congo, and the Joachimsthal mine in Czechoslovakia

See also Atomic Bomb; Becquerel, Henri; Boltwood, Bertram; Cold War; Curie, Marie; Fission; Manhattan Project; Physics; Radiation Protection; Radioactivity; Rutherford, Ernest

References
Badash, Lawrence. *Radioactivity in America: Growth and Decay of a Science.* Baltimore, MD: Johns Hopkins University Press, 1979.
Gustafson, J. K. "Uranium Resources." *The Scientific Monthly* 69:2 (1949): 115–120.
Lanouette, William, with Bela Szilard. *Genius in the Shadows: A Biography of Leo Szilard, The Man behind the Bomb.* Chicago, IL: University of Chicago Press, 1992.
Naimark, Norman M. *The Russians in Germany: A History of the Soviet Zone of Occupation, 1945–1949.* Cambridge. MA: Harvard University Press, 1995.

Urey, Harold

(b. Walkerton, Indiana, 1893; d. La Jolla, California, 1981)

Harold Urey was a U.S. physical chemist who discovered the existence of heavy hydrogen in the 1930s, played a major role in the atomic bomb project, and then developed an influential theory about the composition and origins of the moon and planets. He went to college at the University of Montana, taking a degree in 1917. After working in industry as a research chemist, he entered graduate school in 1921. He took his doctorate in 1923 from the University of California–Berkeley, in physical chemistry. Urey traveled to Europe for a year to work in Niels Bohr's (1885–1962) institute in

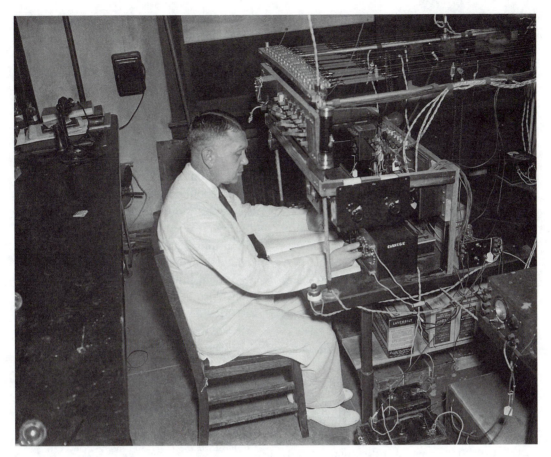

Harold C. Urey is pictured here at work with a mass spectrometer. (Bettmann / Corbis)

Copenhagen. After a brief period at the Johns Hopkins University, he took a position in 1929 at Columbia University.

Urey's renown in the 1930s came from his discovery of heavy hydrogen, more commonly called deuterium. Deuterium is a rare isotope of ordinary hydrogen, but with about twice the mass. In late 1931, Urey and his colleagues Ferdinand Brickwedde (1903–1989) and George Murphy (1903–1968) at Columbia found that "heavy water" consists of one atom of oxygen and two atoms of heavy hydrogen. They isolated it by evaporating four liters of liquid hydrogen, identifying the heavy isotope with spectroscopic methods. One reason for deuterium's importance was that it was immediately seen to be very suitable as a projectile in nuclear exper-

iments. Heavy hydrogen's excess weight would make it a valuable tool in a particle accelerator. Adding energy by speeding deuterons, as Urey called the particles, became a principle way of stimulating nuclear reactions in other elements. For this discovery, Urey received the Nobel Prize in 1932. The following year, he founded the *Journal of Chemical Physics* as a publication to focus on topics such as nuclear reactions. He became its first editor, a position he kept until 1941.

During World War II, Urey was heavily involved in the efforts to build the atomic bomb. He directed a program of isotopic separation, the process by which "enriched" uranium, U-235, was isolated from its more common sibling, U-238. After the war, Urey moved to the Institute for Nuclear Studies at the University of Chicago. His interests shifted to geochemistry, geophysics, and even cosmology. Part of the reason for his change of subject matter, which would lead to work that inspired the scientific agenda of the Apollo missions to the moon, was his desire to distance himself from the weapons programs he helped to create during the war. He did, however, continue to serve in an advisory capacity to the Atomic Energy Commission in the 1950s. Another reason was his fascination with the properties of the moon, reinforced by reading Ralph B. Baldwin's (1912–) 1949 book, *The Face of the Moon*. Urey's own studies of the moon and planets brought his expertise as a chemist to the long-standing questions about planets' (and their satellites') origins.

Urey presented his ideas about the earth's formation in 1949, at a meeting of the National Academy of Sciences. He believed that the earth had condensed from a cloud of dust at relatively low temperatures. He likened the earth's initial core to that of the present-day moon, now altered and masked by the earth's evolution over the millennia. Urey's background in chemistry gave the theory some added credibility. His book *The Planets* appeared in 1952. It is credited with con-

vincing many scientists to enter the field of solar system astronomy, which was an interdisciplinary field embracing the "earth" sciences—geophysics, geochemistry, and meteorology. He was convinced that the moon was still in a primordial state, unaltered as the earth had been by erosion and other dynamic forces. It preserved the historical record better than the earth; in the moon, scientists could understand the earth's own conditions billions of years ago. Sending humans to the moon, which the United States did in 1969, was motivated by national competition more than science; yet Urey's work provided a strong justification. Studying the moon might illuminate the earth's own geophysical history.

See also Earth Structure; Geophysics; Physics; Uranium

References

Brush, Stephen G. "Nickel for Your Thoughts: Urey and the Origins of the Moon." *Science,* New Series, 217: 4563 (3 September 1982): 891–898.

Kragh, Helge. *Quantum Generations: A History of Physics in the Twentieth Century*. Princeton, NJ: Princeton University Press, 1999.

Wilhelms, Don E. *To a Rocky Moon: A Geologist's History of Lunar Exploration*. Tucson: University of Arizona Press, 1993.

V

Vavilov, Sergei

(b. Moscow, Russia, 1891; d. Moscow, USSR, 1951)

Sergei Vavilov spent much of the 1920s in the shadow of his brother, Nikolai Vavilov (1887–1943), whose rising eminence as a geneticist outstripped Vavilov's own as a physicist. Until 1929, he worked at the Institute of Physics and Biophysics in Moscow. He had the opportunity to travel abroad and studied in Berlin in 1926. He was by no means a man of eminence. But in 1932, almost inexplicably, Vavilov became a member of the Academy of Sciences and was asked to direct an institute of physics. His rise in stature partly can be attributed to the instability of the scientific community, as the government increasingly looked for conformity with Communist ideology. Not all scientists fit the mold, but Vavilov appeared to adapt to it very well. Vavilov was one of many new members of the academy intended to instill the spirit of the Communist Party in the academy and in Soviet science generally.

In 1934, the takeover of science by ideologues seemed to be sealed when the academy moved from Leningrad to Moscow in order to be fully subordinated to the government's Council of Ministers. The same year, Vavilov was asked to head a new, well-funded research center for physics, the Fizicheskii Institut Akademii Nauk (FIAN). His students and colleagues later credited him with fostering an atmosphere of pure science, in which he did not interfere with their work, and with protecting them from periodic political upheavals. Most of Vavilov's colleagues appreciated his political skill under Joseph Stalin (1879–1953), which helped them to keep their posts.

In the laboratory, Vavilov was interested in the properties of light, particularly the luminescence of liquids. His graduate student Pavel Cherenkov (1904–1990) discovered in 1935 a faint blue radiation that soon was called Cherenkov radiation. In the Soviet Union, it was called Vavilov-Cherenkov radiation, because of Vavilov's important role in leading Cherenkov to his problem and his participation in the work. Other physicists at the institute, Il'ia Frank (1908–1990) and Igor Tamm (1895–1971), provided the theoretical explanation for the radiation in 1937: It was the result of particles moving faster than the speed of light in a given medium. This work revised the theory of relativity slightly, noting that light is unsurpassed only in a vacuum. Like a sonic boom in the case of sound, exceeding the velocity of light resulted in a special kind of radiation.

This scientific work was not always appreciated. Another Soviet physicist, Piotr Kapitza (1894–1984), derided Vavilov's obsession with liquid fluorescence, saying, "you can play with the apparatus for all your life. . . . He never did anything else" (Kojevnikov 1996, 26). (In Kapitza's estimation, Vavilov did not have the scientific eminence even to belong in the academy and must have achieved his stature by knowing what to say to the right people.)

Much of Sergei Vavilov's notoriety surrounds the death of his brother, Nikolai Vavilov, himself an eminent scientist in the field of genetics. Nikolai was arrested in 1940 for "wrecking" (sabotage) in the field of agriculture. He was a geneticist and had been denounced by Trofim Lysenko (1898–1976), the agronomist who associated genetics with Western, anti-Marxist ideology. Nikolai Vavilov was sentenced to fifteen years in prison; he died in 1943 in Saratov prison. Westerners might have found it surprising that such a high-ranking scientist as Sergei Vavilov should remain in his position of authority after such a close relative was denounced and sentenced as an enemy of the state. Yet this was not so uncommon in the Soviet Union, where hardly any eminent person was untouched by Stalin's purges. One of Vavilov's great successes as a public figure was in navigating such troubled waters without drowning in them himself. Indeed, his brother's death did not bring Vavilov down at all. Nor did it stop him from praising Stalin and dialectical materialism, the official scientific philosophy of the Soviet Union. In 1945, he was "elected" to the presidency of the Academy of Sciences. Lysenko had also been a candidate, but because so many scientists did not respect his ideological enthusiasm, the loyal and respected Vavilov was the logical choice. Many scientists breathed a sigh of relief that the academy did not fall into Lysenko's hands. Vavilov, of course, did not shy from tying science to ideology. In 1949, he helped Stalin celebrate his seventieth birthday by giving a talk entitled "On Stalin's Scientific Genius." Although this was an overt embrace of Stalinist ideology, Vavilov's choice of rhetoric never came close to that of Lysenko's, which drew sharp distinctions between "bourgeois" and "proletarian" science.

Vavilov developed heart disease, and his health deteriorated drastically in 1950. He died in January 1951. Some of his last official actions were to reduce the number of Jews in high-ranking positions, in line with Stalin's anti-Semitic policies. He probably encouraged the resignation of the celebrated physicist Abram Ioffe (1880–1960) from the directorship of the Leningrad Physico-Technical Institute. As historian Alexei Kojevnikov describes him, Vavilov was "a non-Communist and nonsympathizer who happened to become an exemplary Stalinist politician" (Kojevnikov 1986, 26).

See also Academy of Sciences of the USSR;
    Cherenkov, Pavel; Light; Lysenko, Trofim;
    Soviet Science
References
Kojevnikov, Alexei. "President of Stalin's
    Academy: The Mask and Responsibility of
    Sergei Vavilov." *Isis* 87 (1996): 18–50.

Venereal Disease

The turn of the twentieth century was a pivotal moment in the history of medicine, as the study of pathogens by noted scientists Robert Koch (1843–1910), Paul Ehrlich (1854–1915), and others was leading to the identification of bacteria and the successful treatment of the world's most terrible infectious diseases. Epidemics of tuberculosis and diphtheria were being conquered by effective public health measures, such as better sanitation and education about personal hygiene. At the same time, sexually transmitted diseases such as syphilis and gonorrhea were addressed less effectively. As historian Allan M. Brandt has noted, confronting these diseases did not merely entail finding a "magic bullet" to cure them; rather, their existence and spread provoked dilemmas about moral behavior and social control.

Part of the problem of halting diseases such as syphilis was the social stigma attached to them. Progressive-era reformers attempted to prevent such diseases by raising public consciousness about its dangers, not just to one's own life, but also to one's family and community. This disease resulted from sinful behavior and was a mark of social decay. Public health officials capitalized on the fear of infection and promoted the stigmatization through publicity, especially posters, during wartime that emphasized the shame a man would feel if his family and friends discovered he had syphilis. Posters represented women as the carriers of disease, as forbidden fruit that seemed "clean" but in reality were deadly.

Americans in particular made little progress in combating venereal disease in the 1920s, largely because of its social stigma. Americans were more willing to blame the prevalence of the disease on immorality than on insufficient public health measures. The apparent rise in promiscuity in the 1920s seemed directly connected to increasing cases of disease. The key to controlling it would be to control social behavior, such as regulating the actions of couples in dance halls. In the 1930s, Surgeon General Thomas Parran (1892–1968) tried to reverse this emphasis. He waged a campaign against venereal disease in the United States and attempted to put less stress on individual responsibility and more on public health measures. He criticized the Army on the eve of World War II, for example, for failing to eliminate sources of temptation for soldiers. The government should crack down on prostitution, Parran argued, not count on the men to exercise restraint. A successful program would combine education, repression of prostitution, medical treatment, and investigations of contact by infected people.

The issue of confronting venereal disease

Poster for treatment of syphilis, showing a man and a woman bowing their heads in shame (between 1936 and 1938). (Library of Congress)

posed a dilemma of responsibility not found in dealing with other infectious diseases. Health measures such as better sanitation and hygiene certainly seemed reasonable in the case of tuberculosis, cholera, and the like. But why should governments eliminate prostitution or hand out condoms when a person could just as easily avoid disease through a personal choice, namely, abstinence? This question might seem legitimate, but despite moral indignation at human weakness, the probability of human promiscuity remained. During World War II, for example, 53 to 63 percent of U.S. soldiers engaged in sexual intercourse, and about half of the married soldiers had extramarital sex. Insisting on abstinence seemed to be a naive illusion.

Fortunately, the development of penicillin during the war put a halt to the power of syphilis and gonorrhea. Its use cured 90 to 97 percent of those afflicted. By the early 1950s, these venereal diseases seemed to be disappearing. But public health officials would confront their dilemma repeatedly in the ensuing decades, with various sexually transmitted diseases. The AIDS epidemic that began in the 1980s would renew the politicization of a disease that, in some minds, resulted from immoral behavior. The idea that individuals, not societies, should be held responsible for such diseases has not faded entirely.

See also Ehrlich, Paul; Medicine; Public Health; World War II

References
Brandt, Allan M. *No Magic Bullet: A Social History of Venereal Disease in the United States since 1880, with a New Chapter on AIDS.* New York: Oxford University Press, 1987.
Poirier, Suzanne. *Chicago's War on Syphilis, 1937–1940: The Times, the Trib, and the Clap Doctor.* Urbana: University of Illinois Press, 1995.

Von Laue, Max

(b. Pfaffendorf, Germany, 1879; d. Berlin-Dahlem, Germany, 1960)

Max Von Laue conducted early work on X-rays and crystals, pioneering the study of what became known as solid-state physics. He studied under Max Planck (1858–1947) at the University of Berlin, obtaining a doctorate in physics in 1903. After that he worked at the University of Göttingen for two years before returning to becoming Planck's assistant. In 1909, Von Laue attained a position as a privatdozent at the University of Munich. In subsequent years, Von Laue held positions at several universities in Zurich, Frankfurt, Würzburg, and Berlin.

Von Laue attained scientific renown by his work in X-ray diffraction. He proposed that if electromagnetic radiation of very short wavelengths were passed through crystals, one might expect there to be a diffraction effect. He based this idea on the theoretical assumption that if wavelengths were reduced to a magnitude of atomic distances, one should expect some serious disturbances in the direction and intensities of the waves. Theoretically, he reasoned, X-rays should be diffracted by crystalline structures. After this was demonstrated in a laboratory setting to be the case, Von Laue worked out the mathematical reasoning and published both the theory and the evidence for it in 1912. Not only had he shown the existence of diffraction, but he had proven that X-rays were in fact electromagnetic radiation of short wavelengths (at that time, some believed they were particles). For this work, which helped to begin the study of X-ray crystallography, Von Laue won the Nobel Prize in 1914. In later years, Von Laue expanded the field of solid-state physics further through his research on superconductivity.

During World War II, Von Laue moved to Hechingen to be part of a community of physicists to aid the Nazi government in war-related work. Led by Werner Heisenberg (1901–1976), the physicists were charged with the task of developing an atomic bomb. Toward war's end, Von Laue was taken prisoner by the U.S. *Alsos* mission and detained at Farm Hall, a country manor near Cambridge, England, with several other high-ranking German scientists. Their conversations were secretly recorded to learn about the bomb project. They had fallen well short of the goal, having failed to achieve a fission chain reaction, which the Americans accomplished in 1942.

After being released by the authorities, Von Laue returned to Germany and took a position at Göttingen before settling finally in Berlin-Dahlem, where he became the director of the Fritz Haber Institute for Physical Chemistry in 1951. His book on the history of physics, which he wrote during the war, went through several editions. He died from injuries received in an automobile accident when he was eighty years old.

See also Bragg, William Henry; Physics; World War II; X-rays

References

Ewald, P. P., ed. *Fifty Years of X-ray Diffraction.* Utrecht: International Union of Crystallography, 1962.

Goudsmit, Samuel A. *Alsos.* New York: Henry Schuman, 1947.

Hoddeson, Lillian, Ernest Braun, Jürgen Teichmann, and Spencer Weart, eds. *Out of the Crystal Maze: Chapters from the History of Solid-State Physics.* New York: Oxford University Press, 1992.

Von Laue, Max. *History of Physics.* New York: Academic Press, 1950.

Vygotsky, Lev

(b. Orsche, Russia, 1896; d. USSR, 1934)

Russian psychologist Lev Vygotsky helped to establish the science of child development. His work on cognitive disabilities forged an entirely new approach to the subject now known as special education. One of his most important innovations was to differentiate between the physical and developmental aspects of cognitive disabilities. This achievement cannot be overestimated, because it dealt a severe blow to the notion that cognitive disabilities were beyond therapy.

Vygotsky did not begin his career as a psychologist. He took a degree in law from Moscow University in 1917, the year of the Bolshevik Revolution. He turned his studies more fully to psychology in the 1920s, but retained a great affinity for the humanities; he penned *The Psychology of Art* in 1925. Vygotsky worked in a different tradition than most, emphasizing the sociological aspects of psychology such as language and culture, rather than strictly the behavioral aspects. This set him apart from the eminent psychologists of

his time such as Ivan Pavlov (1849–1936) and Jean Piaget (1896–1980). Vygotsky saw culture as the dominant force in cognition. Because he believed the human consciousness was constructed by cultural forces, his brand of psychology was developmental rather than behavioral.

Vygotsky identified signs, symbols, and language as the vehicles for sophisticated cognition. Development did not occur along a linear path, Vygotsky believed; rather, these cultural tools shaped a person as he or she grew from child to adult. Vygotsky was particularly interested in cognitive development in children, child psychology, and learning disabilities. In the Soviet Union, the study of disabilities later was dubbed defectology. Vygotsky recognized that cognitive disabilities were not all cultural in origin, as in the case of what he called primary defects, resulting from biological damage. But he also noted that some disabilities were socially defined, connected to but not resulting from the primary disability. These secondary defects could be delays or distortions in the development of social skills, associated with the interaction of the primary disability with the child's social environment. The goal for defectologists, he believed, was to find ways of compensating: improving communication and facilitating learning to eliminate and prevent common secondary defects.

Vygotsky discarded the notion that cognitive disabilities were permanent and unchanging. Disabilities manifest themselves differently from one person to the next, depending on their education, social interactions, and other factors. Vygotsky applied these insights to his field, called paedology (study of the child) in the Soviet Union. Unfortunately for Vygotsky, his work was criticized as anti-Marxist, despite his efforts to show how it was very much in line with socialist thought. He died of tuberculosis in 1934 while finishing a book, *Thought and Language.* In the

Paedology Decree of the Central Committee of the Communist Party, issued in 1936, paedology was condemned and Vygotsky's work was banned. The suppression of Vygotsky's work was highly successful; it did not come into wide circulation in the West until the 1960s.

See also Mental Retardation; Psychology; Soviet Science

References

Kozulin, Alex. *Vygotsky's Psychology: A Biography of Ideas.* Cambridge, MA: Harvard University Press, 1991.

Van der Veer, René, and Jan Valsiner. *Understanding Vygotsky: A Quest for Synthesis.* Cambridge, MA: Blackwell, 1991.

Vygotsky, L. S. *The Collected Works of L. S. Vygotsky. Volume 2: The Fundamentals of Defectology (Abnormal Psychology and Learning Disabilities),* ed. by Robert W. Rieber and Aaron S. Carton. New York: Plenum Press, 1993.

W

Wegener, Alfred

(b. Berlin, Germany, 1880; d. Greenland icecap, 1930)

Alfred Wegener has earned a reputation as being the father of continental drift and thus the father of the modern theory of plate tectonics. He certainly deserves most of the credit for the former, and almost none for the latter. Trained in astronomy (he received a Ph.D. from the University of Berlin in 1904), Wegener was wounded in World War I and served his country as a meteorologist. He was also part explorer, participating in expeditions to Greenland, one of which ended his life. He became controversial during the interwar years for his geophysical ideas, which drew on biological evidence far more than physical theory.

In 1912, Wegener proposed that the continents moved and that the apparent jigsaw fit of landmasses—most dramatically, the fit of western Africa with eastern South America—was not coincidence. In 1915, he presented his theory of continental drift in a book entitled *The Origins of Continents and Oceans*. Wegener described a supercontinent, "Pangaea," which split apart millions of years ago, the present continents being the broken-up remnants of it. The idea, he said, came to him in 1910 as he was studying a world map.

He found the apparent congruence of coastlines between Africa and South America very striking. That, combined with paleontologists' recent discovery of evidence of a former land bridge between Brazil and Africa, led Wegener to seek out data to corroborate his intuition. Most of his data came from fossil records, because few physicists were willing to accept such a drastic conclusion about continental mobility.

Although the theory of sea-floor spreading and plate tectonics would be a long time in coming, Wegener is often invoked as the father of these theories. But there is no straight line between Wegener and the advocates of sea-floor spreading in the 1960s. Indeed, Wegener was reviled in his own time. His ideas gained little sympathy among scientists, with some notable exceptions, such as South African Alexander Du Toit (1878–1948) and the Briton Arthur Holmes (1890–1965). But for the most part, Wegener's version of continental drift was rejected by contemporary scientists, especially those in the United States. His ideas stood against the prevailing turn-of-the-century explanations of the earth's upheavals, namely, that the earth steadily was contracting from its release of heat. Certainly geologists disagreed on the specifics. The

Austrian Eduard Suess (1831–1914), for example, had viewed mountains as wrinkles on the earth's surface, and the oceans as collapsed portions of the crust. Long ago, he reasoned, there had been a vast continent (he called it Gondwanaland) of which both the present-day continents and oceans were once part. James Dwight Dana (1813–1895), by contrast, had seen continents and oceans as stable and permanently separate. But neither Suess nor Dana had a vision of the earth that allowed for mobile continents floating along the earth's surface. Wegener's ideas, none of which provided a convincing mechanism powerful enough to drive continents apart, persuaded few geophysicists to abandon their beliefs in the earth's fixity.

Wegener admitted that geophysicists generally were arriving at conclusions in opposition to his own. But he also noted that this was a disparity between disciplines: Biologists, tracing the evidence of fossils, readily acknowledged connections among continents in previous epochs. These could be explained through land bridges or, as Wegener believed, through continental drift. Only the physicists seemed to stand in the way. Eventually, Wegener would be vindicated during the revolution in earth science in the 1960s. But unfortunately, the father of continental drift would not see this happen. He died in 1930, returning from a rescue expedition in Greenland; at the time, he was still convinced of the truth of his idea, but he had not succeeded in persuading the skeptics.

See also Continental Drift; Earth Structure; Geology; Geophysics; Jeffreys, Harold

References

Hallam, A. *A Revolution in the Earth Sciences: From Continental Drift to Plate Tectonics.* Oxford: Clarendon Press, 1973.

Oreskes, Naomi. *The Rejection of Continental Drift: Theory and Method in American Earth Science.* Oxford: Oxford University Press, 1999.

Wegener, Alfred. *The Origin of Continents and Oceans.* New York: Dover, 1966.

Women

It is no secret that women have played smaller roles than men in the development of science. Why this should be true has been the subject of considerable debate. The most obvious reason, contested by almost no one, is that prejudice against women excluded many of them from participation in the scientific profession and discouraged their training in scientific disciplines. Even when they did participate, such prejudices made awards, honors, promotions, and other distinctions less likely, and preexisting gender stereotypes reinforced subordinate roles in laboratories and in fieldwork. Some fields, such as oceanography, had almost no women at all and certainly none in positions of leadership, owing to cultural taboos against having women aboard ships, a distinctively masculine work space. Cultural prejudices against women, whether related to science or drawn from society at large, informed the attitudes of many leading scientists. Still, in the early twentieth century, some strides were made in including women in scientific fields, although historians tend to argue about the positive and negative consequences of such inclusion.

In the United States, the numbers of notable women in science grew steadily in the first half of the twentieth century. The reference work *American Men of Science* included about 500 women in its first three editions between 1906 and 1921. That number almost had quadrupled two decades later. The principal obstacle to women was access to education. With the proliferation of women's colleges toward the end of the nineteenth century and the decision by some major universities—especially in the United States and Germany—to grant doctoral degrees to women (by around 1910), such access was less difficult than it had ever been in the past.

Despite access to higher education, involvement of women in science often was limited to "women's work," the relatively dull busywork that scientists delegated to others. Women could be hired to do such work for less money than their male counterparts. One example was the job of cataloguing and classifying specimens, collected by men, deposited in natural history museums. Also, institutions such as the Harvard College Observatory employed large proportions of women to catalogue and analyze photographic plates taken from telescopic observations. Some of the most important findings in astronomy were made by women in such positions, notably the work on Cepheid variables by Henrietta Swan Leavitt (1868–1921). But despite this important role, the work itself was the kind of activity that leading male scientists simply did not wish to do. Genuine leading roles for women scientists were few, with home economics departments being an important exception.

Women's successes in finding work in science led to new divisions in the scientific community as fields became "gendered" according to the proportions of women in them. Those wanting to pursue research often took low-paying (or not paying at all) positions, or worked in laboratories under the title of "secretary." When women were admitted to professional organizations, sometimes a more select category of membership was created at the same time. Part of the reason for this abject sexism, historian Margaret Rossiter has argued, was that male scientists feared that the inclusion of more women into their fields would erode the status of the field as a whole. Professionalizing science meant limiting the degree to which it was feminized and thus presumably softened by a lack of rigor. To give one example, at the turn of the century (1897), an article in *Science* asked whether the rising number of women in the field of botany made it a field that aspiring male scientists should avoid.

Much of the debate about the role of women in science has centered up the excep-

Anatomist Florence Sabin was the first woman elected to the National Academy of Science, in 1925. (Corbis)

tions. There were a few leading women of science who achieved positions of notoriety in the United States: Anatomist Florence Sabin (1871–1953) was the first woman elected to the National Academy of Science, in 1925, and psychologist Margaret Washburn (1871–1939) was elected in 1931. But the academy waited another thirteen years before electing another woman. Polish-born Marie Curie (1867–1934) was not merely one of the most well-known women but was also one of the most celebrated scientists of the early twentieth century. Curie won Nobel Prizes in Chemistry and Physics for her work (in France) on radioactivity, particularly her discovery and isolation of radioactive elements such as polonium and radium. Curie was an international celebrity and might have served as an inspiration to women and a case lesson for men who believed women were less competent in the sciences.

Yet the magnitude of her accomplishments could have had the reverse effect, making her seem an incredible exception rather than an exemplar of the rule.

Despite these examples, most notable women of science worked in relative obscurity for most of their lives. Astronomer Cecilia Payne-Graposhkin once noted how the Harvard College Observatory made her invisible: Her salary was paid out of the equipment budget. Physicist Maria Mayer often worked without pay. Plant geneticist Barbara McClintock (1902–1992), who later won the Nobel Prize, struggled for years against prejudice in an academic setting but found a niche as a researcher at the Cold Spring Harbor Laboratory. Rosalind Franklin (1920–1958) played a crucial yet largely forgotten role in the work leading to the identification of the double-helical structure of DNA.

Some recent work in the history of science from a feminist point of view suggests that, beyond the exclusion of women from scientific practice and recognition, science itself was gendered, reflecting a masculine viewpoint of the world. Authors such as Sandra Harding stress this point, arguing the need for a feminist science that will be less a construction of male ambitions, motivations, and values. Evelyn Fox Keller's biography of Barbara McClintock suggests that McClintock's discoveries came from her particular feminine point of view—she had a "feeling" for the organism. Such authors argue that science itself will change as more women take part in it. These viewpoints take it for granted that knowledge is socially constructed, as is scientific method. However, others argue that this is simply another incarnation of essentialism; to argue that science will become more nurturing and less concerned with rationality when the gender balance is addressed is simply to encourage the stereotypes of women and men. This debate has been enhanced by more studies of important women in science in the early twentieth century.

See also Astronomical Observatories; Curie, Marie; Leavitt, Henrietta Swan; McClintock, Barbara; Meitner, Lise; Nobel Prize; Pickering's Harem

References

Cowan, Ruth Schwartz. "Women in Science: Contested Terrain." *Social Studies of Science* 25 (1995): 363–370.

Harding, Sandra. *Whose Science? Whose Knowledge? Thinking from Women's Lives.* Ithaca, NY: Cornell University Press, 1991.

Keller, Evelyn Fox. *A Feeling for the Organism: The Life and Work of Barbara McClintock.* San Francisco, CA: W. H. Freeman, 1983.

Quinn, Susan. *Marie Curie: A Life.* Reading, MA: Addison-Wesley, 1996.

Rossiter, Margaret. *Women Scientists in America: Struggles and Strategies to 1940.* Baltimore, MD: Johns Hopkins University Press, 1982.

Sicherman, Barbara, John Lankford, and Daniel J. Kevles. "Science and Gender." *Isis* 75:1 (1984): 189–203.

World War I

Although the Great War (1914–1918) was set into motion in the Balkans with the assassination of the heir to Austria's throne, the conflict soon involved all the major powers of Europe and eventually drew in the United States. It was the first major industrialized war, and many of the recent discoveries and inventions in science and technology were put to use. The war saw the birth of what later were called weapons of mass destruction, as both sides of the conflict put scientists to work making chemical weapons. Scientists themselves played an active role in that process, promoting their usefulness and developing innovative defense measures and destructive weapons.

Scientists were no less patriotic than their countrymen, and often no less chauvinistic. Just as the British royal family changed its name from Saxe-Coburg to Windsor to avoid slander against it and to hide its German ancestry, British scientists of German descent risked being isolated and denounced. The secretary of the Royal Society, for example, was German-born physicist Arthur Schuster

British soldiers after being blinded by mustard gas. World War I was known as the "chemist's war," because chemists were deeply involved in developing new explosives, drugs, dyes, and, most memorably, poisonous gas. (Bettmann/Corbis)

(1851–1934); not only did some criticize him, but the local police confiscated his wireless apparatus in fear that he would use it to make contact with the Germans. When he was elected president of the British Association for the Advancement of Science in 1915, scientific colleagues and the public press were deeply critical and were outspokenly doubtful of his loyalty to England. Personal animosity against scientists who lived and worked in Germany was worse, and it increased when a group of them signed a manifesto—"An die Kulturwelt! Ein Aufruf" (often called the Appeal of the Ninety-Three Intellectuals)—defending their country's actions and showing solidarity with what appeared to be German militarism. Many Allied scientists generalized the Germans

unkindly and agreed with physicist William Ramsay (1852–1916), who concluded that German science had done little but inundate the scientific world with mediocrity. Ramsay attributed this to racial qualities (indeed, he believed that most of the genuine productivity of German science was a result of efforts by Jews), whereas others believed that the culture of militarism was responsible. German scientists did the same, claiming as physicist Philipp Lenard (1862–1947) did that the souls of Isaac Newton (1642–1727) and Michael Faraday (1791–1867) had left England's shores and had gone to other countries; the seeming importance of British discoveries in recent years, he said, were because of the British insisting on taking more credit than they deserved.

Scientists became deeply involved in war work. They contributed to research on submarines, aircraft and airship design, wireless communication, artillery weapon construction, medicine, and even geology (for trench warfare). Beyond these, however, the war was known as the chemist's war, because chemists were deeply involved in developing new explosives, drugs, dyes, and, most memorably, poisonous gas. Chemistry was particularly advanced in Germany, where the demands of the dye industry had produced organic chemists in abundance. Trade with Germany ceased during the war, of course, and Britain found itself having to invest heavily in chemical research in order to match German progress. Although the first use of chemical weapons in 1915 was made possible by the efforts of Germans Fritz Haber (1868–1934) and Walther Nernst (1864–1941), both sides of the conflict developed them and used them during the course of the war. Some chemical weapons caused death; others simply harassed soldiers through temporary asphyxiation or even blindness, making them incapable of defending themselves. The early weapons were clouds of chlorine gas intended to drive soldiers from the trenches; in 1917 the Germans integrated arsenic gas and mustard gas into explosives.

The war transformed scientific work in the United States, particularly through efforts to create stronger ties between the government and the scientific elite. After the sinking of the passenger liner *Lusitania* in 1915, the government had asked leading inventor Thomas Edison (1847–1931) to form a committee to help combat the threat of submarines. But the nation's leading scientists in the National Academy of Sciences were not asked to help. Largely because of the efforts of astronomer George Ellery Hale (1868–1938), U.S. scientists struggled to promote their usefulness to the government during the first years of the war. Hale saw the war as an exemplary case of society's misunderstandings of the nature of science and its role in society—scientists could be very useful, and there ought to be a body whose purpose was to harness that potential. In an effort to organize scientific action for the good of the country, Hale in 1918 was instrumental in forming the National Research Council as part of Academy of Sciences. Although some U.S. scientists saw it as an effort to control scientific activity from Washington, this was the first major effort to coordinate scientists during a time of national crisis, and the council continued to perform that function after the war.

The widespread involvement of scientists in the course of the war led some to question the proper role of science in society. Just as the war itself provoked hard questions about the progress of civilization, scientists' role cast doubt on whether one could count on science as a positive agent of change in human life. The salient images of the war were closely tied to science and technology—the use of airplanes, the first gas attacks, and the extended tunnel and trench networks. Allied scientists tended to put the onus of responsibility on the Germans, and in subsequent years "punished" them by disallowing their inclusion in bodies such as the International Research Council. One legacy of the war for science was the closer relationship between scientists and governments; the war also demonstrated the adhesion of scientists to the political currents of their own countries.

See also Chemical Warfare; Haber, Fritz; International Research Council; National Academy of Sciences; Nationalism; Patronage; Royal Society of London

References
Badash, Lawrence. "British and American Views of the German Menace in World War I." *Notes and Records of the Royal Society of London* 34:1 (1979): 91–121.
Haber, L. F. *The Poisonous Cloud: Chemical Warfare in the First World War*. Oxford: Clarendon Press, 1986.
Kevles, Daniel J. "George Ellery Hale, the First World War, and the Advancement of Science in America." *Isis* 59:4 (1968): 427–437.

World War II

World War II changed the practice of science dramatically, and work of scientists during the conflict brought about fundamental technological changes that shaped the postwar world. As in the case of World War I, scientists aided their countries in the cause of war and often mobilized their work efficiently to create new organizations. Most of the major combatants devoted some attention to research and development, whether for short-term or long-term goals. The countries that devoted the most resources to research and development during the war were the United States, Great Britain, and Germany. The most celebrated development of the war, the atomic bomb, was pursued by five of the major combatants. Although Japan did not succeed in developing noteworthy new weapon systems, it did invest in fission research during the war. The Soviet Union also was interested in building an atomic bomb, but the government did not devote much attention to it until after the victory at Stalingrad in early 1943, when it became more realistic to invest in long-term projects. The Soviet Union did, however, have a successful espionage network in the United States that included high-ranking members of the U.S. atomic bomb project, including Klaus Fuchs (1911–1988).

The most significant research projects to have an impact on the war were penicillin, radar, and the atomic bomb. Penicillin was an antibiotic, shown to be extraordinarily successful at killing bacteria. Many of the amputations and deaths, in war or in peace, were not a result of wounds per se, but of the infections that developed in them from the growth of harmful bacteria. Although penicillin had been discovered before the war by British scientist Alexander Fleming (1881–1955), U.S. industry developed techniques to mass-produce it and provide it to medical teams and hospitals, saving thousands of soldiers' lives and limbs from the deleterious effects of these infections. Most of the work was initially accomplished by British scientists, but after 1941 corporations such as Pfizer and Merck worked with the U.S. government to bring the life-saving drug into large-scale production.

Radar was the name given to detection by radio waves. The use of radio waves for communication already existed, and scientists everywhere understood that measuring such waves could also be useful in detecting interfering objects. For example, stations recording radio waves might detect an airplane from the characteristics of the returning radio signal. The task was to develop equipment of sufficient sensitivity and radio waves of sufficiently small wavelengths, or microwaves. Because the technology was well understood (but not yet developed), radar was sought by several countries during the war—Britain, the United States, Germany, the Soviet Union, France, the Netherlands, Italy, and Japan. It was a technology that many believed was simply a matter of "who" and "when" rather than "if." Although the essentials were well known, each country pursued its efforts to develop radar behind closed doors. The important exception was the decision to combine the British and American projects. A result of British physicist Henry Tizard's (1885–1959) visit to the United States in 1940 (also called the Tizard mission), the synthesis of the two was very productive. In fact, it was the beginning of a fruitful scientific and military collaboration among the United States, Great Britain, and Canada. Radar proved the most decisive of all scientific projects; for example, it provided Britain with a strategic advantage during Adolf Hitler's (1889–1945) effort to establish air superiority over the British Isles during the a preliminary phase of his invasion plan. Thanks to radar detection, the German Luftwaffe never dominated the skies during the 1940 "Battle of Britain," and the invasion plan was abandoned. Aside from radar, radio waves also enabled the construction of an important new weapons technology, the proximity fuse. Explosive shells now could be detonated when they were close to a target, rather than waiting for an actual impact.

Radar in operation in World War II. (Library of Congress)

Organizing science for war proved just as important as scientific ideas. In the United States, President Franklin Roosevelt (1882–1945) authorized the creation of the National Defense Research Committee (NDRC) in 1940, and specifically charged it to organize science for war. NDRC's chairman was Carnegie Institution of Washington president Vannevar Bush (1890–1974), who played a leading role in formulating science policies in the United States during and after the war. Even before the bombing of Pearl Harbor, NDRC had given serious attention to two major projects: microwave radar and the atomic bomb. The former seemed much more promising for many reasons, not the least of which was greater familiarity with electronic equipment than with radioactive materials. The main problem was the lack of

an adequate generator of waves at sufficiently small (hence the "micro") wavelengths. This problem was greatly reduced after the Tizard mission, when it became clear that the British already had developed a device, the resonant cavity magnetron, to generate wavelengths in the centimeter range. This kind of radar was known as centimetric radar.

Although NDRC was supportive of radar research, initially it took a more cautious view of the atomic bomb. Although fission had been discovered before the war, the possibility of building a weapon based on splitting the atomic nucleus depended on establishing a chain reaction. Such a reaction would require creating an environment in which a single fission reaction would incite another fission reaction, and so on, until more energy was being released than was required to start it. This was still only a theoretical possibility, and NDRC initially was slow to commit to it. But the enthusiasm of a few leading U.S. physicists, such as cyclotron inventor Ernest Lawrence (1901–1958), and the pressure exerted by British atomic scientists eventually prevailed. But NDRC decided to turn over the responsibility of managing the program to the Army Corps of Engineers. Thus the bomb project in the United States, codenamed the Manhattan District (or the Manhattan Project) fell under the jurisdiction of General Leslie Groves (1896–1970). In 1942, a scientific team led by Enrico Fermi (1901–1954) achieved the first controlled fission chain reaction, in Chicago. Some of the world's leading physicists were assembled in Los Alamos, New Mexico, to design and assemble atomic bombs. Two bombs were dropped, one each on the Japanese cities of Hiroshima and Nagasaki on 6 and 9 August 1945. Although this probably hastened the end of the war, unlike radar it cannot be considered a decisive "war-winning" weapon, because Japan already was near defeat and the United States was planning to invade its home islands.

In the final stages of the war, both the United States and the Soviet Union sent experts into Germany with their armies in order to capture German scientists. One U.S. mission, called *Alsos,* traveled in France and Germany in 1944 and 1945 to discover the details of the German atomic bomb project. The German team, led by Werner Heisenberg (1901–1976), did not advance as far as the American one had; in fact, it never achieved a chain reaction, a feat accomplished in the United States in 1943. But other projects, notably the German rocket program at Peenemunde, were of great interest to the United States. The German government sponsored research during the war on long-range rockets (V-2 rockets) capable of delivering bombs to targets in other countries without the need of airplanes. Their shorter-term goal was developing the Wasserfall ("waterfall"), an anti-aircraft mis-

sile system. Some of the designs of German rockets, such as the V-2, and several of the scientists themselves, such as Wernher Von Braun (1912–1977), were transported to the United States and became important figures in the U.S. rocket and space programs.

World War II increased the interest of the military in funding scientific research, not only for the atomic bomb and radar, but also for other fields. Undersea warfare was a crucial field of research for oceanographers and physicists, and the transmission of sound in sea water appeared to be one of the most pressing scientific subjects during and after the war, as the need to counter the strategic importance of submarines became paramount. Oceanographers and meteorologists also were very active in the United States and Britain in developing swell forecasting. This increased the likelihood of making successful amphibious landings, the first phase of coastal or island invasion. The need for such forecasting became clear during a disastrous assault of the atoll of Tarawa in 1943, where thousands of U.S. Marines were killed or wounded, largely owing to insufficient knowledge of tides. When planning the invasion of the coast of France, in 1944, Allied leaders used the predictions of swell forecasting system developed in the British Isles. Knowledge of accurate meteorological conditions, essential not only for landings, but also for making bombing runs, proved crucial in the war as well.

Science administrators like Vannevar Bush perceived a need to organize not only scientific research, but also its development. In 1941, he orchestrated the creation of the Office of Scientific Research and Development (OSRD) for that very purpose; the new body could request funds from Congress, rather than take its money from the president's emergency funds. The OSRD widened the scope of research and number of projects considerably, and it allowed civilian scientists to work on military projects without interference from the military. When the war ended, Bush attempted to continue this practice, advocating for a national science foundation. Instead, patronage for science continued in bodies such as the Office of Naval Research (created in 1946), and the propriety of basing so much science on military funding provoked lively debate in ensuing years.

See also Artificial Elements; Atomic Bomb; Brain Drain; Computers; Espionage; Great Depression; Hiroshima and Nagasaki; International Cooperation; Manhattan Project; Nazi Science; Oceanography; Office of Naval Research; Patronage; Penicillin; Radar; Rockets

References

Brown, Louis. *A Radar History of World War II: Technical and Military Imperatives.* Bristol: Institute for Physics Publishing, 1999.

Genuth, Joel. "Microwave Radar, the Atomic Bomb, and the Background to U.S. Research Priorities in World War II." *Science, Technology, & Human Values* 13 (1988): 276–289.

Goudsmit, Samuel A. *Alsos.* New York: Henry Schuman, 1947.

Kevles, Daniel J. *The Physicists: The History of a Scientific Community in Modern America.* Cambridge, MA: Harvard University Press, 1995.

Liebenau, Jonathan. "The British Success with Penicillin." *Social Studies of Science* 17 (1987): 69–86.

Neufeld, Michael J. *The Rocket and the Reich: Peenemunde and the Coming of the Ballistic Missile Era.* New York: Free Press, 1995.

Zimmerman, David. *Top Secret Exchange: The Tizard Mission and the Scientific War.* London: Sutton, 1996.

Wright, Sewall

(b. Melrose, Massachusetts, 1889; d. Madison, Wisconsin, 1988)

The U.S. biologist Sewall Wright was one of three individuals—the others were British scientists Ronald A. Fisher (1890–1962) and J. B. S. Haldane (1892–1964)—usually credited with facilitating the synthesis of Mendelian genetics and Darwinian natural selection by the 1930s, through new techniques of population genetics. Population genetics addressed the distribution of genes within large populations of organisms and became a useful tool for understanding genetic variability and inheritance. Wright studied at Harvard University under William Ernest Castle, whose research on hooded rats appeared to demonstrate that natural selection interfered with classical Mendelian laws. Wright's own 1915 doctoral thesis examined the color variability of guinea pigs. He took a research position at the Department of Agriculture, pursuing subjects connected to domestic animal breeding. While there, he met British scientist Ronald Fisher, whose studies of inheritance in whole populations inspired Wright to change his own plans and orient his work toward populations.

After taking a position in the zoology department of the University of Chicago in 1925, Wright slowly developed his own ideas of evolution in populations. He became widely known for his concept of "genetic drift." Genetic drift described the fluctuation in the frequency of genes that occur in isolated populations, owing to variation between generations. Because of the wide variety of genes, a sample of a population will not necessarily have the same frequency of the genes as the entire population might. Wright argued that highly inbred populations would not maintain identical genes. Because of random mutations, even populations not subject to natural selection would change over time. In fact, mutations in these populations were the most susceptible to natural selection. Because of genetic drift, he argued that evolution through natural selection would likely be very successful when large populations were broken up into isolated, highly inbred groups within a species that could adapt to their own, more specialized environments. Natural selection, Wright concluded, should act differently in different local populations within the species. By emphasizing isolated subpopulations, Wright sidestepped the critiques of natural selection that pointed out the statistical difficulties of expecting an entire species to evolve in a single direction.

Wright's objections to the idea of mass natural selection brought him into conflict with Fisher, who took the opposite view. For Fisher, mass selection was the principal vehicle for evolutionary change. He did not believe that genetic drift played as important a role as that ascribed to it by Wright. The clash between Wright and Fisher helped to define research on population genetics in the 1930s. After retiring from the University of Chicago in 1954, Wright took a position at the University of Wisconsin. He continued to work and publish, including his four-volume *Evolution and the Genetics of Populations;* three of the volumes were published while Wright was an octogenarian. He continued an active life as a scientist until his death at the age of ninety-nine.

See also Biometry; Evolution; Genetics; Mutation
References
Bowler, Peter J. *Evolution: The History of an Idea.* Berkeley: University of California Press, 1989.
Mayr, Ernst, and William B. Provine, eds. *The Evolutionary Synthesis: Perspectives on the Unification of Biology.* Cambridge, MA: Harvard University Press, 1980.
Provine, William B. *The Origins of Theoretical Population Genetics.* Chicago, IL: University of Chicago Press, 1971.
————. *Sewall Wright and Evolutionary Biology.* Chicago, IL: University of Chicago Press, 1986.

X-rays

X-rays, discovered at the end of the nineteenth century by German physicist Wilhelm Röntgen (1845–1923), revolutionized both physics and the medical community. The nineteenth century was filled with new discoveries and insights about the nature of electricity; by the 1890s, many leading physicists had experimented with electricity in vacuum tubes. In these tubes, cathodes were used to produce electricity in the form of a beam, or ray. While studying cathode rays at the University of Würzburg, Röntgen by chance found an even more penetrating radiation originating in a metal screen struck by a cathode ray. The new kind of ray emanated from a glowing spot on the screen, and it appeared to affect photographic plates. Soon Röntgen published the first X-ray photographs of human hands, showing shadows of skeletons (and often wedding rings) without skin. Because the metal phosphoresced at the point of origin, many assumed that X-rays emanated from all glowing bodies (in fact, this hypothesis led to the experiments that first revealed the presence of radioactivity in uranium the following year).

The vague term *ray* already was in use to describe poorly understood phenomena connected to the conduction of electricity in cathode ray tubes. The X-ray was given the name *X* for that very reason: Its true nature was unknown. This lack of clarity was compounded by the fact that French physicist Henri Becquerel (1852–1908) discovered radioactivity in 1896, and the disintegrations from unstable elements were also dubbed rays. In fact, there was a plethora of rays studied and allegedly discovered around the turn of the century, and scientists seemed to see them everywhere, even when they did not exist, as in the case of N-rays.

The discovery of X-rays was heralded by news media all over the world, and they soon became the focus of research for physicists. Only after the detonation of the atomic bomb in 1945 would a scientific event receive wider press coverage than X-rays. The thought of photographs produced by rays going through seemingly impenetrable materials captured the public imagination. Some profited from the limited understanding, selling lead-lined undergarments that supposedly would prevent prying eyes from seeing through clothing. Because many physicists already had been experimenting with cathode ray tubes, they were already equipped to produce the new rays, and soon X-rays became the most-studied phenomenon in science. More than a thousand articles and

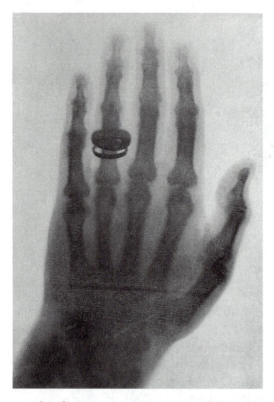

One of the first X-ray photographs made by Wilhelm Röntgen in 1898. (Bettmann/Corbis)

books were published on X-rays in 1896 alone. Röntgen became the first recipient of the Nobel Prize in Physics, awarded in 1901.

The nature of X-rays proved controversial. Many felt that they were akin to light, a kind of electromagnetic radiation with extremely short wavelengths. British physicist Charles Barkla demonstrated in 1905 that X-rays showed signs of polarization, which strengthened the view that the rays were a kind of radiation. But some, such as British physicist William Henry Bragg (1862–1942), continued to believe that they were particles of some kind; the matter would not be resolved until the coming of quantum mechanics in the 1920s, which recognized the wave and particle duality of matter. But even prior to that, German physicist Max Von Laue

(1879–1960) discovered in 1912 that X-rays were diffracted by crystals, which indicated that they behaved like waves; after that, most (including Bragg) accepted that X-rays were a kind of electromagnetic radiation. Bragg and his son, William Lawrence Bragg (1890–1971), soon extended this work, using X-rays to analyze different kinds of metals. Thus began the field of science known as X-ray crystallography.

The medical applications of X-rays as a diagnostic tool appeared obvious from the moment Röntgen published his early photographs. They could be used to look for problems within the human body, such as bone fractures, and to aid in the surgical removal of foreign objects. This proved particularly useful during World War I, when X-rays were used on a large scale to help surgeons removed bomb fragments and bullets from the bodies of wounded soldiers. The dangers of X-rays to the human body were not understood for many years, although leading scientists suspected the dangers were greater than realized. For example, French physical chemist Marie Curie (1867–1934) was instrumental in setting up and maintaining X-ray services during the war; when her health deteriorated later in life, she believed that it was because of overexposure to these harmful rays.

See also Bragg, William Henry; Cancer; Compton, Arthur Holly; Curie, Marie; Light; Medicine; Radiation Protection; Von Laue, Max; World War I

References
Glasser, Otto. *Wilhelm Conrad Röntgen and the Early History of the Roentgen Rays.* Springfield, IL: Thomas, 1934.
Kragh, Helge. *Quantum Generations: A History of Physics in the Twentieth Century.* Princeton, NJ: Princeton University Press, 1999.
Nitzke, W. Robert. *The Life of Wilhelm Conrad Röntgen, Discoverer of the X Ray.* Tucson: University of Arizona, 1971.
Quinn, Susan. *Marie Curie: A Life.* Reading, MA: Addison-Wesley, 1996.

Y

Yukawa, Hideki

(b. Tokyo, Japan, 1907; d. Kyoto, Japan, 1981)

As a young student, Hideki Yukawa's interests lay in literature, not physics. But he excelled in mathematics, and as a teen he became attracted to geometry, which led him ultimately to theoretical physics. In the 1930s, Yukawa held positions at Kyoto University and Osaka University, not gaining his doctorate until 1938, three years after his most important contribution, namely, the prediction of the meson. His theories of forces in the nucleus would dominate his professional interests for most of his career. Yukawa's extraordinary creativity in the 1930s testifies against the notion that science in Japan was inhibited by its authoritarian rigidity and its strict hierarchical structure during this period. Or perhaps Yukawa's experience is the exception to the rule.

When James Chadwick (1891–1974) discovered the neutron in 1932, fundamental particles again took center stage in physics. Before the 1930s, physics accepted only two particles: the electron and the proton. Some, like Chadwick's mentor Ernest Rutherford (1871–1937) and Chadwick himself (for a time), thought that the neutron was in fact a composite of these two, canceling each other's charge. But soon the nucleus of the atom was teeming with particles, including not only neutrons but neutrinos, predicted in 1929 by Wolfgang Pauli (1900–1958), and the antiparticles predicted in 1931 by Paul Dirac (1902–1984): positrons and antiprotons.

Yukawa was far from the centers of activity in Europe and the United States. But a small community of physicists had developed in Japan, strongly influenced by the Copenhagen interpretation of quantum mechanics. Yoshio Nishina (1890–1951) returned to Japan from Niels Bohr's (1885–1962) laboratory in 1929, and Yukawa benefited from his colleague's experience there. His own theories were inspired by European physicists such as Bohr, Werner Heisenberg (1901–1976), and Enrico Fermi (1901–1954). He knew that nuclear forces had not yet been explained adequately and a new theoretical way of keeping the proton and the neutron inside the nucleus was necessary. His expertise in mathematics would provide the means to make major strides along these lines. He gave a ten-minute presentation to the Physico-Mathematical Society in Tokyo in 1934, in which he postulated a new kind of entity, whose existence would help to fix the mathematical problem brought about by the

Professor Hideki Yukawa, noted Japanese physicist and winner of the 1949 Nobel Prize, in his office at Columbia University. (Bettmann / Corbis)

exchange forces in atomic nuclei. Inventing this new particle, Yukawa determined, would help theoretical calculations match the accepted size of the atomic nucleus. Did the particle really exist? Yukawa could not be sure; it was simply a theoretical prediction. He did note that his "heavy quanta," if they existed at all, would be most readily observable in high-energy interactions like cosmic radiation. This theory was published in the 1935 volume of the *Proceedings of the Physico-Mathematical Society of Japan,* not a publication widely read in the West.

Hardly anyone paid attention to Yukawa's prediction. The discovery of a strange anomalous particle by Californians Carl Anderson (1905–1991) and Seth Neddermeyer (1907–1988) in 1937 was not prompted by Yukawa's theoretical predictions. But Yukawa had been right about one thing: Anderson and Neddermeyer's work was on cosmic radiation. Yukawa took notice of their findings and sent a note to the journal *Nature,* pointing out his 1935 prediction. *Nature* denied publication of Yukawa's theory, claiming that it was too speculative. Yukawa's theory came to light in the West only when scientists referred to it in order to reject it. But, as historian Helge Kragh has noted, a negative response is sometimes preferable to no response at all. In this case, Yukawa's prediction soon became widely known, and eventually it became identified with Anderson and Neddermeyer's finding. In fact, some proposed that the particle be dubbed the yukon in honor of Yukawa; eventually it was called the meson, a shortened version of mesotron. The name comes from the Greek root *meso* meaning "middle" or "intermediary." This reflected the meson's role as the middle particle, larger than an electron but smaller in mass than a proton.

Yukawa's prediction and the Anderson-Neddermeyer discovery were not, as it turned out, the same thing. The latter is now called the muon, whereas Yukawa actually had predicted what is now called the charged pion. But this is no detraction from Yukawa's accomplishment, because he had set the pace for studying nuclear particles for the next decade. Much of this period, of course, was interrupted by World War II. Yukawa, being Japanese, was again isolated from many of his colleagues, although much of the important theoretical work in this period was conducted by Japanese physicists such as Sin-Itiro Tomonaga (1906–1979) and Toshima Araki (1897–1978), and also by other Axis scientists, notably a group of experimental physicists in Italy. During this time, Yukawa was honored in his own country with the Impe-rial Prize of the Japan Academy in 1940 and the Decoration of Cultural Merit in 1943.

Yukawa played an important role in the reestablishment of cooperative relations between former enemies after World War II.

Scientists hoped to avoid the long-standing animosities between scientific communities that had occurred after World War I, and Yukawa became a symbol not only of world-class science in Japan but also of the spirit of reconciliation. After the war, most physicists felt that Yukawa's theory was the best effort to describe the forces that held the protons and neutrons together within the nucleus. In 1948, he came to the United States, first at Princeton's Institute for Advanced Study, then to Columbia University as a visiting professor. He received the Nobel Prize in Physics in 1949, the first Japanese ever to receive one.

See also Atomic Structure; Chadwick, James; Cosmic Rays; Physics

References

Kragh, Helge. *Quantum Generations: A History of Physics in the Twentieth Century*. Princeton, NJ: Princeton University Press, 1999.

Yukawa, Hideki. *Tabibito (The Traveler)*, trans. by L. Brown and R. Yoshida. Singapore: World Scientific Publishing, 1982.

Chronology

1895   Wilhelm Röntgen discovers X-rays.

Guglielmo Marconi develops wireless telegraphy.

1896   Henri Becquerel discovers radioactivity in Paris.

1897   J. J. Thomson discovers the electron and develops the "plum pudding" model of the atom.

Henry C. Cowles conducts experiments in Indian Dunes that lead to his concept of plant succession.

Emil Wiechert proposes that the earth has an iron core.

1898   Pierre and Marie Curie announce the discovery of new elements polonium and radium from the radioactive ore pitchblende.

1899   David Hilbert's *Foundations of Geometry* asserts the need to make postulates free of internal inconsistencies.

1900   Hugo DeVries and others revive Gregor Mendel's 1866 work on heredity.

Sigmund Freud publishes *The Interpretation of Dreams,* which becomes a seminal work in the new field of psychoanalysis.

Max Planck introduces notion of quantized energy and founds quantum physics.

Journal *Science* becomes official arm of the American Association for the Advancement of Science.

1901   Karl Pearson founds the journal *Biometrika,* a venue for using statistical methods to discuss inheritance, evolutionary change, and eugenics.

England and Wales adopt fingerprinting techniques for crime detection.

Curt Herbst proposes that embryonic growth occurs through the process of induction.

1901
(cont.)
Robert Koch's public health recommendations help to limit the damage from typhoid fever outbreak in Germany.

Guglielmo Marconi makes first transatlantic communication using wireless technology.

Hugo de Vries proposes "mutation" as the means of creating new characteristics in organisms.

The United States creates the National Bureau of Standards.

First Nobel Prizes awarded.

1902
Emil Fischer determines that amino acids are the units that make up protein.

William Bateson publishes *Mendel's Principles of Heredity*.

Clarence McClung suggests that certain pairs of chromosomes are responsible for determining sex.

Ernest Starling coins the word *hormone* to describe the body's chemical messengers.

George Ellery Hale receives $150,000 to found Mount Wilson Observatory in southern California.

International Council (later International Council for the Exploration of the Sea) is formed.

The Carnegie Institution of Washington is founded and becomes a major patron for science.

1903
Wilhelm Johanssen proposes his "pure lines" theory of variation and inheritance.

Ivan Pavlov develops theory of signalization, connecting sensory input and motor output in organisms.

1904
Jacobus Kapteyn discovers the existence of two distinct "star streams," challenging the notion that stars move about randomly.

1905
Thomas Crowder Chamberlin and Forest Moulton propose the "planetesimal hypothesis" of the earth's (and other planets') formation.

William Bateson coins the term *genetics* to describe the study of heredity based on Mendel's laws.

Albert Einstein publishes major theoretical papers on the photoelectric effect, Brownian motion, and his special theory of relativity.

John Ambrose Fleming develops an "oscillation valve," the first device that allows electric current to be converted to a signal.

Alfred Binet publishes his method of scoring human intelligence, later called the Intelligence Quotient Test (IQ test).

Percival Lowell begins searching for Planet X, beyond Neptune.

1906
Karl Schwarzchild proposes his theory of stellar evolution, based on new findings in quantum physics.

Marie Curie becomes the first woman to teach at the Sorbonne in Paris.

Percival Lowell presents the case for life on Mars in his *Mars and Its Canals.*

Thomas Hunt Morgan begins experiments in the "fly room" at Columbia University, using fruit flies to study the transmission of traits from one generation to the next.

Ernst Haeckel founds the Monist Alliance.

United States passes Pure Food and Drug Act to protect consumers.

Pierre Duhem emphasizes the evolutionary character of scientific change in *La Theorie Physique.*

1907   Bertram Boltwood attempts to revise age of the earth through radioactive dating.

Eugenics Education Society is formed in Britain.

Britain's Imperial College of Science and Technology is founded.

Pope Pius X condemns "modernism" along with the theory of evolution.

1908   Sixty-inch-lens telescope is completed at Mount Wilson Observatory.

1909   Paul Ehrlich develops Salvarsan, a compound useful in treating syphilis.

Wilhelm Johannsen first uses the word *gene* to describe that which carries hereditary information.

Eugene Antoniadi at Paris's Meudon Observatory shows that the "canals" of Mars are an illusion.

Andrija Mohorovičić uses seismic data to discern discontinuity between earth's crust and mantle.

American Robert Peary makes his expedition to the North Pole.

Psychoanalysis gains popularity in United States following lecture tour by Freud, Jung, and Adler.

1910   First published volume of Bertrand Russell and Alfred North Whitehead's *Principia Mathematica* makes its appearance.

Robert Millikan uses oil-drop method to measure charge of an electron.

1911   Thomas Hunt Morgan's fruit flies reveal the chromosome theory of inheritance.

Hertzsprung-Russell diagram of stellar evolution is published.

Ernest Rutherford proposes "planetary" model of atom.

C. T. R. Wilson develops the Cloud Chamber, a device used to see the vapor trails of charged particles.

Marie Curie becomes the first person to receive two Nobel Prizes (1903 in Physics, 1911 in Chemistry).

1911
(cont.)

Kaiser Wilhelm Society is founded, along with its first two institutes (Chemistry and Physical Chemistry).

Norwegian Roald Amundsen's expedition reaches the South Pole.

Frederick Soddy publishes his identification of all radioelements in their proper order of decay.

First Solvay Conference on Physics is convened.

1912

All the members of British explorer Robert Scott's expedition are killed on return trip across Antarctica from South Pole.

Max von Laue demonstrates that X-rays are diffracted by crystals.

Victor Hess flies in a balloon and discovers *cosmic rays,* a term later coined by Robert Millikan.

Henrietta Swan Leavitt notes that the relationships between period and luminosity in Cepheid variables can be a useful tool for calculating the distances of stars.

Henry Herbert Goddard publishes *The Kallikak Family,* arguing that degenerate minds are permanent from one generation to the next.

Piltdown Man is "discovered" in Sussex, England.

Henry Bryant Bigelow initiates long-term intensive oceanographic study of Gulf of Maine.

Britain's Medical Research Committee is formed.

1913

Niels Bohr applies quantum theory to Rutherford's model of the atom.

William Henry Bragg and William Lawrence Bragg conduct pioneering work in X-ray crystallography, using spectra of reflected X-rays to analyze the structure of crystals.

American Society for the Control of Cancer is founded.

John B. Watson makes a case for behaviorist psychology, breaking from structuralism and functionalism.

1914

World War I begins in Europe.

H. G. Wells writes *The World Set Free,* speculating about future "atomic bombs."

Margaret Sanger popularizes the phrase *birth control* and promotes the use of contraception by women.

Magnus Hirschfeld interprets homosexuality as an effect of hormones in *Homosexuality in Man and Woman.*

Beno Gutenberg calculates the depth of the earth's core based on the propagation of seismic waves.

Frtiz Haber begins Germany's efforts to develop chemical weapons.

Carl Jung abandons psychoanalysis and founds analytical psychology.

"To the Civilized World!" is signed by ninety-three German intellectuals, including Max Planck.

Recruiting for World War I reveals widespread problems in basic nutrition in young men.

1915 First use of chemical weapons in war, during World War I, by the Germans at Ieper, Belgium.

Alfred Wegener publishes his theory of Continental Drift in *Origins of Continents and Oceans.*

Einstein completes his general theory of relativity.

Ernest Everett Just receives Spingarn Award from the National Association for the Advancement of Colored People.

1916 The Stanford-Binet IQ Test establishes score of 100 as average at each age level.

National Research Council is formed under the auspices of the National Academy of Sciences.

1917 Bolshevik Revolution in Russia.

United States enters World War I.

Vilhelm Bjerknes returns to Norway and begins the "Bergen school" of meteorology.

1918 Spanish influenza starts in United States and spreads across globe, killing more than 20 million.

First published volume of the *Henry Draper Catalogue* makes its appearance.

1919 Ernest Rutherford assumes the directorship of Cambridge University's Cavendish Laboratory.

Arthur Eddington mounts expedition to island of Principe to witness solar eclipse, recording evidence to affirm Einstein's general theory of relativity.

Joseph Larmor suggests that the earth's magnetism results from convection currents and a dynamo effect.

International Research Council is founded, barring German membership.

1920 "Great Debate" between Harlow Shapley and Heber D. Curtis about the size and uniqueness of the Milky Way takes place at the National Academy of Sciences.

Pacific Science Association is founded to coordinate research among nationszz bordering the Pacific Ocean.

Anti-Semitic campaign against Albert Einstein and his theories begins in Germany.

1921 Frederick Banting and J. J. R. Macleod successfully extract an antidiabetic hormone from a dog's pancreas, later marketed as insulin.

Excavation of Zhoukoudian site begins, unearthing remains of Peking Man throughout 1920s.

1922 J. B. S. Haldane formulates a law of gene linkage, subsequently known as Haldane's Law.

1922
(cont.)
Britain creates Department of Scientific and Industrial Research.

1923
Arthur Compton discovers the "Compton effect," describing the transfer of energy between a photon and an electron upon collision.

Sigmund Freud develops his ideas about repression and the unconscious mind in *The Ego and the Id*.

1924
Harold Jeffreys publishes the first edition of his influential textbook, *The Earth*.

Hans Spemann develops the "organizer" theory, noting that some tissue cells exert an organizing influence on the development of other cells.

Edwin Hubble calculates the vast distance to the Andromeda galaxy, settling the "island universes" controversy.

Max Wertheimer becomes leading proponent of *Gestalt* psychology.

1925
Signing of the Geneva Protocol for the Prohibition of the Use in War of Asphyxiating, Poisonous or Other Gases.

Tennessee passes the Butler Act, banning the teaching of evolution in schools.

Trial of John Scopes, arrested for teaching evolution in school, attracts wide media attention and national debate.

Werner Heisenberg develops matrix mechanics, a version of what becomes known as quantum mechanics.

British vessel *Discovery* begins oceanographic expedition to Antarctic waters.

1926
Arthur Eddington publishes *The Internal Constitution of Stars*.

Scientists from former Central Powers are permitted to join the International Research Council.

Paul Kammerer shoots himself shortly after being denounced as a fraud.

Robert H. Goddard launches the first successful liquid-fueled rocket.

Erwin Schrödinger develops wave mechanics.

Science fiction magazine *Amazing Stories* is founded.

1927
Georges Lemaître proposes theory of universe's origins, later called "Big Bang."

Werner Heisenberg announces his uncertainty principle.

Niels Bohr's complementarity principle asserts the dual nature of matter and sets the foundation of the "Copenhagen interpretation" of quantum mechanics.

Clinton Davisson provides experimental evidence for wave mechanics.

United States Supreme Court upholds eugenics practice of involuntary sterilization for the greater good of society.

1928   Alexander Fleming discovers penicillin.

Subrahmanyan Chandrasekhar identifies a mass limit for stars, beyond which they will not evolve into white dwarfs.

In his *Materials for the Study of Inheritance in Man,* Franz Boas emphasizes the role of the environment in inheritance and development, rather than inherent racial determinism.

John von Neumann provides a proof of his minimax theorem, the cornerstone idea of what becomes known as game theory.

George Gamow explains the release of alpha particles in radioactivity by quantum "tunneling."

Margaret Mead publishes *Coming of Age in Samoa,* observing that many behaviors are culturally rather than biologically determined.

Chandrasekhara Raman discovers a change in wavelength in scattered light, soon known as the "Raman effect."

1929   Crash of the New York Stock Exchange catalyzes worldwide economic depression.

Astronomer Edwin Hubble interprets the red shift in stellar spectra as an indication that the universe is expanding.

1930   At the Cavendish Laboratory, Ernest Walton and John Cockcroft build a linear particle accelerator.

Kurt Gödel announces his "incompleteness" theorem to fellow mathematicians in Germany.

Amateur astronomer Clyde Tombaugh discovers the planet Pluto.

1931   Japan occupies Manchuria.

First analog computer, based on Vannevar Bush's 1927 design for a differential analyzer, is built.

Harold Urey identifies a heavy isotope of hydrogen, deuterium.

1932   Annus mirabilis in physics.

James Chadwick discovers the neutron at the Cavendish Laboratory.

Jan Oort postulates the existence of "dark matter," making up a large proportion of the universe's mass.

Ernest Walton and John Cockcroft use their particle accelerator to create the first disintegration (decay) of a nonradioactive element through artificial means.

1932
(cont.)   Ernest Lawrence improves on previous linear particle accelerators by developing the cyclotron, in which charged particles are accelerated in a spiral.

Second International Polar Year begins.

J. B. S. Haldane's *The Causes of Evolution* demonstrates possible accelerated paces in natural selection.

United States Public Health Service begins syphilis experiment on 400 unwitting African American men in Alabama.

Karl Jansky discovers the principles of radio astronomy.

Aldous Huxley's *Brave New World* depicts a future society dependent on drugs and social engineering.

1933   Adolf Hitler's rise to power in Germany begins a wave of emigration of Jewish scientists out of continental Europe, destined for countries such as Britain, the United States, and the Soviet Union.

Leo Szilard helps to establish in Britain the Academic Assistance Council, to find the means to bring Jewish intellectuals out of Germany.

1934   Academy of Sciences of the USSR moves from Leningrad to Moscow, affirming its subordinate status to the state.

Frédéric Joliot and Irène Joliot-Curie discover artificial radioactivity.

1935   Nuremberg Laws forbidding racial intermarriage are instituted in Nazi Germany.

Hideki Yukawa predicts the existence of "mesons" in the atomic nucleus.

Pavel Cherenkov notes a bluish light from charged particles passing through a medium, soon called Cherenkov radiation.

Arthur Tansley introduces the term *ecosystem* to describe the interdependent exchange of energy and matter.

Karl Popper proposes the importance of "falsifiability" of scientific theories in *Logik der Forschung*.

Charles F. Richter develops the "Richter scale" to measure the magnitude of earthquakes.

1936   Japanese scientist Shiro Ishii's biological warfare unit, later called Unit 731, is formed in Manchuria, and it experiments on humans.

Alan Turing develops theoretical machine using a "tape" with sequential calculations on it, providing a conceptual foundation in the development of computers.

Carl Cori and Gerty Cori study the conversion of the starch glycogen into glucose.

Egas Moniz announces results in alleviating depression and anxiety through radical brain surgery using his "leukotome" device.

Aleksandr Oparin formulates theory of the origin of life as a biochemical process.

Jean Piaget begins to publish theories about the evolution of intelligence and the construction of reality in children.

British scientists at Bawsey Manor develop first airplane detection system based on radar.

1937   Italian Emilio Segrè and C. Perrier believe (incorrectly) that they have created an artificial element in the laboratory.

Carl Anderson and Seth Neddermeyer discover subatomic particles (later called muons) in cosmic rays.

Hans Krebs develops the metabolic cycle later named for him.

U.S. government funds the National Cancer Institute.

Il'ia Frank and Igor Tamm provide the theoretical interpretation of Cherenkov radiation as an effect of charged particles surpassing light's speed in the same medium.

Theodosisu Dobzhansky observes in *Genetics and the Origin of Species* that the vast array of genes in human beings can explain environmental adaptation.

South African Alexander du Toit lends support to the controversial theory of continental drift with his book *Our Wandering Continents.*

Valerii Chkalov is first to fly over the North Pole, piloting his plane from Moscow to Washington state.

1938   Otto Hahn and Fritz Strassman discover nuclear fission in German laboratory.

Johannes Holtfreter demonstrates mechanistic aspects of embryonic growth, diminishing the importance of the "organizer" and of living tissue.

Ugo Cerletti and Lucio Bini experiment on human patients with electric shock therapy.

Orson Welles performs a radio production of *War of the Worlds,* mistaken as a true alien invasion by some.

B. F. Skinner discusses his theory of operant conditioning in *The Behavior of Organisms.*

1939   World War II begins in Europe.

René Dubos isolates tyrothricin, soon to be developed commercially to combat pneumonia.

Ernst Chain and Howard Florey develop techniques to use penicillin for therapeutic purposes.

Albert Einstein signs a letter drafted by Leo Szilard advising President Roosevelt about the possibility of developing an atomic bomb.

1939
(cont.)

Lise Meitner and Otto Frisch interpret the experiment work of Otto Hahn and Fritz Strassman as nuclear fission.

Paul Müller invents DDT.

1940

First truly artificial element, neptunium, identified by Edwin McMillan and Philip Abelson at Berkeley.

Led by Henry Tizard, the "Tizard Mission" of British scientists comes to the United States, intending to facilitate cooperation in science and technology during the war.

Radar gives advantage to the Royal Air Force against the German Luftwaffe in the Battle of Britain.

1941

United States enters World War II.

Glenn T. Seaborg discovers Plutonium, an artificial element.

British MAUD Committee reports positively about the possibility of building an atomic bomb.

George Beadle and Edward Tatum formulate the "one gene, one enzyme" hypothesis.

1942

First fission chain reaction is achieved by a team led by Enrico Fermi, in a squash court underneath the University of Chicago.

Manhattan Engineer District takes over the U.S. atomic bomb project (often called the "Manhattan Project").

Raymond Lindeman develops theoretical means to study ecological change by analyzing trophic (nutritional) relationships as energy flow.

1943

Selman A. Waksman announces development of streptomycin, an antibiotic used to fight tuberculosis.

Victory over Germans at Stalingrad convinces Soviet Union to pursue atomic bomb more vigorously.

First paper on cybernetics is published.

First visualization of DNA achieved through X-ray diffraction techniques.

Josef Mengele arrives at Auschwitz death camp and begins to conduct experiments on inmates.

Manhattan Project enters bomb design phase, and top scientists convene in secret at Los Alamos, New Mexico.

1944

Niels Bohr meets with President Roosevelt and with Prime Minister Churchill to urge them to take steps to avoid a future atomic arms race with the Soviet Union.

George Gaylord Simpson's *Tempo and Mode in Evolution* reconciles the findings of population genetics with the extant fossil record.

John von Neumann and Oskar Morgenstern connect game theory to economic phenomena in *Theory of Games and Economic Behavior.*

Barbara McClintock discovers "jumping genes."

Oswald Avery, Colin MacLeod, and Maclyn McCarty show that the crucial role of DNA is forming cells and suggest that genes are made from DNA.

First V-2 rocket attack by Germany against London, England, is launched.

1945   First successful test of a plutonium-based atomic bomb, codenamed "Trinity," in Alamogordo, New Mexico, is achieved.

Scientists draft Franck Report, which warns of the geopolitical repercussions should the United States use the atomic bomb against Japan.

Hiroshima and Nagasaki become the first two cities to be devastated by atomic weapons, as a result of attack by the United States.

Vannevar Bush writes *Science—the Endless Frontier,* outlining policies for federal patronage of science after World War II.

The Federation of Atomic Scientists is formed.

German atomic scientists are arrested, interned, and put under surveillance at Farm Hall.

U.S. and Soviet occupation authorities begin to smuggle scientists and technicians out of Germany for use in their own defense industries.

1946   Office of Naval Research is founded in the United States, funneling money into basic research in several disciplines.

Unexplained sightings of "ghost rockets" confuse Scandinavians.

Scientists in the United States help defeat May-Johnson Bill, which would have put atomic energy into the military's hands.

Baruch Plan is proposed at the United Nations as a means of establishing international control of atomic energy.

Manhattan Project scientists conduct plutonium injection experiments on unwitting human subjects.

Walter Freeman develops the transorbital lobotomy procedure.

1947   Atomic Energy Commission takes over atomic affairs in the United States.

John Bardeen and Walter Brattain invent the transistor.

Swedish Deep-Sea Expedition begins, using latest coring apparatus.

Largest radio telescope, 218-foot parabolic reflector, is built at Jodrell Bank Observatory.

1948 200-inch-lens telescope is completed at Palomar Observatory.

United States conducts Operation Sandstone, to test atomic bombs and their effects on U.S. soldiers conducting military operations.

Hermann Bondi, Thomas Gold, and Fred Hoyle propose their steady-state cosmological theory.

Willard Libby develops carbon dating, based on measuring the presence of radioactive isotopes of carbon in organic material.

Trofim Lysenko succeeds in banning genetics from the Soviet scientific community, on the grounds that it is incompatible with the Soviet worldview.

George Gamow and Ralph Alpher publish the famous Alpher-Bethe-Gamow article (with Hans Bethe), detailing the creation of light elements in the first few moments after the Big Bang.

The Kaiser Wilhelm Society is renamed the Max Planck Society.

Louis Leakey and Mary Leakey discover fossil remains of hominid named *Proconsul africanus.*

National Bureau of Standards director Edward Condon defends his loyalty before the House Committee on Un-American Activities.

B. F. Skinner publishes his utopia, *Walden Two,* illustrating the potential of a planned society based on the principles of behavioral psychology.

1949 Soviet Union tests its first atomic bomb.

President Harry Truman announces his desire to use science and technology for the good of developing countries and proposes technical assistance programs, later called "Point Four" programs.

Conservationist Aldo Leopold's *A Sand County Almanac* is published posthumously.

University of California institutes loyalty oath for its employees, making rejection of communism a condition of employment and sparking debate about academic freedom.

National Institute of Mental Health is established in United States.

1950 Korean War begins.

Fred Hoyle derisively refers to Georges Lemaître's 1927 theory as "big bang," and the name sticks.

German physicist Klaus Fuchs, part of the British team in the Manhattan Project, is revealed as a Soviet spy.

President Harry Truman announces his decision to proceed with the development of the hydrogen bomb, or "super bomb," with yields thousands of times more powerful than the bombs dropped on Hiroshima and Nagasaki.

State Department committee led by Lloyd Berkner urges effort by the United States to keep ahead in science and technology in order to ensure its national security.

National Science Foundation is founded in the United States, providing a major civilian-operated source of federal funding for scientific research.

Scripps Institution of Oceanography launches first major U.S. postwar deep-sea expedition across the Pacific Ocean.

Jan Oort proposes the existence of a reservoir of debris from which comets are formed, later called the "Oort Cloud."

Selected Bibliography

Please note that this bibliography is pared down considerably, to focus on the books that were the most useful in writing this reference work. They include histories that are both general and specific; in the interest of space, most primary sources and articles have been omitted. For more comprehensive citations, see individual entries.

Akin, William E. *Technocracy and the American Dream: The Technocrat Movement, 1900–1941.* Berkeley: University of California Press, 1977.

Allen, Garland E. *Thomas Hunt Morgan: The Man and His Science.* Princeton, NJ: Princeton University Press, 1978.

Alperovitz, Gar. *The Decision to Use the Atomic Bomb, and the Architecture of an American Myth.* New York: Alfred A. Knopf, 1995.

Alter, Peter. *The Reluctant Patron: Science and the State in Britain, 1850–1920.* New York: St. Martin's Press, 1987.

Argles, Michael. *South Kensington to Robbins: An Account of English Scientific and Technical Education since 1851.* London: Longmans, 1964.

Ash, Mitchell G. *Gestalt Psychology in German Culture, 1890–1967: Holism and the Quest for Objectivity.* New York: Cambridge University Press, 1985.

Aspray, William. *John von Neumann and the Origins of Modern Computing.* Cambridge, MA: MIT Press, 1990.

Badash, Lawrence. *Radioactivity in America: Growth and Decay of a Science.* Baltimore, MD: Johns Hopkins University Press, 1979.

———. *Scientists and the Development of Nuclear Weapons: From Fission to the Limited Test Ban Treaty, 1939–1963.* Atlantic Highlands, NJ: Humanities Press, 1995.

Bailes, Kendall E. *Technology and Society under Lenin and Stalin: Origins of the Soviet Technical Intelligentsia, 1917–1941.* Princeton, NJ: Princeton University Press, 1978.

Bannister, Robert C. *Sociology and Scientism: The American Quest for Objectivity, 1880–1940.* Chapel Hill: University of North Carolina Press, 1987.

Bartholomew, Robert E., and George S. Howard. *UFOs and Alien Contact: Two Centuries of Mystery.* Amherst, NY: Prometheus Books, 1998.

Baümler, Ernest. *Paul Ehrlich: Scientist for Life.* New York: Holmes & Meier, 1984.

Beer, John Joseph. *The Emergence of the German Dye Industry to 1925.* Urbana: University of Illinois Press, 1959.

Bell, E. T. *Men of Mathematics.* New York: Simon & Schuster, 1965.

Benjafield, John G. *A History of Psychology.* Boston: Allyn & Bacon, 1996.

Bertotti, B., R. Balbinot, S. Bergia, and A. Messina, eds. *Modern Cosmology in Retrospect.* New York: Cambridge University Press, 1990.

Bigelow, Henry Bryant. *Memories of a Long and Active Life.* Cambridge, MA: Cosmos Press, 1964.

Biographical Memoirs, National Academy of Sciences. Vols 45–82. Washington, DC: National Academy of Sciences, 1974–2003.

Bjork, Daniel W. *B. F. Skinner: A Life.* New York: Basic Books, 1993.

Blinderman, Charles. *The Piltdown Inquest.* Buffalo, NY: Prometheus Books, 1986.

Bliss, Michael. *The Discovery of Insulin.* Chicago, IL: University of Chicago Press, 1982.

Bocking, Stephen. *Ecologists and Environmental Politics: A History of Contemporary Ecology.* New Haven, CT: Yale University Press, 1997.

Boss, Alan. *Looking for Earth: The Race to Find New Solar Systems.* New York: John Wiley & Sons, 1998.

Bowler, Peter J. *Evolution: The History of an Idea.* Berkeley: University of California Press, 1989.

———. *The Mendelian Revolution: The Emergence of Hereditarian Concepts in Modern Science and Society.* Baltimore, MD: Johns Hopkins University Press, 1989.

———. *Theories of Human Evolution: A Century of Debate, 1844–1944.* Baltimore: Johns Hopkins University Press, 1986.

Brandt, Allan M. *No Magic Bullet: A Social History of Venereal Disease in the United States since 1880, with a New Chapter on AIDS.* New York: Oxford University Press, 1987.

Braun, Ernest, and Stuart MacDonald. *Revolution in Miniature: The History and Impact of Semiconductor Electronics.* Cambridge: Cambridge University Press, 1978.

Brock, Thomas D. *Robert Koch: A Life in Medicine and Bacteriology.* Madison, WI: Science Tech Publishers, 1988.

Brown, Andrew P. *The Neutron and the Bomb: A Biography of Sir James Chadwick.* Oxford: Oxford University Press, 1997.

Brown, Louis. *A Radar History of World War II: Technical and Military Imperatives.* Bristol: Institute for Physics Publishing, 1999.

Brush, Stephen G. *Nebulous Earth: The Origin of the Solar System and the Core of the Earth from Laplace to Jeffreys.* New York: Cambridge University Press, 1996.

Bryder, Linda. *Below the Magic Mountain: A Social History of Tuberculosis in Twentieth-Century Britain.* Oxford: Oxford University Press, 1998.

Buderi, Robert. *The Invention that Changed the World: How a Small Group of Radar Pioneers Won the Second World War and Launched a Technological Revolution.* New York: Simon & Schuster, 1996.

Burchfield, Joe D. *Lord Kelvin and the Age of the Earth.* New York: Science History Publications, 1975.

Campbell, John. *Rutherford: Scientist Supreme.* Christchurch: AAS, 1999.

Campbell-Kelly, Martin, and William Aspray. *Computer: A History of the Information Machine.* New York: Basic Books, 1996.

Carpenter, Kenneth J. *Protein and Energy: A Study of Changing Ideas in Nutrition.* New York: Cambridge University Press, 1994.

Carter, Paul A. *The Creation of Tomorrow: Fifty Years of Magazine Science Fiction.* New York: Columbia University Press, 1977.

Cassidy, David. *Einstein and Our World.* Atlantic Highlands, NJ: Humanities Press, 1995.

———. *Uncertainty: The Life and Science of Werner Heisenberg.* New York: W. H. Freeman, 1992.

Casti, John L., and Werner DePauli. *Gödel: A Life of Logic.* New York: Perseus, 2001.

Chandrasekhar, S. *Eddington: The Most Distinguished Astrophysicist of His Time.* New York: Cambridge University Press, 1983.

Chapman, Michael. *Constructive Evolution: Origins and Development of Piaget's Thought.* New York: Cambridge University Press, 1988.

Chapman, Paul Davis. *Schools as Sorters: Lewis M. Terman, Applied Psychology, and the Intelligence Testing Movement, 1890–1930.* New York: New York University Press, 1988.

Childs, Herbert. *An American Genius: The Life of Ernest Orlando Lawrence, Father of the Cyclotron.* New York: Dutton, 1968.

Christianson, Gale E. *Edwin Hubble: Mariner of the Nebulae.* New York: Farrar, Strauss, and Giroux, 1995.

Clark, Claudia. *Radium Girls: Women and Industrial Health Reform, 1910–1935.* Chapel Hill: University of North Carolina Press, 1997.

Clark, Ronald W. *Einstein: The Life and Times.* New York: World Publishing Co., 1971.

Cochrane, Rexmond C. *The National Academy of Sciences: The First Hundred Years, 1863–1963.* Washington, DC: National Academy of Sciences, 1978.

Cole, Sonia. *Leakey's Luck: The Life of Louis Seymour Bazett Leakey, 1903–1972.* New York: Harcourt Brace Jovanovich, 1975.

Collard, Patrick. *The Development of Microbiology.* New York: Cambridge University Press, 1976.

Cortada, James W. *Before the Computer: IBM, NCR, Burroughs, and Remington Rand and the Industry They Created, 1865–1956.* Princeton, NJ: Princeton University Press, 1993.

Crawford, Elisabeth. *The Beginnings of the Nobel Institution: The Science Prizes, 1901–1915.* New York: Cambridge University Press, 1984.

———. *Nationalism and Internationalism in Science, 1880–1939: Four Studies of the Nobel Population.* New York: Cambridge University Press, 1992.

Crowther, J. G. *The Cavendish Laboratory, 1874–1974.* New York: Science History Publications, 1974.

Crowther, M. Anne, and Brenda White. *On Soul and Conscience: The Medical Expert and Crime: 150 Years of Forensic Medicine in Glasgow.* Aberdeen, Scotland: Aberdeen University Press, 1988.

Dahl, Per F. *Flash of the Cathode Rays: A History of J. J. Thomson's Electron.* London: Institute of Physics, 1997.

Davis, E. A., and Isabel Falconer. *J. J. Thomson and the Discovery of the Electron.* Bristol, PA: Taylor & Francis, 1997.

Dawson, John W., Jr. *Logical Dilemmas: The Life and Work of Kurt Gödel.* Wellesley, MA: A. K. Peters, 1997.

Dick, Steven J. *The Biological Universe: The Twentieth-Century Extraterrestrial Life Debate and the Limits of Science.* New York: Cambridge University Press, 1996.

Douglas, Susan J. *Inventing American Broadcasting, 1899–1922.* Baltimore, MD: Johns Hopkins University Press, 1987.

Dowling, Harry F. *Fighting Infection: Conquests of the Twentieth Century.* Cambridge, MA: Harvard University Press, 1977.

Dunlap, Thomas R. *DDT: Scientists, Citizens, and Public Policy.* Princeton, NJ: Princeton University Press, 1981.

Dupree, A. Hunter. *Science in the Federal Government: A History of Policies and Activities.* Cambridge: Belknap Press, 1957.

Edge, David O., and Michael J. Mulkay. *Astronomy Transformed: The Emergence of Radio Astronomy in Britain.* New York: John Wiley & Sons, 1976.

England, J. Merton. *A Patron for Pure Science: The National Science Foundation's Formative Years, 1945–57.* Washington, DC: National Science Foundation, 1982.

Ewald, P. P., ed. *Fifty Years of X-Ray Diffraction.* Utrecht: International Union of Crystallography, 1962.

Farley, John. *The Spontaneous Generation Controversy from Descartes to Oparin.* Baltimore, MD: Johns Hopkins University Press, 1977.

Faye, Jan. *Niels Bohr: His Heritage and Legacy.* Dordrecht: Kluwer, 1991.

Ferguson, Kyle E., and William O'Donahue. *The Psychology of B. F. Skinner.* Thousand Oaks, CA: Sage Publications, 2001.

Fine, Reuben. *A History of Psychoanalysis.* New York: Columbia University Press, 1981.

Finn, Bernard, Robert Bud, and Helmuth Trischler, eds. *Exposing Electronics.* Amsterdam: Harwood, 2000.

Flamm, Kenneth. *Creating the Computer: Government, Industry, and High Technology.* Washington, DC: Brookings Institution, 1988.

Fleming, James Rodger, ed. *Historical Essays on Meteorology, 1919–1995.* Boston, MA: American Meteorological Society, 1996.

Florkin, Marcel. *A History of Biochemistry.* New York: Elsevier, 1972.

Foucault, Michel. *Madness and Civilization: A History of Insanity in the Age of Reason.* New York: Vintage, 1988.

Friedman, Robert Marc. *Appropriating the Weather: Vilhelm Bjerknes and the Construction of a Modern Meteorology.* Ithaca, NY: Cornell University Press, 1989.

Galishoff, Stuart. *Safeguarding the Public Health: Newark, 1895–1918.* Westport, CT: Greenwood Press, 1975.

Gamwell, Lynn, and Nancy Tomes. *Madness in America: Cultural and Medical Perceptions of Mental Illness before 1914.* Ithaca, NY: Cornell University Press, 1995.

Gay, Peter. *Freud: A Life for Our Time.* New York: W. W. Norton, 1988.

Gilbert, Scott, ed. *A Conceptual History of Modern Embryology.* New York: Plenum Press, 1991.

Gillispie, Charles Coulston, ed. *Dictionary of Scientific Biography.* 17 vols. New York: Charles Scribner's Sons, 1970–1990.

Goldsmith, Maurice. *Frédéric Joliot-Curie: A Biography.* Atlantic Highlands, NJ: Humanities Press, 1976.

Goldstine, Herman H. *The Computer from Pascal to von Neumann.* Princeton, NJ: Princeton University Press, 1972.

Good, Gregory A., ed. *Sciences of the Earth: An Encyclopedia of Events, People, and Phenomena.* New York: Garland Publishing, 1998.

Goodstein, Judith R. *Millikan's School: A History of the California Institute of Technology.* New York: W. W. Norton, 1991.

Goran, Morris. *The Story of Fritz Haber.* Norman: University of Oklahoma Press, 1967.

Gordon, Linda. *Woman's Body, Woman's Right: A Social History of Birth Control in America.* New York: Grossman, 1976.

Gould, Stephen Jay. *The Mismeasure of Man.* New York: W. W. Norton, 1996.

Gowing, Margaret. *Britain and Atomic Energy, 1939–1945.* London: Macmillan, 1964.

Graham, Loren R. *Science and Philosophy in the Soviet Union.* New York: Alfred A. Knopf, 1972.

Greenaway, Frank. *Science International: A History of the International Council of Scientific Unions.* New York: Cambridge University Press, 1996.

Grene, Marjorie. *Dimensions of Darwinism: Themes and Counterthemes in Twentieth Century Evolutionary Theory.* Cambridge: Cambridge University Press, 1983.

Gribbin, John. *In Search of Schrödinger's Cat: Quantum Physics and Reality.* New York: Bantam, 1984.

Grob, Gerald N. *From Asylum to Community: Mental Health Policy in Modern America.* Princeton, NJ: Princeton University Press, 1991.

———. *Mental Illness and American Society, 1875–1940.* Princeton, NJ: Princeton University Press, 1983.

Grossman, Atina. *Reforming Sex: The German Movement for Birth Control and Abortion Reform, 1920–1950.* New York: Oxford University Press, 1995.

Groves, Leslie R. *Now It Can Be Told: The Story of the Manhattan Project.* New York: Harper, 1962.

Guerlac, Henry E. *Radar in World War II.* New York: American Institute of Physics, 1987.

Haber, L. F. *The Poisonous Cloud: Chemical Warfare in the First World War.* New York: Clarendon Press, 1986.

Hacker, Barton C. *Elements of Controversy: The Atomic Energy Commission and Radiation Safety in Nuclear Weapons Testing, 1947–1974.* Berkeley: University of California Press, 1994.

Hagen, Joel B. *An Entangled Bank: The Origins of Ecosystem Ecology.* New Brunswick, NJ: Rutgers University Press, 1992.

Hall, A. Rupert. *Science for Industry: A Short History of the Imperial College of Science and Technology and Its Antecedents.* London: Imperial College, 1982.

Hallam, A. *A Revolution in the Earth Sciences: From Continental Drift to Plate Tectonics.* Oxford: Clarendon Press, 1973.

Hamburger, Viktor. *The Heritage of Experimental Embryology: Hans Spemann and the Organizer.* New York: Oxford University Press, 1988.

Harding, Sandra. *Whose Science? Whose Knowledge? Thinking from Women's Lives.* Ithaca, NY: Cornell University Press, 1991.

Harris, Sheldon H. *Factories of Death: Japanese Biological Warfare, 1932–1945, and the American Cover-Up.* New York: Routledge, 2002.

Headrick, Daniel R. *The Invisible Weapon: Telecommunications and International Politics, 1851–1945.* Oxford: Oxford University Press, 1991.

Heilbron, J. L. *The Dilemmas of an Upright Man: Max Planck and the Fortunes of German Science.* Cambridge, MA: Harvard University Press, 2000.

Heilbron, John L., and Robert W. Seidel. *Lawrence and His Laboratory: A History of the Lawrence Berkeley Laboratory,* vol. 1. Berkeley: University of California Press, 1989.

Heims, Steve Joshua. *The Cybernetics Group.* Cambridge, MA: MIT Press, 1991.

Herf, Jeffrey. *Reactionary Modernism: Technology, Culture and Politics in Weimar and the Third Reich.* New York: Cambridge University Press, 1984.

Herken, Gregg. *Brotherhood of the Bomb: The Tangled Lives and Loyalties of Robert Oppenheimer, Ernest Lawrence, and Edward Teller.* New York: Henry and Holt, 2000.

Herrmann, Dieter B. *The History of Astronomy from Herschel to Hertzsprung.* New York: Cambridge University Press, 1984.

Hersey, John. *Hiroshima.* New York: Vintage, 1989.

Hewlett, Richard G., and Oscar E. Anderson, Jr. *The New World: A History of the United States Atomic Energy Commission, vol. 1, 1939–1946.* Berkeley: University of California Press, 1990.

Hewlett, Richard G., and Francis Duncan. *Atomic Shield: A History of the United States Atomic Energy Commission, vol. II, 1947–1952.* University Park: Pennsylvania State University Press, 1969.

Hoddeson, Lillian, Ernest Braun, Jürgen Teichmann, and Spencer Weart, eds. *Out of the Crystal Maze: Chapters from the History of Solid-State Physics.* New York: Oxford University Press, 1992.

Hodges, Andrew. *Alan Turing: The Enigma.* New York: Simon & Schuster, 1983.

Hofstadter, Richard. *Social Darwinism in American Thought.* Boston, MA: Beacon, 1955.

Holloway, David. *Stalin and the Bomb: The Soviet Union and Atomic Energy, 1939–1956.* New Haven, CT: Yale University Press, 1994.

Howard, Jane. *Margaret Mead: A Life.* New York: Simon & Schuster, 1984.

Isham, Chris J. *Lemaître, Big Bang, and the Quantum Universe.* Tucson, AZ: Pachart, 1996.

Jacobs, David Michael. *The UFO Controversy in America.* Bloomington: Indiana University Press, 1975.

Jammer, Max. *The Conceptual Development of Quantum Mechanics.* New York: McGraw-Hill, 1966.

Johnson, Jeffrey Allan. *The Kaiser's Chemists: Science and Modernization in Imperial Germany.* Chapel Hill: University of North Carolina Press, 1990.

Jones, Bessie Z., and Lyle Boyd. *The Harvard College Observatory: The First Four Directorships, 1839–1919.* Cambridge, MA: Harvard University Press, 1971.

Jones, Ernest. *The Life and Work of Sigmund Freud, vol 2: Years of Maturity, 1901–1919.* New York: Basic Books, 1955.

Jones, James H. *Bad Blood: The Tuskegee Syphilis Experiment.* New York: The Free Press, 1981.

Joravsky, David. *The Lysenko Affair*. Cambridge, MA: Harvard University Press, 1970.

Josephson, Paul R. *Red Atom: Russia's Nuclear Power Program from Stalin to Today*. New York: W. H. Freeman and Company, 2000.

Jungk, Robert. *Brighter than a Thousand Suns: A Personal History of the Atomic Scientists*. San Diego, CA: Harvest, 1956.

Kargon, Robert H. *The Rise of Robert Millikan: Portrait of a Life in American Science*. Ithaca, NY: Cornell University Press, 1982.

Karier, Clarence J. *Scientists of the Mind: Intellectual Founders of Modern Psychology*. Urbana: University of Illinois Press, 1986.

Karoe, G. M. *William Henry Bragg, 1862–1942: Man and Scientist*. New York: Cambridge University Press, 1978.

Kater, Michael H. *Doctors under Hitler*. Chapel Hill: University of North Carolina Press, 1989.

Keller, Evelyn Fox. *The Century of the Gene*. Cambridge, MA: Harvard University Press, 2000.

———. *A Feeling for the Organism: The Life and Work of Barbara McClintock*. San Francisco, CA: W. H. Freeman, 1983.

Kendall, Edward C. *Cortisone: Memoirs of a Hormone Hunter*. New York: Charles Scribner's Sons, 1971.

Kennedy, David M. *Birth Control in America: The Career of Margaret Sanger*. New Haven, CT: Yale University Press, 1970.

Kevles, Daniel J. *In the Name of Eugenics: Genetics and the Uses of Human Heredity*. New York: Alfred A. Knopf, 1985.

———. *The Physicists: The History of a Scientific Community in Modern America*. Cambridge, MA: Harvard University Press, 1995.

Kirwan, L. P. *A History of Polar Exploration*. New York: W. W. Norton, 1960.

Koestler, Arthur. *The Case of the Midwife Toad*. New York: Random House, 1972.

Kohler, Robert E. *From the Medical Chemistry to Biochemistry: The Making of a Biomedical Discipline*. New York: Cambridge University Press, 1982.

———. *Lords of the Fly:* Drosophila *Genetics and the Experimental Life*. Chicago, IL: University of Chicago Press, 1994.

———. *Partners in Science: Foundations and Natural Scientists, 1900–1945*. Chicago, IL: University of Chicago Press, 1991.

Kozulin, Alex. *Vygotsky's Psychology: A Biography of Ideas*. Cambridge, MA: Harvard University Press, 1991.

Kraft, Victor. *The Vienna Circle: The Origins of Neo-Positivism*. New York: Philosophical Library, 1953.

Kragh, Helge. *Cosmology and Controversy: The Historical Development of Two Theories of the Universe*. Princeton, NJ: Princeton University Press, 1996.

———. *Quantum Generations: A History of Physics in the Twentieth Century*. Princeton, NJ: Princeton University Press, 1999.

Kramish, Arnold. *The Griffin: The Greatest Untold Espionage Story of World War II*. Boston, MA: Houghton Mifflin, 1986.

Kühl, Stefan. *The Nazi Connection: Eugenics, American Racism, and German National Socialism*. New York: Oxford University Press, 1993.

Kuklick, Henrika. *The Savage Within: The Social History of British Anthropology, 1885–1945*. New York: Cambridge University Press, 1992.

Lanouette, William, with Bela Szilard. *Genius in the Shadows: A Biography of Leo Szilard, the Man behind the Bomb*. Chicago, IL: University of Chicago Press, 1992.

Larson, Edward J. *Summer for the Gods: The Scopes Trial and America's Continuing Debate over Science and Religion*. New York: Basic Books, 1997.

Leahy, Thomas Hardy. *A History of Psychology: Main Currents in Psychological Thought*. Englewood Cliffs, NJ: Prentice-Hall, 1987.

Leakey, L. S. B. *By the Evidence: Memoirs, 1932–1951*. New York: Harcourt Brace Jovanovich, 1974.

Lederer, Susan E. *Subjected to Science: Human Experimentation in America before the Second World War*. Baltimore, MD: Johns Hopkins University Press, 1995.

LeGrand, H. E. *Drifting Continents and Shifting Theories*. Cambridge: Cambridge University Press, 1988.

Lewis, Robert. *Science and Industrialization in the USSR: Industrial Research and Development, 1917–1940*. London: Macmillan, 1979.

Lifton, Robert Jay. *The Nazi Doctors: Medical Killing and the Psychology of Genocide*. New York: Basic Books, 1986.

Lightman, Alan, and Roberta Brawer. *Origins: The Lives and Worlds of Modern Cosmologists*. Cambridge, MA: Harvard University Press, 1990.

Lindee, M. Susan. *Suffering Made Real: American Science and the Survivors at Hiroshima*. Chicago, IL: University of Chicago Press, 1994.

Lomask, Milton. *A Minor Miracle: An Informal History of the National Science Foundation*. Washington, DC: National Science Foundation, 1976.

Losee, John. *A Historical Introduction to the Philosophy of Science* New York: Oxford University Press, 1972.

Macfarlane, Gwyn. *Alexander Fleming: The Man and the Myth*. Cambridge, MA: Harvard University Press, 1984.

MacKenzie, Donald A. *Statistics in Britain, 1865–1930: The Social Construction of Scientific Knowledge*. Edinburgh: Edinburgh University Press, 1981.

Macrae, Norman. *John Von Neumann*. New York: Pantheon, 1992.

Macrakis, Kristie. *Surviving the Swastika: Scientific Research in Nazi Germany*. New York: Oxford University Press, 1993.

Maddox, Robert James. *Weapons for Victory: The Hiroshima Decision Fifty Years Later*. Columbia: University of Missouri Press, 1995.

Magner, Lois N. *A History of the Life Sciences*. New York: Marcel Dekker, 1979.

Malphrus, Benjamin K. *The History of Radio Astronomy and the National Radio Astronomy Observatory: Evolution toward Big Science*. Malabar, FL: Krieger Publishing, 1996.

Manning, Kenneth R. *Black Apollo of Science: The Life of Ernest Everett Just*. New York: Oxford, 1983.

Marage, P., and G. Wallenborn, eds. *The Solvay Council and the Birth of Modern Physics*. Basel, Switzerland: Birkhäuser, 1998.

Marconi, Degna. *My Father, Marconi*. New York: Guernica, 2002.

Mayr, Ernst. *The Growth of Biological Thought: Diversity, Evolution, and Inheritance*. New York: Belknap Press, 1985.

Meadows, A. J. *Greenwich Observatory, vol. 2, Recent History (1836–1975)*. New York: Scribner, 1975.

Mehra, Jagdish. *The Solvay Conferences on Physics: Aspects of the Development of Physics since 1911*. Dordrecht: Reidel, 1975.

Meine, Curt. *Aldo Leopold: His Life and Work*. Madison: University of Wisconsin Press, 1988.

Mills, Eric L. *Biological Oceanography: An Early History, 1870–1960*. Ithaca, NY: Cornell University Press, 1989.

Minton, Henry L. *Lewis M. Terman: Pioneer in Psychological Testing*. New York: New York University Press, 1988.

Moore, Walter. *Schrödinger: Life and Thought*. New York: Cambridge University Press, 1989.

Morell, Virginia. *Ancestral Passions: The Leakey Family and the Quest for Humankind's Beginnings*. New York: Simon & Schuster, 1995.

Moss, Norman. *Klaus Fuchs: The Man Who Stole the Atom Bomb*. New York: St. Martin's Press, 1987.

Müller-Hill, Benno. *Murderous Science: Elimination by Scientific Selection of Jews, Gypsies, and Others, Germany 1933–1945*. Oxford: Oxford University Press, 1988.

Nasar, Sylvia. *A Beautiful Mind: A Biography of John Forbes Nash, Jr., Winner of the Nobel Prize in Economics, 1994*. New York: Simon & Schuster, 1999.

Needell, Allan A. *Science, Cold War and the American State: Lloyd V. Berkner and the Balance of Professional Ideals*. Amsterdam: Harwood Academic Publishers, 2000.

Neufeld, Michael J. *The Rocket and the Reich: Peenemunde and the Coming of the Ballistic Missile Era*. New York: Free Press, 1995.

Nitzke, W. Robert. *The Life of Wilhelm Conrad Röntgen, Discoverer of the X Ray*. Tucson: University of Arizona, 1971.

Nobel Lectures, Chemistry, 1901–1970. 4 vols. Amsterdam: Elsevier Publishing Company, 1964–1970.

Nobel Lectures, Physiology or Medicine, 1901–1970. 4 vols. Amsterdam: Elsevier Publishing Company, 1964–1970.

North, John. *The Norton History of Astronomy and Cosmology*. New York: W. W. Norton, 1995.

Nye, Robert D. *The Legacy of B. F. Skinner: Concepts and Perspectives, Controversies and Misunderstandings*. New York: Wadsworth, 1992.

Oldroyd, David R. *Thinking about the Earth: A History of Ideas in Geology*. Cambridge, MA: Harvard University Press, 1996.

Oreskes, Naomi. *The Rejection of Continental Drift: Theory and Method in American Earth Science*. Oxford: Oxford University Press, 1999.

Osterbrock, Donald E. *Yerkes Observatory, 1892–1950: The Birth, Near Death, and Resurrection of a Scientific Research Institution*. Chicago, IL: University of Chicago Press, 1997.

Oudshoorn, Nelly. *Beyond the Natural Body: An Archaeology of Sex Hormones*. London: Routledge, 1994.

Overy, Richard. *Why the Allies Won*. New York: W. W. Norton, 1995.

Pais, Abraham. *Niels Bohr's Times: In Physics, Philosophy, and Polity*. New York: Oxford University Press, 1991.

———. *"Subtle is the Lord . . .": The Science and Life of Albert Einstein*. New York: Oxford University Press, 1982.

Parascandola, John, ed. *The History of Antibiotics: A Symposium*. Madison, WI: American Institute of the History of Pharmacy, 1980.

Patterson, James T. *The Dread Disease: Cancer and Modern American Culture*. Cambridge, MA: Harvard University Press, 1987.

Paul, Erich Robert. *The Milky Way Galaxy and Statistical Cosmology, 1890–1924*. New York: Cambridge University Press, 1993.

Petruccioli, Sandro. *Atoms, Metaphors, and Paradoxes: Niels Bohr and the Construction of a New Physics*. New York: Cambridge University Press, 1993.

Philmus, Robert M. *Into the Unknown: The Evolution of Science Fiction from Francis Godwin to H. G. Wells*. Berkeley: University of California Press, 1970.

Pickens, Donald K. *Eugenics and the Progressives*. Nashville, TN: Vanderbilt University Press, 1968.

Pinkett, Harold T. *Gifford Pinchot: Private and Public Forester*. Urbana: University of Illinois Press, 1970.

Poirier, Suzanne. *Chicago's War on Syphilis, 1937–1940: The Times, the Trib, and the Clap Doctor*. Urbana: University of Illinois Press, 1995.

Poundstone, William. *Prisoner's Dilemma: John Von Neumann, Game Theory, and the Puzzle of the Bomb*. New York: Anchor Books, 1992.

Powers, Thomas. *Heisenberg's War: The Secret History of the German Bomb*. New York: Alfred A. Knopf, 1993.

Proctor, Robert N. *The Nazi War on Cancer*. Princeton, NJ: Princeton University Press, 1999.

———. *Racial Hygiene: Medicine under the Nazis*. Cambridge, MA: Harvard University Press, 1988.

Provine, William B. *The Origins of Theoretical Population Genetics*. Chicago, IL: University of Chicago Press, 1971.

————. *Sewall Wright and Evolutionary Biology*. Chicago: University of Chicago Press, 1986.

Pyenson, Lewis. *Civilizing Mission: Exact Sciences and French Overseas Expansion, 1830–1940*. Baltimore, MD: Johns Hopkins University Press, 1993.

————. *Cultural Imperialism and Exact Sciences: German Expansion Overseas, 1900–1930*. New York: Peter Lang, 1985.

Quinn, Susan. *Marie Curie: A Life*. Reading, MA: Addison-Wesley, 1996.

Raphael, Frederic. *Popper*. New York: Routledge, 1999.

Rawlins, Dennis. *Perry at the North Pole: Fact or Fiction?* New York: Robert B. Luce, 1973.

Reich, Leonard S. *The Making of American Industrial Research: Science and Business at GE and Bell, 1876–1926*. New York: Cambridge University Press, 1985.

Reid, Robert. *Marie Curie*. New York: Saturday Review Press, 1974.

Rhodes, Richard. *The Making of the Atomic Bomb*. New York: Simon & Schuster, 1995.

Richards, Pamela Spence. *Scientific Information in Wartime: The Allied-German Rivalry, 1939–1945*. Westport, CT: Greenwood Press, 1994.

Rife, Patricia. *Lise Meitner and the Dawn of the Nuclear Age*. Boston, MA: Birkhäuser, 1999.

Riordan, Michael, and Lillian Hoddeson. *Crystal Fire: The Invention of the Transistor and the Birth of the Information Age*. New York: W. W. Norton, 1997.

Rose, Paul Lawrence. *Heisenberg and the Nazi Atomic Bomb Project: A Study in German Culture*. Berkeley: University of California Press, 1998.

Rosen, George. *A History of Public Health*. New York: MD Publications, 1958.

————. *Preventive Medicine in the United States, 1900–1975: Trends and Interpretations*. New York: Science History, 1975.

Rossiter, Margaret. *Women Scientists in America: Struggles and Strategies to 1940*. Baltimore, MD: Johns Hopkins University Press, 1982.

Rowe, David E., and John McCleary, eds. *The History of Modern Mathematics*. San Diego, CA: Academic Press, 1989.

Rowney, Don K. *Transition to Technocracy: The Structural Origins of the Soviet Administrative State*. Ithaca, NY: Cornell University Press, 1989.

Rozwadowski, Helen. *The Sea Knows No Boundaries: A Century of Marine Science under ICES*. Seattle: University of Washington Press, 2002.

Russell, Bertrand. *The Autobiography of Bertrand Russell*. New York: Routledge, 2000.

Sapolsky, Harvey M. *Science and the Navy: The History of the Office of Naval Research*. Princeton, NJ: Princeton University Press, 1990.

Schlee, Susan. *The Edge of an Unfamiliar World: A History of Oceanography*. New York: Dutton, 1973.

————. *On Almost Any Wind: The Saga of the Oceanographic Research Vessel "Atlantis."* Ithaca, NY: Cornell University Press, 1978.

Schwartz, Joseph. *Cassandra's Daughter: A History of Psychoanalysis*. New York: Penguin, 2001.

Searle, Geoffrey. *Eugenics and Politics in Britain, 1900–1914*. Leyden: Noordhoff International Publishing, 1976.

Seitz, Frederick, and Norman G. Einspruch. *Electronic Genie: The Tangled History of Silicon*. Urbana: University of Illinois Press, 1998.

Sekido, Yataro, and Harry Elliot, eds. *Early History of Cosmic Ray Studies: Personal Reminiscences with Old Photographs*. Dordrecht: D. Reidel, 1985.

Shapiro, Harry L. *Peking Man*. New York: Simon & Schuster, 1975.

Sheehan, John C. *The Enchanted Ring: The Untold Story of Penicillin*. Cambridge, MA: MIT Press, 1982.

Sheehan, William. *The Planet Mars: A History of Observation and Discovery*. Tucson: University of Arizona Press, 1996.

Sherwin, Martin J. *A World Destroyed: The Atomic Bomb and the Grand Alliance*. New York: Alfred A. Knopf, 1975.

Sime, Ruth Lewin. *Lise Meitner: A Life in Physics*. Berkeley: University of California Press, 1996.

Smith, Alice Kimball. *A Peril and a Hope: The Scientists' Movement in America, 1945–1947*. Chicago, IL: University of Chicago Press, 1965.

Smith, Robert W. *The Expanding Universe: Astronomy's 'Great Debate' 1900–1931*. Cambridge: Cambridge University Press, 1982.

Smocovitis, V. Betty. *Unifying Biology: The Evolutionary Synthesis and Evolutionary Biology*. Princeton, NJ: Princeton University Press, 1996.

Sorell, Tom. *Scientism: Philosophy and the Infatuation with Science*. New York: Routledge, 1991.

Soyfer, Valery N. *Lysenko and the Tragedy of Soviet Science*. New Brunswick, NJ: Rutgers University Press, 1994.

Spencer, Frank, and Ian Langham. *Piltdown: A Scientific Forgery*. London: Oxford University Press, 1990.

Stepan, Nancy. *The Idea of Race in Science: Great Britain, 1800–1960*. Hamden, CT: Archon, 1982.

Stocking, George W., Jr. *After Tylor: British Social Anthropology, 1888–1951*. Madison: University of Wisconsin Press, 1995.

———. *The Ethnographer's Magic and Other Essays in the History of Anthropology*. Madison: University of Wisconsin Press, 1992.

———, ed. *Bones, Bodies, Behavior: Essays on Biological Anthropology*. Madison: University of Wisconsin Press, 1988.

Sullivan, W. T., III, ed. *The Early Years of Radio Astronomy: Reflection Fifty Years after Jansky's Discovery*. Cambridge: Cambridge University Press, 1984.

Sulloway, Frank J. *Freud, Biologist of the Mind: Beyond the Psychoanalytic Legend*. New York: Basic Books, 1979.

Swann, John P. *Academic Scientists and the Pharmaceutical Industry: Cooperative Research in Twentieth-Century America*. Baltimore, MD: Johns Hopkins University Press, 1988.

Szilard, Leo, Spencer R. Weart, and Trude Weiss Szilard, eds. *Leo Szilard: His Version of the Facts*. Cambridge, MA: MIT Press, 1980.

Temin, Peter. *Taking Your Medicine: Drug Regulation in the United States*. Cambridge, MA: Harvard University Press, 1980.

Theunissen, Bert. *Eugène Dubois and the Ape-Man from Java: The History of the First "Missing Link" and Its Discoverer*. Dordrecht: Kluwer, 1989.

Thomson, George P. *J. J. Thomson and the Cavendish Laboratory in His Day*. London: Thomas Nelson and Sons, 1964.

Tomes, Nancy. *The Gospel of Germs: Men, Women, and the Microbe in American Life*. Cambridge, MA: Harvard University Press, 1998.

Trenn, Thaddeus J., ed. *Radioactivity and Atomic Theory*. New York: Halsted Press, 1975.

Trent, James W., Jr. *Inventing the Feeble Mind: A History of Mental Retardation in the United States*. Berkeley: University of California Press, 1994.

Tyor, Peter L., and Leland V. Bell. *Caring for the Retarded in America: A History*. Westport, CT: Greenwood Press, 1984.

Valenstein, Elliot S. *Great and Desperate Cures: The Rise and Decline of Psychosurgery and Other Radical Treatments for Mental Illness*. New York: Basic Books, 1986.

Van Woerden, Hugo, Willem N. Broew, and Hendrik C. Van de Hulst, eds. *Oort and the Universe: A Sketch of Oort's Research and Person*. Dordrecht: D. Reidel, 1980.

Von Braun, Wernher, and Frederick I. Ordway II. *History of Rocketry and Space Travel*. New York: Thomas Y. Crowell, 1969.

Vucinich, Alexander. *Empire of Knowledge: The Academy of Sciences of the USSR, 1917–1970.* Berkeley: University of California Press, 1984.

Wali, Kameshwar C. *Chandra: A Biography of S. Chandrasekhar.* Chicago, IL: University of Chicago Press, 1991.

Walker, J. Samuel. *Permissible Dose: A History of Radiation Protection in the Twentieth Century.* Berkeley: University of California Press, 2000.

Wang, Jessica. *American Science in an Age of Anxiety: Scientists, Anticommunism, and the Cold War.* Chapel Hill: University of North Carolina Press, 1999.

Weart, Spencer R. *Scientists in Power.* Cambridge, MA: Harvard University Press, 1979.

Weintraub, E. Roy, ed. *Toward a History of Game Theory.* Durham, NC: Duke University Press, 1992.

Weiss, Sheila Faith. *Race Hygiene and National Efficiency: The Eugenics of Wilhelm Schallmayer.* Berkeley: University of California Press, 1987.

Welsome, Eileen. *The Plutonium Files: America's Secret Medical Experiments in the Cold War.* New York: Delta, 1999.

Werskey, Gary. *The Visible College: The Collective Biography of British Scientific Socialists in the 1930s.* London: Allen Lane, 1978.

Whorton, James. *Before Silent Spring: Pesticides and Public Health in Pre-DDT America.* Princeton, NJ: Princeton University Press, 1975.

Williams, Robert Chadwell. *Klaus Fuchs, Atom Spy.* Cambridge, MA: Harvard University Press, 1987.

Williams, Vernon J., Jr. *Rethinking Race: Franz Boas and His Contemporaries.* Lexington: University Press of Kentucky, 1996.

Willier, Benjamin H., and Jane M. Oppenheimer, eds. *Foundations of Experimental Embryology.* Englewood Cliffs, NJ: Prentice-Hall, 1964.

Winks, Robin W. *Laurence S. Rockefeller: Catalyst for Conservation.* Washington, DC: Island Press, 1997.

Winzer, Margaret A. *The History of Special Education: From Isolation to Integration.* Washington, DC: Gallaudet University Press, 1993.

Worster, Donald. *Nature's Economy: The Roots of Ecology.* San Francisco, CA: Sierra Club Books, 1977.

Wright, Helen. *Explorer of the Universe: A Biography of George Ellery Hale.* New York: Dutton, 1966.

Zachary, G. Pascal. *Endless Frontier: Vannevar Bush, Engineer of the American Century.* Cambridge, MA: MIT Press, 1999.

Zenderland, Leila. *Measuring Minds: Henry Herbert Goddard and the Origins of American Intelligence Testing.* New York: Cambridge University Press, 1998.

Zimmerman, David. *Top Secret Exchange: The Tizard Mission and the Scientific War.* London: Sutton, 1996.

Index

About the Author

Jacob Darwin Hamblin is a historian of science. He is the author of *Oceanographers and the Cold War: Disciples of Marine Science* (2005) and has published several articles on science and politics in the twentieth century. He received his Ph.D. in history from the Program in History of Science, Technology, and Medicine at the University of California, Santa Barbara. A former postdoctoral fellow at the Muséum National d'Histoire Naturelle, Paris, he has taught at Loyola Marymount University and California State University, Long Beach.

45/7630

hill